Advanced Hybrid Nanomaterials for Energy Storage

Integrating nanotechnology and sustainable energy frontiers, *Advanced Hybrid Nanomaterials for Energy Storage* explores the groundbreaking field of material design at the nanoscale for next-generation energy storage solutions. This comprehensive text delves into the synthesis, characterization, and optimization of hybrid nanomaterials developed by combining the advantageous properties of diverse materials. This diverse range of materials includes metal oxides, carbon nanostructures, biopolymers, and functionalized surfaces. These materials have the potential to revolutionize energy storage technologies such as batteries and supercapacitors due to their synergistic properties and innovative applications.

FEATURES

- Explores the latest advances in hybrid nanomaterial design for energy storage applications
- Discusses the benefits of combining different materials at the nanoscale range, exhibiting their combined properties which significantly outperform those of individual components
- Defines the various types of hybrid nanomaterials, including metal oxide/carbon nanocomposites, metal-doped composites, and biopolymer-based materials
- Focuses on the real-world implications of hybrid nanomaterials in battery electrodes, supercapacitor electrodes, and other energy storage devices
- Summarizes the important role of these materials in transitioning to a clean and sustainable energy environment

This book serves as an important resource for both industry professionals and academic researchers and is ideal for scientists and engineers working in advanced materials for energy storage applications.

Emerging Materials and Technologies

Series Editor:
Boris I. Kharissov

The *Emerging Materials and Technologies* series is devoted to highlighting publications centered on emerging advanced materials and novel technologies. Attention is paid to those newly discovered or applied materials with potential to solve pressing societal problems and improve quality of life, corresponding to environmental protection, medicine, communications, energy, transportation, advanced manufacturing, and related areas.

The series takes into account that, under present strong demands for energy, material, and cost savings, as well as heavy contamination problems and worldwide pandemic conditions, the area of emerging materials and related scalable technologies is a highly interdisciplinary field, with the need for researchers, professionals, and academics across the spectrum of engineering and technological disciplines. The main objective of this book series is to attract more attention to these materials and technologies and invite conversation among the international R&D community.

Nanofluids: Fundamentals, Applications, and Challenges
Shriram S. Sonawane and Parag P. Thakur

MXenes: From Research to Emerging Applications
Edited by Subhendu Chakroborty

Biodegradable Polymers, Blends and Biocomposites: Trends and Applications
Edited by A. Arun, Kunyu Zhang, Sudhakar Muniyasamy and Rathinam Raja

Bioinspired Materials and Metamaterials: A New Look at the Materials Science
Edward Bormashenko

Computational Studies: From Molecules to Materials
Edited by Ambrish Kumar Srivastava

2D Semiconductors for Environmental Remediation
*Edited by Honey John, Nisha T Padmanabhan, Sona Stanly
and Jith C Janardhanan*

Materials from Natural Sources: Structure, Properties, and Applications
Edited by Ramesh Gardas, Neha Patni and Amita Chaudhary

Dielectric Materials for Capacitive Energy Storage
Edited by Haibo Zhang and Hua Tan

Multifunctional Coordination Materials for Green Energy Technologies
*Edited by Ghulam Yasin, Anuj Kumar, Sajjad Ali, Tuan Anh Nguyen
and Saira Ajmal*

Advancements in Nanomaterials for Energy Conversion and Storage
Edited by Piyush Kumar Sonkar and Vellaichamy Ganesan

For more information about this series, please visit: www.routledge.com/
Emerging-Materials-and-Technologies/book-series/CRCEMT

Advanced Hybrid Nanomaterials for Energy Storage

Edited by Won-Chun Oh
and Suresh Sagadevan

CRC Press
Taylor & Francis Group
Boca Raton London New York

CRC Press is an imprint of the
Taylor & Francis Group, an **informa** business

Designed cover image: Shutterstock

First edition published 2025
by CRC Press
2385 NW Executive Center Drive, Suite 320, Boca Raton FL 33431

and by CRC Press
4 Park Square, Milton Park, Abingdon, Oxon, OX14 4RN

CRC Press is an imprint of Taylor & Francis Group, LLC

© 2025 by selection and editorial matter, Won-Chun Oh and Suresh Sagadevan; individual chapters, the contributors

Reasonable efforts have been made to publish reliable data and information, but the author and publisher cannot assume responsibility for the validity of all materials or the consequences of their use. The authors and publishers have attempted to trace the copyright holders of all material reproduced in this publication and apologize to copyright holders if permission to publish in this form has not been obtained. If any copyright material has not been acknowledged please write and let us know so we may rectify in any future reprint.

Except as permitted under U.S. Copyright Law, no part of this book may be reprinted, reproduced, transmitted, or utilized in any form by any electronic, mechanical, or other means, now known or hereafter invented, including photocopying, microfilming, and recording, or in any information storage or retrieval system, without written permission from the publishers.

For permission to photocopy or use material electronically from this work, access www.copyright.com or contact the Copyright Clearance Center, Inc. (CCC), 222 Rosewood Drive, Danvers, MA 01923, 978-750-8400. For works that are not available on CCC please contact mpkbookspermissions@tandf.co.uk

Trademark notice: Product or corporate names may be trademarks or registered trademarks and are used only for identification and explanation without intent to infringe.

Library of Congress Cataloging-in-Publication Data
Names: Oh, Won-Chun, editor. | Sagadevan, Suresh, editor.
Title: Advanced hybrid nanomaterials for energy storage / edited by Won-Chun Oh and Suresh Sagadevan.
Description: First edition. | Boca Raton : CRC Press, 2025. |
Series: Emerging materials and technologies | Includes bibliographical references and index.
Identifiers: LCCN 2024030897 (print) | LCCN 2024030898 (ebook) |
ISBN 9781032817279 (hardback) | ISBN 9781032910840 (paperback) |
ISBN 9781003561262 (ebook)
Subjects: LCSH: Energy storage—Materials. | Nanostructured materials. |
Nanocomposites (Materials)
Classification: LCC TK2945.N36 A377 2025 (print) | LCC TK2945.N36 (ebook) | DDC 621.31/260284—dc23/eng/20241023
LC record available at https://lccn.loc.gov/2024030897
LC ebook record available at https://lccn.loc.gov/2024030898

ISBN: 978-1-032-81727-9 (hbk)
ISBN: 978-1-032-91084-0 (pbk)
ISBN: 978-1-003-56126-2 (ebk)

DOI: 10.1201/9781003561262

Typeset in Times
by codeMantra

Contents

Editors

Won-Chun Oh is a Full Professor in the Department of Advanced Materials and Engineering at Hanseo University in Korea and at the School of Materials Science and Engineering at Anhui University of Science and Technology in China. He is a Guest Professor at various other universities throughout China, Thailand, and Indonesia. He obtained the Research Front Award from the Korean Carbon Society, the Yangsong Award from the Korea Ceramic Society, the Excellent Paper Award from the *Korea Journal of Material Research*, as well as the Best Paper Award from the *Journal of Industrial and Engineering Chemistry* for his pioneering work. He has also received the Award of Appreciation from ICMMA2011, ICMMA2014, and ICMMA2019. He is an ICMMA Committee Board Member, and from 2007 to the present, he has served as a Conference Chairman and Vice-Chairman for ICMMA. He has authored or coauthored 806 papers in domestic and international journals, and he currently serves as the Editor-in-Chief of the *Journal of Multifunctional Materials and Photo Science* and the *Asian Journal of Materials Chemistry*. He is an Advisory Board Member of the *Asian Journal of Chemistry* and *Nanomaterials*.

Suresh Sagadevan is an Associate Professor at the Nanotechnology and Catalysis Research Centre at the University of Malaya. He has published more than 450 research papers in ISI top-tier journals and Scopus and has authored 12 international book series and 60 book chapters. He is an Editor, Guest Editor, and Editorial Board Member of many reputed ISI journals and a member of many professional bodies at the national and international levels. He is a recognized reviewer for many reputed journals, and his work spans various fields, such as nanofabrication, functional materials, graphene, polymeric nanocomposite, glass materials, thin films, bioinspired materials, drug delivery, tissue engineering, cell culture, supercapacitor, optoelectronics, photocatalytic, green chemistry, and biosensor applications.

Contributors

Safia Ali
Department of Applied Physical and
  Material Sciences
University of Swat
Khyber Pakhtunkhwa, Pakistan

Amalia Kurnia Amin
Research Center for Chemistry
National Research and Innovation
  Agency (BRIN)
The B.J. Habibie Science and
  Technology Area
South Tangerang, Banten, Indonesia

Aneu Aneu
Faculty of Education
Universitas Muhammadiyah
Cirebon, West Java, Indonesia

Adyatma Bhagaskara
Department of Chemistry
Faculty of Mathematics and Natural
  Sciences
Universitas Gadjah Mada
Yogyakarta, Indonesia

Yujie Ding
School of Chemical and Environmental
  Engineering
Anhui Polytechnic University
Anhui Wuhu, China

Ganjar Fadiilah
Department of Chemistry
Faculty of Mathematics and Natural
  Sciences
Universitas Islam Indonesia
Yogyakarta, Indonesia

Is Fatimah
Department of Chemistry
Faculty of Mathematics and Natural
  Sciences
Universitas Islam Indonesia
Yogyakarta, Indonesia

Riska Astin Fitria
Department of Chemistry
Faculty of Mathematics and Natural
  Sciences
Universitas Gadjah Mada
Yogyakarta, Indonesia

Khushnuma Gul
Department of Applied Physical and
  Material Sciences
University of Swat
Khyber Pakhtunkhwa, Pakistan

Latifah Hauli
Research Center for Chemistry
National Research and Innovation
  Agency (BRIN)
The B.J. Habibie Science and
  Technology Area
South Tangerang, Banten, Indonesia

Rahmat Hidayat
Department of Applied Chemistry
Politeknik Negeri Lampung
Lampung, Indonesia

Zuhong Ji
Department of Chemical Engineering
Hanseo University
Seosan-si, Chungnam, Korea

Latiful Kabir
Department of Advanced Materials
    Science and Engineering
Hanseo University
Seosan, Chungnam, Korea

Minh-Vien Le
Faculty of Chemical Engineering
Ho Chi Minh City University of
    Technology (HCMUT)
Ho Chi Minh City, Vietnam

Dewi Yuanita Lestari
Department of Chemistry
Faculty of Mathematics and Natural
    Sciences
Universitas Negeri Yogyakarta
Yogyakarta, Indonesia

Van Hoang Luan
Faculty of Chemical Engineering
Ho Chi Minh City University of
    Technology (HCMUT)
Ho Chi Minh City, Vietnam

Dat Ly
Faculty of Chemical Engineering
Ho Chi Minh City University of
    Technology (HCMUT)
Ho Chi Minh City, Vietnam
and
Toyoda Masahiro
Oita University
Oita, Japan

Won-Chun Oh
Department of Advanced Materials
    Science and Engineering
Hanseo University
Seosan, Chungnam, Korea

Zambaga Otgonbayar
Department of Chemical and Biological
    Engineering
Hanbat National University
Yuseong-gu, Daejeon, Korea

Remi Ayu Pratika
Department of Chemistry
Faculty of Mathematics and Natural
    Sciences
Universitas Palangka Raya
Palangka, Indonesia

Suresh Sagadevan
Nanotechnology and Catalysis Research
    Centre
University of Malaya
Kuala Lumpur, Malaysia

Wahyu Dita Saputri
Research Center for Quantum Physics
National Research and Innovation
    Agency (BRIN)
The B.J. Habibie Science and
    Technology Area
South Tangerang, Banten, Indonesia

Aldino Javier Saviola
Department of Chemistry
Faculty of Mathematics and Natural
    Sciences
Universitas Gadjah Mada
Yogyakarta, Indonesia

Xiangfeng Shu
School of Materials Science and
    Engineering
Anhui University of Science and
    Technology
Huainan, Anhui, China

Iqmal Tahir
Department of Chemistry
Faculty of Mathematics and Natural
    Sciences
Universitas Gadjah Mada
Yogyakarta, Indonesia

Kefayat Ulla
Department of Applied Physical and
Material Sciences
University of Swat
Khyber Pakhtunkhwa, Pakistan

Jing Wang
School of Materials Science and
Engineering
Anhui University of Technology
Huainan, Anhui, China

Karna Wijaya
Department of Chemistry
Faculty of Mathematics and Natural
Sciences
Universitas Gadjah Mada
Yogyakarta, Indonesia

Ika Yanti
Department of Chemistry
Faculty of Mathematics and Natural
Sciences
Universitas Islam Indonesia
Yogyakarta, Indonesia

Chang-Min Yoon
Department of Chemical and Biological
Engineering
Hanbat National University
Yuseong-gu, Daejeon, Korea

Asadullah Zaffar
Department of Applied Physical and
Material Sciences
University of Swat
Khyber Pakhtunkhwa, Pakistan

Zhuanzhe Zhao
School of Artificial Intelligence
Anhui Polytechnic University
Anhui Wuhu, China

Preface

The ever-growing demand for sustainable and efficient energy solutions has propelled the development of advanced energy storage technologies. In this rapidly evolving landscape, hybrid nanomaterials have emerged as a class of promising materials with the potential to revolutionize energy storage capabilities.

This book, *Advanced Hybrid Nanomaterials for Energy Storage*, delves into the exciting world of these novel materials, exploring their design, properties, and applications in various energy storage systems.

This book addresses the critical need for a comprehensive resource that bridges the gap between fundamental research on hybrid nanomaterials and their practical application in energy storage devices. We aim to provide readers with a thorough understanding of:

- The latest advancements in solid-state electrolytes based on 2D MXenes (Chapter 1).
- The promising potential of two-dimensional MXenes for supercapacitors (Chapter 2).
- The development of carbon-based hybrid materials for high-performance supercapacitors (Chapter 3).
- The exploration of carbon derivatives for enhanced lithium-ion battery anodes (Chapter 4).
- The diverse applications of carbon-based composites in energy storage (Chapter 5).

This book progresses systematically, covering the fundamental principles of nanomaterials for electrode development (Chapter 6) and the specific role of carbon nanomaterials in supercapacitors (Chapter 7). We then delve into the exciting possibilities of g-C_3N_4-based nanocomposites for hydrogen production and battery applications (Chapter 8).

The proper understanding of the low-dimensional MXenes and their functionalization for energy storage is crucial. Chapter 9 explores these concepts in detail. The broader application of nanomaterials in electrochemistry is addressed in Chapter 10, while Chapter 11 focuses specifically on electric double-layer capacitors utilizing carbon materials.

Finally, Chapter 12 brings the book full circle by revisiting the concept of advanced hybrid nanomaterials for energy storage, summarizing the key points, and highlighting the future prospects of this rapidly evolving field.

This book is a valuable resource for researchers, scientists, engineers, and students working in the field of energy storage materials and devices. It provides a comprehensive overview of the latest advancements in hybrid nanomaterials and their potential to shape the future of sustainable energy solutions.

Won-Chun Oh
Suresh Sagadevan

1 Latest Technologies in Solid-State Electrolytes Related to Energy Storage Applications Based on 2D MXenes

Latiful Kabir, Karna Wijaya and Won-Chun Oh

1 INTRODUCTION

Technological advances, often referred to as "modern electronic devices," have dramatically evolved over the past couple few centuries, becoming increasingly important in daily life. The need for highly advanced energy storage devices has rapidly grown, especially with the increasing use of household devices like wearables and smartphones [1–3]. In recent times, a supercapacitors (SCs) have emerged as the best energy storage devices, bridging the gap between capacitors and batteries. SCs offer several advantages including long cycle life, high conductivity, large specific capacitance, fast charge and discharge rates, flexible and thin-film form factors, environmental capability, and potential for high energy density. SCs are categorized into three categories according to their energy storage mechanisms: electrochemical double-layer capacitors (EDLCs), hybrid capacitors, and pseudocapacitors (PCs). EDLCs store energy through electrostatic ion adsorption/desorption at the interface [4–7]. Pseudocapacitors accumulate charge via the Faraday process, transfer of charge between electrodes and electrolytes, whereas EDLCs accumulate charge through an electrochemical double layer. By selecting a suitable combination of electrode materials and a proper electrolyte, which in turn enhances both power and energy density [7–12]. Pseudocapacitive materials typically have larger specific capacities than EDLC, which is explained by the possibility that fast surface redox processes can store more charge than EDLC [13, 14]. Porous carbon nano-capsules have shown potential for enhancing SC performance[15,16]. These NCs can be integrated with of porous carbon with MXene materials [18], porous carbon modified through nitrogen doping [17], MXene/polymer hybrids, and transition metal compounds [19]. SCs can be classified into the following subgroups based on their characteristics:

DOI: 10.1201/9781003561262-1

1. Extremely high electrical capacity: SCs have an impressive capacity, reaching up to 6000F, which is thousands of times greater than regular capacitors [20,21].
2. Remarkably superior strength: SCs can handle current capacities ranging from hundreds to thousands of amperes over short time intervals [22,23].
3. Rapidly charging capability: Due to their physical and chemical properties, SCs can be charged and discharged rapidly. This fast current flow results in loss density and minimal capacity [24,25].
4. Prolonged service life: SCs have an extended lifespan due to their excellent reversibility in electrochemical reactions and resilience to material degradation [26,27].
5. Outstanding performance in cold conditions: Operating on electrode surfaces, SCs maintain stable charge transfer under varying temperatures. In contrast, batteries can experience a capacity drop of over 70% in cold conditions [28,29].

The summaries of the SCs, batteries, and capacitors are given in Table 1.1.

SC electrode materials represent a novel innovation, offering stability and high energy at an affordable price. It is important to further improve this technology to enhance energy storage in electronic devices. High-performance electrode materials necessary for developing SCs because they enhance the capabilities and applicability of SCs, making then more competitive with other energy storage systems. Electrode materials such as carbon nanotubes (CNTs) and graphene are particularly promising due to their high surface area, excellent conductivity, mechanical strength, and chemical stability. For example, graphene has led to the development of EDLCs.

PCs can store additional energy on their surfaces at high speeds through faradaic processes. Furthermore, the redox reactions occurring on the surface allow PCs to achieve a higher specific capacitance (sp) than EDLCs due to the rapid surface redox processes. Electrode materials such as layered double hydroxides, transition metal

TABLE 1.1

Summarized factors of supercapacitors, batteries and capacitors

No.	Parameter	SCs	Battery	Capacitor
1	Functioning voltage	2.3–2.75 V each section	1.2–4.2 V each section	6–800 V
2	Temperature range of functionality	−40 to 85	−20 to −40	−20 to 100
3	Size	Small size	Large size	Small to large size
4	Duration cycles	Greater than 100 cycles	Among 20–1500 cycles	Greater than 100 cycles
5	Energy concentration (W/Kg)	High >4000	Low >100–3000	High >5000
6	Electrically charged/neutralized	Milliseconds to seconds	1–10 h	Picosecond to millisecond
7	Charging procedure	Charge among conductors	Current and voltage	Charge between terminals
8	Power reserve	Watt-second energy	Watt-hours of energy	Watt-second energy

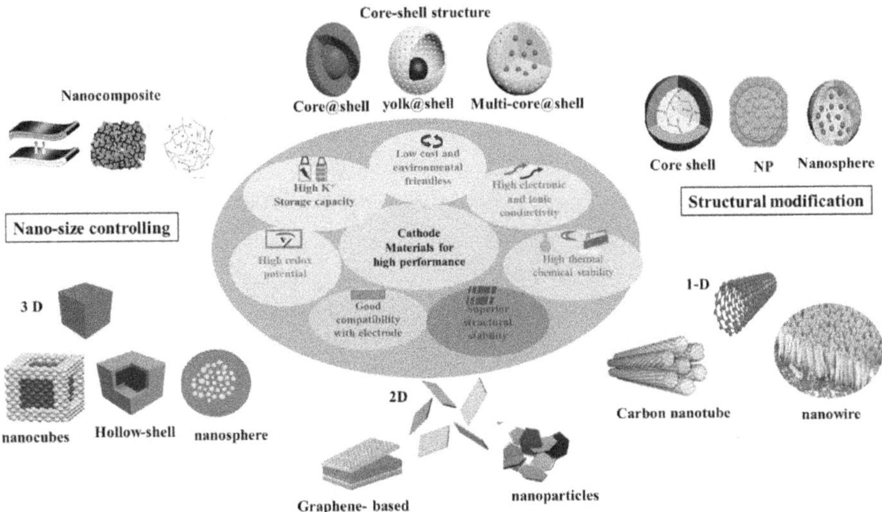

FIGURE 1.1 Enhanced performance through various cathode material types and modification approaches [30–32].

oxides, and conductive polymers have been used in SCs, offering certain advantages. However, they also have limitations; for instance, metal oxides have low conductivity and are expensive, while conductive materials lack stability during charge insertion and extraction. These challenges create barriers to their use in SC electrodes. Therefore, there is a growing focus on exploring unique materials for SC electrodes that provide high power capacity, increased energy concentration, cost-effectiveness, and long-term recyclability. The development of advanced energy storage methods and materials is essential to meet the increasing power demands of electronic devices.

Figure 1.1 illustrates a type of cathode used to achieve energy storage with exceptional performance. Moreover, the basic characteristics of "nanomaterials," along with methods for both functional and morphological modifications, are employed in constructing the cathode material, allowing the SC to operate at high performance levels. As shown in Figure 1.1, demonstrates electrolyte performance can be enhanced, such as the storage mechanism and charge carriers dynamics, by modifying the structure and shape of the nanocomposite. Furthermore, electrode preparation techniques can be modified to improve the effectiveness of SCs by reducing electron flow resistance at the electrode–electrolyte interface. Consequently, This can decrease time constants and enhance the performance of the SCs. One key factor influencing the electrochemical performance of SCs is the surface characteristics of carbonaceous materials, such as their porous structure and hydrophilicity.

Studies that used various methods for preparing electrodes and applied them in various applications have significantly improved the efficacy of SCs. During the preparation of electrodes, the morphology of nanomaterials can be altered, making it more difficult to achieve superior performance. Figure 1.2 illustrates how electrode performance is affected by different modification protocols and strategies. As also shown in Figure 1.2, achieving high performance in nanocomposites involves controlling morphology, reducing dimensions, and using composite materials in

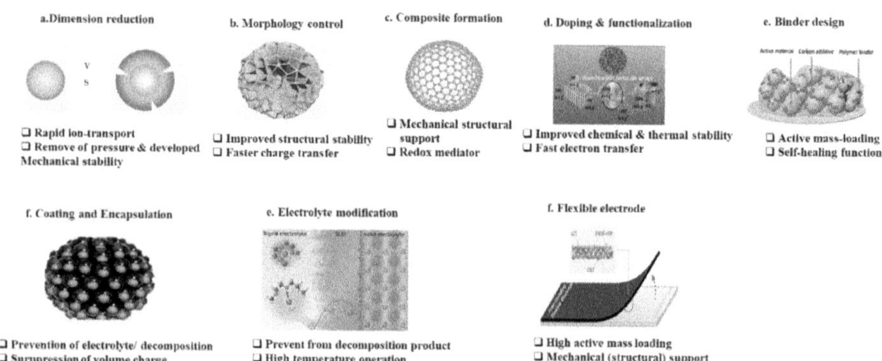

FIGURE 1.2 Technique for enhancing overall performance and justification: It is possible to achieve greater compactness in active materials by reducing dimensions, composite fabrication, electrode softness, electrode modification, binding design, modification of morphology and binding, and active material coating [34–37].

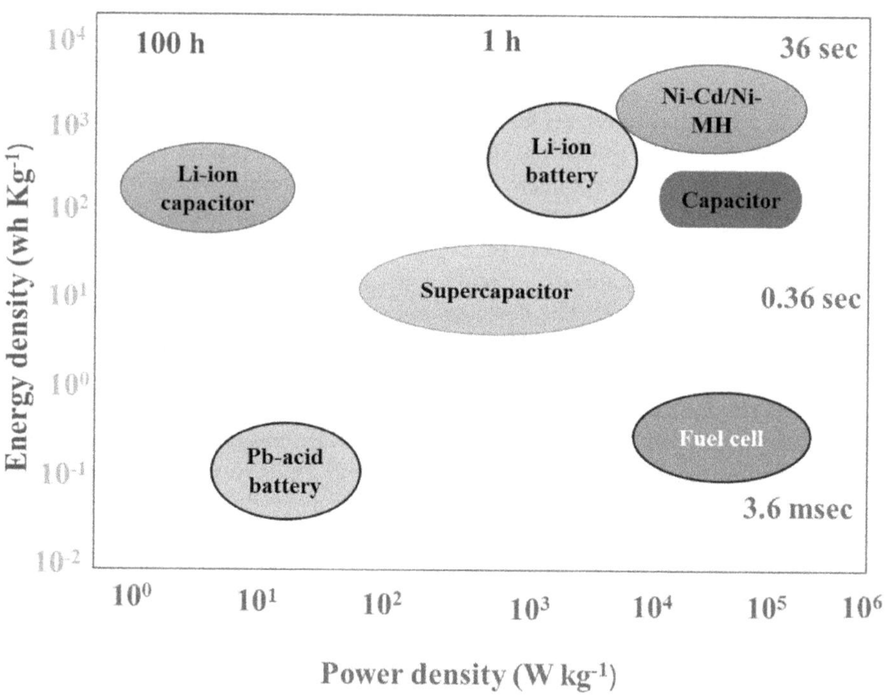

FIGURE 1.3 Comparison of current power capacity and reserves in electrochemical devices with modern technology [38–40].

cathode preparation. The coating and binder design are flexible electrode characteristics. These types of electrodes can be applied in SCs and communication devices, offering excellent electrical conductivity, flexibility, and efficient current collection. Furthermore, the functionalization and modification of electrolytes involve the transitioning from a liquid to a solid electrochemical state. Figure 1.3 illustrates energy preservation systems and their energy and power capacities. Emerging technologies, including wearable and flexible electronics, require durable and secure power sources such as SCs, which are safe and eco-friendly energy storage devices. For the safety of users, the (quasi)-solid electrolytes in SCs ensure high-performance electronics by reducing decay, flammability, and leakage [33]. These electrolytes are available in solid, quasi-solid, and liquid states.

MXene is a group of two-dimensional (2D) transition metals that contain carbonitrides, nitrides, or carbides. The name "MXene" is derived from its chemical formula, where M represents a transition material (such as titanium, tantalum, or niobium), X represents carbon or nitrogen, and T_x denotes the surface functional groups. MXenes are produced by selectively etching away the A layer from a MAX phase, which consists of layers of crystal structure consisting of M, A, and X layers. This process leaves behind a stack of 2D sheets, making MXene a subclass of 2D materials like graphene. MXenes exhibit several remarkable properties: (1) Conductivity: They are excellent electrical conductors, making them suitable for use in electronics and power storage devices. (2) Chemical stability: MXenes are stable in a wide range of chemical environments, enhancing their durability and suitability for various applications. (3) Large surface area: Due to their 2D structure, MXenes have a large surface area, which is valuable for applications in SCs, batteries, and catalysts. Additionally, MXenes have found applications in a range of fields, including energy storage, catalysis, sensors, and additives in composites and coatings. In the context of SCs, the electrochemical performance of 2D MXene materials can be hindered by their dense stacking. To address this, nanomaterials and three-dimensional (3D) structural design may be required. For instance, MXene films treated using a sulfuric acid oxidation method show practical performance with high surface-specific capacitance [41]. MXenes at low temperatures in an argon atmosphere improves the capacitance performance [42]. After 5,000 cycles, this process leads to an active C-Ti-O sites and expanded interlayer voids, resulting in a significant increase in capacitive energy and an 89% retention of capacitance. As a result, it provides excellent mechanical performance, impressive capacitance retention, and specific capacitance. Cai et al. have developed a layer-by-layer column structure using MXene pearls for structurally flexible energy storage device (ESD) and all-solid-state SCs [43]. Due to significant improvements in electrochemical energy storage devices, $Ti_3C_2T_x$ MXene-based SC electrodes are highly sought after. The $Ti_3C_2T_x$ MXene/electrolyte offers important electrochemical efficacy, contributing to advancements in electrodes and electrolytes for SCs.

2 SYNTHESIS OF MXENES

The remarkable and beneficial qualities of MXenes are closely linked to their synthesis mechanism. Typically, factors such as lateral size of flakes, termination surface area, chemical composition, and etching activity significantly depend on the etchant solution and processing conditions. MXenes are 2D inorganic compounds that contain carbonates, nitrides, or carbides. They vary widely in the formation of hydrophilic terminations. It is known that the MAX phase of aluminum is commonly used for synthesizing MXenes, although it is less frequently reported compared to other elements. The importance of MXene lies in their high electrical and thermal conductivity, large surface area, and good hydrophilicity, which make $Ti_3C_2T_x$-based SC electrodes particularly in demand. These properties contribute to an extended life cycle and high effectiveness. Moreover, $Ti_3C_2T_x$ used as SC electrodes offers many advantages over other MXene elements. Most 2D structural SC materials have a low flow ratio.

The 3D MAX phase consists of ternary layered nitrides, metal carbides, and carbonitrides that new MXenes, are expressed the layer of sp element [44]. In the general formula for MAX phases, M represents a d-block transition metal (such as Nb, Ta, Cr, Mo, Zr, Sc, or Hf), A is an sp element (from groups 13 and 14), and X is made up of C or N atoms. A long stability MXene is involved with a broad n. M-X characteristics display robust bonding that is covalent, metallic, and ionized. On the other hand, M-A layers are loosely connected and have entirely metallic character, leading to the easy separation of M-A bonds at high temperatures. There are three types of MAX phases: M_2AX, M_3AX_2, and M_4AX_3. The pH of these phases can be adjusted through centrifugation, washing, and sonication [45–47]. MAX phases are a part of a broad category of 2D materials, with MXene compounds found in over 30 distinct varieties. These materials are divided into three types of MAX phases [48,49]. A comprehensive set of first principles has been employed to theoretically explore over 40 into other MXenes (as shown in Figure 1.4) [50]. To achieve high-performance yields and good-quality MXenes, experimental conditions such as temperature, particle size of the MAX phase, etching times, and hydrofluoric acid (HF) concentrations must be carefully controlled [51–53].

More than 20 different MXenes have been obtained through various etching solutions and methods, as reported in Table 1.2. In majority of instances, the etchant involves two parts, namely, (1) acidic solutions (NH_4HF_2, HF, and mixture solutions) and (2) fluorine-containing salts (NaF, LiF, and NH_3F).

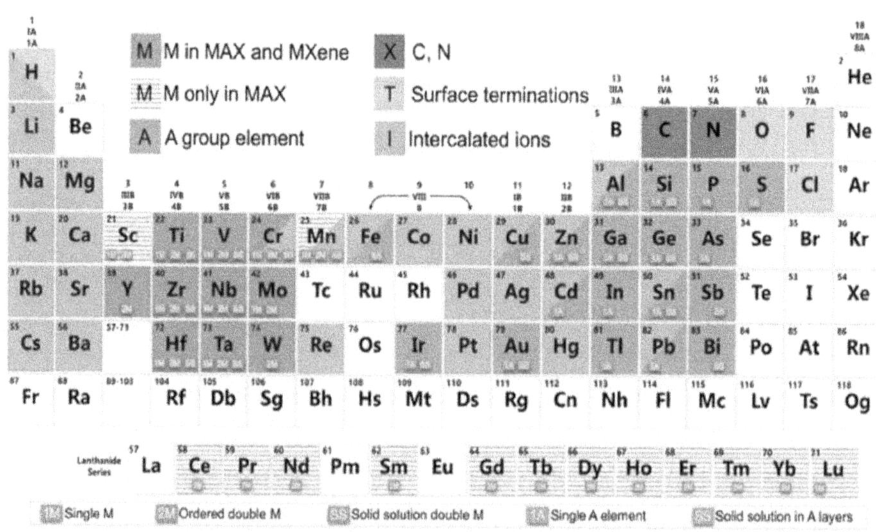

FIGURE 1.4 MXene nanocomposite and MAX phases: a list of the elements applied in the composition of MAX phases and MXenes [46–51].

TABLE 1.2

Classification of the MXene, MAX, Etchant, and Circumstances for Synthesis

№	MXene	MAX	Etchant	Conditions			Ref.
				Time (hr)	T (°C)	Yield (%)	
1	Nb_2CT_x	Nb_2Alc	50% HF	90	RT	100	[54]
2	$Zr_3C_2T_x$	$Zr_3Al_3C_5$	50% HF	60	RT	NA	[62]
3	$Ti_3C_2T_x$	Ti_3AlC_2	NH_3F	24	150	NA	[73]
4	$Ti_3C_2T_x$	Ti_3AlC_2	50% HF	2	RT	100	[57]
5	Ti_2CT_x	Ti_2AlC	10% HF	10	RT	80	[58]
6	Ti_3CNT_x	Ti_3AlCN	30% HF	18	RT	80	[59]
7	$Mo_2Ti_2C_3T_x$	$Mo_2Ti_2AlC_3$	50% HF	90	55	100	[60]
8	$Mo_2TiC_2T_x$	Mo_2TiAlC_2	50% HF	48	RT	100	[61]
9	Ti_2NT_x	Ti_2AlN	5% HF	24	RT	NA	[55]
10	$Nb_{4/3}CT_x$	$(Nb_{2/3}Sc_{1/3})_2AlC$	48% HF	30	RT	NA	[63]
11	$W_{4/3}CT_x$	$(W_{2/3}Sc_{1/3})_2AlC$	48% HF	30	RT	NA	[64]
12	$Ti_4N_3T_x$	Ti_4AlN_3	59% KF + 29% LiF + 12% NaF	0.5	550	NA	[65]
13	Mo_2CT_x	Mo_2Ga_2C	3 M LiF + 12 M HCl	384	35	NA	[66]
14	$Cr_2TiC_2T_x$	Cr_2TiAlC_2	5 M LiF + 6 M HCl	42	55	80	[63]
15	$W_{4/3}CT_x$	$(W_{2/3}Sc_{1/3})_2AlC$	4 g LiF + 12 M HCl	48	35	NA	[67]
16	Ti_2CT_x	Ti_2AlC	0.9 M LiF + 6 M HCl	15	40	NA	[68]
17	V_2CTx	V_2AlC	2 g LiF + 40 M HCl	48	90	NA	[69]
18	Ti_3CNT_x	Ti_3AlCN	0.66 g LiF + 6 M HCl	12	35	NA	[70]
19	$Ti_3C_2T_x$	Ti_3AlC_2	3 M LiF + 6 M HCl	45	40	100	[71]
20	$(Nb_{0.8}Zr_{0.2})_4C_3T_x$	$(Nb_{0.8}Zr0.2)_4AlC_3$	LiF + 12 M HCl	50	168	NA	[72]
21	$Ti_3C_2T_x$	Ti_3AlC_2	NH_3F	24	150	NA	[73]
22	V_2AlC	V_2CT_x	50% HF	RT	90	60	[56]

After applying an acidic solution, particularly HF, which disrupts M-A chemical bonds, MXene layers are formed. Naguib et al. synthesized $Ti_3C_2T_x$ by submerging Ti_3AlC_2 powder in 50% concentrate HF for two hours at room temperature. This process produced MXenes with different functional groups and layered structures. Different ions and chemicals are used for product treatment via sonication [74–77]. The HF etching process has been applied to various MAX phases to produce MXene phases, including Ti_3CNT_x, $Nb_{4/3}CT_x$, $(Ti_{0.5}Nb_{0.5})_2CT_x$, $W_{4/3}CT_x$, $Mo_{4/3}CT_x$, $(V_{0.5}Cr_{0.5})_3C_2T_x$, V_2CT_x, Ti_2NT_x, Nb_2CT_x, and $Ti_3C_2T_x$. Despite its effectiveness, HF etching has specific disadvantages—they impact the environment and human health. To address these issues, a harmless etchant—a mixture of LiF and HCl—has been employed for synthesizing $Ti_3C_2T_x$ from the Ti_3AlC_2 MAX phase at 35°C for 24 hours. MXene multilayers have also been synthesized using salts formed with fluorine ions (e.g., NaF, CaF_2, and KF) and using acids (e.g., H_2SO_4). While mixed etchants emit minor quantities of hazardous HF gas into the surroundings, alternatives like ammonium hydroxide (NH_4HF_2) produce gases that are harmless to the environment. MXenes synthesized using these methods exhibit large surface areas and functional groups such as OH, O, and T-F. The synthesis using a fluorine-containing molten Lewis acids adjusts the surface chemistry and characteristics of MXenes. This method involves removing the "A" atom from the MAX precursor and the cation removal from the Lewis acid. The synthesis method destroys the "A" layer from the MAX phase, which is made by MXene, following a "top–down" strategy. MXenes typically range from several hundred nanometers to 10 micrometers in lateral size. In contrast, the "bottom–up" approach includes chemical vapor deposition (CVD), electrochemical techniques, and lithiation-expansion micro-explosion methods. For instance, a CVD method was developed in 2015 for ultrathin Mo_2C MXene synthesis [78], and in 2019, an improved lithiation-expansion micro-explosion process for introduced for MAX phases, such as Ti_3SiC_2 and Ti_3AlC_2 [79]. MXenes synthesis is accepted for the few layer. In 2020, Zhi et al. introduced a synthesis method involving ion preservation of MXenes with LiTFSI and $Zn(OTf)_2$ and in situ etching of MAX phases using a combined ionic electrolyte solution [80].

Figure 1.5 provides a brief overview of the methods explored for producing MXenes. It shows that Ti_3AlC_2 (MAX) granules were immersed in a 50% HF solution to produce Ti_3C_2 (MXene), which was accomplished in 2011. Since then, numerous methods have been described involving etching durations and HF concentrations [81]. Advances in the synthesis method of MXene have potential to facilitate scaled-up industrial production to some extent. Researchers have also focused on improving the efficiency of MXenes by altering surface area. In addition, the generation of one or several layers of $Ti_3C_2T_x$ can be achieved by introducing dimethyl sulfoxide (DMSO) molecules and employing ultrasound to disrupt or expand the interlayer bonds after the initial HF-based production of $Ti_3C_2T_x$ [82]. Initial SC electrode materials were formed from Ti_3AlC_2 and exfoliated, with successful applications in 2013. In 2014, Gogotsi and Barsoum described the HF-use of HF-strong etching solution for the treatment of MAX [83]. Additionally, HCl/LiF combinations and other fluoride salt/HCl mixtures such as KF, NaF, NH_4F, and FeF_3/HCl have been employed to create V_2CT_x, $Ti_3C_2T_x$, and Ti_2CT_x. This technique offers the advantage of achieving exfoliation and etching in a single stage, thus reducing the overall consumption of HF. The process employed to produce MXene derivatives from MAX phases is illustrated in Figure 1.5.

Nowadays, various approaches for synthesizing MXenes from MAX phases have been suggested. The use of supercritical fluids (SCFs) is commonly employed to break down and intercalate closely stacked materials. Furthermore, supercritical carbon dioxide is favored for its non-toxic, inexpensive, non-flammable properties, with a critical pressure (T_p) of 7.38 MPa and a critical temperature (T_c) of 31.04°C. Chen et al. used supersonic etching to prepare the large quantities of $Ti_3C_2T_x$ MXene, producing about 1 kilogram in 4 hours. Their research identified MXenes in five distinct categories, namely, Nb_2CT_x, Ti_2CT_x, Mo_2CT_x, Ti_3CNT_x, and $Ti_3C_2T_x$, with yields of 1 kg achieved

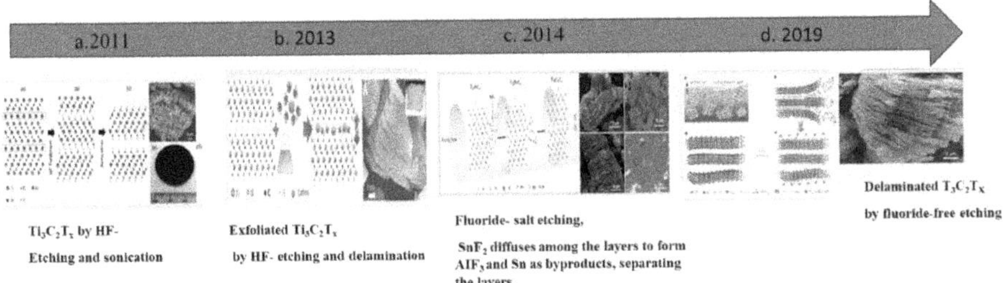

FIGURE 1.5   (a) $Ti_3C_2T_x$–MXene [84], (b) MXene-based electrode for SCs [85], (c) $Ti_3C_2T_x$–MXene–clay [83], and (d) $Ti_3C_2T_x$–MXene by molten salt [86].

within a time frame of 2–5 hours. After improving the MXene materials were applied to combine the final capacity, and Na-ion batteries were 100 mAh g^{-1} conversely, the columbic activity was approximately 100%, and the 70 mAh g^{-1} at 1000 mA g^{-1} occurred. These recent advancements in supercritical techniques represent a promising approach for generating MXene materials with 2D-layered designs [87].

In 2023, Wang et al. improved the CVD method using graphite, titanium, and titanium tetrachloride on the surface of titanium to synthesize Ti_2CCl_2 MXenes. During the synthesis, the MXene layers formed vesicles, with outer layer buckling [88]. The key advantage of the CVD method is that it uses a molten salt etchant containing Lewis acid, eliminating the need for MAX materials. This approach holds potential for more direct synthesis, making it suitable for practical applications. The CVD process requires a temperature of 950°C a reaction time of 2 hours. The morphology of Ti_2CCl_2 MXenes remained unchanged, resembling vertical plates of titanium MXenes . Moreover, variations in contact heating affected the manufacturing process, with preliminary investigations showing the MXene production increased as the temperature raised to 950°C. However, above 950°C, the primary product became TiC_x, reducing the yield of Ti_2CCl_2 MXenes. Furthermore, the product formation ratios were not time-dependent, with durations ranging from 2 hours to 10 days below 950°C. Following this, Ding et al. introduced a chemical scissor-mediated dimensional modification procedure [89] consisting of four distinct phases. In the first phase, Lewis acidic molten salt scissors open the external van der Waals force. In the second phase, the chemical strength of the MAX phases is reduced. In the third stage, MXenes is terminated with a metal scissor via electron injection and the surface functional groups are removed. In the final stage, anions interact to oxidize metal molecules to create terminated MXenes. This method has several advantages, including the ability to reconstruct MAX phases into metal-intercalated 2D carbides. The advancements drive innovation in fields ranging from energy storage to printed circuit boards via metal scissor terminations and combining together 2D carbide nanosheets via molecular intercalation.

3   CHARACTERISTICS OF MXENE MATERIALS

The ability to customize the exceptional properties of the 2D MXene family stems from adjustments in the stoichiometric ratio of M and X atoms. Moreover, MXenes is a potential candidate for simulating high-performance SCs. The unique features of MXenes, including conductivity, hydrophilicity, physical features, potential range, and mechanisms for holding voltage, make them versatile materials for various applications. Certain characteristics of MXenes can be enhanced by incorporating various components and adjusting the interfacial terminations of 2D structures. Over the past few years, instruments along with models for computer and instruments have found spread use in features the favorable characteristics of MXene.

3.1　ATTRIBUTES OF MXENE BODY

In emerging applications, the mechanical properties of MXene electrodes may significantly influence their electrochemical performance, particularly in sensors, thermomechanical systems, flexible electronics, and devices that must endure bending, twisting, and mechanical stresses. From observing the MXene functional attributes, functional groups, composition of the theoretical models, and thickness were used. For future applications, especially in electrodes subject to twisting, stress, and bending, these mechanical characteristics will impact their electrochemical efficiency. MXene measures of the operational groupings technically, thickness as well as the formation were employed in their mechanical qualities. Notable, a single MXene nanosheet exhibits exceptional pliability. Surface functionalization can further enhance the mechanical stability of MXenes [90].

3.2　HYDROPHILICITY

$Ti_3C_2T_X$ is the major property of the SC electrode. However, $Ti_3C_2T_X$ activation is highly dependent on the quality of MXenes, which includes strong conductivity in both heat and electricity, significant hydrophilicity alteration, and surface area. During the synthesis of MXenes, surface termination is managed by preferred methods, with MXene hydrophilicity primarily influenced by -O and -OH groups. MXenes have good affinity for their hydrophilic nature, allowing water to spread across their surfaces [91,92].

3.3　CONDUCTIVITY

A fast discharge rate and high energy density in SCs require highly conductive electrodes. MXenes, similar to their MAX phase precursor, exhibit metallic properties. The individual MXenes conductivities are affected during the concluded by effective groupings on the surface. On the other hand, -O functional groups, for example Ti_2CO_2, the "d" the Ti_2C bare reveal the metallic functions. Furthermore, the titanium layers containing F-functional groups in MXenes, for instance Ti_2CF_2, the fermi energy located in "d" band. The advantages of MXenes stem from their ability to achieve these features by developing surface conductivity and selecting M and X elements, which can changed them into materials ranging from fully metallic to semi-insulating. Over time, issues related to concentration and large lateral widths of MXenes are consistently linked to high conductivity. To facilitate rapid electron transport, the high metallic conductivity of MXenes is necessary for them to continue to be potential SC materials [93].

3.4 Categories of Potential with Capacitor as well as the System for Keeping Charges

MXenes, particularly $Ti_2C_2T_X$ MXenes, exhibit high capacitance because of continuous oxidation state changes and ample redox sites. The capacitance characteristics rely on factors such as surface termination, morphology, cation size, and interlayer spacing. Smaller cations intercalate into the interlayers, contributing to capacitance and fast redox reactions, while larger cations face electrostatic repulsion due to capacitance and adsorption, which leads to MXenes becoming individual materials for energy storage in acidic solutions. The extended durability and high capacity of MXenes stem from its charge storage mechanism. The 2D layered structure of MXenes is essential for the fast ion intercalation and effective charge preservation, without an electrolyte fluid. Significantly, the method of collecting charges and the capacitance of MXene and MXene-based electrodes can be altered for SCs. Typically, the capacitance of MXenes ranges between 300 and 500 F g^{-1}, which is lower than that of EDLCs [94].

3.5 MXene Surface-Level Activation

MXene surface activation involves hydroxide, fluorine, chloride, and oxygen, which enhances the activity and versatility of MXenes with permanency. The properties of MXene membranes can be adjusted according to the type of MXenes and the Tx groups formed during synthesis. This activation process can be categorized into non-covalent and covalent modifications. Non-covalent modifications involve changes in van der Waals efficiency, electrostatic attractions, and hydrogen bonding. Covalent modifications include three subgroups: (1) polymerization (attachment of macromolecules to the outer layer), (2) small molecules, and (3) distinct heteroatoms. The functionality of MXene surface has the advantage, which is the features of MXene are individually raised, for instance: optical and mechanical characteristics improved between 77% and 90%. In contrast, the surface modification techniques leave the concluding sections, resulting in the edges and titanium atoms on the surface happen able of fast ion diffusion, due to, its increased capacitance [95].

3.6 Tunable Interlayer Space of MXenes

It's significant importance to improve a general approach for adjusting surface termination and interlayer spacing of MXene materials, but difficult to mention the improving a technique which can continuously be recognizing this rule. Apply Lewis basic halides, a generic method able to use terminate MXenes and monitoring the intermediary layer distance in the same duration. On the other hand, the raised outcome of the cooperative relationship among the several desolvated k$^+$ then Na$^+$ and intercalation of cation. Moreover, MXene interlayer space is changed so that it can impact the intercalation of ions, due to this, the efficiency of the electrodes can be developed. The modulation of the efficacy features in the foreign element as it is being charge preserve is instrumental in influencing energy storage, and this is achieved through ion intercalation. There have been numerous host molecules used among charge-holding devices in the ion-intercalation compounds with transport approach. Furthermore, the chemical transformation of the foreign particles happen with atom intercalation, due to this alternate in the electrical configuration [96].

4 IMPRESSION OF ELECTROLYTES

MXene is a crucial candidate for SC electrodes because of its excellent electrical conductivity, specific flake morphology, and various surfactant-terminated groups. Among MXenes, $Ti_3C_2T_x$ has the greatest impact on mechanical and electrochemical performance. SCs depend on electrolytes for their performance, as they facilitate charge balancing and transfer between two electrodes. Two types of electrolytes are used in SCs, as shown in Figure 1.6. The electrochemical processes within SCs are profoundly influenced by the interaction between electrolytes and electrodes, affecting both the electrode–electrolyte interface and the internal structure of the active materials. Several different electrolytes have been studied for their application in SCs. Aquatic electrolytes demonstrate good capacitance and high conductivity. However, at high-performance, it has leakage issues and poor cycling permeability. Ionic liquids (ILs) and organic electrolytes can be used at high voltages, but they decrease ionic candidates yet; For instance, "water-in-salt" electrolytes could be used in SCs, but they have reduced temperature precipitation of salt though they have lower conductivity and high viscosity. Solid-state electrolytes (SSEs) eliminate leakage problems but have lower ionic conductivity. Therefore, SCs completely rely on the development of compatible electrolytes that ensure high performance and high capacitance while operating at high voltages. Figure 1.6 provides a comparative analysis of the use of various electrolytes across different countries, illustrating the distribution and focus of research in this area.

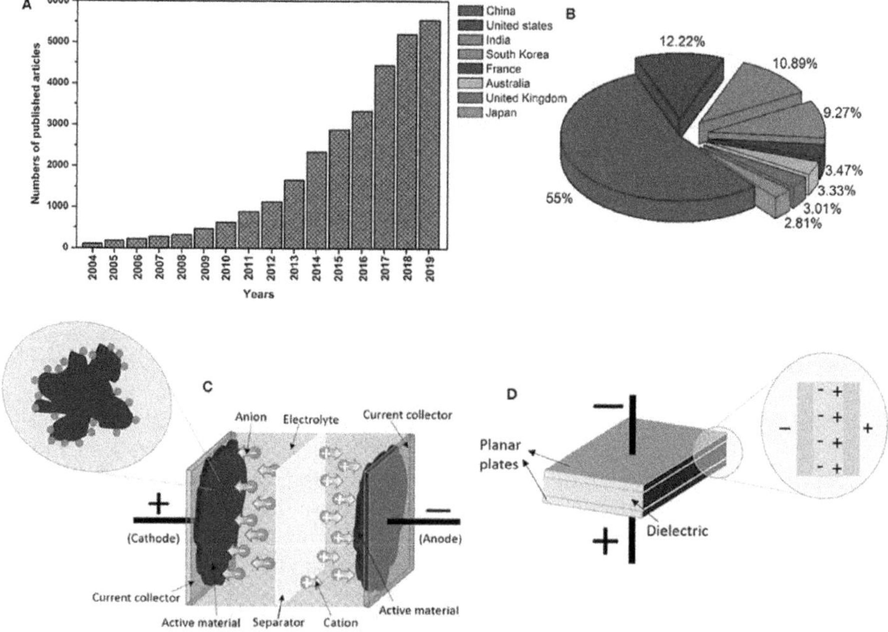

FIGURE 1.6 Distribution of journals discussing PCs, hybrid capacitors, SC electrolytes, and the spread of articles by country.

The electrical entrance and transport of electrons activities are significantly influenced by the choice of electrolyte. In general, $Ti_3C_2T_x$ demonstrates notably high capacitance when used with aqueous electrolytes, particularly H_2SO_4, and it also exhibits a wider operating voltage range when paired with organic electrolytes [97,98]. Each type of electrolyte has unique functionalities, such as flexibility, non-flammability, suitable viscosity, and high conductivity. However, solid electrolyte solutions (SSEs) face a notable feedback in terms of poor conductance [99]. On the other hand, an electrolyte made of gel polymer compounds exhibit high ion transport properties [100,101]. Combining solid and liquid electrolytes can be effective as it leverages the advantages of both states [102,103]. Additionally, redox additives and salts containing small ions can enhance the capacity of the device [104–106]. The solid state of SCs connceted connected to the portable nanoelectronics device [107]. Moreover, this electrode is divided into two types, such as (quasi)-solid and liquid states [108]. Liquid-state electrolytes can be further divided into two types: inorganic and polymer-based electrolytes. Figure 1.8 shows the redox-active groups, while Figure 1.7 shows pseudocapacitive materials used in these electrodes.

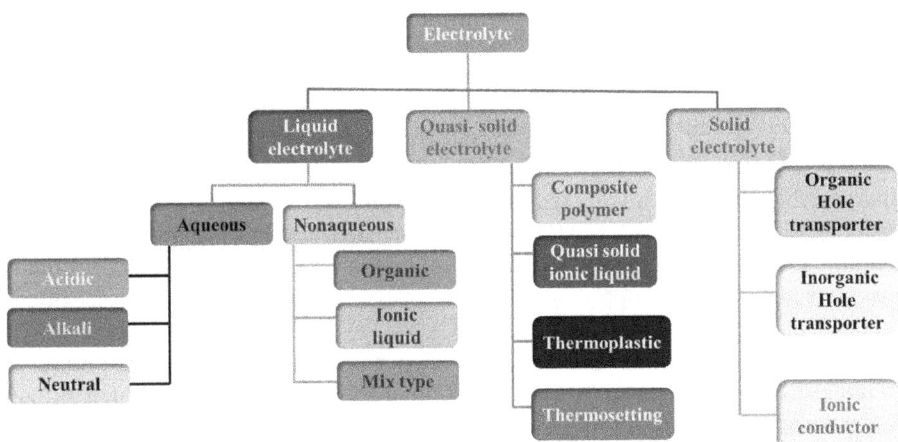

FIGURE 1.7 Category of electrolytes.

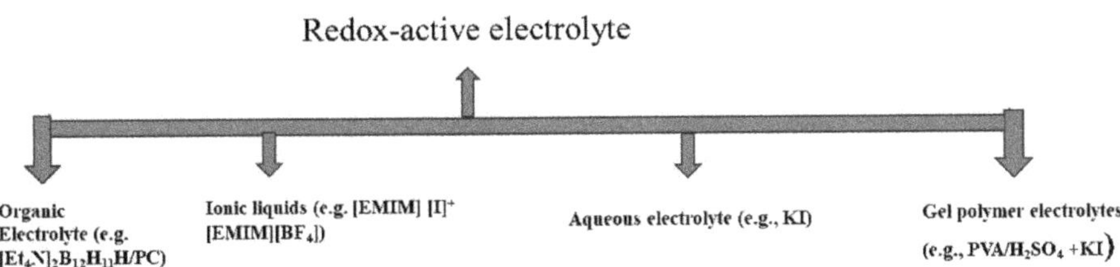

FIGURE 1.8 Redox-active electrolytes.

4.1 ELECTROLYTES IN SOLID STATES

Research on electrolytes utilized in SCs is summarized in Figure 1.9. The optimal electrolyte ensures reliability in electrochemistry, rapid functionality, and cost-effectiveness while achieving the necessary intermediary performance. SCs have utilized a variety of electrolytes, as seen in Figures 1.8 and 1.9. Small molecule diameter, high ionic mobility, and low inner resistance enable electrolytes in water used in SCs to move through micropores with ease. Aqueous electrolytes serve as the primary medium for electrochemical reactions in PCs and EDLCs. The neutral, alkaline, and acidic aqueous solutions such as H_2SO_4 and KOH are the most frequently used in research on water-based electrolytes. A major drawback of SSEs in comparison to liquid electrolytes is their low conductivity [109]. However, gel polymer electrolytes (GPEs) exhibit remarkable ion transport properties. Polymer-based electrolytes may have limited interaction with the electrode due to accessibility issues with micropores, which could lead to poor specific capacitance. In a (quasi)-SSE(QSSE) SC, electrodes, electrolytes, and at a time individually make up the structure. Because of its ionic conductivity and charged polymer chains, a pliable and extensible SC needs a cautious choice of electrolyte. These monomers, which are made up of multiple simple building blocks, serve as the polymer structural echoes. Dry solid-polymer electrolytes (dSPEs), also known as SPEs, have been used to study the conduction of materials like poly(ethylene oxide) (PEO) since the 1970s. However, limited contact between polymer-based electrolytes and electrodes can result in poor specific capacitance. Quasi-solid-state SCs occasionally contain separators. Due to polymeric molecules that carry electricity in pliable and stretchable SCs, high ionic conductivity and careful electrolyte selection are necessary. Moreover, SCs are versatile, secure, and portable. Building a GPE requires the use of salt, solvent, and a polymer substrate, which expands in connection with the materials. Furthermore, to enhance the efficiency of the electrochemical device, the solvent might also include a redox mediator in addition to supplying electron transport. Figure 1.10 depicts a schematic representation of redox-active GPEs.

Among QSSEs, GPEs exhibit the highest electron conduction due to the presence of fluid solvents. It is caused by the interaction of solvent-based atoms, which utilizes a plasticizer combination that enhances molecular ionization and the durability of

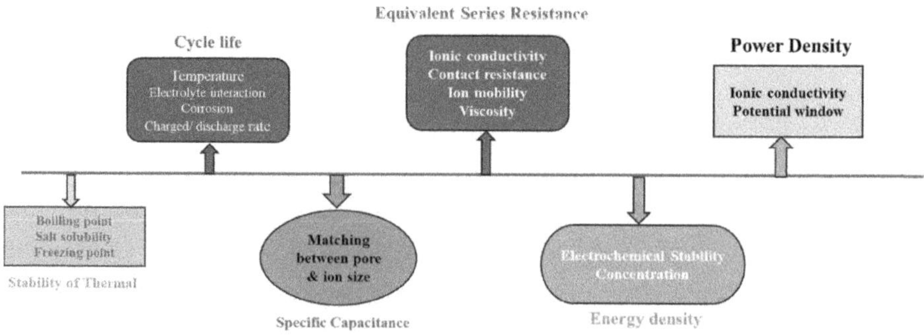

FIGURE 1.9 How electrolytes affect SC performance?

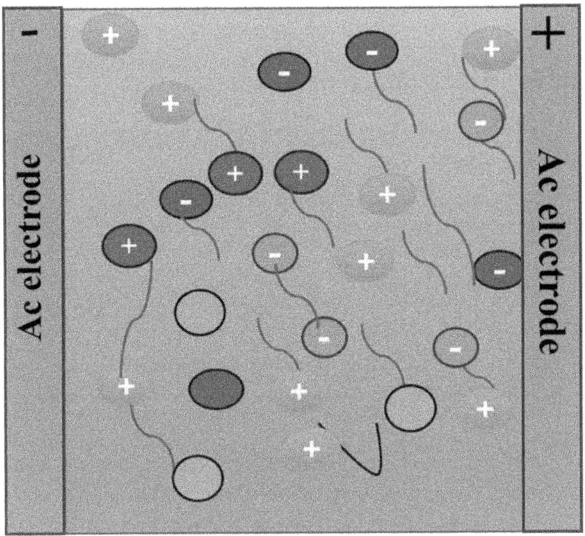

FIGURE 1.10   Schematic of the redox-active GPE.

TABLE 1.3
List of the Solvents' Physicochemical Characteristics

No	Solvent	Dielectric constant	Viscosity (cP)	Melting Point (°C)	Density (g/cm³)	Solubility	Boiling Point (°C)
1	Dimethyl carbonate	3.1	0.59	2.4	1.06	20.3	90
2	Water	77.46	1	0	1	–	100
3	Dimethylacetamide	37.8	0.93	−20	0.95	10.8	166.1
4	Glycerol	42.5	945	17.8	1.25	–	290
5	Propylene carbonate	66.1	2.51	−48.8	1.20	27.2	242
6	Ethylene carbonate	89.8	1.90	36.4	1.32	30.1	248

metals. Furthermore, these interactions rearrange the coordinating centers of polymers, increasing amorphicity and lowering the crystallization temperature. Due to their flexibility and ability to form devices in various shapes, GPEs are ideal for use in wearable technology. Electrolytes need a pliable plasticizer to function well due to their low viscosity, strong diffusion, a high dielectric constant, and significant transparency. Unlike liquids, ions conduct electricity differently solid-state systems. The most studied SSEs are GPEs, which come in the three forms: IL-based GPEs, hydrogels, and organogels. Hydrogels use water as a plasticizer, whereas organogels incorporate different molecules with high dielectric constants. Table 1.3 summarizes the physicochemical characteristics of these solvents. Organogels' high viscosity gives them a solid-like appearance and boosts their barrier properties; however,

TABLE 1.4

Attributes of the Holy Description of Polymer

No	Host Polymer	T_{melt} (°C)	Glass Transition Temp (°C)	Modules of Repeat	Ref.
1	Poly aniline	175	97	–	[116]
2	Polyacrylonitrile	335–362	109–129	$-(CH_2-CH(-CN))_n-$	[112]
3	Polyvinyl alcohol	228	85	$[CH_2CH(OH)]_n$	[113]
4	Nafion	155	110	–	[114]
5	Polyacrylic acid	–	118	$-(CH_2-CH(COOH))_n$	[115]
6	Polyethylene oxide	62	–59	–	[111]
7	Polyvinylpyrrolidine	–	54–175	–	[117]

it also reduces the flow of ions throughout the medium. Organic electrolytes, on the other hand, have lower capacitance because their higher concentration of solvated ions limits their ability to penetrate electrochemical pores.

The amount of plasticization depends on the plasticizer-to-polymer ratio in the mixture. Moreover, the glass transition temperature is impacted by this ratio. Table 1.4 provides a list of the polymer characteristics. A GPE has a unique ability that enables it to be customized for various applications. However, pure electrolytes are necessary for the effective functioning of organogels because they are highly susceptible to humidity and moisture [110]. The manufacturing process involves vacuum dehydration or using an inert environment to eliminate pollutants. This process ensure that performance-critical sections remain unaffected, preserving the lifetime of these energy storage devices.

4.2 Hybrid Electrolytes

Hybrid electrolytes are formed by combining conventional electrolytes such as hydrophilic and organic electrolytes, IL and organic compounds, IL and gel polymer compounds, and water-based and gel polymers. This hybridization introduces new reaction mechanisms, expands applications, and provides new prospects in energy storage systems . The range of voltage, ion mobility, tolerance to temperature variations, and other features are enhances, raising power density. Table 1.5 lists hybrid electrolytes and their comparisons.

Adaptable solid-state SCs (ASSCs), which use flexible electrodes and solid-state gel electrolytes, offer advantages over conventional SCs and show greater potential for use in flexible and versatile wearable electronics. Among the many benefits of SSEs are their comparatively high ionic conductivity, ease of manufacturing, simple and cost-effective packaging, and the absence of leakage issues. SC devices do not require dividers, unlike planar micro-SCs. For the development of highly effective ASSCs, selecting the right SSE (gel) is crucial. The four main subcategories are IL GPEs, organic GPEs, redox-active gel electrolytes, and aqueous GPEs. Table 1.6 lists the special hybrid electrolytes that were created by combining two or more compounds, along with details on their performance. The electrochemical performance

TABLE 1.5

Hybrid Electrolyte Comparison Results

No	Electrode	Cell Voltage (V)	Temp. (°C)	Safety	Electrolytes	Remarkable Point	Ref.
1	Carbon-based	3.7	−100 to 25	No corrosion	ILs and organic	Elevated melting with low	[118]
2	Carbon-based	2.5	−20 to 100	Non-flammable	Aqueous and organic	High conductivity	[119]
3	Carbon-based and pseudocapacitive	2.2	−20 to 120	Direct contact on human skin	Aqueous and gel polymer	Wearable device, microscale device	[120]
4	Carbon-based and pseudocapacitive	3.5	25–90	Non-flammable	ILs and gel polymer	Wearable electronic device	[121]

TABLE 1.6

Shows an Overview of the Performance of SCs Employing Hybrid Electrolytes

No	Electrode	Cell Voltage	Electrolyte	Particular Conductance	Power Density (W kg⁻¹)	Energy Density (Wh kg⁻¹)	Ref.
1	AC-based carbon	1.8	Aquatic and gel polymer	175.8 F g⁻¹ (1 A g⁻¹)	225	19.8	[122]
2	Honeycomb porous carbon	3.5	ILs and gel polymer	48.4 F g⁻¹ (0.5 A g⁻¹)	350	94.1	[123]
3	Rose petal-based porous carbon	2.4	Aqueous and organic	63 F g⁻¹ (0.5 A g⁻¹)	564 (60°C)	44 (60⁰c)	[124]
4	1.5 M EMI-PC-DME	3.0	ILs and organic	131 F g⁻¹ (1 A g⁻¹)	900	38.5	[125]

of SCs is influenced by several factors. However, hybrid electrolytes made of multiple elements can work in concert to provide a wide voltage window, reduce viscosity, increase electron conductivity , improve permeability, reduce ionic particles, and other benefits.

4.3 AQUATIC GPEs

Aqueous GPEs are widely used due to their low cost. GPEs utilize organic solvents and their combination as plasticizers. Gel electrolyte plasticization is regulated by the polymer-to-plasticizer ratio, which also impacts the glass transition temperature. To distinguish between the amorphous zones and ion aggregates, the plasticizer content can be altered in an aquatic gel polymer electrolyte. The 3D polymeric frameworks trap water particles, and the electrode materials determine the electrolyte salts used. As numerous distribution pores can be damaged by acidic or alkaline solutions, aqueous electrolyte printing should be done cautiously. Peracetic acid, PEO, and polyvinyl alcohol are frequently found in polymer matrices. Water acts as a typical plasticizer, and the electrolytic salt can be an effective acid, basic, or neutral salt. PVA is the preferred polymer for SC applications because of its strong chemical stability, biocompatibility, and safety. It is produced by complete or partial hydrolysis of polyvinyl acetate in an alkaline solution to eliminate acetate groups. Additional hydrolysis of the polymer reduces the number of acetate groups. Although PVA contains hydroxyl groups, the polymer crystallization capacity is unaffected by them due to their small size compared to the other acetate groups. Hydrogen ions interacting with water molecules and polymeric chains make up the linear polymer PVA. The -OH groups in PVA also absorb a substantial amount of water, increasing the ionic conductivity of electrolytes. On flexible stainless steel substrates (FSSS), Patil et al. [126] produced porous -MnO2 nanospheres and O-SnS nanoflowers on FSSS. The electrodes used to create ($-MnO_2$/SS//O-SnS/SS) asymmetric solid-state SCs demonstrated exceptional cycling performance, maintaining 95.3% efficiency after 5000 Galvanostatic Charge Discharge (GCD) rotations at 10 mA. The huge capacitance and extreme individual asymmetric sodium-ion SCs (ASSCs) allow for the achievement of significant amounts of both power and energy of 29.8 and 1.3 kW kg^{-1}, correspondingly.

4.4   IL-CENTERED GPEs

Key features of ILs, a subclass of GPEs, are their nonvolatility, wide operating temperature ranges, mechanical qualities, chemical, thermal, and electrical conductivity, and non-flammability. ILs are categorized into three classes (aprotic, protic, and zwitterionic) and can be employed in SCs, fuel cells, lithium batteries, and membranes. These materials hold considerable promise for the pliability and adaptability of SCs. IL-based gel electrolytes are ideal for use in fabricated solid-state capacitors due to their high permeability and resistance to nozzle corrosion. They have been explored in numerous polymer hosts, including organic and aqueous GPEs. The defining properties of ILs are large voltage windows, low vapor pressure, high electrochemical and thermal stability, nonvolatility, and inflammability. An organic salt is mixed with extra chemical compounds that compete with one another to reduce the boiling point and stop crystallizing in eutectic combinations. It is essential to correctly combine the identical anion and cation to avoid ordered structures and crystallization. However, the working temperature of GPEs may be constrained due to their moisture content and low mechanical strength.

4.5   ORGANIC GPEs

Organic GPEs address the limited electrical range within in which electrolytes made of hydrophilic gel polymers can operate. An electrolyte composed of organic gel polymers is produced by mixing salt with a nonaqueous solvent solution containing a high-molecular weight polymer. Plasticizers including ethylene carbonate (EC), dimethyl formamide (DMF), and propylene carbonate (PC) are used to increase the operating voltage range. To boost SC energy density, the cell voltage window needs to be extended. The approach improved by Lee et al., involving the cautious selection of electrolyte and electrode materials and the formation of a pliable layer, resulted in a slim, rectangular, high-performing SC with dynamic flexibility. Mn/Mo-mixed oxide@Multi-walled carbon nanotube electrodes were applied after the organic GPEs containing lithium, succinonitrile, and adiponitrile were spread across a porous elastomer sheet. In Huang et al.'s study [127], an OGE for an AC-based SC, a PEO–PAN mix (PAN-b-PEG-b-PAN) was used as the host, with DMF as the plasticizer and $LiClO_4$ as the electrolytic salt. With a voltage window of 2.1 V and a capacitance of 101 F g^{-1} at 0.125 A g^{-1}, the integrated cell generated 11.5 W h kg^{-1} of energy at a maximum output of 10 kW kg^{-1}. Over 30,000 cycles, the cell demonstrated exceptional stability and showed no loss of capacitance. The electrical conductivity of ions was significantly impacted by the salt content and the salt-to-host polymer ratio in the electrolyte. The electrode–electrolyte interface design of SCs is crucial for achieving higher ionic conductivity, cycling, inner resistance, voltage density, and intensity of energy lifetime. The following are the electrolyte primary characteristics: relationships between electrons and the electrode–electrolyte include (a) ion concentration, (b) ion interactions with the electrode and electrolyte, (c) ion dimension, (d) ion interaction with the solution, and (e) ion type. For QSSEs, the structure of the SC electrode–electrolyte interface is crucial. Enhancing the integration of the electrolyte and the electrode's outer layer is vital [128], as it reduces the inner barrier [129] and extends the lifespan by reducing hydrogel dehydration.

5 SCS USING MXENE AND MXENE-DERIVED COMBINATIONS IN FLUID ELECTROLYTES

MXene-derived materials have large surface areas, 2D structures, and promising electrical characteristics suitable for SCs. After 10,000 cycles, Lukatskaya et al. showed no deterioration while demonstrating the caustic integration that occurs spontaneously within $Ti_3C_2T_x$ the surfaces of MXene, obtaining a volumetric capacitance of 340 F cm^{-3} at 2 mV s^{-1} [130]. These results indicates promising prospects for the development and use of MXenes in SCs. $Ti_3C_2T_x$, a well-researched MXene material, exhibits pseudocapacitive electrochemical activity in sulfuric acid [131]. Hu et al. studied the method by which power works in aqueous electrolytes containing sulfate ions, only discovering electrical dual-layer power in $(NH_4)_2SO_4$ and $MgSO_4$ [132]. The research on MXenes in SCs is summarized in the following paragraphs:

Ghidiu et al. used clay-like $Ti_3C_2T_x$ MXene electrodes for SCs to achieve 900 F cm^3 at 2 mV s^{-1} [133]. After 10,000 cycles, the electrode showed no loss in capacitance, which is due to the reduced size of H^+ and improved interlayer spacing. Binder-free $Ti_3C_2T_x$-based electrodes for SCs were developed by using a modified electrophoretic deposition process, which produced good electrochemical performance [134]. Zhang et al. produced solid-state SCs with highly transparent and conductive $Ti_3C_2T_x$, attaining an outstanding volumetric capacitance of 676 F cm^{-3} [135]. The creation of few-layered MXenes on paper through an interaction with physical sputtering, vacuum-assisted accumulation, and laser printing resulted in outstanding electrochemical performance and high electrical conductivity. The macroporous $Ti_3C_2T_x$ MXene hydrogels proposed by Lukatskaya et al. have volumetric capacitance and high capacity of 210 F g^{-1} at 10 V s^{-1} and 100 F g^{-1} at 40 V s^{-1} [136]. The effect of $Ti_3C_2T_x$ MXenes on SCs' electrochemical performance in H_2SO_4 electrolytes was also studied by Dall'Agnese et al [137].

Furthermore, composites based on MXenes have been researched for use in SCs. When Ling et al. found that $Ti_3C_2T_x$/polymer composite films demonstrated excellent volumetric capacitance with KOH electrolytes [138]. Additionally, a mixture of $Ti_3C_2T_x$ and conductive polypyrrole (PPy) was used to create SCs, which retained 92% of their capacitance after 25,000 cycles [139]. Conductive and free-standing PPy/$Ti_3C_2T_x$ hybrid films were also developed, showing good electrochemical performance. To create hybrid designs for power holding facilities with better efficiency, polyfluorene derivatives (PFDs) were employed [140]. $Ti_3C_2T_x$ MXenes were initially proposed for their significant volumetric sensitivity when used in SC electrodes with ionic fluid electrolytes [141]. After $Ti_3C_2T_x$, Ti_2CT_x is the MXene that has been most extensively researched. Rakhi et al. investigated the effects of the post-etch annealing on Ti_2CT_x, examining its structural characteristics and electrical functionality [142]. They found that N_2/H_2 annealing produced the highest performance, with a specific capacitance of 51 F g^{-1}. Moreover, due to its large energy storage capacity, the prepared Ti_2CT_x was also employed in wire-type SCs [143]. The capabilities of MXene-based electrochemical systems for SCs are summarized in Table 1.7.

TABLE 1.7

An Investigation of the Electrochemical Characteristics of Dissimilar MXene Using SCs

No	Electrode	Electrolyte	FC	CN	SC	Rate	Ref.
1	Ti_2CT_x	30 wt% KOH	93%	6,000	–	93%	[142]
2	$Ti_3C_2T_x$	6 M KOH	~100%	5,000	–	5 A g^{-1}	[145]
3	PPy/$Ti_3C_2T_x$	0.5 M H_2SO_4	~ 400 F cm^{-3}	10,000	406 F cm^{-3}	1 mA cm^{-2}	[140]
4	N-$Ti_3C_2T_x$	1 M H_2SO_4	314 F cm^{-3}	10,000	–	50 mV s^{-1}	[146]
5	MnO_x/$Ti_3C_2T_x$	1 M Li_2SO_4	–	–	602 F cm^{-3}	2 mV s$^-$	[152]
6	$Ti_3C_2T_x$/rGO	1M $MgSO_4$	370 F cm^{-3}	10,000	340 F cm^{-3}	10 A g^{-1}	[150]
7	$Ti_3C_2T_x$/PVA	1 M KOH	~314 F cm^{-3}	10,000	370 F cm^{-3}	5 A g^{-1}	[138]
8	$Ti_3C_2T_x$	1 M H_2SO_4	~ 100%	10,000	–	10 A g^{-1}	[144]
9	LDH/$Ti_3C_2T_x$	6 M KOH	70%	4,000	–	4 A g^{-1}	[147]
10	$Ti_3C_2T_x$/CNT	6 M KOH	~390 F g^{-1}	10,000	384 F g^{-1}	50 mV s^{-1}	[148]
11	$Ti_3C_2T_x$/TiO_2	6 M KOH	~ 96%	3,000	143 F g^{-1}	5 mV s^{-1}	[149]
12	$Ti_3C_2T_x$	1 M KOH	330 F cm^{-3}	10,000	350 F cm^{-3}	1 A g^{-1}	[130]
13	$Ti_3C_2T_x$/paper	1 M H_2SO_4	92%	10,000	–	2 mA cm^{-2}	[151]

SC = starting capacitance

CN = cycle number

FC = capacity following cycles

6   NEW MXENES AND MXENES COMBINED WITH A SOLID-STATE ELECTROLYTES FOR THE DEVELOPMENT OF SC

The electrochemically active regions between the electrode and electrolyte have a major impact on the specific capacitance of SCs. The performance of energy storage systems and steady mechanical flexibility have been significantly improved by interfacial engineering. It is important to note that low moisture content can result in insufficient solid–solid heterogeneity interaction, inappropriate small structures, and stress fractures, which can significantly diminish the electrolyte/electrode interface. Beginning carbides of transition metals and carbonitrides: a unique class of 2D materials, which is called MXene, has shown significant prospects as an electrode compound for SCs. $Ti_3C_2T_x$, the most extensively studied MXene, is produced by etching Al from the multilayer carbide Ti_3AlC_2 host. The outer layers are frequently terminated with O, OH, and/or F, where T_x stands for the end of the fundamental area. MXenes are generally desirable for energy storage applications due to their diverse chemistry, customizable surface terminations, metallic conductivity, and surface hydrophilicity, particularly as SC electrode materials. Superior chemical permeability, high ionic conductivity, excellent mechanical properties, ideal interface, significant transduction capability, electrical stability, dendrite prevention, strong adhesion to electrode substrates, high electrical resistivity, and eco-friendliness are the major properties of state-of-the-art SSEs. A good junction having an excellent capacitance dimension could result from the combination of the characteristics of MXene and SSEs. SCs can be divided into hybrid, asymmetric, and symmetric SCs based on their energy storage mechanisms. The following paragraphs will discuss electrodes derived from MXene used in various SCs.

Symmetrical SCs: In 2021, Sun et al. produced MXene nanosheets on a cobalt sulfide nickel/carbon fabric, which demonstrated high cycling stability and high specific capacitance [153]. The nanosheets improved conductivity and ion penetration, making them appropriate for adaptable and efficient energy storage applications. To improve the mechanical properties of poly(3,4-ethylenedioxythiophene) (PEDOT), particularly conductive polymers, Liu and colleagues created $Ti_3C_2T_x$@PEDOT composites, which have better electrical characteristics, electrochemical stability, and capacitive performance [154]. $Ti_3C_2T_x$ and PEDOT worked together synergistically to increase interlayer conductivity and charge transport, which prevented restacking. Yan et al. produced self-supporting PPy-$Ti_3C_2T_x$ MXene films, demonstrated good electrochemical characteristics [155]. Recently, cobalt and nickel oxides have gained popularity as pseudocapacitive materials because of their low price, low toxicity, and high theoretical capacitance. Zhang et al. coated cobalt–nickel bimetallic oxides on MXene nanolayers , which led to better electrochemical performance [156]. Although $Ti_3C_2T_x$ MXene has strong electrolytic movement, electrical connectivity, and pliability, self-assembly remains a significant challenge. Zhang et al. addressed this issue by incorporating layered hierarchical porous carbon (HPC) into $Ti_3C_2T_x$ films, which avoided self-assembly and enhanced ion movement both laterally and longitudinally [157]. Following charge/discharge cycles, the symmetric SC with 60% HPC in a quasi-solid condition exhibited outstanding reliability and retained 86% of its capacitance.

Asymmetric SCs: The power window for MXene-based symmetric SCs is only 0.6 V due to oxidation taking place at anodes. Asymmetric SCs, which are made of pseudocapacitive compounds and MXene electrodes, can efficiently widen the voltage window. For example, Xu et al. developed an adaptable asymmetric SC in 2021 featuring a PVA/sulfuric acid hydrogel electrolyte and an IDT@rGo structural heterojunction [158]. This device exhibited a high capacitance of 60 F g^{-1} and an output voltage of 1.6 V, achieving 17 Wh kg^{-1} energy efficiency up to 8 kW kg^{-1}. The introduced ESD had good mechanical flexibility, cycle stability, and multiplicity, making it promising for wearable and smart devices. Ma et al. enhanced the MXene surface using lignosulfonate-modified GO 3D porous aerogels [159]. This modification prevented restacking and offered better electrochemical characteristics. To obtain a high power content of 142 Wh m^{-2} and capacitance retention of 96.3% following 10,000 cycles, the asymmetric SC took advantage of the elevated level of pseudocapacitance and chemical sensitivity properties of lignosulfonates. Zhao et al. used an electrostatic self-assembly technique to produce Ti_3C_2/FeOOH polymer hybrids with quantum dots (QDs) [160]. The Ti_3C_2 nanosheet repacking is prevented by the amorphous FeOOH QDs, which also have a large capacitance. The asymmetric device achieved a maximum energy density of 8.2 W cm^{-2} and surface conductance 2.3 times greater than pure Ti_3C_2 sheets. Conductive polymers offer unique advantages such as low cost, simple production, compatibility, and customizable intrinsic features. Recent advances have increased their use in ESDs . Wang et al. produced PAN/MXene polymer films with density thin-film conductors by incorporating Polyaniline (PANI) nanoparticles into MXene interlayers [161]. These electrodes featured a flexible substrate, high conductivity, well-dispersed MXene nanosheets, and effective binding. PANI nanoparticles provided synergistic effects for interlayer conductivity and powerful pseudocapacitive elements.

Flexible SCs: In the last few years, there has been an increase in interest in the production of active electrode materials for flexible SCs employing MXene substrates. Yu et al. created a one-step procedure to produce flexible SCs using $Ti_3C_2T_x$ MXenes in organic electrolytes, which exhibited conductivity and electrochemically activity [162]. Activated charcoal grains were sandwiched between MXene layers, eliminating the necessity for an insulating polymer binder. The electrode apply of MXene for various purposes, such as a binder, flexible support, conductive additive, and additional active substances. The combination of MXenes with activated carbon resulted in a 3D network that enhanced the MXenes between layers and significantly increased the electrode capacitance and multiplicity capacity, with a specific capacity of 126 F g^{-1} in 0.1 to 100 A g^{-1} organic electrolytes. Zhou et al. created an elastic, self-sustaining electrode using a carbon cloth (CC) composite of MnO_2 nanorods and MXene [163]. The electrode enhanced of the electrochemical performance, cycling stability, and supercapacitor flexibility. Zheng et al. laminated $Ti_3C_2T_x$ MXene sheets and PEDOT films using vapor phase polymerization (VPP) and spraying [164]. The PEDOT/MXene-coated cotton materials exhibited superior electrochemical, Joule heating, EMI shielding, and strain sensing capabilities, paving the way for wearable electronics and multifunctional fibers . The rate performance and mass loading of MXene nanosheets are significantly impacted by their self-stacking. Fan et al. addressed this issue by adding $Fe(OH)_3$ nanoparticles with diameters of 3–5 nm to MXene films, producing a flexible nano-porous MXene film [165]. This modification resulted in high bulk capacitance and good flexibility, making

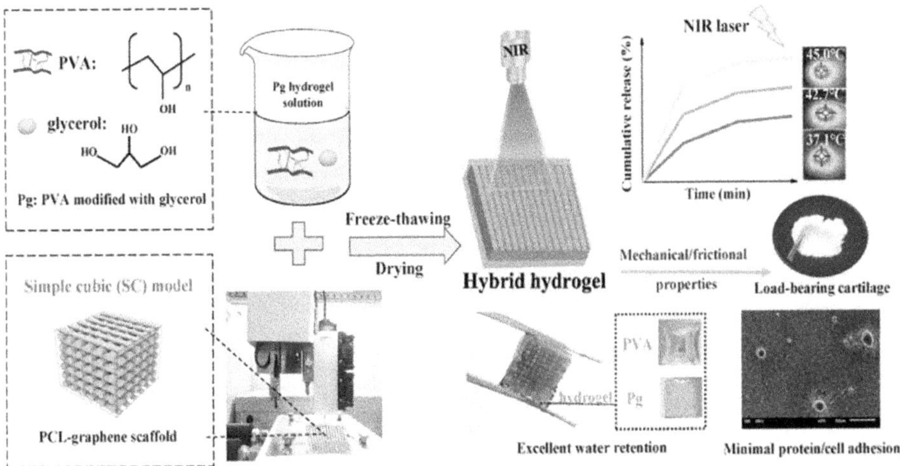

FIGURE 1.11 Graphical representation of SC film and hydrogel synthesis, including a picture.

it a potential electrode material for small, lightweight, and flexible energy storage systems. The film is a successful material design for flexible ESD because of its remarkable flexibility under high-quality stresses. Conducting polymers and their composites provide excellent flexible electrode materials, but the degradation of the polymer structure during charge and discharge can reduce their capacitance. The two most common conducting polymers are PANI and PPy. The following sentences provide a summary of the research on conducting polymers utilizing flexible electrodes. For example, a pliable PANI film electrode developed by Qi Wang et al. had a real capacitance of 143 mF cm^{-2}, while a flexible CNT/PPy electrode constructed by Liu et al. had a fiber-shaped areal specific capacitance of 58.82 mF cm^{-2}. The elimination of doping ions during the charge and discharge cycles may have caused the polymer structure to be destroyed [104]. Flexible SCs with a traditional design often have a multilayer laminated construction; however, when subjected to physical deformation, this design is susceptible to structural damage. An integrated electrode–electrolyte–electrode (all-in-one) configuration is needed in a new manufacturing technique to create high-performance flexible SCs. GPEs can act as both flexible substrates and electrolytes in integrated devices. The synthesis and physical characteristics of hydrogel electrolytes are shown in Figure 1.11. The cell design affects electrical and mechanical properties of QSSE-flexible SC [127], as depicted in Figure 1.12. Achieving high-performance electrical and mechanical qualities for SCs significantly depends on the cell design and the electrolyte/electrode selection. Conductive polymers have the potential for reverse doping–dedoping, which can enhance their conductivity; in contrast, SCs based on Conductive polymer (CP) have limited cycle stability [166]. The apply of hybrid electrodes, such as carbon-based materials and blending metal oxides carbon-based materials among only one electrode, focus to raise the sluggish energy densities of some SCs [167]. Flexible substrates include polymers, metal sheets, and paper made of carbon. QSSEs must be used in building flexible SCs [168], [169], necessitating new and enhanced QSSEs and SSEs. As stated by An and Cheng [170], PVA-based GPEs are used in most stretchable SCs. An electrolyte with a broad voltage window, excellent ion transport, and good stability is needed [171].

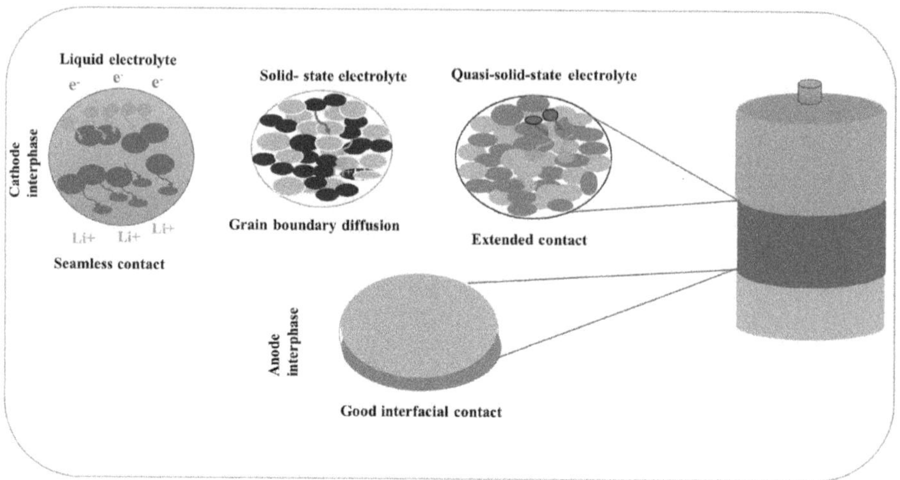

FIGURE 1.12 Design of a QSSE.

TABLE 1.8
An Overview of the Flexible SCs

No	Electrode Substances	Retention	Cycle Number	Sp. cap	Rate	Flexibility	Ref.
1	rGO/Ti$_3$C$_2$T$_x$	100%	10,000	370 F cm^{-3}	1 A g^{-1}	Bending	[172]
2	MXene/CNTs	86.3%	10,000	74.1 F g^{-1}	5 mV s^{-1}	Bending/ folding	[173]
3	MXene/carbon cloth	94.2%	1,000	413 mF cm^{-2}	1 mA cm^{-2}	Bending	[174]
4	MXene/MnO$_2$/ CC	83%	10,000	511.2 F g^{-1}	5 A g^{-1}	Bending/ folding	[163]
5	Ti$_3$C$_2$T$_x$@ PEDOT	96.5%	10,000	564 F g^{-1}	1 A g^{-1}	Twisting	[175]
6	MXene/rGO (20%)	90%	40,000	260.1 F g^{-1}	1000 mV s^{-1}	Bending/ twisting	[176]
7	MXene/MnO$_2$ (NWs)	85%	10,000	216 2 mF cm^{-1}	10 mV s^{-1}	Stretching/ bending	[177]

Table 1.8 summarizes a few typical MXene-based flexibility SCs. The following sentences present a summary of the latest research on MXene as well as SSEs for SCs. The first study carried out by Zhang et al. focused on a cation-induced Ti$_3$C$_2$T$_x$ MXene hydrogel for energy storage applications. Although Ti$_3$C$_2$T$_x$ MXenes have been a strong candidate in the energy storage market, it is still challenging to make 3D MXene hydrogels due to inherent hydrophilicity and excellent conductivity. A quick and easy cation-induced approach was used to successfully create cation-induced Ti$_3$C$_2$T$_x$ MXene hydrogels. Studies on the hydrogel characteristics revealed that it

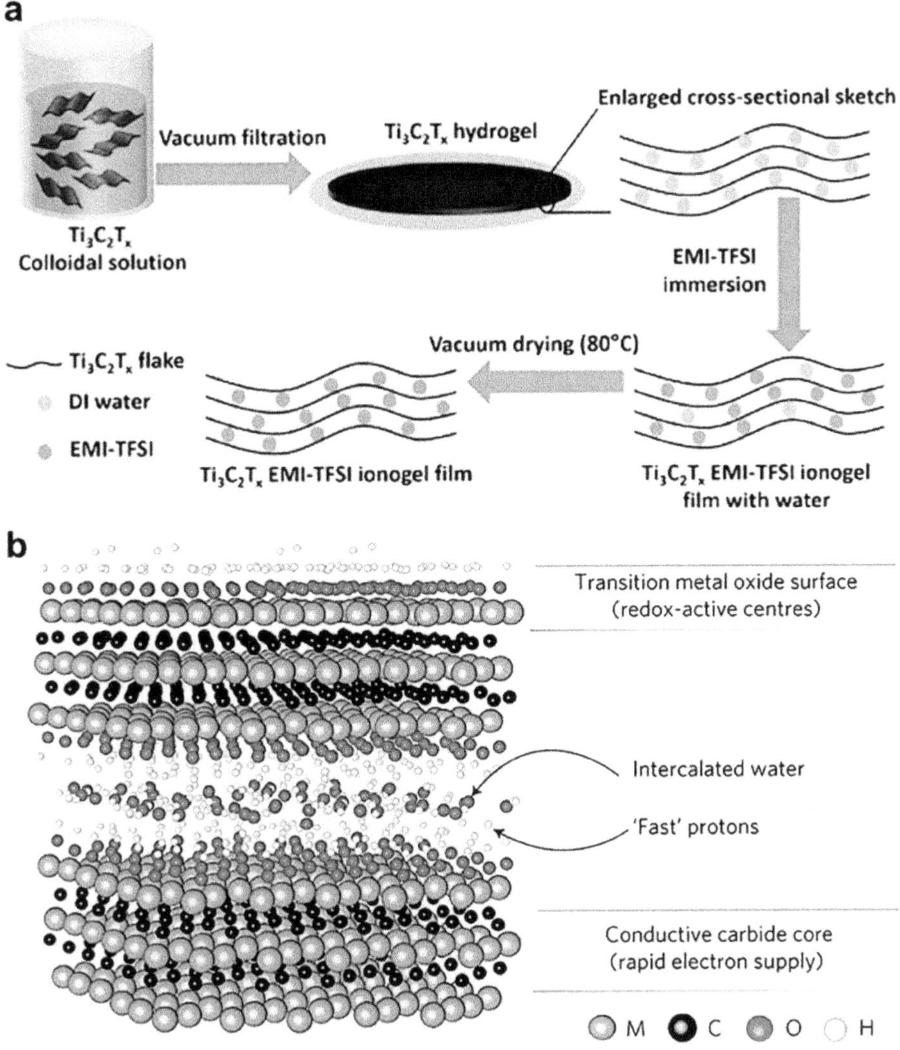

FIGURE 1.13 (a) Diagrammatic representation of the synthesis of the all-MXene hydrogel and ionogel (b) Pre-intercalated water molecules within the framework of nanosheets constructed from MXene enable proton transport to redox-active sites [178].

had a configurable basal spacing and a 3D porous structure, contributing to its good capacitance and high-rate capability. According to computation utilizing density functional theory (DFT), a higher number of intercalated cations and larger hydrated radii are advantageous for expanding the basal space. Therefore, high-rate capabilities and excellent capacitance were attained in cation-induced Ti$_3$C$_2$T$_x$ hydrogels. This study shows a straightforward method for creating MXene electrodes for preserving capacitance charge that uses a unique morphological control standard. Figure 1.13a illustrates the purification procedure used in this study [178].

To facilitate the removal process, the hydrogel coating was subsequently immersed in acetone. The gel-like structure was caused by an energetic physical interaction within the MXene film caused by a significant amount of residual pre-intercalated water molecules. However, the van der Waals intercellular interactions that are now in place cannot be sustained by such a type of force. Therefore, to prevent the MXene particles from restacking, each of the aforementioned studies employed a distinct approach to inhibit the hydrogels from collapsing. Lin et al. allowed solvents to interchange by submerging the non-dried hydrogel sheet in a thermally resistant ionic solution, such as 1-ethyl-3-methylimidazolium bis(trifluoromethylsulfonyl)imide. The ionized solution stayed in the MXene sheet upon drying under vacuum at 80°C, which inhibited reassembling and enhanced the layer-to-layer gap, generating an ionogel, $Ti_3C_2T_x$ rather than a hydrogel. Lukatskaya et al. employed the produced MXene hydrogel directly for electrochemical experiments after immersing it in a H_2SO_4 solution for a period of time to retain its structure. Protons can move quickly in the MXene configuration because of the pre-intercalated water molecules, as shown in Figure 1.13b.

MXene/MO composites can be produced through field assembly or in situ processing. Growing metal oxides directly on liquid-phase MXene nanosheets chemically is a straightforward method to promote significant interactions between components during in situ synthesis. The synthesis process of CoS_2 nanoparticles growing on an MXene surface is depicted in Figure 1.14a. A straightforward, one-step solvent thermal technique was employed for the synthesis. Scanning electron microscopy (SEM) images show that the composite exhibits two different morphologies: a lamellar structure and a spherical structure. This combination has the potential to have a synergistic impact, which will improve its electrochemical performance. The electrochemical performance of the MXene/CoS_2 (CCH) composition in a liquid electrolyte with 2 M KOH is shown in Figure 1.14b–g. The maximum operational voltage window that the MXene/CoS_2 (CCH)/rGO asymmetric SCs (ASCs) can achieve is 1.6 V, as illustrated in Figure 14b–g. Furthermore, after 5000 cycles of charging and discharging, the specific capacitance maintenance of the device is approximately 98%, indicating exceptional long-time cyclic efficiency. It is important to note that the permeability gradually decreases during the first 1,000 cycles before rising to 98% within the next 1,000–5,000 cycles. The wettability between active materials and electrolyte ions is significantly enhanced, which could be the reason for the activation of electrode materials.

Figure 1.15a illustrates the process used to create oxygen-functionalized Ti_3C_2 MXene nanosheets. According to SEM images, compact Ti_3AlC_2 granules were etched in an aquatic HF mixture to create the multilayer $Ti_3C_2T_x$ MXene containing a packed accordion-like structure. The polarization curve of the sample is displayed in Figure 1.15b. This verifies the low performance of E-$Ti_3C_2T_x$ for the hydrogen evolution reaction (HER). Remarkably, at a voltage density of 10 mA cm^{-2}, the overpotentials of E-$Ti_3C_2T_x$ heated at 450°C (E-$Ti_3C_2T_x$-450) and E-Ti_3C_2 (OH)$_x$ were calculated to be approximately 266 and 217 mV, respectively. These values are significantly lower than those of E-$Ti_3C_2T_x$. E-$Ti_3C_2O_x$ had the most effective HER catalytic performance (~190 mV overpotential at 10 mA cm^{-2}) when compared to E-$Ti_3C_2T_x$, E-$Ti_3C_2T_x$-450, and E-Ti_3C_2(OH)$_x$. According to the theoretical conclusions, the higher efficiency of E-$Ti_3C_2O_x$ compared to E-Ti_3C_2(OH)$_x$ implies that

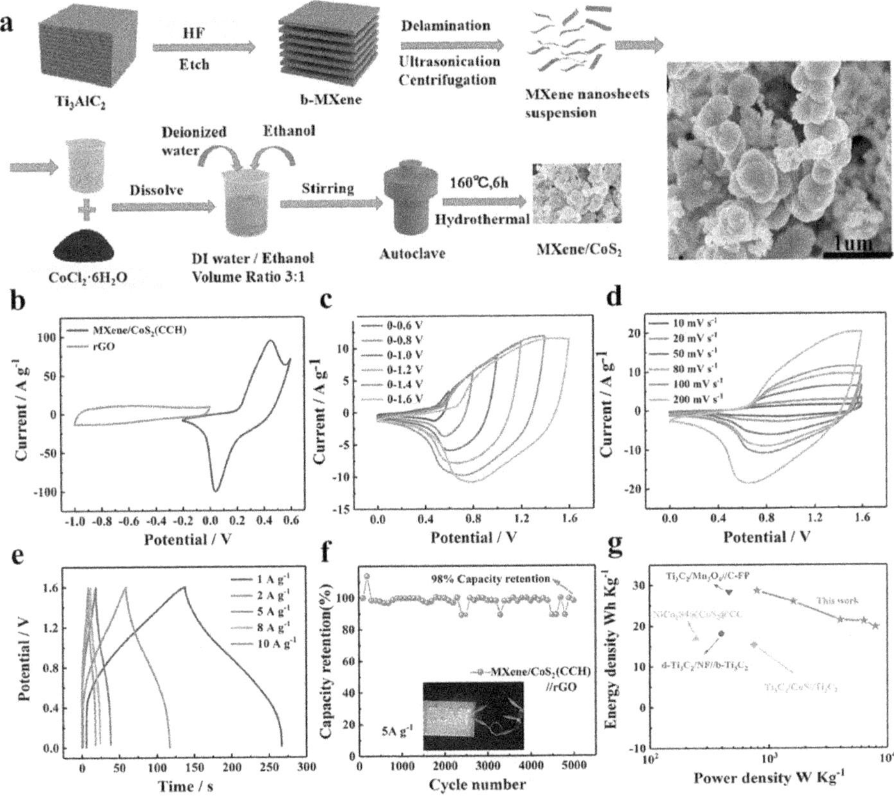

FIGURE 1.14 (a) Schematic diagram of the synthesis process for the MXene/CoS$_2$ (CCH) composite. (b) SEM image and the (b–g) electrochemical performance of MXene/CoS$_2$ (CCH)//rGO ASCs in a 2 M KOH aqueous electrolyte [187].

extremely active regions for the HER instead of the OH terminations are found on the basal plane.

Additionally, the development of two-electrode symmetric and asymmetric devices based on Mo$_{1.33}$C MXenes demonstrates the potential of this material for real-world applications. One such device is a two-electrode symmetric SC [180], which utilized composites made of a polymer and Mo$_{1.33}$C. A high-performance ultrathin flexible Mo$_{1.33}$C MXene/PEDOT:poly(styrene sulfonic acid) (PSS) composite film was created by combining Mo$_{1.33}$C MXene with PEDOT:PSS. This film acted as an electrode for solid-state SCs, delivering a high volumetric capacity of 568 F cm^{-3}, an ultrahigh power consequence of 33.2 mWh cm^{-3}, and a voltage density of 19,470 mW cm^{-3}. Moreover, the Mo$_{1.33}$C MXene/PEDOT:PSS composite material exhibited a high volumetric capacitance of 1310 F cm^{-3} following treatment with H$_2$SO$_4$. The rate capability of the Mo$_{1.33}$ MXene/PEDOT:PSS composite sheet was also increased by treatment with H$_2$SO$_4$ as it contains conductive PEDOT and PSS. In addition to Mo-based i-MXenes, a preliminary evaluation of W$_{1.33}$C i-MXene, which is created from (W$_{2/3}$Sc$_{1/3}$)$_2$AlC and (W$_{2/3}$Y$_{1/3}$)$_2$AlC i-MAX phases, is discussed. Free-standing

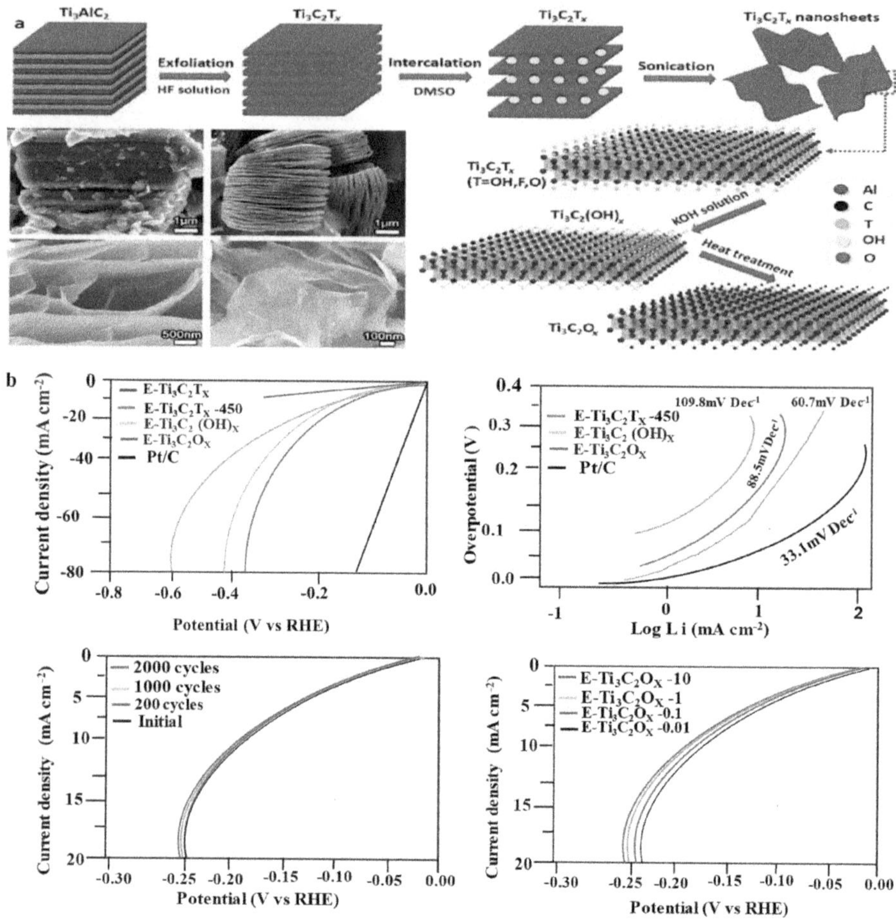

FIGURE 1.15   SEM images of Ti_3AlC_2, $Ti_3C_2T_x$ coatings, layers with $Ti_3C_2T_x$ incorporated with DMSO, and E-$Ti_3C_2T_x$. (a) Fabrication of $Ti_3C_2T_x$ MXene and E-$Ti_3C_2O_x$. (b) Tafel slopes from polarization graphs of E-$Ti_3C_2O_x$, E-$Ti_3C_2(OH)_x$, E-$Ti_3C_2O_x$, E-$Ti_3C_2T_x$-450, and 20 weight percent Pt/C on the glassy carbon electrode at 5 mV s^{-1} in 0.5 m H_2SO_4; Polarization curves for E-$Ti_3C_2O_x$ before and after potential sweeps (0 to −0.3 V vs. reversible hydrogen electrode (RHE) at a scan rate of 50 mVs^{-1}); E-$Ti_3C_2O_x$ −10, E-$Ti_3C_2O_x$ −1, E-$Ti_3C_2O_x$ −0.1, and E-$Ti_3C_2O_x$ −0.01 [179].

$W_{1.33}C$ electrodes with a thickness of 2–4 m were made as composites with 10 wt.% PEDOT:PSS, and they were assessed in a solution of 1 m H_2SO_4 electrolyte. The normalized capacitances were calculated to be 610 F cm^{-3} (116 F g^{-1}) and 591 F cm^{-3} for 2-m-thick $W_{1.33}C$ (Sc)/PEDOT:PSS and $W_{1.33}C$ (Y)/PEDOT:PSS electrodes, respectively. $W_{1.33}C$ i-MXene, generated from the parent phase containing Y, has a slightly lower charge storage capacitance than its Sc-containing counterpart, comparable to Mo-based i-MXene. It was found that the charge storage mechanism in W-based i-MXenes was dominated by surface-controlled capacitive behavior (Figure 1.16).

Chu et al. developed additional gel electrolytes and SCs based on MXene. Limited cell potential windows have made it difficult for aqueous MXene-based

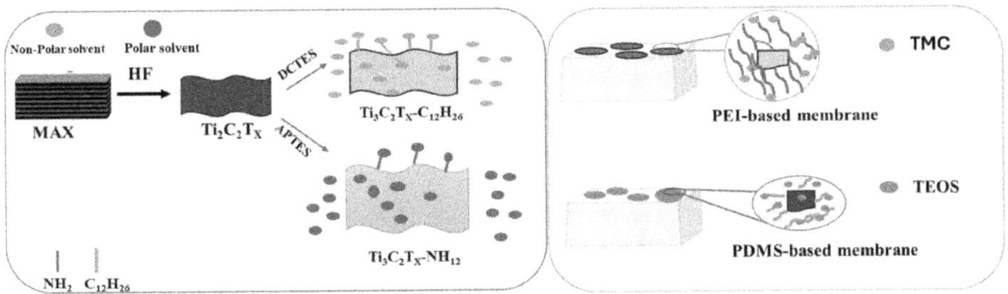

FIGURE 1.16 Schematic representation of galvanostatic charge–discharge profiles for the electrode preparation and actual sp.

SCs to proliferate. There have been suggestions for ASCs to extend potential windows to solve this problem. Building conducting polymer hydrogels (CPHs) with coupled 3D nanostructures and substantial PCs is a step in supramolecular techniques. The power density could be increased to 16.6 Wh cm^{-2}, and the CPH/Ti$_3$C$_2$T$_x$ ASC in its as-assembled state could function in a wide potential window of 1.15 V. To increase the efficiency and possible openings of aquatic MXene SCs, the use of a positive CPH electrode is a promising method [181]. Wang et al. introduced a new MXene-based SC. Due to its flexibility, metallic conductivity, and excellent capacitance, MXene is an ideal electrode for flexible SCs. However, reassembling within MXene concentrations limits the ion kinetics and reduces the number of ion storage sites. The restacking problem of traditional MXene films can be resolved using a 3D intercalation method. For effective ion transport in the Magnesium (MG) film electrodes, graphdiyne nanotubes (GDY-NTs), which feature in-plane pores, allow for a greater cross-sectional flow area. Graphene nanotubes were used to create a composite film that has increased capacitance and rate capabilities. GDY-NTs with built-in in-plane pores enabling horizontal–vertical intercalation among MXene layers were used to create a free-standing, flexible, structurally 3D-connected, and hydronium ion-penetrable MXene/GDY-NT composite film. With a high energy density of 19.7 Wh kg^1, the as-assembled MG-based SC has good wearability. The asymmetric solid-state flexible SC has high energy density and capacitance retention after 10,000 cycles. This enhanced capacitance and rate performance resulted in enhanced electrochemical performance, making it a promising candidate for wearable, portable, and adaptable electronics [182]. Table 1.9 summarizes some common solid electrolytes and MXene-based electrodes for SCs.

Despite numerous advancements and successes, research on MXenes for SCs is still in its early stages, as seen by their limited capacity to create novel MXenes and enhance their usefulness. The next steps include controlling the surface chemistry, comprehending energy storage methods that use capacitance, and developing more efficient SCs. One crucial factor is the electrolyte, which is especially important because it determines how better the Flexible supercapacitors (FSCs) preserve energy, second only to electrode materials in terms of importance. Although aquatic electrolytes offer strong ionic conductivity, SCs face significant limitations in improving their energy and power densities. Despite the high water-based electrolytes' electrical conductivity, SCs have a very limited ability to increase the concentrations of their power and

TABLE 1.9

Summary of the SCs Based on MXene

No	Electrode Material	Electrolyte	Power Density	Capacitance	Energy Density	Capacitance Retention	Ref.
1	rGO/$Ti_3C_2T_x$	PVA-KOH	60 mW cm^{-3}	–	63 mW h cm^{-3}	100% over 10,000 (5 A g^{-1})[3]	[23]
2	$Mo_{1.33}$ C MXene/PEDOT:PSSS	PVA-H_2SO_4	19470 mW cm^{-3}	1310 F cm^{-3}	33.2 mWh cm^{-3}	90% over 10,000 (2 mV s^{-1})	[180]
3	MXene-WO_3 nanorod/rGO sponge	KOH/ $K_4[Fe(CN)_6]$ /PVA gel	1450 W kg^{-1}	376.8F g^{-1}	34 W h kg^{-1}	86% over 3,000 (5 A g^{-1})	[183]
4	2D SnO_2/$Ti_3C_2T_x$	PVA-H_2SO_4	1,206 µW cm^{-2}	190 F g^{-1}	6.7 µWh cm^{-2}	87.4% over 4,000 (10 mV s^{-1})	[43]
5	$Mo_{1.33}$C/$Ti_3C_2T_x$/Polymer	Na_2SO_4/PPy/ SDBS	32.9 W cm^{-3}	69.5 mF cm^{-2}	250.1 mWh cm^{-3}	89% above 4,000 (10 mV s^{-1})	[184]
6	$Ti_3C_2T_x$/RuO_2	PVA-H_2SO_4	40 mW cm^{-2}	50 mF cm^{-2}	37 µW h cm^{-2}	86% over 20,000 (20 mV s^{-1})	[185]
7	MXene/Ag NWs/cellulose	PVA-H_2SO_4	–	505 F g^{-1} at 10 mV s^{-1}	–	90.9% over 500 (10 mV s^{-1})	[186]

energy. Due to the growing power demand for wearable, portable, printable, small, and highly flexible electronic devices, SSE-based SCs have received considerable attention. For this reason, they are necessary for transferring and maintaining the voltages between the two electrodes in SCs. SSEs are made up of inorganic solid and organic polymer electrolytes, including single crystals, amorphous materials, and polycrystalline materials, and an organic polymer solid electrolyte made up of an organic polymer matrix and salt. Furthermore, most SSEs for SCs are constructed using polymers, such as solid polymers, gel polymers (quasi-solid states), and polyelectrolytes. The high ionic conductivity of GPEs, which is result of their liquid phase, has led to their application in SCs. As the performance of the MXene–electrolyte interface determines the electrochemical outcomes of MXene, the choice of the electrolyte has a substantial impact on the electrochemical results of MXene electrodes. Additionally, flexible and printable MXene-based SCs can enhance their mechanical features without altering their electrochemical outcomes by using reinforced materials in the MXene matrix.

7 CONCLUSION

This study discusses the latest advancements in MXene-based electrode functions involving preserving energy. Key topics include synthesis methods, shape design, chemical modifications, electrochemical properties, and a basic understanding of the charge-holding process in various electrolytes. MXenes were produced using three main synthetic methods: HF etching, cation intercalation and delamination, moderately enriched F solutions, and F-free electrolytes. Energy storage is one of the primary applications for MXene materials because of their ability to alter their surface chemistry. Enhancing the electrochemical characteristics of MXenes through surface chemical modifications is a crucial technique. MXenes can be altered chemically and doped to improve performance. For reinforced materials to have better mechanical properties, MXene matrices can be added. Although the development of MXenes for SCs is still in its early stages, the following process entails understanding the mechanisms of capacitive energy storage, managing surface chemistry, and creating architectures that improve the efficacy of SCs. Recent advancements in energy preservation strategies include the development of hybrid gel electrolytes by analyzing their interactions, enhancing their compatibility, and incorporating redox-active components. These advancements can enhance the overall efficacy of SCs, which are essential for flexible, lightweight, and wearable capacitive electronics. Due to the improved ionic conductivity of QSSEs like GPEs, a rising number of researchers are exploring their potential. When developing novel QSSEs for SCs, it is essential to choose the right polymers and optimize the power region to prevent the breakdown of solvents.

REFERENCES

1. L. Zhang, D. Shi, T. Liu, M. Jaroniec, J. Yu, Nickel-based materials for supercapacitors, *Mater. Today*, 25 (2019), 35–65.

2. B. S. Singu, E. S. Goda, K. R. Yoon, Carbon nanotube–Manganese oxide nanorods hybrid composites for high-performance supercapacitor materials, *J. Ind. Eng. Chem.*, 97 (2021), 239–249

3. J. Sung, C. Shin, Recent studies on supercapacitors with next-generation structures, *Micromachines*, 11 (12) (2020), 1125.

4. B. Pal, S. Yang, S. Ramesh, V. Thangadurai, R. Jose, Electrolyte selection for supercapacitive devices: A critical review, *Nanoscale Advances*, 1 (10) (2019), 3807–3835.

5. Y. Shao, J. Li, Y. Li, H. Wang, Q. Zhang, R. B. Kaner, Flexible quasi-solid-state planar micro-supercapacitor based on cellular graphene films, *Mater. Horizons*, 4 (6) (2017), 1145–1150.

6. Q. Meng, K. Cai, Y. Chen, L. Chen, Research progress on conducting polymer based supercapacitor electrode materials, *Nano Energy*, 36 (2017), 268–285.

7. W. Qin, N. Zhou, C. Wu, M. Xie, H. Sun, Y. Guo, L. Pan, Mini-review on the redox additives in aqueous electrolyte for high performance supercapacitors, *ACS Omega*, 5 (8) (2020), 3801–3808.

8. Y. Zhu, O. Fontaine, *Most modern supercapacitor designs advanced electrolyte and interface*, IntechOpen (2021).

9. K. Sun, E. Feng, G. Zhao, H. Peng, G. Wei, Y. Lv, G. Ma, A single robust hydrogel film based integrated flexible supercapacitor, ACS Sustain. *Chem. Eng.*, 7 (2018), 165–173.

10. B. S. Yin, S. W. Zhang, K. Ke, Z. B. Wang, Advanced deformable all-in-one hydrogel supercapacitor based on conducting polymer: Toward integrated mechanical and capacitive performance, *J. Alloys Compd.*, 805 (2019), 1044–1051.

11. B. Zhang, J. Li, F. Liu, T. Wang, Y. Wang, R. Xuan, G. Zhang, R. Sun, C. P. Wong, Self-healable polyelectrolytes with mechanical enhancement for flexible and durable supercapacitor, *Chemistry*, 25 (2019), 11715–11724.

12. Y. Lu, H. Mi, C. Ji, F. Guo, Z. Bai, Y. Liu, C. Yu, J. Qiu, Synergizing layered carbon and gel electrolyte for efficient energy storage, *ACS Sustain. Chem. Eng.*, 8 (10) (2020), 4207–4215.

13. L. Li, Z. Hu, Q. Liu, J.-Z. Wang, Z. Guo, H.-K. Liu, Cathode materials for high-performance potassium-ion batterie, *Cell Rep. Phy. Sci.*, 2 (2021) 100657, https://doi.org/10.1016/j.xcrp.2021.100657.

14. J. Cao, Z. Sun, J. Li, Y. Zhu, Z. Yuan, Y. Zhang, D. Li, L. Wang and W. Han, Microbe-assisted assembly of $Ti_3C_2T_x$ MXene on fungi-derived nanoribbon heterostructures for ultrastable sodium and potassium ion storage, *ACS Nano*, 15 (2021), 3423–3433.

15. B. Yan, L. Feng, J. Zheng, S. Jiang, C. Zhang, Y. Ding, J. Han, W. Chen, S. He, High performance supercapacitors based on wood-derived thick carbon electrodes synthesized via green activation process, Inorg. *Chem. Front.*, 9 (2022), 6108–6123.

16. J. Zheng, B. Yan, Q. Zhang, C. Zhang, W. Yang, J. Han, S. Jiang, S. He, Potassium citrate assisted synthesis of hierarchical porous carbon materials for high performance supercapacitors, *Diam. Relat. Mater.*, 128 (2022), 109247.

17. S. Zheng, J. Zhang, H. Deng, Y. Du, X. Shi, Chitin derived nitrogen-doped porous carbons with ultrahigh specific surface area and tailored hierarchical porosity for high performance supercapacitors, *J. Bioresour. Bioprod.*, 6 (2) (2021), 142–151.

18. L. Wei, W. Deng, S. Li, Z. Wu, J. Cai, J. Luo, Sandwich-like chitosan porous carbon Spheres/MXene composite with high specific capacitance and rate performance for supercapacitors, *J. Bioresour. Bioprod.*, 7 (1) (2022), 63–72.

19. J. Xiao, H. Li, H. Zhang, S. He, Q. Zhang, K. Liu, S. Jiang, G. Duan, K. Zhang, Nanocellulose and its derived composite electrodes toward supercapacitors: Fabrication, properties, and challenges, *J Bioresour Bioprod.*, 7 (4) (2022), 245–269.

20. S. Z. Butler, S. M. Hollen, L. Cao, Y. Cui, J. A. Gupta, H. R. Guterrez, T. F. Heinz, S. S. Hong, J. Huang, A. F. Ismach, E. J. Halperin, M. Kuno, V. V. Plashtina, R. D. Robinson, R. S. Ruoff, S. Salahuddin, J. Shan, L. Shi, M. G. Spencer, M. Terrones, W. Windi, J. E. Goldberger, Progress, challenges, and opportunities in two-dimensional materials beyond graphene. *ACS Nano*, 7 (4) (2013), 2898–2926.

21. Y. Huang, H. Li, Z. Wang, M. Zhu, Z. Pei, Q. Xue, Y. Huang, C. Zhi, Nanostructured polypyrrole as a flexible electrode material of supercapacitor. *Nano Energy.* 22 (2016), 422–438.
22. Y. Wang, W. Lai, N. Wang, Z. Jiang, X. Wag, P. Zou, Z. Lin, H. J. Fan, F. Kang, C. P. Wong, C. Yang, A reduced graphene oxide/mixed-valence manganese oxide composite electrode for tailorable and surface mountable supercapacitors with high capacitance and super-long life, *Energy Environ. Sci.*, 10 (4) (2017), 941–949.
23. S. Xu, G. Wei, J. Li, W. Han, Y. Gogotsi, Flexible MXene-graphene electrodes with high volumetric capacitance for integrated co-cathode energy conversion/storage devices, *J. Mater. Chem. A*, 5 (33) (2017), 17442–17451.
24. B. Ahmed, D. H. Anjum, Y. Gogotsi, H. N. Alshareef, Atomic layer deposition of SnO_2 on MXene for Li-ion battery anodes, *Nano Energy*, 34 (17) (2017), 249–256.
25. L. Wen, F. Li, H. Cheng, Carbon nanotubes and graphene for flexible electrochemical energy storage: From materials to devices, *Adv. Mater.*, 28 (22) (2016), 4306–4337.
26. Z. Su, C. Yang, B. Xie, Z. Lin, Z. Zhang, J. Liu, B. Li, F. Kang, C. P. Wong, Scalable fabrication of MnO_2 nanostructure deposited on free-standing Ni nanocone arrays for ultrathin, flexible, high-performance micro-supercapacitor, *Energy Environ. Sci.*, 7 (8) (2014), 2652–2659.
27. D. Er, J. Li, M. Naguib, Y. Gogotsi, V. B. Shenoy, Ti_3C_2 MXene as a high capacity electrode material for metal (Li, Na, K, Ca) ion batteries, *ACS Appl Mater Interfaces*, 6 (2014), 11173–11179.
28. D. Cai, H. Huang, D. Wang, B. Liu, L. Wang, Y. Liu, Q. Li, T. Wang, High-performance supercapacitor electrode based on the unique $ZnO@Co_3O_4$ core/shell heterostructures on nickel foam, *ACS Appl. Mater. Interfaces*, 6 (18) (2014), 15905–15912.
29. C. Si K. H. Jin, J. Zhou, Z. Sun, F. Liu, Large-gap quantum spin hall state in MXenes: d-band topological order in a triangular lattice, *Nano Lett.*, 16 (10) (2016), 6584–6591.
30. K. O. Oyedotun, J. O. Ighalo, J. F. Amaku, C. Olisah, A. O. Adeola, K. O. Iwuozor, K. G. Akpomie, J. Conradie, K. A. Adegoke, Advances in supercapacitor development: Materials, processes, and applications, *J. Electron. Mater.*, 52 (2023), 96–129.
31. H. Zhang, J. Zhang, X. Gao, L. Wen, W. Li, D. Zhao, Advances in materials and structures of supercapacitors, *Ionics*, 28 (2022), 515–531.
32. D. P. Chatterjee, A. K. Nandi, A review on the recent advances in hybrid supercapacitors, *J. Mater. Chem. A*, 9 (2021), 15880–15918.
33. D. Kim, G. Lee, D. Kim, J. S. Ha, Air-stable, high-performance, flexible micro supercapacitor with patterned ionogel electrolyte, *ACS Appl. Mater. Interfaces*, 7 (8) (2015), 4608–4615.
34. J. Chen, P. S. Lee, Electrochemical supercapacitors: From mechanism understanding to multifunctional applications, *Adv. Energy Mater.*, 11 (6) (2021), 2003311.
35. X. He, X. Zhang, A comprehensive review of supercapacitors: Properties, electrodes, electrolytes and thermal management systems based on phase change materials, *J. Energy Storage*, 56 (2022), 106023.
36. Y. Zhang, H. Mei, Y. Cao, Z. Yan, J. Yan. H. Gao, H. Luo, S. Wang, X. Jia, L. Kachalova, J. Yang, S. Xue, C. Zhou, L. Wang, Y. Gui, Recent advances and challenges of electrode materials for flexible supercapacitors, *Coord. Chem. Rev.*, 438 (2021), 213910.
37. P. Forouzandeh, V. Kumaravel, S. C. Pillai, Electrode materials for supercapacitors: A review of recent advances, *Catalysts*, 10 (9) (2020), 969.
38. A. Riaz, M. R. Sarker, M. H. Md Saad, R. Mohamed, Review on comparison of different energy storage technologies used in micro-energy harvesting, WSNs, low-cost microelectronic devices: Challenges and recommendations, *Sensors*, 21 (15) (2021), 5041.
39. M. Chen, Y. Zhang, G. Xing, S. L. Chou, Y. Tang, Electrochemical energy storage devices working in extreme conditions, *Energy Environ. Sci.*, 14 (2021), 3323–3351.

40. K. Brinkert, P. MAndin, Fundamentals and future applications of electrochemical energy conversion in space, *NPJ Microgravity*, 8 (52) (2022), 6730, https://doi.org/10.1038/s41526-022-00242-3.

41. J. Tang, T. Mathis, X. Zhong, X. Xiao, H. Wang, M. Anayee, F. Pan, B. Xu, Y. Gogotsi, optimizing ion pathway in titanium carbide MXene for practical high-rate supercapacitor, *Adv. Energy Mater.*, 11 (4) (2021), 1–8.

42. Z. Zhang, Z. Yao, X. Zhang, Z. Jiang, 2D carbide MXene under postetch low-temperature annealing for high-performance supercapacitor electrode, *Electrochim Acta.*, 359 (2020), 136960.

43. C. Cai, W. Zhou, Y. Fu, Bioinspired MXene nacre with mechanical robustness for highly flexible all-solid-state photothermo-supercapacitor, *Chem. Eng. J.*, 418 (2021), 129275.

44. M. W. Barsoum, The $M_{n+1}AX_n$ phases: A new class of solids, Prog. *Solid St. Chem.*, 28 (2000), 201–281.

45. Y. Zhao, K. Watanabe, K. Hashimoto, Self-supporting oxygen reduction electrocatalysts made from a nitrogen-rich network polymer, *J. Am. Chem. Soc.*, 134 (2012), 19528–19531.

46. A. A. Shamsabadi, Z. Fakhraai, M. Soroush. Chapter 18- MXene-based molecular sieving membranes for highly efficient gas separation. https://doi.org/10.1016/B978-0-12-823361-0.00017-4

47. K. Deshmukh, T. Kovarik, S. K. Khadheer Pasha, State of the art recent progress in two dimensional MXenes based gas sensors and biosensors: A comprehensive review, *Coord. Chem. Rev.*, 424 (2020), 213514.

48. Y. Gogotsi, B. Anasori, The rise of MXenes, *ACS Nano*, 13 (2019), 8491–8494

49. R. Qin, G. Shan, M. Hu, W. Huang, Two-dimensional transition metal carbides and/or nitrides (MXenes) and their applications in sensors, *Mater. Today Phys.*, 21 (2021), 100527.

50. B. Anasori, M. R. Lukatskaya, Y. Gogotsi, 2D metal carbides and nitrides (MXenes) for energy storage, *Nature Rev. Mater.*, 2 (2017), 16098.

51. A. Sinha, Dhanjai, H. Zhao, Y. Huang, X. Lu, J. Chen, R. Jain, MXene: An emerging material for sensing and biosensing, *Trends. Anal. Chem.*, 5 (2018), 424–435.

52. B. Lalmi, H. Oughaddou, H. Enriquez, A. Kara, S. Vizzini, B. Ealet, B. Aufray, Epitaxial growth of a silicone sheet, *Appl. Phys. Lett.*, 97 (2010), 223109.

53. P. K. Kalambate, N. S. Gadhari, X. Li, Z. Rao, S. T. Navale, Y. Shen, V. R. Patil, Yunhui Huang, Recent advances in MXene-based electrochemical sensors and biosensors, *Trends Anal. Chem.*, 120 (2019), 115643.

54. M. Naguib, J. Halim, J. Lu, K. M. Cook, L. Hultman, Y. Gogotsi, M. W. Barsoum, New two-dimensional Niobium and Vanadium carbides as promising materials for Li-Ion batteries, *J. Am. Chem. Soc.*, 135 (2013), 15966–15969.

55. B. Soundiraraju, B. K. George, Two-dimensional Titanium Nitride (Ti_2N) MXene: Synthesis, characterization, and potential application as surface-enhanced Raman scattering substrate, *ACS Nano*, 11 (2017), 8892–8900.

56. Vahid Mohammadi, A. Hadjikhani, S. Shahbazmohamadi, M. Beidaghi, Two-dimensional Vanadium Carbide (MXene) as a High-capacity cathode material for rechargeable aluminum batteries, *ACS Nano*, 11 (2017), 11135–11144.

57. M. Naguib, M. Kurtoglu, V. Presser, J. Lu, J. Niu, M. Heon, L. Hultman, Y. Gogotsi, M. W. Barsoum, Two-dimensional nanocrystals produced by exfoliation of Ti_3AlC_2, *Adv. Mater.*, 23 (2011), 4248–4253.

58. K. Zhu, Y. Jin, F. Du, S. Gao, Z. Gao, X. Meng, G. Chen, Y. Wei, Y. Gao, Flexible Ti_2C MXene film: Synthesis, electrochemical performance and capacitance behavior, *J. Mater. Chem.*, 31 (2018), 1–8.

59. M. Naguib, O. Mashtalir, J. Carle, V. Presser, J. Lu, L. Hultman, Y. Gogotsi, M. W. Barsoum, Two-dimensional transition metal carbides, *ACS Nano*, 6, (2012), 1322–1331.

60. B. Anasori, Y. Xie, M. Beidaghi, J. Lu, M. W. Barsoum, Two-dimensional, ordered, double transition metals carbides (MXenes), *ACS Nano*, 9 (2015), 9507–9516.

61. J. Zhou, X. Zha, F. Y. Chen, Q. Ye, P. Eklund, S. Du, Q. Huang, A two-dimensional Zirconium carbide by selective etching of Al_3C_3 from Nanolaminated $Zr_3Al_3C_5$, *Angew. Chem., Int. Ed.*, 55 (2016), 5008–5013.

62. J. Halim, J. Palisaitis, J. Lu, J. Thörnberg, E. J. Moon, M. Precner, P. Eklund, P. O. Å. Persson, M. W. Barsoum, J. Rosen, Synthesis of two-dimensional $Nb_{1.33}C$ (MXene) with randomly distributed vacancies by etching of the quaternary solid solution $(Nb_2/3Sc_{1/3})_2AlC$ MAX Phase, *ACS Appl. Nano Mater.*, 1 (2018), 2455–2460.

63. R. Meshkian, M. Dahlqvist, J. Lu, B. Wickman, J. Halim, J. Thornberg, Q. Tao, S. Li, S. Intikhab, J. Snyder, M. W. Barsoum, M. Yildizhan, J. Palisaitis, L. Hultman, P. O. A. Persson, J. Rosen, W-Based atomic laminates and Their 2D Derivative $W_{1.33}C$ MXene with vacancy ordering, *Adv. Mater.*, 30 (2018), e1706409.

64. P. Urbankowski, B. Anasori, T. Makaryan, D. Er, S. Kota, P. L. Walsh, M. Zhao, V. B. Shenoy, M. W. Barsoum, Y. Gogotsi, Synthesis of two-dimensional titanium nitride Ti_4N_3 (MXene), *Nanoscale*, 8 (2016), 11385–11391.

65. J. Halim, S. Kota, M. R. Lukatskaya, M. Naguib, M. Q. Zhao, E. J. Moon, J. Pitock, J. Nanda, S. J. May, Y. Gogotsi, M. W. Barsoum, Synthesis and characterization of 2D Molybdenum Carbide (MXene), *Adv. Funct. Mater.*, 26 (2016), 3118–3127.

66. B. Anasori, Y. Xie, M. Beidaghi, J. Lu, B. C. Hosler, L. Hultman, P. R. C. Kent, Y. Gogotsi, M. W. Barsoum, Two-dimensional, ordered, double transition metals carbides (MXenes), *ACS Nano*, 9 (2015), 9507–9516.

67. S. Kajiyama, L. Szabova, H. Iinuma, A. Sugahara, K. Gotoh, K. Sodeyama, Y. Tateyama, M. Okubo, A. Yamada, Enhanced Li-Ion accessibility in MXene Titanium Carbide by Steric Chloride termination, *Adv. Sci. News*, 7 (2016), 1601873.

68. F. Liu, J. Zhou, S. Wang, B. Wang, C. Shen, L. Wang, Q. Hu, Q. Huang, A. Zhou, Preparation of High-Purity V_2C MXene and electrochemical properties as Li-Ion batteries, *J. Electrochem. Soc.*, 164 (2017), A709–A713.

69. F. Du, H. Tang, L. Pan, T. Zhang, H. Lu, J. Xiong, J. Yang, C. J. Zhang, Environmental friendly scalable production of colloidal 2D Titanium Carbonitride MXene with minimized nanosheets restacking for excellent cycle life Lithium-Ion batteries, *Electrochim. Acta*, 235 (2017), 690–699.

70. M. Ghidiu, M. R. Lukatskaya, M. Q. Zhao, Y. Gogotsi, M. W. Barsoum, Conductive two-dimensional titanium carbide 'clay' with high volumetric capacitance, *Nature*, 516 (2014), 78–81.

71. J. Yang, M. Naguib, M. Ghidiu, L.-M. Pan, J. Gu, J. Nanda, J. Halim, Y. Gogotsi, M.W. Barsoum, Two-dimensional Nb-Based M_4C_3 solid solutions (MXenes), *J. Am. Ceram. Soc.*, 99 (2015), 660–666.

72. L. Wang, H. Zhang, B. Wang, C. Shen, C. Zhang, Q. Hu, A. Zhou, B. Liu, Synthesis and electrochemical performance of $Ti_3C_2T_x$ with hydrothermal process, *Electron. Mater. Lett.*, 12 (2016), 702–710.

73. J. Halim, M. R. Lukatskaya, K. M. Cook, J. Lu, C. R. Smith, L. A. Naslund, S. J. May, L. Hultman, Y. Gogotsi, P. Eklund, M. W. Barsoum, Transparent Conductive Two-Dimensional Titanium Carbide Epitaxial Thin Films, *Chem. Mater.*, 26 (2014), 2374–2381.

74. O. Mashtalir, M. Naguib, V. N. Mochalin, Y. Dall'Agnese, M. Heon, M. W. Barsoum, Y. Gogotsi, Intercalation and delamination of layered carbides and carbonitrides, transparent conductive two-dimensional titanium carbide Epitaxial thin films, *Nat. Commun.*, 4 (2013), 1716.

75. M. Naguib, R. R. Unocic, B. L. Armstrong, J. Nanda, Large-scale delamination of multi-layers transition metal carbides and carbonitrides "MXenes", *Dalton Trans.*, 44 (2015), 9353–9358.

76. K. Arole, J. W. Blivin, S. Saha, D. E. Holta, X. Zhao, A. Sarmah, H. Cao, M. Radovic, J. L. Lutkenhaus, M. J. Green, Water-dispersible $Ti_3C_2T_z$ MXene nanosheets by molten salt etching, *iScience*, 24 (12) (2021), 103403.

77. M. Naguib, O. Mashtalir, J. Carle, V. Presser, J. Lu, L. Hultman, Y. Gogotsi, M. W. Barsoum, Two-dimensional transition metal carbides, *ACS Nano*, 6 (2012), 1322–1331.

78. C. Xu, L. Wang, Z. Liu, L. Chen, J. Guo, N. Kang, X. L. Ma, H. M. Cheng, W. Ren, Large-area high-quality 2D ultrathin Mo_2C superconducting crystals, *Nat Mater.*, 14 (11) (2015), 1135–1141.

79. Z. Sun, M. Yuan, L. Lin, H. Yang, C. Nan, H. Li, G. Sun, X. Yang, Selective lithiation-expansion–microexplosion synthesis of two-dimensional fluoride-free Mxene, *ACS Mater Lett.*, 1 (6) (2019), 628–632.

80. Z. Li, Y. Ren, L. Mo, C. Liu, K. Hsu, Y. Ding, X. Zhang, X. Li, L. Hu, D. Ji, G. Cao, Impacts of oxygen vacancies on zinc ion intercalation in VO_2, *ACS Nano*, 14 (5) (2020), 5581–5589.

81. M. Naguib, O. Mashtalir, J. Carle, V. Presser, J. Lu, L. Hultman, Y. Gogotsi, M. W. Barsoum, Two-dimensional transition metal carbides, *ACS Nano*, 6 (2012), 1322–1331

82. J. Halim, S. Kota, M. R. Lukatskaya, M. Naguib, M.-Q. Zhao, E. J. Moon, J. Pitock, J. Nanda, S. J. May, Y. Gogotsi, M. W. Barsoum, Synthesis and characterization of 2D Molybdenum Carbide (MXene), *Adv. Funct. Mater.* 26 (2016), 3118–3127.

83. K. Arole, J. W. Blivin, S. Saha, D. E. Holta, X. Zhao, A. Sarmah, H. Cao, M. Radovic, J. L. Lutkenhaus, M. J. Green, Water-dispersible Ti3C2Tz MXene nanosheets by molten salt etching, *iScience*, 24 (12) (2021), 103403.

84. M. Naguib, M. Kurtoglu, V. Presser, J. Lu, J. Niu, M. Heon, L. Hultman, Y. Gogotsi, M. W. Barsoum, Two-dimensional nanocrystals produced by exfoliation of Ti_3AlC_2, *Adv. Mater.*, 23 (2011), 4248–4253.

85. M. R. Lukatskaya, O. Mashtalir, C. E. Ren, Y. Dall'Agnese, P. Rozier, P. L. Taberna, M. Naguib, P. Simon, M. W. Barsoum, Y. Gogotsi, Cation intercalation and high volumetric capacitance of two-dimensional titanium carbide, *Science*, 341 (2013), 1502–1505.

86. Z. Otgonbayar, S. Yang, I. J. Kim, W. C. Oh, Recent Advances in Two-Dimensional MXene for Supercapacitor Applications: Progress, Challenges, and Perspectives, Nanomaterials, 13 (2023) 919, https://doi.org/10.3390/nano13050919.

87. N. Chen, Z. Duan, W. Cai, Y. Wang, B. Pu, H. Huang, Y. Xie, Q. Tang, H. Zhang, W. Yang, Supercritical etching method for the large-scale manufacturing of MXenes, *Nano Energy*, 107 (2023), 108147.

88. D. Wang, C. Zhou, A. S. Filatov, W. J. Cho, F. Lagunas, M. Wang, S. Vakikuntanathan, C. Liu, R. F. Klie, D. V. Talapin, Direct synthesis and chemical vapor deposition of 2D carbide and nitride MXenes, *Science*, 379 (2023), 1242–1247.

89. H. Ding, Y. Li, M. Li, K. Chen, K. Liang, G. Chen, J. Lu, J. Palisatis, P. O. A. Persson, P. Eklund, L. Hultman, S. Du, Z. Chai, Y. Gogotsi, Q. Huang, Chemical scissor–Mediated structural editing of layered transition metal carbides, *Science*, 379 (2023), 1130–1135.

90. M. Alhabeb, K. Maleski, T.S. Mathis, A. Sarycheva, C.B. Hatter, S. Uzun, A. Levitt, Y. Gogotsi, Selective etching of silicon from Ti_3SiC_2 (MAX) to obtain 2D titanium carbide (MXene), *Angew Chemie.*, 130 (19) (2018), 5542–5546.

91. J. Luo, C. Wang, H. Wang, X. Hu, E. Matios, X. Lu, W. Zhang, X. Tao, W. Li, Pillared MXene with ultralarge interlayer spacing as a stable matrix for high performance sodium metal anodes, A*dv. Funct. Mater.*, 29 (3) (2019), 1–12.

92. C. Fang, J. Luo, C. Jin, H. Yuan, O. Sheng, H. Huang, Y. Gan, Y. Xia, C. Liang, J. Zhang, W. Zahng, X. Tao, Enhancing catalyzed decomposition of Na_2CO_3 with Co_2MnO_x nanowire-decorated carbon fibers for advanced Na-CO_2 batteries, *ACS Appl. Mater. Interfaces*, 10 (20) (2018), 17240–17248.

93. V. N. Borysiuk, V. N. Mochalin, Y. Gogotsi, Molecular dynamic study of the mechanical properties of two-dimensional Titanium carbides $Ti_{n+1}C_n$ (MXenes), *Nanotechnology*, 26 (26) (2015), 1–10.

94. J. H. Chen, C. Jang, S. Xiao, M. Ishigami, M. S. Fuhrer, Intrinsic and extrinsic performance limits of graphene devices on SiO_2, *Nat. Nanotechnol.*, 3 (4) (2008), 206–209.

95. O. Mashtalir, M. R Lukatskaya, A. I. Kolesnikov, E. R. Pinero, M. W. Barsoum, Y. Gogotsi, The effect of hydrazine intercalation on the structure and capacitance of 2D titanium carbide (MXene), *Nanoscale*, 8 (17) (2016), 9128–9133.

96. S. Kulandaivalu, N. H. N. Azman, Y. Sulaiman, Advances in Layered Double Hydroxide/ Carbon Nanocomposites Containing Ni^{2+} and $Co^{2+/3+}$ for Supercapacitors, Front. Mater., 7 (2020) 147, https://doi.org/10.3389/fmats.2020.00147.

97. M. Z. Iqbal, S. Zakar, S. S. Haider, Role of aqueous electrolytes on the performance of electrochemical energy storage device, J. Electri. Chem., 858 (2020) 113793, https://doi.org/10.1016/j.jelechem.2019.113793.

98. B. Pal, S. Yang, S. Ramesh, V. Thangadurai, R. Jose, Electrolyte selection for supercapacitive devices: A critical review, *Nanoscale Adv.*, 1 (2019), 3807–3835.

99. Sumboja, J. Liu, W. G. Zheng, Y. Zong, H. Zhang, Z. Liu, Electrochemical energy storage devices for wearable technology: A rationale for materials selection and cell design, *Chem. Soc. Rev.*, 47 (2018), 5919–5945.

100. C. W. Huang, C. A. Wu, S. S. Hou, P. L. Kuo, C. Te Hsieh, H. Teng, Gel electrolyte derived from Poly(ethylene glycol) Blending Poly(acrylonitrile) Applicable to roll-to-roll assembly of electric double layer capacitors, *Adv. Funct. Mater.*, 22 (2012), 4677–4685.

101. Y. J. Kang, S. J. Chun, S. S. Lee, B. Y. Kim, J. H. Kim, H. Chung, S. Y. Lee, W. Kim, All-solid-state flexible supercapacitors fabricated with bacterial nanocellulose papers, carbon nanotubes, and Triblock-Copolymer Ion Gels, *ACS Nano*, 6 (2012), 6400–6406.

102. Osada, H. De Vries, B. Scrosati, S. Passerini, Ionic-Liquid-based polymer electrolytes for battery applications, *Angew. Chem. - Int. Ed.*, 55 (2016), 500–513.

103. J. Zhang, B. Sun, X. Huang, S. Chen, G. Wang, Honeycomb-like porous gel polymer electrolyte membrane for lithium ion batteries with enhanced safety, *Sci. Rep.*, 4 (2014), 6007.

104. J. Zhou, Y. Yin, A. N. Mansour, X. Zhou, Experimental Studies of Mediator-Enhanced Polymer Electrolyte Supercapacitors, *Electrochem. Solid-State Lett.*, 14 (2011), A25–A28.

105. G. Ma, J. Li, K. Sun, H. Peng, J. Mu, Z. Lei, High performance solid-state supercapacitor with PVA-KOH-K 3[Fe(CN)6] gel polymer as electrolyte and separator, *J. Power Sources*, 256 (2014), 281–287.

106. F. Yu, M. Huang, J. Wu, Z. Qiu, L. Fan, J. Lin, Y. Lin, A redox-mediator-doped gel polymer electrolyte applied in quasi-solid-state supercapacitors, *J. Appl. Polym. Sci.*, 131 (2014), 39784.

107. Z. Wu, L. Li, J. M. Yan, X. B. Zhang, Materials design and system construction for conventional and new-concept supercapacitors, *Adv. Sci.*, 4 (2017), 1600382.

108. R. Kotz, M. Carlen, Principles and applications of electrochemical capacitors, *Electrochim. Acta*, 45 (2000), 2483– 2498.

109. Z. Song, H. Duan, L. Li, D. Zhu, T. Cao, Y. Lv, W. Xiong, Z. Wang, M. Liu, L. Gan, High-energy flexible solid-state supercapacitors based on O, N, S-tridoped carbon electrodes and a 3.5 V gel-type electrolyte, *Chem. Eng. J.*, 372 (2019), 1216–1225.

110. G. A. Snook, P. Kao, A. S. Best, Conducting-polymer-based supercapacitor devices and electrodes, *J. Power Sources*, 196 (1) (2011), 1–12.

111. M. Avella, E. Martuscelli, Poly-d-(–)(3-hydroxybutyrate)/poly(ethylene oxide) blends: Phase diagram, thermal and crystallization behaviour, *Polymer*, 29 (1988), 1731–1737.

112. Y. Furushima, M. Nakada, H. Takahashi, K. Ishikiriyama, Study of melting and crystallization behavior of polyacrylonitrile using ultrafast differential scanning calorimetry, *Polymer*, 55 (2014), 3075–3081.

113. J. S. Park, J. W. Park, E. Ruckenstein, On the viscoelastic properties of poly(vinyl alcohol) and chemically crosslinked poly(vinyl alcohol), *J. Appl. Polym. Sci.*, 82 (2001), 1816–1823.

114. G. Alberti, R. Narducci, M. L. Di Vona, S. Giancola, Annealing of Nafion 1100 in the presence of an annealing agent: A powerful method for increasing ionomer working temperature in PEMFCs, *Fuel Cells*, 13 (2013), 42–47.

115. G. S. Haldankar, H. G. Spencer, Properties of bound water in poly(acrylic acid) and its sodium and potassium salts determined by differential scanning calorimetry, *J. Appl. Polym. Sci.*, 37 (1989), 3137–3146.

116. C. Srinivas, IOSR, Synthesis and characterization of nano size conducting polyaniline, *J. Appl. Phys.*, 1 (2012), 12–15.

117. D. T. Turner, A. Schwartz, The glass transition temperature of poly(N-vinyl pyrrolidone) by differential scanning calorimetry, *Polymer*, 26 (1985), 757–762.

118. J. Tian, C. Cui, Q. Xie, W. Qian, C. Xue, Y. Miao, Y. Jin, G. Zhang, B. Guo, EMIMBF4–GBL binary electrolyte working at − 70°C and 3.7 V for a high performance graphene-based capacitor, *J. Mater. Chem. A*, 6 (8) (2018), 3593–3601.

119. S. W. Xu, M. C. Zhang, G. Q. Zhang, J. H. Liu, X. Z. Liu, X. Zhang, D. D. Zhao, C. L. Xu, Y. Q. Zhao, Temperature-dependent performance of carbon-based supercapacitors with water-in-salt electrolyte, *J. Power Sources*, 441 (2019), Article 227220.

120. L. Liu, Q. Dou, Y. Sun, Y. Lu, Q. Zhang, J. Meng, X. Zhang, S. Shi, X. Yan, A moisture absorbing gel electrolyte enables aqueous and flexible supercapacitors operating at high temperatures, *J. Mater. Chem. A*, 7 (35) (2019), 20398–20404.

121. Z. Song, L. Li, D. Zhu, L. Miao, H. Duan, Z. Wang, W. Xiong, Y. Lv, M. Liu, L. Gan, Synergistic design of a N, O co-doped honeycomb carbon electrode and an ionogel electrolyte enabling all-solid-state supercapacitors with an ultrahigh energy density, *J. Mater. Chem. A*, 7 (2) (2019), 816–826.

122. Y. Lu, H. Mi, C. Ji, F. Guo, Z. Bai, Y. Liu, C. Yu, J. Qiu, Synergizing layered carbon and gel electrolyte for efficient energy storage, *ACS Sustain. Chem. Eng.*, 8 (10) (2020), 4207–4215.

123. Z. Song, L. Li, D. Zhu, L. Miao, H. Duan, Z. Wang, W. Xiong, Y. Lv, M. Liu, L. Gan, Synergistic design of a N, O co-doped honeycomb carbon electrode and an ionogel electrolyte enabling all-solid-state supercapacitors with an ultrahigh energy density, *J. Mater. Chem. A*, 7 (2) (2019), 816–826.

124. S. W. Xu, M. C. Zhang, G. Q. Zhang, J. H. Liu, X. Z. Liu, X. Zhang, D. D. Zhao, C. L. Xu, Y. Q. Zhao, Temperature-dependent performance of carbon-based supercapacitors with water-in-salt electrolyte, *J. Power Sources*, 441 (2019), Article 227220.

125. H. Lu, L. He, X. Li, W. Zhang, J. Che, X. Liu, Z. Hou, H. Du, Y. Qu, Ionic liquid-solvent mixture of propylene carbonate and 1, 2-dimethoxyethane as electrolyte for electric double-layer capacitor, *J. Mater. Sci. Mater. Electron.*, 30 (15) (2019), 13933–13938.

126. A. M. Patil, V. C. Lokhande, U. M. Patil, P. A. Shinde, C.D. Lokhande, High performance all-solid-state asymmetric supercapacitor device based on 3D nanospheres of β-MnO$_2$ and nanoflowers of O-SnS, *ACS Sustain. Chem. Eng.*, 6 (1) (2018), 787–802.

127. L. Fenglin Lai, Zeming Fang, Lin Cao, Wei Li1, Zhidan Lin, Peng Zhang, Self-healing flexible and strong hydrogel nanocomposites basedon polyaniline for supercapacitors. Ionics 26 (2020) 3015-3025, https://doi.org/10.1007/s11581-020-03438-3.

128. L. Kou, T. Huang, B. Zheng, Y. Han, X. Zhao, K. Gopalsamy, H. Sun, C. Gao, Coaxial wet-spun yarn supercapacitors for high-energy density and safe wearable electronics, *Nat. Commun.*, 5 (2014), 3754.

129. R. Lin, P. Huang, J. Ségalini, C. Largeot, P. L. Taberna, J. Chmiola, Y. Gogotsi, P. Simon, Solvent effect on the ion adsorption from ionic liquid electrolyte into sub-nanometer carbon pores, *Electrochim. Acta*, 54 (2009), 7025–7032.

130. M. R. Lukatskaya, O. Mashtalir, C. E. Ren, Y. Dall'Agnese, P. Rozier, P. L. Taberna, M. Naguib, P. Simon, M. W. Barsoum, Y. Gogotsi, Cation intercalation and high volumetric capacitance of two-dimensional Titanium carbide, *Science*, 341 (2013), 1502–1505.

131. M. R. Lukatskaya, S. M. Bak, Yu X., Yang X. Q., M. W. Barsoum, Y. Gogotsi, probing the mechanism of high capacitance in 2D Titanium carbide using in situ x-ray absorption spectroscopy, *Adv. Energy Mater.*, 5 (2015), 1500589.

132. M. Hu, Z. Li, T. Hu, S. Zhu, C. Zhang, X. Wang, High-capacitance mechanism for Ti$_3$C$_2$T$_x$ MXene by in situ electrochemical Raman spectroscopy investigation, *ACS Nano*, 10 (2016), 11344–11350.

133. M. Ghidiu, M. R. Lukatskaya, Zhao M. Q., Y. Gogotsi, M. W. Barsoum, Conductive two-dimensional Titanium carbide 'clay' with high volumetric capacitance, *Nature*, 516 (2014), 78–81.

134. S. Xu, G. Wei, J. Li, Y. Ji, N. Klyui, V. Izotov, W. Han, Binder-free $Ti_3C_2T_x$ MXene electrode film for supercapacitor produced by electrophoretic deposition method, *Chem. Eng. J.*, 317 (2017), 1026–1036.

135. C. J. Zhang, B. Anasori, A. Seral-Ascaso, S. H. Park, N. McEvoy, A. Shmeliov, G. S. Duesberg, J. N. Coleman, Y. Gogotsi, V. Nicolosi, Transparent, flexible, and conductive 2D Titanium Carbide (MXene) films with high volumetric capacitance, *Adv. Mater.* 29 (2017), 1702678.

136. M. R. Lukatskaya, S. Kota, Lin Z., Zhao M. Q., N. Shpigel, M. D. Levi, J. Halim, P. L. Taberna, M. W. Barsoum, P. Simon, Y. Gogotsi, Ultra-high-rate pseudocapacitive energy storage in two-dimensional transition metal carbides, *Nat. Energy*, 6 (2017), 17105.

137. Y. Dall'Agnese, M. R. Lukatskaya, K. M. Cook, P. L. Taberna, Y. Gogotsi, P. Simon, High capacitance of surface-modified 2D Titanium carbide in acidic electrolyte, *Electrochem. Commun.*, 48 (2014), 118–122.

138. Z. Ling, C. E. Ren, M. Q. Zhao, J. Yang, J. M. Giammarco, J. Qiu, M. W. Barsoum, Y. Gogotsi, Flexible and conductive MXene films and nanocomposites with high capacitance, *Proc. Natl. Acad. USA*, 111 (2014), 16676–16681.

139. M. Boota, B. Anasori, C. Voigt, M. Q. Zhao, M. W. Barsoum, Y. Gogotsi, Pseudocapacitive electrodes produced by oxidant-free polymerization of pyrrole between the layers of 2D Titanium Carbide (MXene), *Adv. Mater.*, 28 (2016), 1517–1522.

140. M. Zhu, Y. Huang, Q. Deng, J. Zhou, Z. Pei, Q. Xue, Y. Huang, Z. Wang, H. Li, Q. Huang, C. Zhi, Highly flexible, freestanding supercapacitor electrode with enhanced performance obtained by hybridizing polypyrrole chains with MXene, *Adv. Energy Mater.*, 6 (2016), 1600969.

141. Z. Lin, D. Barbara, P. L. Taberna, K. L. Van Aken, B. Anasori, Y. Gogotsi, P. Simon, Capacitance of $Ti_3C_2T_x$ MXene in ionic liquid electrolyte, *J. Power Sources*, 326 (2016), 575–579.

142. R. B. Rakhi, B. Ahmed, M. N. Hedhili, D. H. Anjum, H. N. Alshareef, Effect of postetch annealing gas composition on the structural and electrochemical properties of Ti_2CT_x MXene electrodes for supercapacitor applications, *Chem. Mater.*, 27 (2015), 5314–5323.

143. K. Krishnamoorthy, P. Pazhamalai, S. Sahoo, S. J. Kim, Titanium carbide sheet based high performance wire type solid state supercapacitors, *J. Mater. Chem. A*, 5 (2017), 5726–5736.

144. M. Hu, Z. Li, H. Zhang, T. Hu, C. Zhang, Z. Wu, X. Wang, Self-assembled $Ti_3C_2T_x$ MXene film with high gravimetric capacitance, *Chem. Commun.*, 51 (2015), 13531–13533.

145. Y. Tang, J. Zhu, C. Yang, F. Wang, Enhanced capacitive performance based on diverse layered structure of two-dimensional Ti_3C_2 MXene with long etching time, *J. Electrochem. Soc.*, 163 (2016), A1975–A1982.

146. Y. Wen, T. E. Rufford, X. Chen, N. Li, M. Lyu, L. Dai, L. Wang, Nitrogen-doped $Ti_3C_2T_x$ MXene electrodes for high-performance supercapacitors, *Nano Energy*, 38 (2017), 368–376.

147. Y. Wang, H. Dou, J. Wang, B. Ding, Y. Xu, Z. Chang, X. Hao, Three-dimensional porous MXene/layered double hydroxide composite for high performance supercapacitors, *J. Power Sources*, 327 (2016), 221–228.

148. P. Yan, R. Zhang, J. Jia, C. Wu, A. Zhou, J. Xu, X. Zhang, Enhanced super capacitive performance of delaminated two-dimensional titanium carbide/carbon nanotube composites in alkaline electrolyte, *J. Power Sources*, 284 (2015), 38–43.

149. J. Zhu, Y. Tang, C. Yang, F. Wang, M. Cao, Composites of TiO_2 nanoparticles deposited on Ti_3C_2 MXene nanosheets with enhanced electrochemical performance, *J. Electrochem. Soc.*, 163 (2016), A785–A791.

150. M. Q. Zhao, C. E. Ren, Z. Ling, M. R. Lukatskaya, C. Zhang, K. L. Van Aken, M. W. Barsoum, Y. Gogotsi, Flexible MXene/Carbon nanotube composite paper with high volumetric capacitance, *Adv. Mater.*, 27 (2015), 339–345.

151. N. Kurra, B. Ahmed, Y. Gogotsi, H. N. Alshareef, MXene-on-paper coplanar microsupercapacitors, *Adv. Energy Mater.*, 6 (2016), 1601372.

152. Y. Tian, C. Yang, W. Que, X. Liu, X. Yin, L. B. Kong, Flexible and free-standing 2D Titanium carbide film decorated with manganese oxide nanoparticles as a high volumetric capacity electrode for supercapacitor, *J. Power Sources*, 359 (2017), 332–339.

153. L. Sun, Q. Fu, C. Pan, Hierarchical porous "skin/skeleton"-like MXene/biomass derived carbon fibers heterostructure for self-supporting, flexible all solid-state supercapacitors, *J Hazard Mater.*, 410 (2021), 124565.

154. Z. Liu, L. Wang, Y. Xu, J. Guo, S. Zhang, Y. Lu, A $Ti_3C_2T_x$@PEDOT composite for electrode materials of supercapacitors, *J. Electroanal Chem.*, 881 (2021), 114958.

155. J. Yan, Y. Ma, C. Zhang, X. Li, W. Liu, S. Yao, S. Luo, Polypyrrole-MXene coated textile-based flexible energy storage device, *RSC Adv.*, 8 (69) (2018), 39742–39748.

156. X. Zhang, B. Shao, A. Guo, Z. Gao, Y. Qin, C. Zahng, F. Cui, X. Yang, Improved electrochemical performance of CoO_x-NiO/$Ti_3C_2T_x$ MXene nanocomposites by atomic layer deposition towards high capacitance supercapacitors, *J. Alloys. Compd.*, 862 (2021), 158546.

157. D. Zhang, M. Luo, K. Yang, P. Yang, C. Liu, W. Chen, X. Zhou, Porosity-adjustable MXene film with transverse and longitudinal ion channels for flexible supercapacitors, *Microporous Mesoporous Mater.*, 326 (159) (2021), 111389.

158. Y. Xu, B. Pan, W.S. Li, L. Dong, X. Wang, F. G. Zhao, High-performance flexible asymmetric supercapacitor paired with indanthrone@graphene heterojunctions and MXene electrodes, *ACS Appl. Mater. Interfaces*, 13 (35) (2021), 41537–41544.

159. L. Ma, T. Zhao, F. Xu, T. You, X. Zhang, A dual utilization strategy of lignosulfonate for MXene asymmetric supercapacitor with high area energy density, *Chem. Eng. J.*, 405 (35) (2021), 126694.

160. K. Zhao, H. Wang, C. Zhu, S. Lin, Z. Xu, X. Zhang, Free-standing MXene film modified by amorphous FeOOH quantum dots for high-performance asymmetric supercapacitor, *Electrochim. Acta.*, 308 (2019), 1–8.

161. Y. Wang, X. Wang, X. Li, Y. Bai, H. Xiao, Y. Liu, G. Yuan, Scalable fabrication of polyaniline nanodots decorated MXene film electrodes enabled by viscous functional inks for high-energy-density asymmetric supercapacitors, *Chem. Eng. J.*, 405 (2021), 126664.

162. L. Yu, L. Hu, B. Anasori, Y.T. Liu, Q. Zhu, P. Zhang, Y. Gogotsi, B. Xu, MXene-bonded activated carbon as a flexible electrode for high-performance supercapacitors, *ACS Energy Lett.*, 3 (7) (2018), 1597–1603.

163. H. Zhou, Y. Lu, F. Wu, L. Fang, H. Luo, Y. Zhang, M. Zhou, MnO_2 nanorods/MXene/CC composite electrode for flexible supercapacitors with enhanced electrochemical performance, *J. Alloys. Compd.*, 802 (2019), 259–268.

164. X. Zheng, J. Shen, Q. Hu, W. Nie, Z. Wang, L. Zou, C. Li, Vapor phase polymerized conducting polymer/MXene textiles for wearable electronics, *Nanoscale*, 13 (3) (2021), 1832–1841.

165. Z. Fan, Y. Wang, Z. Xie, X. Xu, Y. Yuan, Z. Cheng, Y. Liu, A nanoporous MXene film enables flexible supercapacitors with high energy storage, *Nanoscale*, 10 (20) (2018), 9642–9652.

166. Quasi-solid-state electrolytes - strategy towards stabilising Lilinorganic solid electrolyte interfaces in solid-state Li metal batteries Lucia Mazzapioda, Akiko Tsurumaki, Graziano Di Donato, Henry Adenusi[2], Maria Assunta Navarra, Stefano Passerini.

167. R. B. Ambade, S. B. Ambade, R. R. Salunkhe, V. Malgras, S. H. Jin, Y. Yamauchi, S. H. Lee, Flexible-wire shaped all-solid-state supercapacitors based on facile electropolymerization of polythiophene with ultra-high energy density, *J. Mater. Chem. A*, 4 (2016), 7406–7415.

168. T. Lv, M. Liu, D. Zhu, L. Gan, T. Chen, Nanocarbon-based materials for flexible all-solid-state supercapacitors, *Adv. Mater.*, 30 (2018), 1705489.

169. L. Li, Z. Lou, D. Chen, K. Jiang, W. Han, G. Shen, Recent advances in flexible/stretchable supercapacitors for wearable electronics, *Small*, 14 (2018), 1–23.

170. T. An, W. Cheng, Recent progress in stretchable supercapacitors, *J. Mater. Chem. A*, 6 (2018), 15478–15494.

171. C. Cao, Y. Chu, Y. Zhou, C. Zhang, S. Qu, Recent advances in stretchable supercapacitors enabled by low-dimensional nanomaterials, *Small*, 14 (2018), 1–26.

172. L. Wang, C. Lin, F. Zhang, J. Jin, Phase transformation guided single-layer β-$Co(OH)_2$ nanosheets for pseudocapacitive electrodes, *ACS Nano*, 8 (4) (2014), 3724–3734.

173. P. Zhang, Q. Zhu, R. A. Soomro, S. He, N. Sun, N. Qiao, B. Xu, In situ ice template approach to fabricate 3D flexible MXene film-based electrode for high performance supercapacitors, *Adv. Funct. Mater.*, 30 (47) (2020), 1–10.

174. C. Ma, W. T. Cao, W. Xin, J. Bian, M. G. Ma, Flexible and free-standing reduced graphene oxide and polypyrrole coated air-laid paper-based supercapacitor electrodes, *Ind. Eng. Chem. Res.*, 58 (27) (2019), 12018–12027.

175. M. Zhang, F. Héraly, M. Yi, J. Yuan, Multitasking tartaric-acid-enabled, highly conductive, and stable MXene/conducting polymer composite for ultrafast supercapacitor, *Cell Reports Phys. Sci.*, 2 (6) (2021), 100449.

176. J. Miao, Q. Zhu, K. Li, P. Zhang, Q. Zhao, B. Xu, Self-propagating fabrication of 3D porous MXene-rGO film electrode for high-performance supercapacitors, *J. Energy Chem.*, 52 (2021), 243–250.

177. X. Li, H. Li, X. Fan, X. Shi, J. Liang, 3D-printed stretchable micro-supercapacitor with remarkable areal performance, *Adv. Energy Mater.*, 10 (14) (2020), 1–12.

178. Yi-Zhou Zhang, Jehad K. El-Demellawi, Qiu Jiang, Gang Ge, Hanfeng Liang, Kanghyuck Lee, Xiaochen Dong, Husam N. Alshareef, MXene hydrogels: Fundamentals and applications. Chem. Soc. Rev., 49 (2020) 7229-7251, https://doi.org/10.1039/D0CS00022A.

179. Yanan Jiang, Tao Sun, Xi Xie, Wei Jiang, Jia Li, Bingbing Tian, Chenliang Su, Oxygen-Functionalized ultrathin Ti3C2Tx MXene for enhanced electrocatalytic hydrogen evolution. ChemSusChem,12 (2019) 1368–1373, https://doi.org/10.1002/cssc.201803032.

180. Xia Sun, Hong Zhang, Caideng Yuan, Yuping Wei, Junjie Li, Preparation Strategies and Applications of MXene-Polymer Composites: A Review, Macromole. Rap. Commun., 42 (2021) 2100324, https://doi.org/10.1002/marc.202100324.

181. X. Chu, Y. Wang, L. Cai, H. Huang, Z. Xu, Y. Xie, C. Yan, Q. Wang, H. Zhang, H. li, W. Yang, Boosting the energy density of aqueous MXene-based supercapacitor by integrating 3D conducting polymer hydrogel cathode, *Sus. Mat.*, 2 (3) (2022), 379–390.

182. Y. Wang, N. Chen, Y. Liu, Z. Zhou, B. Pu, Y. Qing, M. Zhang, Z. Jiang, J. Huang, Q. Tang, B. Zhou, W. Yang, MXene/Graphdiyne nanotube composite films for free-standing and flexible Solid-State supercapacitor, *Chem. Eng. J.*, 450 (2022), 138398.

183. R. S. Karmur, D. Gogoi, A. Biswas, C. Prahthibha, M. R. Das, N. N. Ghosh, Nanocomposite having hierarchical architecture of MXene-WO_3 nanorod@rGOsponge and porous carbon for cathode and anode materials for high-performance flexible all-solid-state asymmetric supercapacitor device, *Appl. Surf. Sci.*, 623 (2023), 1570442.

184. L. Qin, Q. Tao, A. E. Ghazaly, J. Fernandez-Rodriquez, Per O. A. Persson, J. Rosen, F. Zhang, High-performance ultrathin flexible solid-state supercapacitors based on solution processable $Mo_{1.33C}$ MXene and PEDOT:PSS, *Adv. Funct. Mater.*, 28 (2018), 170380.

185. Q. Jiang, N. Kurra, M. Alhabeb, Y. Gogotsi, H. N. Alshareef, All pseudocapacitive MXene-RuO_2 asymmetric supercapacitors, *Adv. Energy Mater.*, 8 (13) (2018), 1703043.

186. H. Tang, R. Chen, Q. Huang, W. Ge, X. Zhang, Y.Yang, X. Wang, Scalable manufacturing of leaf-like MXene/Ag NWs/cellulose composite paper electrode for all-solid-state supercapacitor, *EcoMat*, 4 (6) (2022), e12247.

187. Z. Zhang, Z. Yao, Y. Li, S. Lu, X. Wu, Z. Jiang, Cation-induced $Ti_3C_2T_x$ MXene hydrogel for capacitive energy storage, *Chem. Eng. J.*, 433 (1) (2022), 134488.

2 Two-Dimensional MXenes for Supercapacitor Applications and Prospects

Zuhong Ji, Kefayat Ullah and Won-Chun Oh

1 INTRODUCTION

The type of technology known as "modern electronic devices" has become increasingly important in daily life and has developed significantly over the past few decades [1–5]. The widespread use of household appliances such as smartphones and wearable devices has led to a sharp increase in demand for high-quality energy storage accessories. The use of supercapacitors (SCs) (also known as electrochemical capacitors) in various energy storage devices has attracted attention due to its many advantages, such as very long cycle life and rapid discharge and charge capabilities [6–12]. The development of SCs requires more in-depth research on electrode materials, as electrode materials are known to be critical to SCs. According to the charge storage mode, electrode materials used for SCs can be divided into pseudocapacitors and electric double-layer capacitors (EDLCs) [13–16]. EDLCs are the most widely used SCs on the market, in which charge is stored primarily through reversible electrochemical desorption/adsorption of electrolyte ions at the electrolyte/electrode interface. However, the specific capacitance of EDLCs based on pure carbon materials is limited to 250 F g^{-1} in practical applications [17]. At the same time, pseudocapacitors allow charge storage via Faradaic processes involving fast ion intercalation and/or redox reactions at/near the surface. More charge can be stored due to rapid surface redox processes; pseudocapacitive materials generally have higher specific capacities than EDLCs [18]. The development of SCs benefits from the use of various pseudocapacitive materials as electrode materials, including transition metal oxides (TMOs), transition metal dichalcogenides (TMDs), layered double hydroxides (LDHs), and conductive polymers [19]. In addition, MXene/polymer compounds and transition metal compounds [20] and the combination of MXene with porous carbon [21], as well as N-doped porous carbon [22], porous carbon NCs [23, 24] have become promising candidates for SCs.

DOI: 10.1201/9781003561262-2

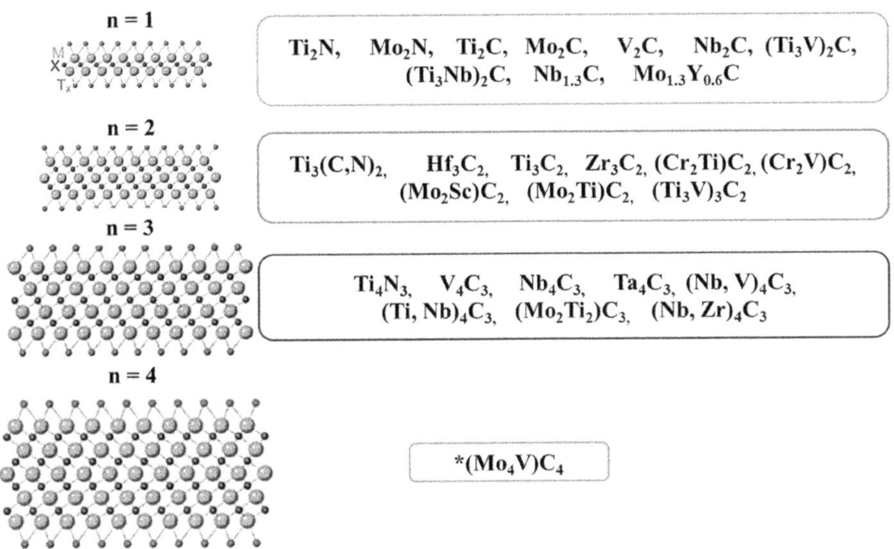

FIGURE 2.1 The list of studied MAX and MXene nanocomposites.

These pseudocapacitive materials present bottlenecks due to these limitations, but also have their own advantages. In contrast, pseudocapacitive polymers often exhibit structural fluctuations during interpolation and extraction and have shorter lifetimes when used in SCs. For example, typical pseudocapacitive oxides require high production costs and have low electrical conductivity. Their usefulness is largely limited by this drawback when used as SC electrodes [25]. In order to develop new energy storage technologies and meet the growing energy demands of electronic devices, new SC electrode materials with stability and high strength at low cost must be found. MXenes, first reported in 2011, consist of atomically thin layers of carbonates or metal nitrides, and carbides and belong to the class of 2D inorganic compounds, with various hydrophilic termini composing the synthetic MXenes [26]. The MAX phase of aluminum produces MXenes, but there are few reports on the production of MXenes with other A elements [27, 28] (e.g., Si and Ga) (Figure 2.1).

Although the most attractive SC electrode material is $Ti_3C_2T_X$; its performance largely depends on the demanding properties of MXenes, such as compatibility of functional groups, strong hydrophilicity, surface area, and high thermal and electrical conductivity. $Ti_3C_2T_X$-based SC electrodes have been demonstrated to have relatively high volumetric capacitance, excellent rate results at high scan rates, and excellent cycle life. $Ti_3C_2T_X$ may have many advantages when used as SC electrodes compared to other forms of MXenes. Most of the MXene materials used in SCs generally belong to 2D structures. From a structural perspective, the restacking and horizontal aggregation of MXene nanosheets lead to the accessibility and use of electrolyte ions due to the strong van der Waals interactions between adjacent layers. The entire 2D MXene surface limits these possibilities. Designing open structures of MXene nanosheets has proven to be an advanced approach to overcome this limitation, which offers several advantages; examples include increasing the electrochemically accessible surface area of MXene and tailoring properties to improve the rate or morphology of ion transport to active redox sites. Various strategies for tuning morphology, such as vertical line design, three-dimensional (3D) porous structures, interlayer spacing expansion, and particle size control, have been proposed in research publications targeting high-power and high-capacity performance MXenes. Designing a porous/3D electrode structure with a large active surface area to facilitate the connection between ions and ion transport channels can more effectively improve the high-rate performance of MXenes. Various methods, such as hard template strategies [29], different types of freeze-drying methods [30,31], chemical cross-linking [32], oxidative etching [33], and self-assembly methods [34], have been used to prepare 3D porous MXene materials, and most 3D porous MXene materials have macropores/mesopores with a wide size distribution.

To meet the growing demand for wearable technology, wearable and flexible electronics are currently undergoing rapid development. Supercapacitor materials are important energy supply components and will inevitably undergo different deformations such as folding, curving, and bending when used in practice. The synthesis of novel SC electrode materials offers outstanding electrochemical achievements and remarkable versatility, thereby greatly meeting the growing energy demands of wearable technology. Taken together, $Ti_3C_2T_X$ shows great potential as SC electrode materials in real-world energy storage applications. However, despite many reviews mentioning the advantages of MXene in electrochemical energy storage, the use of $Ti_3C_2T_X$ (MXenes) as an SC electrode material has not been extensively studied. In view of the rapid progress in this field and the advantageous use of $Ti_3C_2T_x$ (MXene) in attractive flexible/wearable energy storage devices, a thorough analysis of MXene ($Ti_3C_2T_x$)-based SC electrodes is urgently needed. For practical applications and improved performance, it is necessary to examine the precise gold properties of MXenes materials, including the retention of physical and mechanical properties of MXenes and the interactions between MXenes and guest materials. Therefore, this article provides an in-depth discussion of the recent developments in $Ti_3C_2T_X$-based SC electrode materials, focusing on the relevant concepts and the important role played by $Ti_3C_2T_X$ in achieving excellent electrochemical results. First, the production process of $Ti_3C_2T_X$ (MXene) is closely related to the generation of special physicochemical characteristics, and we will study how $Ti_3C_2T_X$ (MXene) affects the electrochemical properties of the obtained materials and their synthesis techniques. Second, the influence of electrode shape, surface treatment group, and electrolyte on the electrochemical results is discussed, as well as whether pure $Ti_3C_2T_X$ (MXene) is suitable for SC electrode use. Third, since composites of $Ti_3C_2T_X$ with materials such as metal compounds, polymers, and carbon have been used to fabricate SC electrodes (Figure 2.2), the last section discusses the potential of $Ti_3C_2T_X$-based materials in SC applications and their current challenges. In conclusion, this chapter provides a detailed discussion of the applications and associated advantages and disadvantages, mechanical properties, fabrication processes, and fundamentals of SC applications, configuration, fabrication, and design of $Ti_3C_2T_X$-based SC electrodes.

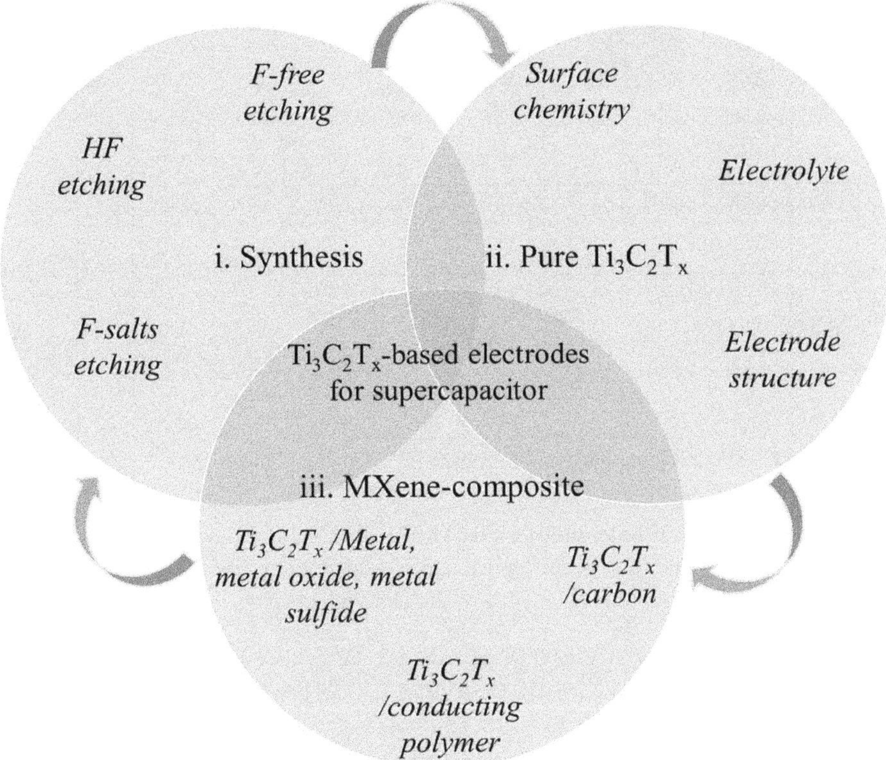

FIGURE 2.2 Overview of typical synthesis and variables affecting electrochemical achievement and $Ti_3C_2T_x$-based composite electrode for SC.

2 SYNTHESIS OF 2D MXENE

Since $Ti_3C_2T_x$ MXene was first reported in 2011, an increasing number of single- and multi-element bidirectional metal nitrides, carbides, and MXenes are under investigation. MXene generation typically uses top-down selective recording. This aggregation route does not change or lose properties as the batch size increases and has been proven to be scalable. As shown in Figure 2.3, to generate elusive MXenes, many techniques have been exploited, including wet-etching [35–37], electrochemical etching [38], chemical vapor deposition [39, 40], Lewis acidic molten-salt etching [41] and approaches using molten salt [42, 43].

For example, when etching Ti_3AlC_2 in HF solution at ambient temperatures, the surface of the carbide layer bonds with O, OH, and/or F atoms, resulting in the selective removal of A(Al) atoms. In addition to the treatment of HF solvents, recent studies have reported the use of various ions and reagents, alkali metal ions [45], TBAOH [46], and DMSO [47], as well as sonication methods. The synthesis method of Ti_3C_2 (MXene) is to attack Ti_3AlC_2 (MAX) powder with 50% concentrated HF solution (Figure 2.4a), which was first reported in 2011. The accordion-like layered structure with multiple functional groups is what the final MXene exhibits. Since MXene was first reported,

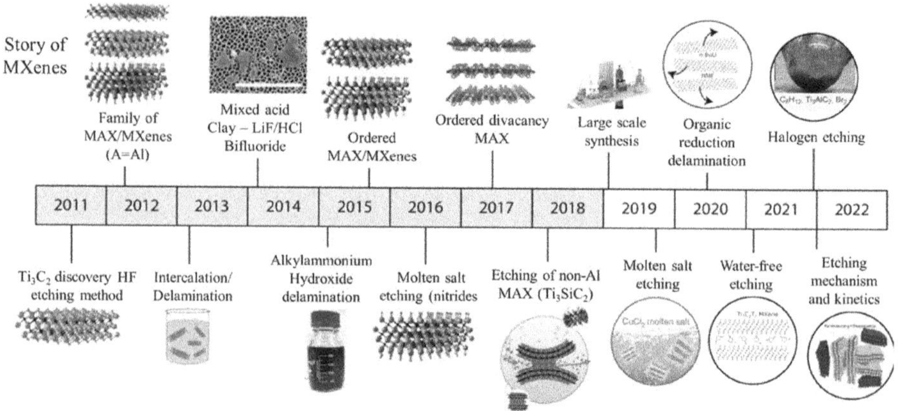

FIGURE 2.3 Synthesis methods of MXenes from 2011 to 2022 [35–43].

the forms of MXene include multifarious Ti$_3$CN, (V0.5,Cr0.5)$_3$C$_2$, TiNbC, Ta$_4$C$_3$, and Ti$_2$CT$_x$ using different HF concentrated solutions and etching times [48]. The following equations express the overall chemistry of the etching process:

$$M_{n+1}\text{AlX}_n + 3\text{HF} = \text{AlF}_3 + 3/2\text{H}_2 + M_{n+1}X_n \qquad [1]$$
$$M_{n+1}X_n + 2\text{H}_2\text{O} = M_{n+1}X_n(\text{OH})_2 + \text{H}_2 \qquad [2]$$
$$M_{n+1}X_n + 2\text{HF} = M_{n+1}X_nF_2 + \text{H}_2 \qquad [3]$$

In addition, by using HF solution to prepare Ti$_3$C$_2$T$_X$, mixing it with DMSO molecules, and then using ultrasound to relax the interlayer interaction or increase the layer spacing, single or multilayer Ti$_3$C$_2$T$_X$ can be obtained [36]. Exfoliated Ti$_3$C$_2$T$_X$ prepared by HF etching (Ti$_3$AlC$_2$) was first reported as an SC electrode material in 2013 (Figure 2.4b) [49]. To form fewer layers, DMSO was also used to decompose multilayer Ti$_3$C$_2$T$_X$. In the next step, filter-free paper electrodes were fabricated using MXene particles as raw materials [49]. DMSO has been proven to be an effective cross-linker in retarding Ti$_3$C$_2$T$_x$ [50] and (Mo$_{2/3}$Ti$_{1/3}$)$_3$C$_2$T$_x$ [51]. As shown for V$_2$CT$_x$ and Ti$_3$CNT$_x$, other organic molecules such as TBAOH can also be prepared using overlapping samples [46]. In addition to HF strong etchant treatment, Gogotsi and Barsoum reported the indirect use of HF etching treatment for MAX in 2014 [45]. In addition to LiF/HCl mixed solvents, HCl/fluoride salt mixed solvents (such as NH$_4$F, KF, NaF, and HCl/FeF$_3$) and fluorine-containing ammonium salts have also been used to prepare V$_2$CT$_x$, Ti$_2$CT$_x$, and Ti$_3$C$_2$T$_X$. The advantage of this method is that the entire synthesis path reduces the high dosage of the HF agent, and the stripping and etching process is performed in one step. Due to the intercalation of Li$^+$ and NH^{4+} ions, the interlayer distance of MXene also increases. As a result, the product can be layered directly into single-layer flakes using only ultrasonic or manual stirring. However, since single-layer MXene is susceptible to oxidation in both aqueous solutions and air, methods to prevent MXene oxidation and low-temperature storage, such as using additives that do not interfere with MXene use, have been reviewed. All these methods overuse HF-based, F-based etchants, or F-containing etchants, which

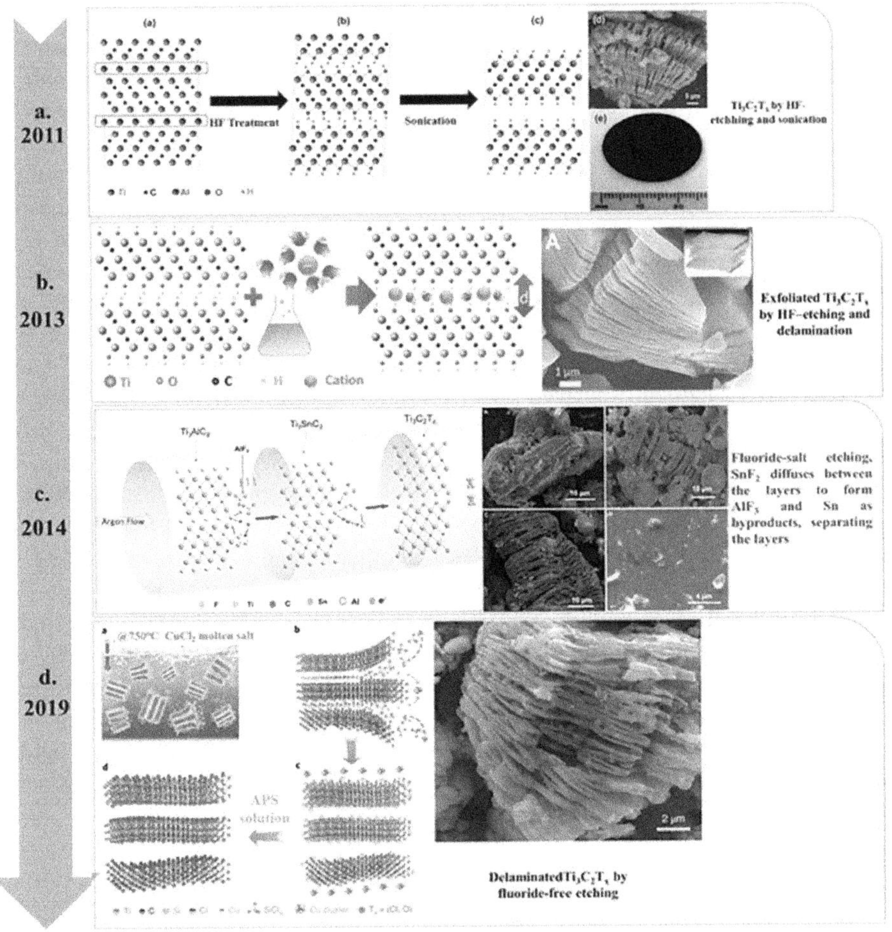

FIGURE 2.4 (a) $Ti_3C_2T_x$-MXene [44], (b) MXene-based electrode for SCs [49], (c) $Ti_3C_2T_x$-MXene-clay [47], and (d) $Ti_3C_2T_x$-MXene by molten salt [37].

leads to considerable environmental and safety issues. In order to solve these major problems, research on developing MXenes using fluorine-free etching solutions has gradually become more and more popular. In 2018, scientists synthesized MXenes using a fluorine-free hydrothermal alkali etching method [52]. Therefore, the formation of MXenes from MAX (Ti_3SiC_2) with non-Al using a mixed HF/H_2O_2 etchant solution was reported [53]. Thereafter, studies on different synthesis methods, such as employing MXene-carbon NCs, MXene-polymers, and MXene-metal oxides, were reported in 2019, demonstrating that composite MXene is a state-of-the-art hybrid material suitable for multiple applications [54–56]. In 2016, Urbankowski et al. provided an effective, short-cycle etching method, using a molten fluorine-containing salt mixture of KF, NaF, and LiF as an etchant, and obtained the first 2D transition metal nitride MXene [57]. As shown in Figure 2.4d, the preparation of MXene using Lewis acid molten salts with various A elements and MAX is reported, such as Ga (Ti_2GaC), Si (Ti_3SiC_2), Al (Ti_2AlC), and Zn (Ti_3ZnC_2). This etching technique provides an excellent opportunity to tune the properties and surface chemistry of MXenes and potentially generate new MXenes, including those unavailable for etching processes. Lewis acid molten salts have proven to be an effective tool for etching the MAX phase in addition to fluoride-containing salt melts. The cations in the Lewis acid and the "A" atoms in the MAX precursor are removed through a redox reaction. In 2020, Talapin et al. provided a general strategy and synthesized bare MXene without surface termination as well as a variety of MXenes with single surface termination (O, S, Te, NH, Br, Cl, and Se) and found some Nb_2C series of MXenes, including Nb_2CCl MXene, are superconducting at low temperatures [58].

The etching method described above is based on removing the "A" layer from the MAX and forming MXene without or with surface terminations, with lateral dimensions of chemically derived MXene ranging from a few hundred nanometers to about ten micrometers. Bottom-up approaches for MXene synthesis list three different synthesis methods, namely, *in situ* electrochemical synthesis method, lithiation-expansion-microexplosion mechanism, and chemical vapor deposition (CVD). In 2015, Xu et al. provided a CVD method for the synthesis of large-scale and ultra-thin MXenes (Mo_2C) [59]. Two years later, another research group provided a CVD method to prepare MXenes with lateral dimensions on the centimeter scale [60]. In 2020, Buke et al. studied whether the formation of Mo2C crystals is affected by impurity factors [61]. For the lithiation-expansion-microexplosion mechanism, Sun et al. discovered a new method to prepare single or several layers of MXenes through Ti_3AlC_2 MAX materials. Apart from that, this method (2019) also uses Ti_3SiC_2 MAX as the precursor material [62]. Recently, in 2020, Zhi et al. published an integrated process that combines ion storage of MXene and *in situ* etching of MAX using $Zn(OTF)_2$ mixed ions and LiTFSI electrolyte as etchant solution [63].

3 KEY PROPERTIES OF 2D MXENES FOR SUPERCAPACITORS

Members of the 2D-MXene family possess outstanding properties that can be altered by controlling the stoichiometric ratio of the X and M elements. Subsequently, MXene's tunable and unique properties, including an electrochemically active surface, metal-like conductivity, hydrophilicity, dispersibility in water, and good mechanical properties, make it a potential candidate for high energy density and high power capacitor designs. The following subsections briefly summarize the unique properties of MXenes.

3.1 EXCELLENT MECHANICAL FLEXIBILITY OF MXENES

MXene 2D structures exhibit excellent mechanical properties, making MXenes well suited for use in flexible supercapacitor devices. The mechanical properties of a material include its processability, ease of workability, elasticity, and strength, where the material's flexible properties enable it to be formed into any shape. In recent years, the demand for retainable and compact "microelectronic" system devices, such as micro-SCs and flexible SCs, has increased significantly, and the 2D layered structure and ductility of MXenes make it a promising candidate for such devices. Most

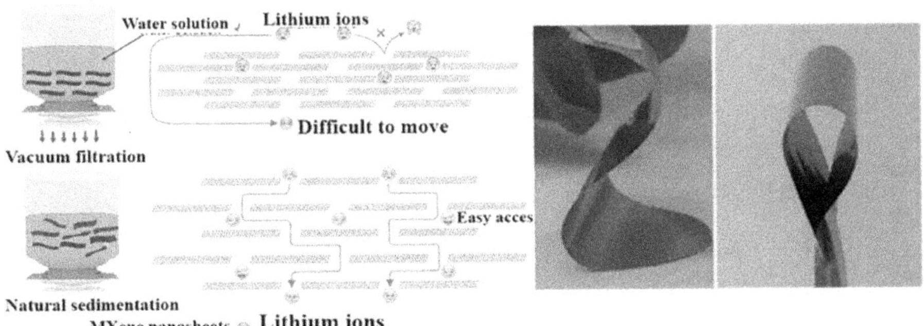

FIGURE 2.5 Comparison of enhanced ion availability for digital photography of naturally deposited MXene films [64].

research so far has focused on further improving the structural properties of MXenes (as shown in Figure 2.5), and 2D-MXene nanosheets can be easily assembled into thin film electrodes [64–66]. At present, these research directions have shown certain progress.

3.2 HYDROPHILICITY AND DISPERSIBILITY OF MXENES

MXene has a hydrophilic surface with a strong affinity for water, and the spreading of water on this surface is preferred. However, crucial to the fabrication of electrode inks that are subsequently used in composite electrodes or fabricated electrodes is the dispersion ability of the electrode material in various solvents. In the preparation of MXene, the choice of synthesis method provides control over the surface termini (-O, -Cl, -F, -OH); therefore, the -OH and -O groups are important in imparting stable hydrophilicity to MXene in aqueous solutions (Figure 2.6). Furthermore, the electrochemical results of MXene-based electrodes are mainly driven by the surface end groups of MXenes [67].

3.3 CONDUCTIVITY OF MXENES

Typically, SCs have fast discharge rates and high energy densities; therefore, they require electrodes like MXene with high conductivity, and the advantages of MXenes in facilitating the above properties include the selection of X and M and its surface end groups' adjustments to increase the conductivity of MXenes, which can become fully insulating, semi-insulating, or metalloid [68–70]. Therefore, low defect concentration and large lateral width in MXenes are generally associated with higher electrical conductivity [71–72]. As shown in Figure 2.7, severe defects exist on fiber-molded $Ti_3C_2T_X$-MXene films etched on cold-pressed $Ti_3C_2T_X$ (MXene) films (1000 S cm^{-1}). In comparison, after etching with LiF/HCl solution, spin-cast $Ti_3C_2T_X$ (MXene) films can achieve extremely high conductivity (6500 S cm^{-1}) [73–74]. Achieving high power densities requires MXenes to be very conductive, but this also reduces the need for conductive materials to make electrodes and even current collectors [45], thus increasing the overall energy density of the device.

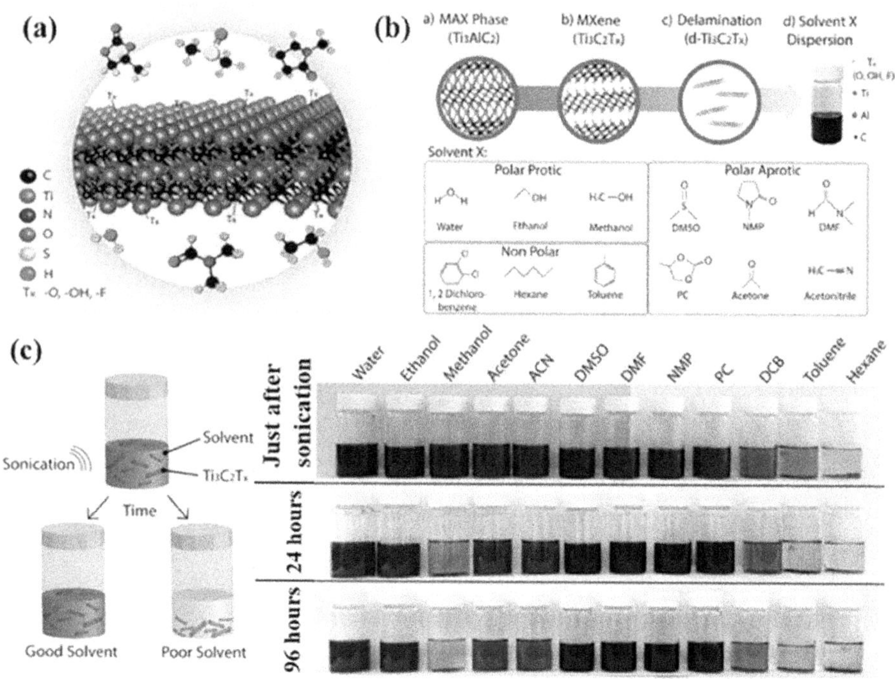

FIGURE 2.6 (a) An exemplification of the MXene, (b) description of the synthesis and dispersion of $Ti_3C_2T_x$ in organic solvents, and (c) dispersibility of $Ti_3C_2T_x$ in 12 different solvents [67].

3.4 CHARGE STORAGE MECHANISM

MXene's pseudocapacitive ion gap charge storage mechanism contributes to its long lifetime and high capacity. Capacity and longevity are key components of SCs. The 2D layered structure of MXene allows rapid intercalation of ions (e.g., Na+, Li+, or H+), enabling efficient operation in electrolytes – free media as well as in acidic aqueous electrolytes and (as recently demonstrated) fast pseudocapacitive charge storage. Therefore, MXene has high capacitance (300–500 F g^{-1}) in aqueous electrolytes or is more or less comparable to EDLC carbon materials; moreover, after thousands of operating cycles, MXene outperforms pseudocapacitive materials, and there is very little or no capacitance loss. Furthermore, by controlling the composition of MXene-containing NCs, the capacitance of MXene-based electrodes in SCs can be improved. For example, MnO_2@V_2C-MXene had a capacitance of approximately 551.8 F g^{-1} and a retention of about 96.5% after 5,000 cycles [75]; $Ti_3C_2T_x$-AuNPs film has a capacitance of 696.67 F g^{-1} at 5 mVs^1 in 3 M H_2SO_4 electrolyte [76], and the Ni-MOF/$Ti_3C_2T_x$ nanocomposite has a high specific capacitance of 497.6 F g^{-1} at 0.5 A g^{-1} in the KOH/PVA electrolyte [77]. To fabricate high-performance metal oxide symmetric capacitors, an efficient approach is summarized from these results, and a simple method to improve the design of MXene-based electrodes is proposed.

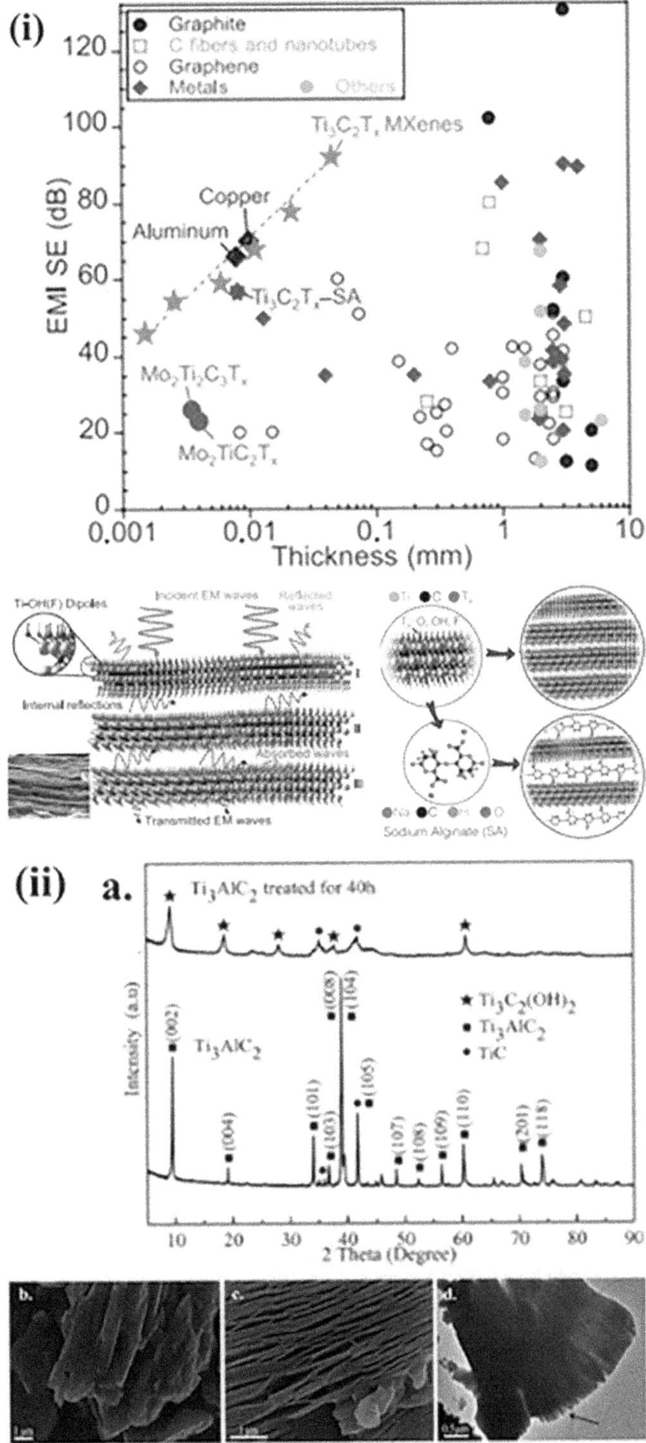

FIGURE 2.7 (i) An exemplification of TC and TC-SA composite with EMI SE, (ii) (a) XRD and (b–d) SEM/TEM images of before and after HF treatment on MAX [73].

3.5　High Density and Gravimetric Capacitance of MXene

High-capacity MXenes produce high-gravity, high-density MXenes. In practical applications, the increase in energy density and device power is highly dependent on the volumetric capacitance of the material [72, 45]. Vacuum-filtered $Ti_3C_2T_X$ (MXene) membranes are also included, and their volume capacitance can reach 900 F cm^{-3}, while MXene hydrogel membranes can achieve a volume capacitance of 1500 F cm^{-3} [78]. Figure 2.8 illustrates all the data about Ti3C2TX MXene. These results are superior to those obtained with other supercapacitor electrode materials, such as porous AC (60–200 F cm^{-3})[79, 80] or optimally activated graphene electrodes (200–350 F cm^{-3}), while achieving the same the results of.

4　DESIGNING 2D MXENE ELECTRODE

Electrochemical capacitors and batteries, two examples of energy storage technologies, have always used layered materials formed by binding ions together, and few

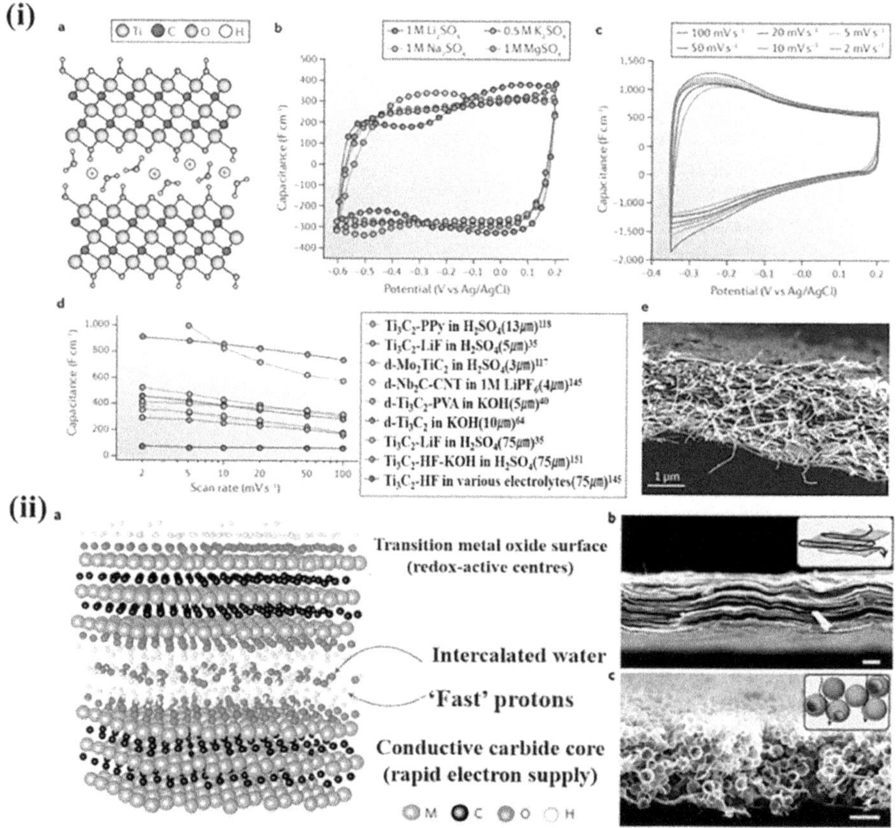

FIGURE 2.8　(i) (a) An exemplification of MXenes, (b–c) Cyclic voltammograms of $Ti_3C_2T_x$ paper electrode. (d) Comparison study of MXene electrodes. (e) Cross-sectional image [72]. (ii). An exemplification and SEM images of $Ti_3C_2T_x$ MXene [79].

core materials contain ions relatively larger than lithium. As already mentioned, MXene exhibits rapid ion intercalation behavior, resulting in a broad active surface that can be used in electrochemistry and is suitable for use as a pseudocapacitive mechanism in acidic nonaqueous and aqueous electrolytes. In 2013, the first report on MXenes in capacitive electrodes was published [49]. This study shows that there is spontaneous intercalation of cations from various salt solutions between MXene layers. Furthermore, MXene blends a 2D conductive carbide layer with a hydrophilic (mainly hydroxyl-terminated) surface. Therefore, the electrochemical interactions of many cations such as Al^{3+}, Mg^{2+}, NH^{4+}, K^+, and Na^+ can indeed provide capacitances exceeding 300 farads per cubic centimeter (relatively higher than the capacitance of porous carbon). Therefore, in this study, not only MXene ($Ti_3C_2T_X$) flakes were investigated, but also $Ti_3C_2T_X$ binder-free paper was prepared, which was used as a capacitor electrode. According to their research, cations of various sizes and charges in aqueous solutions can be successfully embedded into exfoliated $Ti_3C_2T_X$ multi-layer films and MXene papers composed of multiple $Ti_3C_2T_X$ layers. Both pH and cationic composition influence this phenomenon. Complete testing of the electro-chemical properties of $Ti_3C_2T_X$ in these aqueous electrolytes illustrates its significant embedded capacitance. In conclusion, this first report of MXene in monovalent and multivalent ionic capacitors provides a basis for exploring the applicability of 2D superfamilies in electrochemical energy storage applications. The above study used HF solvent as an etchant for $Ti_3C_2T_X$ synthesis, while another study used a mixture of LiF/HCl as an etchant. According to their research, LiF/HCl-etched Ti_3C_2 (MXene) clay exhibits an excellent gravimetric capacitance of 245 F g^{-1} with minimal capaci-tance loss after 10,000 cycles at 10 A g^{-1} current. Furthermore, when using hydrogel $Ti_3C_2T_X$ (MXene) electrodes in 3M H_2SO_4 electrolyte, capacitances up to 380 F g^{-1} or 1,500 F cm^{-3} were achieved using glassy carbon current collectors [81]. The research-ers tested F-containing reagents other than LiF/HCl and HF etchants; however, there was no improvement in the energy storage capacity of MXene-based electrodes over the above results.

Figure 2.9 shows different patterning methods for MXene-based SCs. Generally speaking, the fabrication methods include two categories: direct patterning of MXene on the substrate/current collector through reactive ion etching or laser scribing; and transferring MXene ink into different patterns through the following printing meth-ods. There are more preparation methods for MXene-based SCs, such as laser cut-ting freeze-drying, direct writing, laser cutting vacuum filtration, and laser cutting spraying methods.

5 CHARGE STORAGE AND ITS INFLUENCE ON THE SURFACE GROUP OF MXENES

There are three main processes that exist in the energy storage process of MXene: collection of electrons by the collector, propagation of ions from the electrolyte to the interlayer, and electron transfer to the surface. Therefore, in order to store energy at high speeds, the conductivity of MXene-based electrodes should be as high as possible. The conduction of electrons is responsible for the TMC core formed by the C and M layers dispersed in the M_3C_2 MXene layer. According to experimental

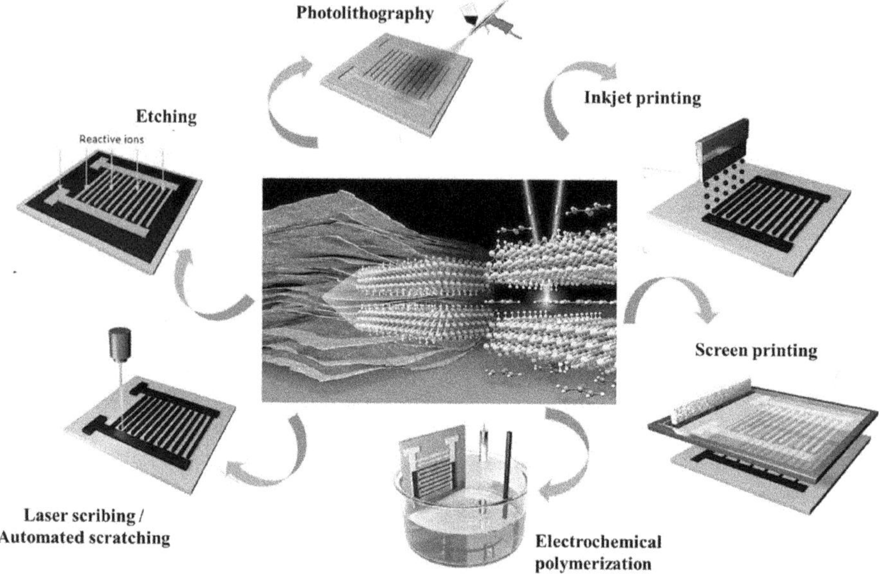

FIGURE 2.9   The fabrication methods for interdigital MXene SCs.

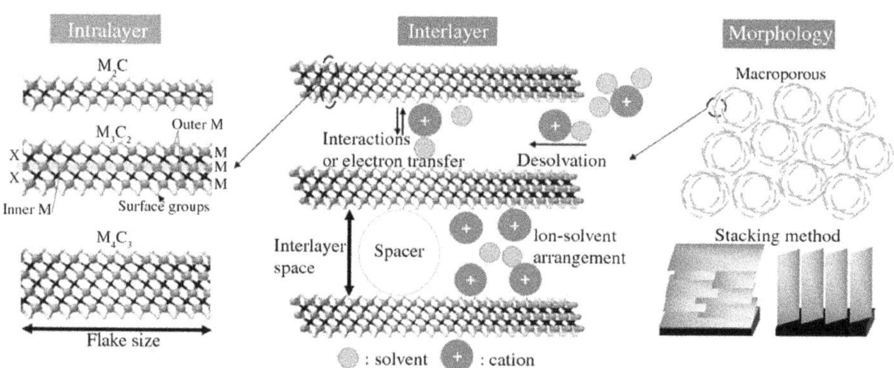

FIGURE 2.10   Rate determination of 2D materials at three levels: Intra and interlayer, and electrode morphology [82].

reports, $Ti_3C_2T_X$ has better conductivity than Ti_2CT_x, which may be due to the lack of carbide core in M_2C MXene. Replacing the outer layer of the first metal with ordered dual TM-MXene to form the second metal results in the structures $M1_2M_2C_2$ and $M1_2M_2C_3$, where M_1 and M_2 represent two separate initial transition metals, which can be Ti, V, Nb, Ta, Cr, or Mo (as shown in Figure 2.10) [82].

$Ti_3C_2T_x$, which has the highest electrical conductivity of all known MXenes -20,000 S cm^{-1}, is made by extracting additional aluminum from the Ti_3AlC_2 (MAX) phase. The metal-like properties of MXene will change from metals to semiconductors as the outermost layer changes to Mo. Pseudocapacitive ions can be embedded through surface redox reactions, and EDLCs can be fabricated on the surface of MXenes. As shown in Figure 2.11a, the MXene surface groups interact with the interacting ions and transfer part of the electrons to the material, thereby initiating surface oxidation reactions on the MXene. Therefore, the pseudocapacitive mode is the charge storage mode of the MXene electrode, and MXene relies directly on this quality to transfer and hold more capacitance than Electric Double Layer Capacitors (EDLC). $Ti_3C_2T_X$ powder was mixed with MXene powder, 15 wt, 80 wt carbon black, and 5 wt PTFE binder, processed into individual electrode films, and tested in 3M H_2SO_4 aqueous electrolyte to evaluate the effect of surface modification on electrochemical performance. As shown in Figure 2.11b, nitrogen surface functionalization can unlock the full potential of MXenes through a high-rate, highly reversible approach. Higher voltage windows extend these measurements, and even higher voltages cannot drive more ions to electrosorb in current mode; in other words, we expect the curve to bend downward at higher voltages.

The surface chemistry of MXene affects the capacity of MXene-based charge storage electrodes due to surface redox charging patterns. After etching the MAX-A layer in fascia solution, functional groups were added to the surface of the MXene. Treatment with different etchants has been shown to produce different electrochemical properties, and Ti_3C_2X (X = Br, Cl...) has been shown to exhibit different improved capacities. Additionally, surface voids and defects in MXene can serve as active sites for ion adsorption and other interfacial reactions, thereby affecting the material's capacitance, potential window, and rate-to-velocity capabilities. In most cases, titanium-deficient defects are clustered in the same sublayer (Figure 2.11c–e), and by changing the amount of etchant used during the etching process, the concentration of surface defects can be adjusted. Figure 11c shows a single-layer $Ti_3C_2T_X$ flake, with its typical defect configuration: Ti vacancies (VTi) and Ti vacancy clusters, with the number of clusters ranging from 2 to 17 VTi. Using different concentrations of HF etchants to adjust the defect concentration will affect the Ti vacancies on the surface functional groups, and the electronic conductivity will also be adjusted accordingly. Critical to controlling the functional properties of the MXene phase is the ability to tune the defect concentration. Figure 2.11d shows the microscopic crystal structure was further characterized by high-resolution TEM (HRTEM).

6 MXENES AS ELECTRODE MATERIALS FOR SC

Synthetic methods for MXenes show a variety of surface terminations and morphologies with different energy storage properties, and there have been several studies examining electrode composition control, stoichiometry, and surface modification. The following sections summarize these studies.

6.1 Control of Size of MXene Flakes

The corrosiveness of synthesis conditions greatly affects the structure and morphology of MXenes. For example, extending the etching time results in a more open structure than HF etching of Ti_3AlC_2, thereby resulting in an accordion-shaped $Ti_3C_2T_X$ (MXene). As shown in Figure 2.12, Ti_3C_2 with a corrosion time of 216 hours has a deformed octahedral structure. Compared with Ti_3C_2 with a corrosion time of

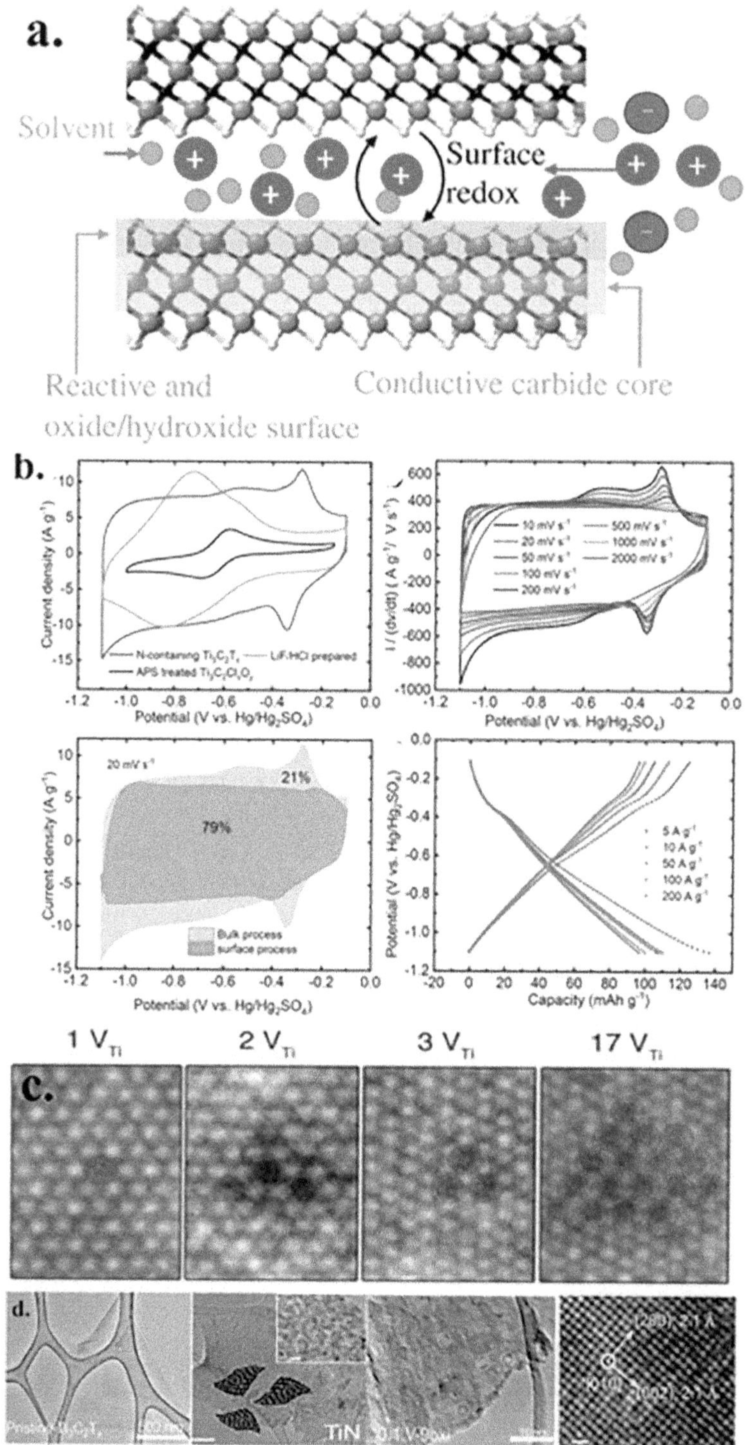

FIGURE 2.11   (a) An exemplification of MXene, (b) charge storage rate [83]. (c) Dark-filed TEM [84] and (d–e) TEM image TM-nitride nanocrystals at 700°C [85, 86].

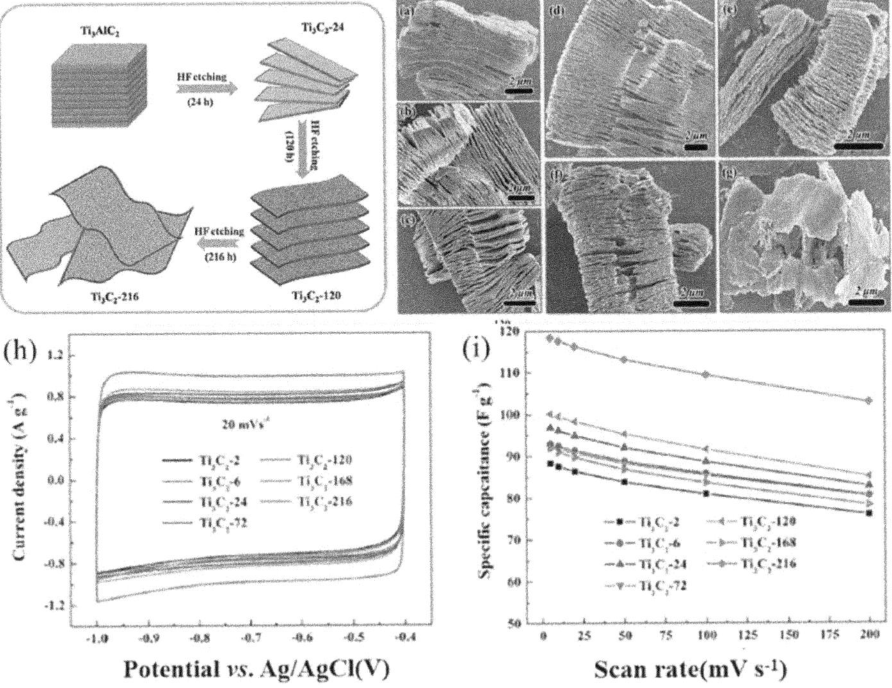

FIGURE 2.12 Schematic illustration for the typical structures of Ti_3C_2 with the evolution of the etching time, SEM images of the typical as-fabricated MXene with the CV curves and specific capacitance vs. scan rate for the electrodes with different etching times [86, 87].

24 hours, the surface loses more Ti and C contacts, allowing it to obtain significantly better capacitive performance [86, 87]. However, smaller MXene fragments have lower electrical conductivity due to higher surface-to-surface contact resistance, despite higher ionic conductivity and easier access to ion diffusion paths. Kayali et al. [88] studied the influence of the side diameter of MXene on its electrochemical performance. Furthermore, Mustafa et al. [89] demonstrated the fabrication of rationally designed electrodes by mixing MXene with aqueous chloroauric acid ($HAuCl_4$) solution. As shown in Figure 2.13, the symmetric supercapacitor made of AuNP and MXene exhibits an excellent specific capacitance of 696.67 F g^{-1} in 3 M H_2SO_4 electrolyte at 5 mV s^{-1} and can maintain 90% of the original capacitance after 5,000 cycles. The device operates within a 1.2 V potential window, achieving maximum power and energy densities of 2076 Wh kg^{-1} and 138.4 W kg^{-1}, respectively. Morphology analysis clearly shows that Au NPS are uniformly distributed on the MXene nanosheets; therefore, Au NPs are the best intercalators in the hybrid structure [89].

6.2 CATEGORY OF COMPOSITION

Table 2.1 summarizes most of the previous studies examining $Ti_3C_2T_X$ and other MXenes for use in supercapacitor electrodes or energy storage applications. Ti_2CT_x

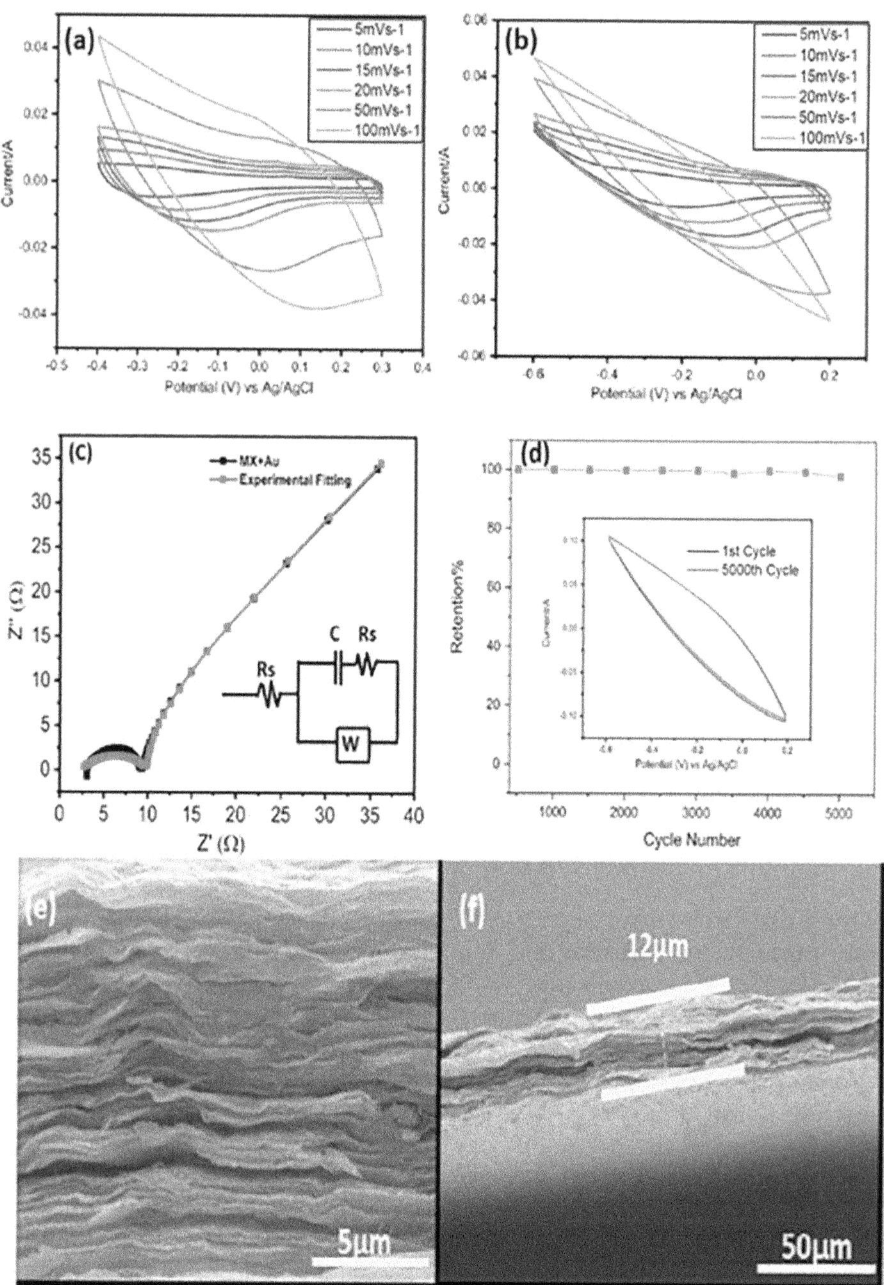

FIGURE 2.13 (a) CV curves of MXene at different scan rates in a three-electrode setup; (b) CV curves of MX/AuNPs in 1 M H_2SO_4 at different scan rates in a three-electrode setup; (c) Nyquist plots of MX/AuNPs; inset shows an equivalent circuit model for the Nyquist plots; (d) MX/AuNPs electrode showing excellent cyclic stability with 98% capacitance retention at 100 mV s^{-1} 1M H_2SO_4 after 5,000 cycles; (e) cross-sectional image of the MX/AuNPs composite film demonstrating stacked layered structure; (f) cross-sectional image showing the thickness of the composite film [89].

TABLE 2.1

Capacitive Capabilities of Different MXene Electrode Materials for SCs

Versatility	Synthesis	Electrode Assembly	Electrolyte	Capacitance	Cycle Stability	Ref.
$Ti_3C_2T_x$	HF etching Ti_3AlC_2 and DMSO delamination	Filtrating into film	1M KOH	350 F cm^{-3}	No degradation (10,000 cycles)	43
$Ti_3C_2T_x$	HCl/HF etching Ti_3AlC_2	Rolling	1M H_2SO_4	900 F cm^{-3} or 245 F g^{-1}	No degradation (10,000 cycles)	41
$Ti_3C_2T_x$	Lewis acidic molten salt etching Ti_3SiC_2	Rolling	1M $LiPF_6$-EC/DMC	738 C g^{-1} or 205 mAh g^{-1}	90% (2,400 cycles)	31
Ti_3CT_x	HF etching Ti_2AlC	Rolling	1M KOH	517 F cm^{-3}	No degradation (3,000 cycles)	90
Ti_3NT_x	Oxygen-assisted molten salt etching Ti_2AlN_2 and HCl treatment	Coating	1M $MgSO_4$	201 F g^{-1}	140% (1,000 cycles)	93
V_2C	HF etching V_2AlC and TMAOH delamination	Rolling	1M H_2SO_4	487 F g^{-1}	83% (10,000 cycles)	94
$V_4C_3T_x$	HF etching V_4AlC_3	Coating	1M H_2SO_4	~209 F g^{-1}	97.23% (10,000 cycles)	95
$Nb_4C_3T_x$	HF etching Nb_4AlC_3 and TMAOH delamination	Rolling	1M H_2SO_4	1075 F cm^{-3}	76% (5,000 cycles)	96
Ta_4C_3	HF etching Ta_4AlC_3	Coating	0.1M H_2SO_4	481 F g^{-1}	89% (2,000 cycles)	97
Mo_2CT_x	HF etching Mo_2Ga_2C and TBAOH delamination	Filtrating into film	1M H_2SO_4	700 F cm^{-3}	No degradation (10,000 cycles)	36
$Mo_2TiC_2T_x$	HF etching and DMSO delamination Mo_2TiAlC_2	Rolling	1M H_2SO_4	413 F cm^{-3}	No degradation (10,000 cycles)	45

MXenes [90, 91] provided excellent results. Nitride MXene exhibits higher electrical conductivity compared to carbides [92]. Accordion-structured, -O/-OH-terminated Ti_2NT_x nanolayers were obtained by oxygen-assisted molten salt synthesis of the Ti_2AlN (MAX) phase followed by HCl treatment.

The scan rate of nanolayer Ti_2NT_x MXene with -O/-OH surface termination in 1 M $MgSO_4$ electrolyte at 2 mV s^{-1}, exhibiting a capacitance of 200 F g^{-1}, and this MXene material was treated with HCl prepared. In addition to traditional titanium-based MXene materials, by using M-center metal alloys (such as Mo-[42] and Ta-[104]), the members of the MXene family have been infinitely expanded to include N- and C-bonded transition metal species, based on MXenes. The material is considered a reliable electrode for supercapacitor applications. Electrochemical results, stability, and conductivity are all significantly affected by the inherent properties of the X and M elements, and surface finish also strongly affects these qualities. On the other hand, since surface terminals can be managed, and their performance can be modified by selecting different X and M components, there is great potential for the development of new MXenes with good capacity performance. Therefore, new MXenes with higher electrochemical efficiency are needed considering their potential use in supercapacitors. Due to its unique atomic bonds, appropriate synthesis methods, including etching and separation processes, should be optimized for each MAX precursor. This is necessary to achieve the desired achievements and structure. Exploring new MXenes and their preparation technology requires a lot of research and is an important and arduous task.

6.3   HETEROATOMS DOPING AND THE CONTROL OF SURFACE-TERMINUS GROUP

Heteroatom doping can change the surface properties and electron acceptability of materials. N atoms are considered to be active heteroatoms affecting MXene electrochemistry, and many studies have confirmed this relationship. The reaction mechanism is roughly as follows: by replacing C atoms in the C layer with N atoms, it has been proven that the interlayer distance can be increased, thereby effectively improving the conductivity [98]. The most commonly used nitrogen doping methods can be divided into liquid stripping, solvothermal treatment, hydrothermal treatment, liquid intermediate doping, and ammonia heat treatment [92,99–102]. For example, as shown in Figure 2.14, a Ti_3C_2 thin film with a flexible N-doped layer was prepared through *in situ* solvothermal treatment and static adsorption, using a urea-saturated alcohol solution as the N atom source [101].

In addition, Ti_2CT_x-MXene undergoes nitrogen chemical doping and liquid-gas phase layering simultaneously [102]. In an inert environment, NH_2CN serves as an intercalation agent and nitrogen source combined with Ti_2CT_x nanosheets, and the condensation reaction generates polymerized carbon nitride (p-C_3N_4), which is required during heat treatment. It has been demonstrated that solid-state doping is possible. S and N co-doped $Ti_3C_2T_X$-MXene was prepared by pyrolysis of thiourea in Li_2SO_4 electrolyte, which exhibited 75 F g^{-1} at 2 mV s^{-1} [103, 104]. Historically, the primary focus of most experiments investigating MXene doping has been nitrogen. The potential for modified MXene could also be demonstrated by using other types of dopant atoms (such as heteroatoms and metals). Calculated results show that B

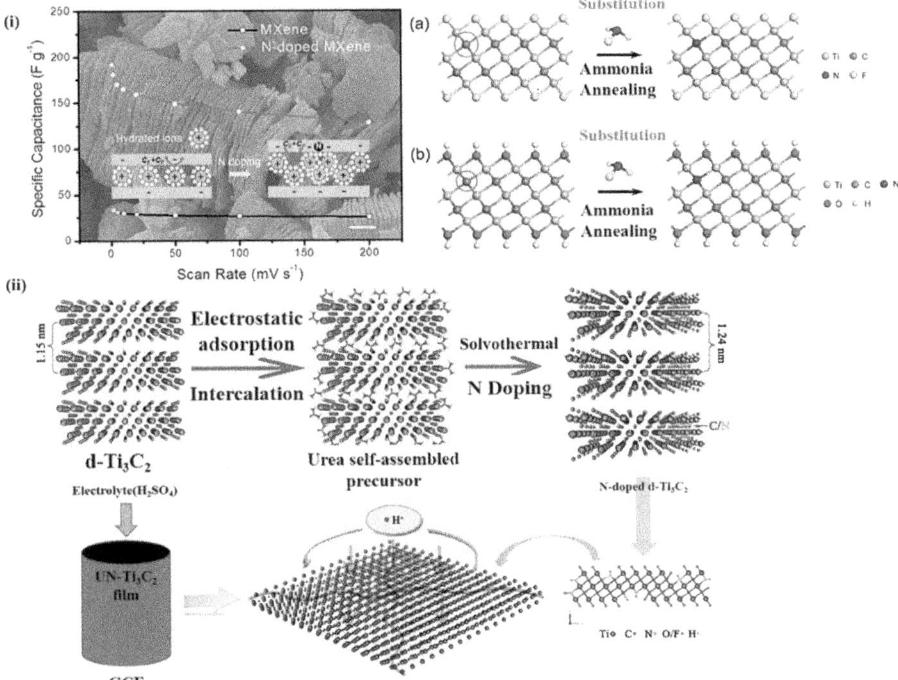

FIGURE 2.14 (i) An exemplification of charge-storage and film-electrode fabrication of UN-Ti$_3$C$_2$. (ii) Schematic illustration of film preparation and electrode fabrication of the UNTi$_3$C$_2$' [101].

doping increases the stretchability of MXene, making it ideal for flexible electronic devices [105]. The capacitance, conductivity, and interlayer spacing of MXene can also be increased by adding doping with metal atoms such as Nb and V [106,107]. As mentioned above, chemical modifications can be used to increase the yield of MXenes and fine-tune their properties. Since vacancies can cause defects in the MXene matrix, affecting the activity, surface properties, and structure of the MXene, this may also be beneficial in modifying its properties. With the continuous deepening of research on chemical modification technology, new terminals and doping atoms are constantly added to MXenes [108–112].

6.4 Fabricating Vertical Alignments

Figure 2.15 shows a schematic diagram regarding the fabrication of RGO/Ti$_3$C$_2$T$_X$ MXene-modified fabric by simple dip coating and spray coating methods. It characterized the electrochemical performance of MXenes and RGO-modified fabrics by conducting CV tests using 1 M H$_2$SO$_4$/PVA gel electrolyte in a two-electrode configuration. Figure 2.15a shows the CV curves of the RMC supercapacitor and all-solid-state MCF at a scan rate of 10 mV s^{-1}. MCF-based supercapacitors demonstrate poor charge storage capabilities as they display a minimal closed CV curve (almost a line). At the same time, after adding RGO, the charge storage capacity is

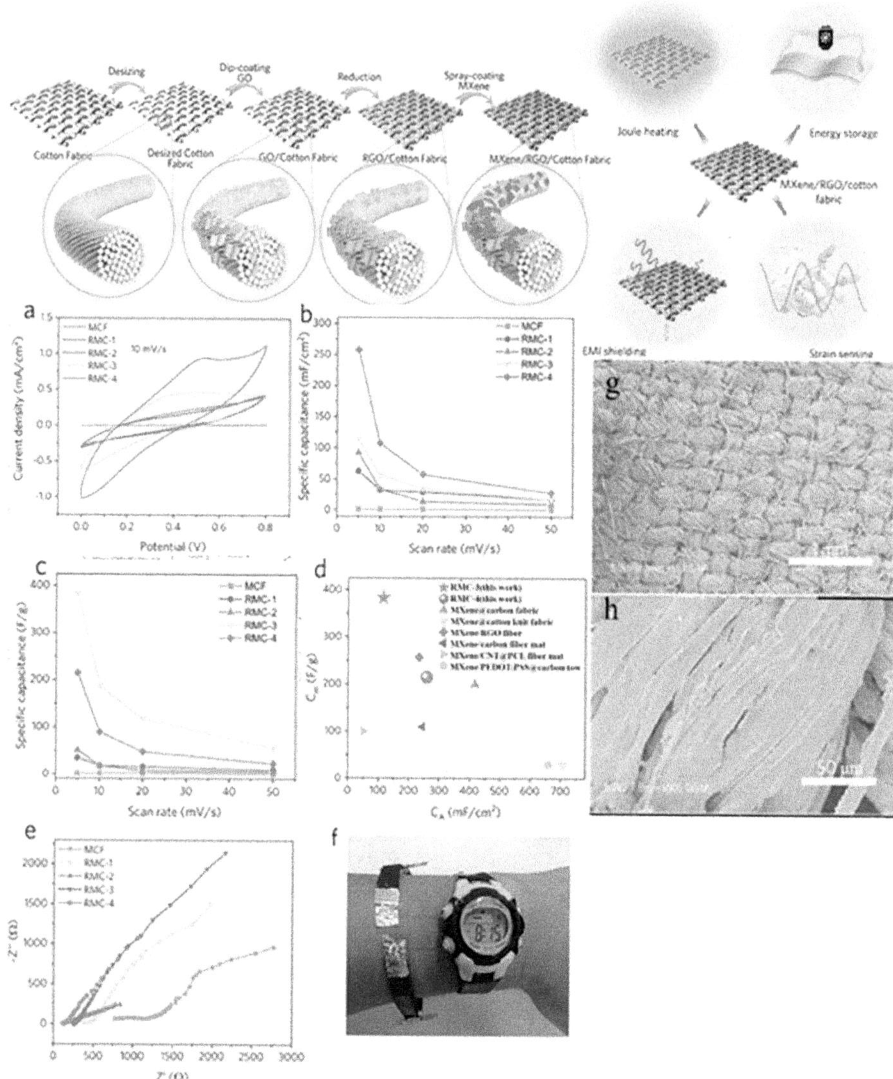

FIGURE 2.15 Schematic illustration of multifunctional RGO/Ti$_3$C$_2$T$_x$ MXene fabrics, Electrochemical performance of the MCF and RMC fabrics. (a) CV curves of the MCF and RMC at 10 mV s^{-1}. Areal specific capacitance (b) and gravimetric specific capacitance (c) of MCF and RMC. (d) Areal and gravimetric specific capacitance of RMC compared with the reported MXene-modified fibers, yarns, and fabrics. (e) Nyquist plots of MCF and RMC. (f) Four RMC-4 supercapacitors connected in series as a wristband to power a watch, (g and h) SEM images of fabrics [113].

significantly improved, and the CV area of RMC_{-1} is increased. Furthermore, the addition of RGO significantly increased the CA of RMC_{-1} to 62.9 mF cm^{-2}, and the CA of RMC_{-2} further increased to 92.5 mF cm^{-2}, 115.0 mF cm^{-2}, and 258.0 mF cm^{-2}, respectively, RMC_{-3} and RMC_{-4}. The increase in C_A is attributed to the synergistic effect between the dominant pseudocapacitance and the increased conductivity of MXene. The morphology image of $RGO/Ti_3C_2T_X$ MXene was analyzed by SEM, and the obtained image is shown in Figure 2.15g–h. The SEM image shows that MXene and RGO sheets are densely filled on the fiber surface and in the fiber gaps, and in the gaps between warp and weft yarns, graphene sheets clearly appear.

By fabricating MXene membrane electrodes from "T-shaped" fragments; thus achieving efficient longitudinal ion transfer [114]. The capacity of the $Ti_3C_2T_X$ thin film electrode in H_2SO_4 electrolyte ranges from 2 A g^{-1} to 361 F g^{-1}, and the capacity increases by 76% when the current density increases to 20 A g^{-1}. Table 2.2 summarizes the results and designs of electrodes that can meet the current efficiency and supercapacitor capacity requirements. As shown in Table 2.2, to approach the theoretical limit of MXene capacity, we can design hydrogel frameworks. Simultaneously using porous MXene structures enables high-speed results at scan rates up to 10 V s^{-1}, and using vertically oriented MXene particles enables high-capacity power with film thicknesses up to 200 μm. This confirms the output. These designs and their combinations offer exciting new possibilities for MXenes in supercapacitor applications.

TABLE 2.2

Capacitive Outcome of MXene Electrodes with Optimized Structural Design

Electrode	Preparation	Electrolyte	Capacitance	Cyclability	Ref.
		Controlling flake size			
$Ti_3C_2T_x$ film	Mixing large and small flakes	3M H_2SO_4	435–86 F g^{-1}	10,000 cycles	74
		Adding spacer between the MXene interlayer			
75 μm $Ti_3C_2T_x$ pillared by hydrazine	Suspending in hydrazine	1M H_2SO_4	250–210 F g^{-1}	No decay (10,000 cycles)	115
$Ti_3C_2T_x$/ graphene-3% film	Mixing and filtration	3M H_2SO_4	438–302 F g^{-1}	No decay	116
Sandwiched $Ti_3C_2T_x$/CNT-5% film	Alternative filtration	1M $MgSO_4$	390–280 F cm^{-3}	No decay	117
V_2CT_x/alkali metal cations film	Cation-driven assembly	3M H_2SO_4	1315–>300 F cm^{-3}	106 cycles	118
$Ti_3C_2T_x$ ionogel film	Immersing into EMITFSI	EMITFSI	70–52.5 F g^{-1}	1,000 cycles	119
Ti_3C_2/CNTs film	Electrophoretic deposition	6M KOH	134–55 F g^{-1}	No decay	120

(Continued)

**TABLE 2.2
(Continued)**

Electrode	Preparation	Electrolyte	Capacitance	Cyclability	Ref.
Carbon-intercalated $Ti_3C_2T_x$	*In situ* carbonization	$1M\ H_2SO_4$	364.3–$193.3\ F\ g^{-1}$	10,000 cycles	121
$Ti_3C_2T_x$@rGO film	Plasma exfoliation	PVA/H_2SO_4	54–$35\ mF\ cm^{-2}$	1,000 cycles	122
Designing 3D/porous structure					
13 μm $Ti_3C_2T_x$ film	1–2 μm PMMA template	$3M\ H_2SO_4$	310–$100\ F\ g^{-1}$	–	63
MXene/CNTs film	Ice template	$3M\ H_2SO_4$	375–$92\ F\ g^{-1}$	10,000 cycles	123
Ti_3C_2 aerogel	Assembly and freeze-drying	$1M\ KOH$	87.1–$66.7\ F\ g^{-1}$	10,000 cycles	124
$Ti_3C_2T_x$ hydrogel	Assembly and freeze-drying	H_2SO_4	370–$165\ F\ g^{-1}$	10,000 cycles	125
$Ti_3C_2T_x$/carbon cloth	Freeze-drying with KOH treatment	$1M\ H_2SO_4$	312–$200\ mF\ cm^{-2}$	8,000 cycles	126
$Ti_3C_2T_x$/Ni foam	Electrophoretic deposition	$1M\ KOH$	140–$110\ F\ g^{-1}$	10,000 cycles	127
Compact and nanoporous $Ti_3C_2T_x$ film	Freeze-drying and mechanically pressing	$3M\ H_2SO_4$	932–$462\ F\ cm^{-3}$	5,000 cycles	128
3D porous $Ti3C_2T_x$ film	Reduced-repulsion freeze casting assembly	$3M\ H_2SO_4$	358.8–$207.9\ F\ g^{-1}$	10,000 cycles	129
Fabricating vertical alignments					
Anti-T $Ti_3C_2T_x$ film	Filtrating through an entwined metal mesh	$1M\ H_2SO_4$	361–$275\ F\ g^{-1}$	10,000 cycles	111
200 μm $Ti_3C_2T_x$ film	Mechanical shearing of a discotic lamellar liquid-crystal MXene	$3M\ H_2SO_4$	$>200\ F\ g^{-1}$	20,000 cycles	110

6.5   3D Microporous Sphere/Tube $Ti_3C_2T_x$

Both Ti-based and V-based MXenes tend to restack due to strong van der Waals interactions, resulting in potassium ion dynamics and decreased surface area. In an attempt to solve this problem, many methods have been developed to fabricate 3D architectures using Photonic

Photonic Integrated Circuits (PICs). In order to make 3D microporous tubes or spheres using $Ti_3C_2T_X$ as the raw material, Fang et al. designed a simple spray freeze-drying process [130]. Figures 2.16a and 2.16b show a 3D Ti_3C_2 structure consisting of some tubes and several spheres with smooth surfaces. The solution to the stacking problem of MXene nanosheets relies on the development of 3D morphology and enables rapid ion transport (Figure 2.16c). The constructed PIC has significant power density and energy density ($7015.7\ W\ kg^{-1}$ and $98.4\ Wh\ kg^{-1}$). According to

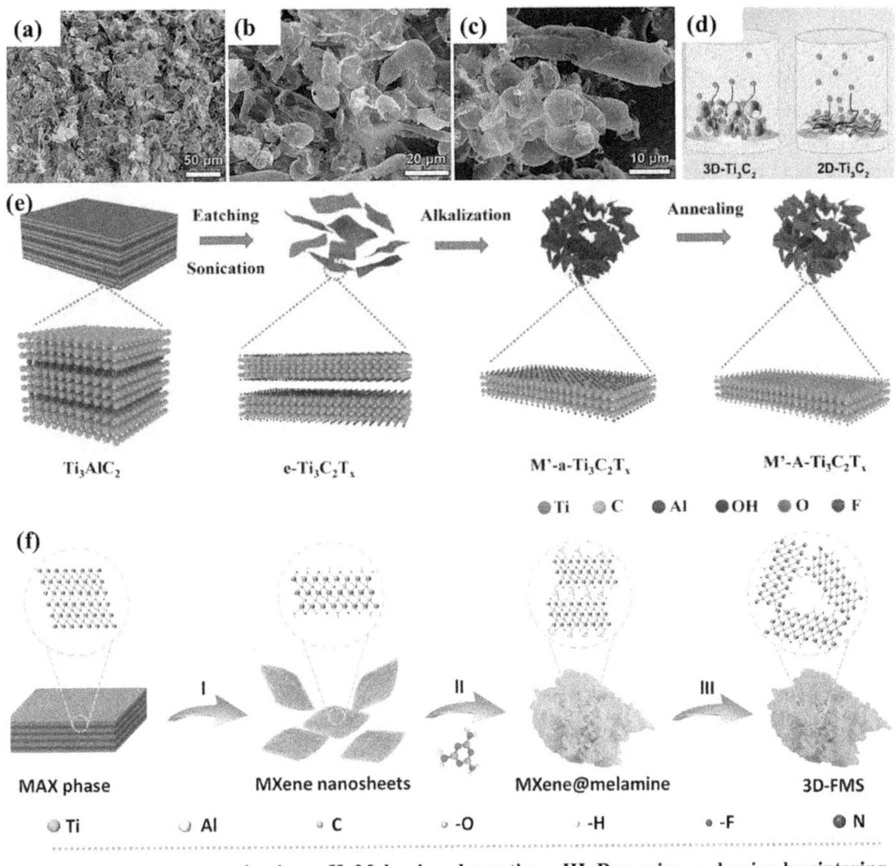

FIGURE 2.16 (a–b) SEM images, (c) an exemplification of ion transport in 3D and 2D-Ti$_3$C$_2$ electrodes [132], (d) an exemplification of fabrication [133], and (e) the synthesis of 3D-FMS [134].

Figure 2.16d and e, Zhao et al. synthesized 3D K-pre-intercalated Ti$_3$C$_2$T$_X$ (MXene) (K-Ti$_3$C$_2$T$_X$) through KOH treatment, freeze-drying, and electrostatic flocculation. After K-intercalation, the MXene layer gap increased to 1.32 nm, which is beneficial to K+ dynamics. Through the combination of AC cathode and 3D K-Ti$_3$C$_2$T$_X$ anode, the proposed PIC exhibits an energy density of 163 Wh kg^{-1} and a high power density of 8.7 kW kg^{-1}. In addition to the above studies, the research group of Zhang et al. prepared 3D foam-like 3D-FMS by electrostatically neutralizing Ti$_3$C$_2$T$_X$ with positively charged melamine and subsequent calcination [131]. Due to the surface area of 89.5 m^2 g^{-1}, there is an acceleration in the rate of K conductivity.

6.6 DESIGN OF 3D POROUS MXENE ELECTRODE

The performance of MXene materials in energy storage applications can be improved by assembling 2D MXene wafers into 3D electrode structures. The design of MXene

3D foams is accomplished using a variety of techniques. Therefore, it is beneficial to increase the interlayer distance and reduce the fragment size, thereby improving the accessibility and transport of ions to the active sites, thus increasing the efficiency of MXene. However, the high-flux potential of MXenes can be further enhanced by forming 3D/porous electrode structures with large active surfaces accessible to ions as well as interconnected pores for ion transport channels. The achievements of MXenes are expected to increase as appropriate porous structures can provide pathways for ion transport and electrolyte wetting, thereby shortening diffusion distances and reducing electrical resistance. We can generate porous MXene electrodes through various methods, such as electrodeposition, hydrazine-induced foaming, template method, and chemical etching [135–138]. Recently, we have used small ice particles as self-sacrificial templates to form flexible MXene/CNT films with nanopores and macropores. The controlled porous structure allows the realization of capacities of 251 and 375 F g-1 at 1000 and 5 mV s-1, respectively. Carbon assembly, corrugation, and foaming have also been considered as effective methods to create porous MXene frameworks [139–142].

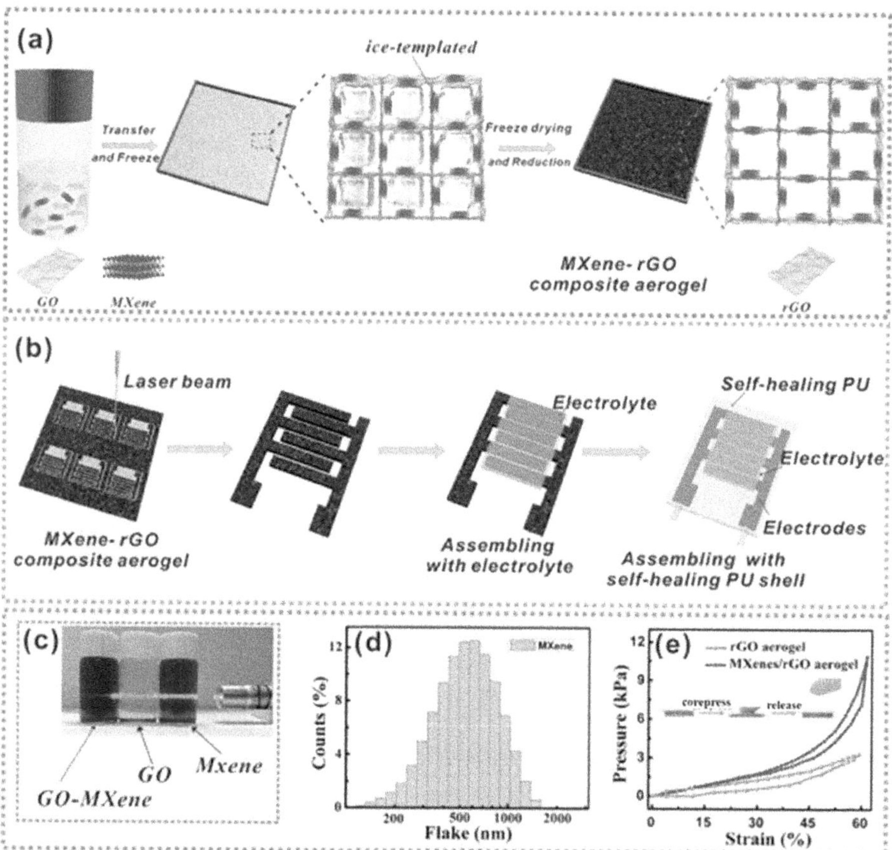

FIGURE 2.17 An exemplification of fabrication of (a) aerogel, (b) MSCs, (c) photo of NCs in the dispersion of water, (d) size of MXene flakes, and (e) pressure-strain curve [125].

Graphene can also be combined with 3D MXene aerogels [131]. As shown in Figure 2.17, during the freeze-drying process of the MXene and GO mixed solution, the GO/MXene nanosheets are forced to gradually align along the ice crystal boundaries and are eventually cross-linked through interactions to form a 3D lattice porous structure.

7 MXENE-BASED COMPOSITE MATERIALS FOR CAPACITOR ELECTRODE

Due to its unique 2D wafer structure and excellent electrical conductivity, MXene is considered a potential building block of composite materials for energy storage applications. MXene is combined with multiple active ingredients, including conductive polymers and metal oxides, to create synergistic effects. However, these elements reduce MXene rearrangements and increase interlayer spacing, which is expected to accelerate ion transfer and improve ion accessibility. Therefore, to demonstrate the potential use of MXene in supercapacitor technology, MXene-based composites often exhibit enhanced electrochemical capabilities. The following subsections provide a review of the studied MXene composites and their electrochemical properties in the field of capacitors.

7.1 MXene/Conducting Polymers

Those polymers that conduct electricity are called organic polymers, also known as conductive polymers, or more accurately, intrinsically conductive polymers (ICPs). The substance in question may be metallic in conductivity or semiconducting. Charge transfer complexes were the first conducting polymers because they were the first highly conductive organic compounds, such as polyaniline, which was first discovered in the mid-nineteenth century. Superconductivity was first demonstrated in 1980; however, researchers first demonstrated in the early 1970s that tetrathiafulvalene salts exhibit almost metallic conductivity. Compared to other pseudocapacitive materials, conductive polymers offer significant advantages for wearable supercapacitors due to their inherent conductivity and flexibility. In acidic solutions, composites of conductive polymers and MXene can also achieve high capacitance. Various conductive polymers, including PFD, PDA, PPy, PEDOT, and PANI [143–148], have been combined with MXene to produce hybrid materials with excellent electrochemical properties.

The conductive polymer usually combined with MXenes is highly conductive PANI [143, 149, 150], which is polymerized to form a CP@rGO electrode with a $Ti_3C_2T_X$ film electrode, as shown in Figure 2.18a [143]. Figure 2.18b shows the strategy to extend the voltage window of 2D $Ti_3C_2T_X$ sheets in 3 M H_2SO_4 by pairing them with CP-containing supercapacitors. That the MXene film consists of well-aligned stacked sheets was confirmed by the morphology images of the MXene electrode film. Figure 2.18a–c shows the CV curves (5 mV s^{-1}) of individual electrodes (CP@rGO and Ti3C2Tx) in a three-electrode setup, where both electrodes exhibit redox activity at completely different potentials, indicating the pseudocapacitive complete fabrication of asymmetric devices in aqueous electrolytes. Due to the formation of $Ti_3C_2T_x$//PEDOT@rGO, it confirms facile proton transport within the hybrid device, exhibiting stable pseudocapacitive properties even at a scan rate of 100 mV s^{-1}.

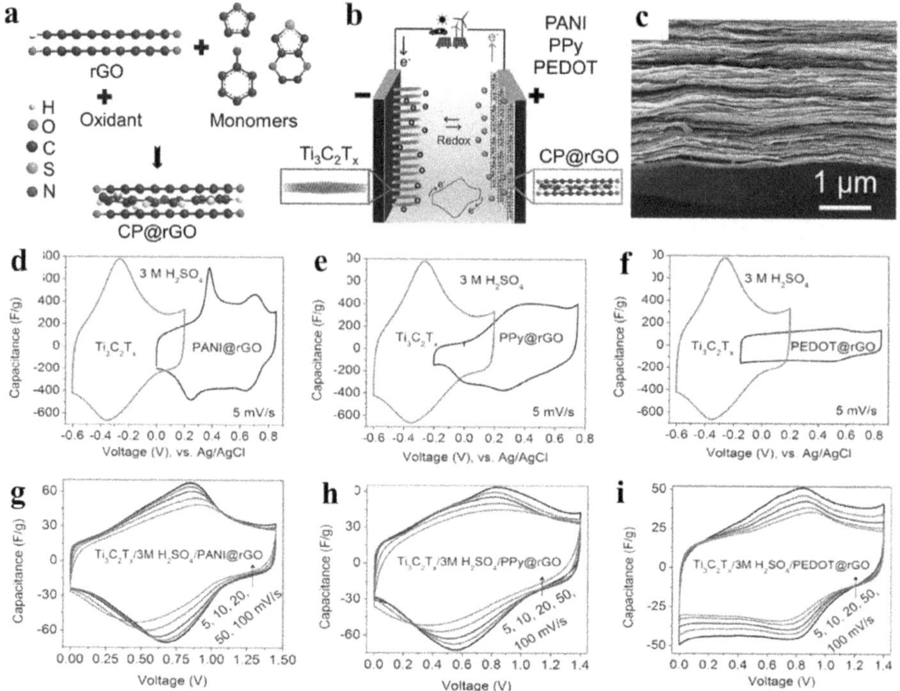

FIGURE 2.18 (a) An exemplification of MXene-synthesis and the (b) CV-curve of the MXene-based-electrodes [143].

7.2 MXENE/METAL AND MXENE/METAL-OXIDES

Pseudocapacitive materials including a variety of metal-based compounds are used in supercapacitors. Although these materials have high theoretical potential, their electronic conductivities are generally below average. Since MXene has metal-like conductivity, composite materials composed of metals and MXene have been studied to provide the possibility to solve this problem. For ZnO, WO_3, NiO, TiO_2, SnO_2, RuO_2, MoO_3, MnO_2, and MO are often used in combination with MXene to obtain excellent capacitance [151, 152, 153, 154]. MXene/MO composites can be generated by on-site assembly or *in-situ* processing [157, 158]. Chemical growth of metal oxides directly on liquid phase MXene nanosheets is a simple technique to induce meaningful interactions between *in-situ* synthesized components [155]. Figure 2.19a shows the synthesis and growth of CoS_2 nanoparticles on the MXene surface using a simple one-step solvothermal method.

It can be seen from the SEM images that the composite material has two morphologies: one is a layered structure and the other is a spherical structure. Their electrochemical performance can be promoted as synergistic effects that can be produced between them. The results of electrochemical testing of MXene/CoS_2 (CCH) composites in 2M KOH aqueous electrolyte are shown in Figure 2.19b–g. As shown in Figure 2.19b–g, the MXene/CoS_2 (CCH)//rGO ASCs can reach a maximum operating voltage window of 1.6 V. In addition, the device has a specific capacitance

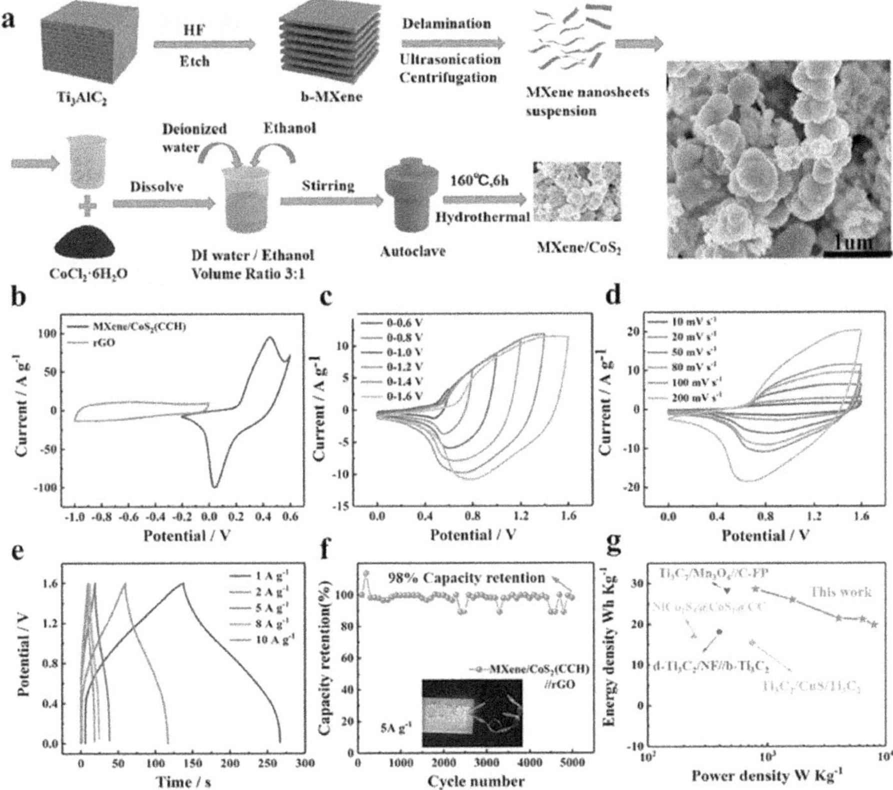

FIGURE 2.19   (a) Schematic diagram for the synthesis process of MXene/CoS$_2$ (CCH) composite with the (b) SEM image, and the (b–g) electrochemical performance of MXene/CoS$_2$ (CCH)//rGO ASCs in 2 m KOH aqueous electrolyte. [159].

retention rate of approximately 98% after 5,000 charge-discharge cycles, showing excellent long-term cycle performance. Notably, the capacitance decreases slowly during the first 1,000 cycles and increases to 98% in the 1,000–5,000 cycle range. Possible explanations are the significant enhancement of wettability between electrolyte ions and active materials and the activation of electrode materials.

TiO$_2$/Ti$_3$C$_2$ composites can be produced *in situ* by directly oxidizing Ti$_3$C or by adding TiO$_2$ precursors, as TiO$_2$ can be prepared by converting Ti$_3$C$_2$ MXene, which is a metal oxide. Incorporation of TiO$_2$ NPs into Ti$_3$C$_2$ nanolayers using tetrabutyl titanate as the precursor and *in situ* hydrolysis yielded open MXene ion transport paths, large MXene interlayer distance and high specific surface area [154]. Then a less difficult one-step room temperature oxidation process was proposed to synthesize Ti$_3$C$_2$/TiO$_2$ nanocomposites (Figure 2.20).

Ti$_3$C$_2$ MXene was coupled with monoclinic WO$_3$ nanorods and hexagonal WO$_3$ nanoparticles using HCl and HNO$_3$ respectively under various hydrothermal conditions [156]. In the potential range of −0.5 to 0 V vs. Ag/AgCl, the capacitance of the hexagonal WO$_3$/Ti$_3$C$_2$ composite (566 F g^{-1}) is almost twice that of pure hexagonal

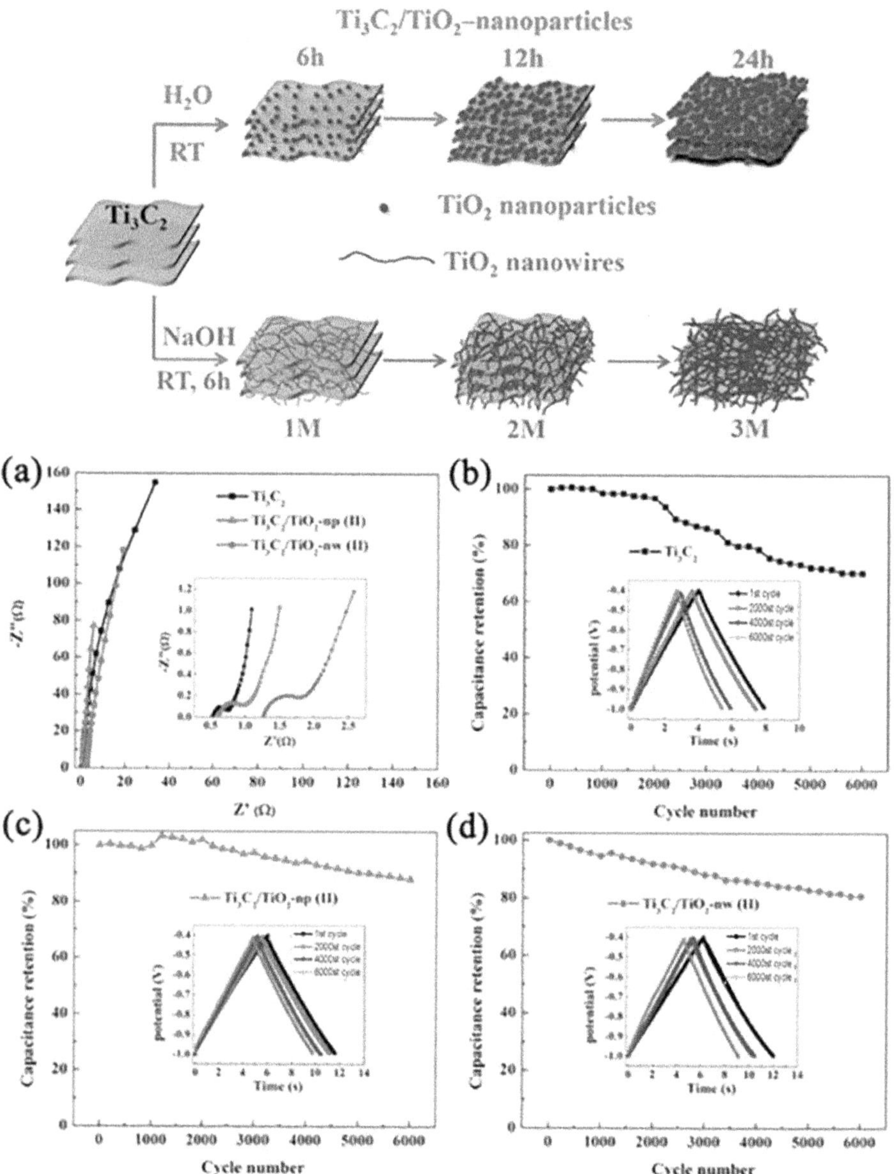

FIGURE 2.20 An exemplification of Ti_3C_2/TiO_2 composites fabrication, (a) EIS, (b–d) cyclability retention of various Ti_3C_2/TiO_2-electrodes [155].

WO_3 in 0.5M H_2SO_4 electrolyte (Figure 2.21). In order to combine MXenes and metal oxides, microwave processing is also used [153]. MXene is the base electrolyte composite that primarily promotes sporadic flexibility and fast electron transfer with minimal capacitance contribution. Because MXene can provide high pseudocapacitance based on redox reactions starting from H^+ intercalation, it can also be used as an active material in acidic electrolytes.

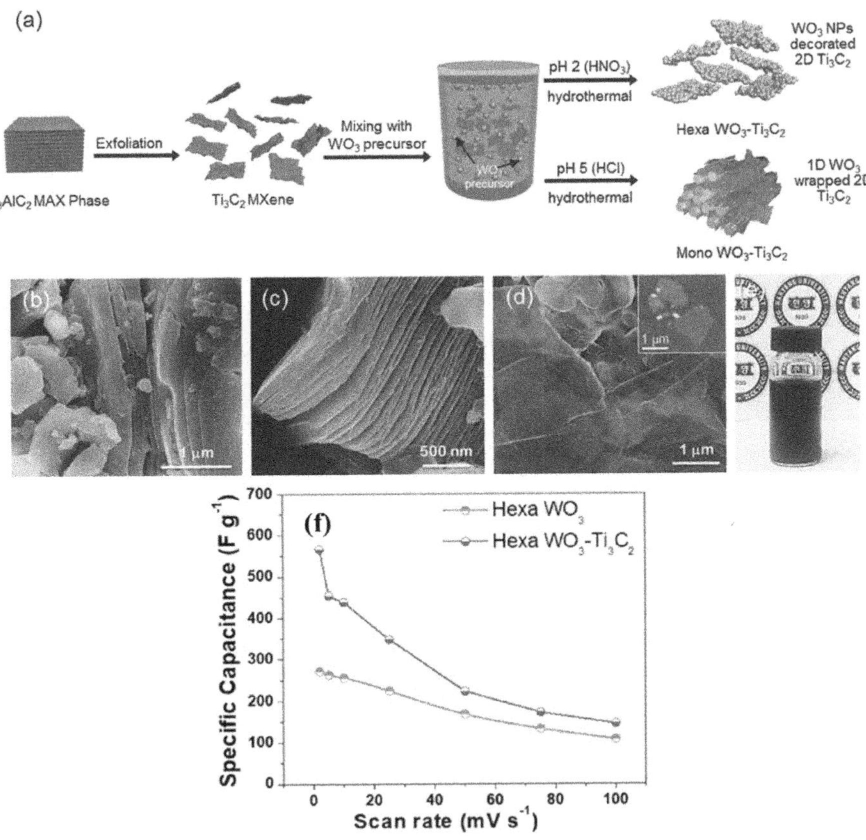

FIGURE 2.21   (a) An exemplification of fabrication, (b–d) FESEM, AFM, and vial images with the (f) spec. capacitance rate of Hexa-WO₃ electrodes [156].

8   CAPACITIVE MECHANISM OF MXENES IN ELECTROLYTES

The realization of high capacities of MXene-based electrodes for energy storage applications and the development of improved or new methods are based on the fundamental understanding of the MXene energy storage mechanism. Furthermore, in acidic electrolytes in high scan rate environments, the electrochemical characteristics of MXenes show a broad redox peak at the top of a rectangular CV shape, indicating redox pseudocapacitance. Intercalation pseudocapacitance is characterized by the presence of H⁺ insertion reaction sites between MXene layers in acidic electrolytes without phase changes and is a storage mechanism [160]. It has a long cycle life (over 10,000 cycles) and is in contrast to the slow ion intercalation observed in battery-type layered electrode materials such as graphite. This section discusses and reviews recent advances in the research area investigating the charge storage processes and capacitive properties of MXenes in various electrolytes. For capacitors, various types of electrolytes have been used, including nonaqueous electrolytes, aqueous electrolytes (neutral and alkaline), and acidic electrolytes. Because acidic electrolytes adopt a different charge storage mechanism than neutral and alkaline electrolytes, MXene electrodes have

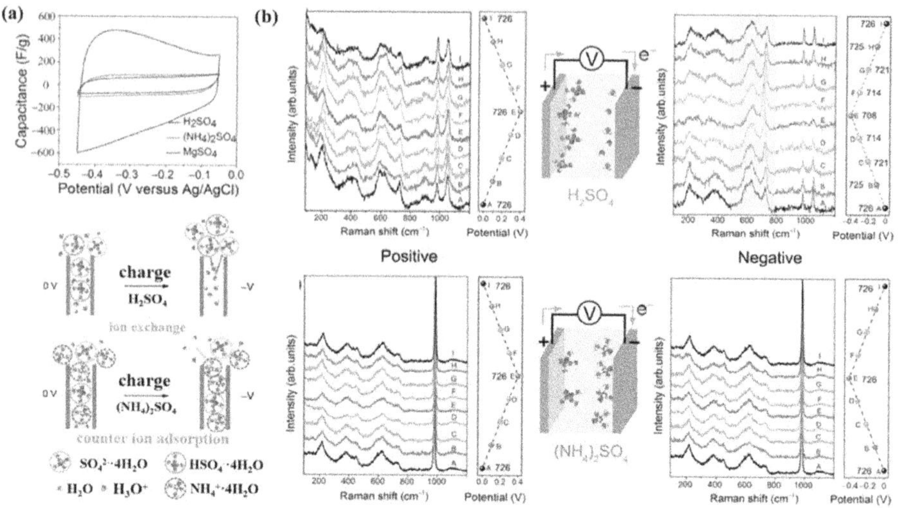

FIGURE 2.22 (a) CV-curves. (b) *In situ* Raman spectra of $Ti_3C_2T_x$ MXene recorded on: positive electrode and negative electrode [161].

excellent productivity and have the largest capacitance among H_2SO_4 electrolytes in acidic electrolytes. Figure 2.22a shows a comparison of the CV curves of $Ti_3C_2T_x$ thin film electrodes in neutral and acidic electrolytes [161]. By looking at the large redox bump in the CV curve from −0.4 to −0.3 V, it can be seen that pseudocapacitive behavior exists in the H_2SO_4 electrolyte. H_2SO_4 electrolytes produce larger capacitances than neutral electrolytes ($MgSO_4$ and $(NH_4)_2SO_4$) [161]. *In situ* X-ray absorption spectroscopy (XAS) measurements reveal the changes in Ti oxidation state during the electrochemical discharge of $Ti_3C_2T_X$ in H_2SO_4. The pseudocapacitive behavior based on redox reactions was confirmed [112]. Figure 2.22b demonstrates the use of *in situ* electrochemical Raman spectroscopy to represent the discharge process of the Ti3C2TX thin film electrode in H_2SO_4. According to the observed reversible voltage-dependent changes after discharge [161], it is suggested that the electrochemical reaction occurs:

$$(M = Ox) + 1/2 \times H+ + 1/2 \times e- \rightarrow M - O(1/2) \times (OH)(1/2)x \quad (5)$$

Compared to neutral and alkaline electrolytes, acidic electrolytes exhibit pseudocapacitive behavior of redox reactions, resulting in faster kinetics and higher capacitance due to fast proton transport. For example, at a potential scan rate of 1,000 mV s^{-1}, they can continue to deliver 500 F cm^{-3} [75]. There is considerable evidence that electrode materials that exhibit fast Faradaic reactions (e.g., metal oxides) and acidic electrolytes are compatible [162]. The higher capacitance of MXene (more than 1500 F cm^{-3}) can be fully exploited in acidic electrolytes. However, although MXenes can have high rate and high capacitance in acidic aqueous electrolytes based on the intercalation pseudocapacitive process, the practical use of such devices is limited by their limited operating voltage range (approximately 1 V).

Water molecules must be present between the MXene layers to achieve high capacity. They support the formation of the electric double layer (EDL) in the MXene interstitial

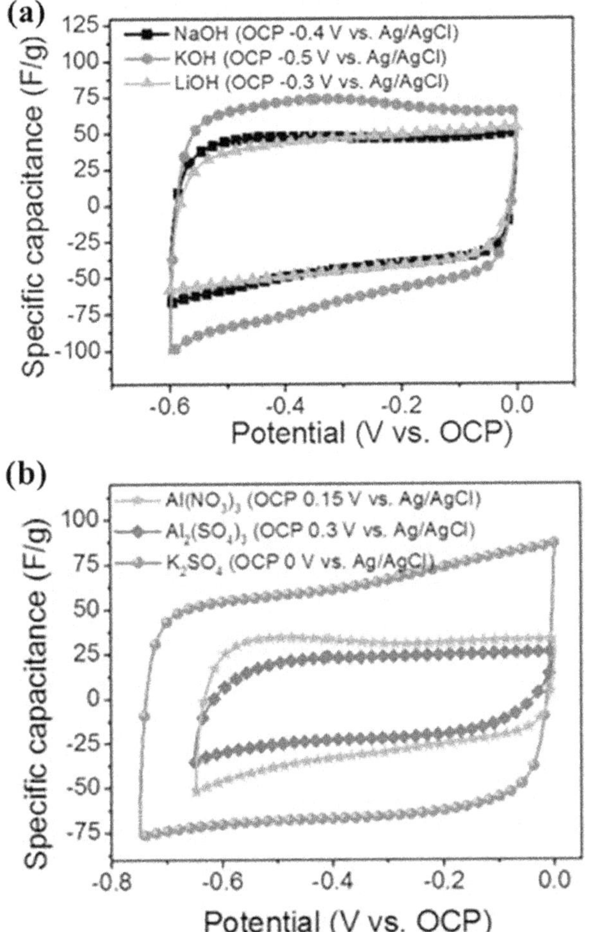

FIGURE 2.23 (a and b) CV profiles of $Ti_3C_2T_x$-based SCs in basic and neutral electrolytes [38].

space as well as ion adsorption and intercalation [163–165]. As shown in Figure 2.23, hydrated cations (e.g., Li+) are formed from water molecules surrounding the cations inserted into the MXene layer without dehydration. The hydration shell isolates the atomic orbitals and cations of MXene, thereby separating negative and positive charges and preventing orbital hybridization. Internal Electric Double Layer Capacitors (EDLC) occurs due to the potential difference in the interlayer spacing. Therefore, despite embedding hydrated cations in the interstitial space, MXene has a near-constant capacity in the potential window of the aqueous electrolyte and exhibits most of the typical properties of EDLCs.

Solid-state electrolyte (SSE)-based SCs have recently attracted great interest due to the rapidly growing power demands of highly flexible electronics, microelectronics, printable electronics, portable electronics, and wearable electronics. The electrolyte is an important component of supercapacitors and plays an important role in balancing and transferring charge between the two electrodes. In fact, the interaction between electrodes and electrolytes during electrochemical processes widely affects the internal structure of active materials and the state of the electrolyte-electrode interface, which deserves attention for advanced applications of flexible supercapacitors. A valuable approach is to design MXene materials and a suitable MXene/solid electrolyte interface. SCs composed of aqueous electrolyte ionic liquids can operate at high voltages and exhibit capacitance and high conductivity although there are leakage issues. Using solid-state electrolytes (SSEs) may help avoid leakage problems [166]. SSE is composed of an organic polymer electrolyte and an inorganic solid electrolyte. Organic polymer solid electrolytes are composed of salts and organic polymer matrices, while inorganic solid electrolytes include polycrystalline, amorphous compounds, and single crystals. Most SSEs of SCs are polymer-based, including polyelectrolytes, gel polymers (quasi-solids), and solid polymers. Among these three SSEs, gel polymer electrolytes have received special attention in SCs due to their high ionic conductivity derived from the liquid phase. For the preparation of gel polymer electrolytes, various types of polymers such as PVDF-HFP, PMMA, and PVA, as well as some organic solvents (PC, EC, DMF) and water, are used as plasticizers [167–170]. Figure 2.24 represents a diagram of a hydrogel polymer composed of an aqueous electrolyte (KOH, H_2SO_4, etc.) and a polymer host (PEG, PVA, or PEO) or a conductive salt dissolved in a solvent [171].

Table 2.3 summarizes the comparative study results of MXene composite SC, pure MXene-SC, and other SCs. As shown in Table 2.3, MXene-based SCs possess high weight capacitance and long-term reusability properties, which further strongly confirms the theoretical reasons why MXene is a candidate for SCs.

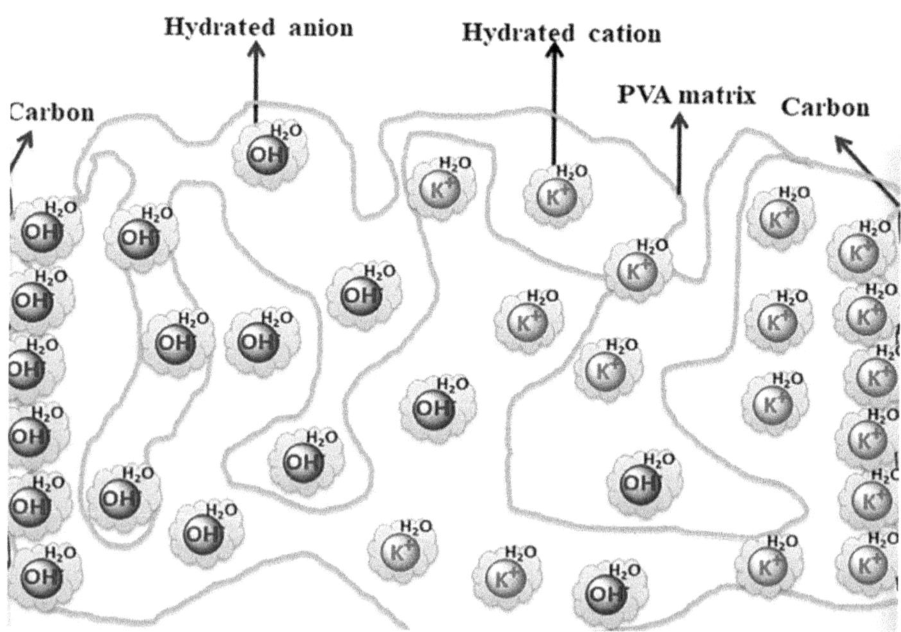

FIGURE 2.24 Graphical representation of the hydro-gel polymer electrolyte (PVA/KOH/H_2O) [171].

TABLE 2.3

Comparative Study of Pure MXene-SCs, MXene Composite SCs, and other SCs

Electrode	Electrolyte	Gravimetric Capacitance (F g^{-1})	Cycling Stability	Flexibility	Ref.
Ti$_3$C$_2$T$_x$	1M H$_2$SO$_4$	314 (2 mV s^{-1})	89.1% retention (5 mA g^{-1}), 10,000 cycles	–	[172]
Ti$_3$C$_2$T$_x$	3M H$_2$SO$_4$	380 (10 mV s^{-1})	Over 90% retention (10 mA g^{-1}), 10,000 cycles	Flexible film	[173]
Ti$_3$C$_2$T$_x$	1M KOH	130 (2 mV s^{-1})	100% retention (1 mA g^{-1}), 10,000 cycles	Flexible film	[174]
Ti$_3$C$_2$T$_x$/rGO	3M H$_2$SO$_4$	335 (2 mV s^{-1})	100% retention (1 mA g^{-1}), 20,000 cycles	Flexible film	[175]
Ti$_3$C$_2$T$_x$/CNT	MgSO$_4$	125 (2 mV s^{-1})	100% retention (5 A g^{-1}), 10,000 cycles	Flexible film	[176]
Ti$_3$C$_2$T$_x$/PPy	1M H$_2$SO$_4$	416 (5 mV s^{-1})	92% retention (100 mV s^{-1}), 25,000 cycles	Flexible film	[177]
Ti$_3$C$_2$T$_x$/NiCo$_2$S$_4$	3 M KOH	1147.47 (1 A g^{-1})	91.1% retention (10 A g^{-1}), 3,000 cycles	–	[178]
Ti$_3$C$_2$T$_x$/Graphene	H$_2$SO$_4$	542 (5 mV s^{-1})	~52% retention (1000 mV s^{-1}) 5,000 cycles	–	[179]
Ti$_3$C$_2$T$_x$/hydrogel	H$_2$SO$_4$	370–165 (5–1,000 A g^{-1})	~98% retention (10 A g^{-1}) 10,000 cycles	–	[180]
CuO@ AuPd@ MnO$_2$ core-shell Whiskers	1M KOH	1376 (5 mV s^{-1})	99% retention (5 mVs^{-1}) 5,000 cycles	–	[181]
Ni-MOF	3M KOH	988 (1.4 A g^{-1})	96.5% retention(1.4 A g^{-1}) 5,000 cycles	Flexible film	[182]
PANI-ZIF67	3M KCl	2146 (10 mV s^{-1})	–	Flexible	[183]
α-Fe$_2$O$_3$@ C	1M Na$_2$SO$_4$	1232.4 (2 mA cm^{-2})	97.6% retention(2 mA cm^{-2}) 4,000 cycles	Flexible (97.16% retention after 4000 bending cycles)	[184]
MOFC/CNT	6M KOH	381.2 (5 mV s^{-1})	95% retention (5 mV s^{-1}) 10,000 cycles	Flexible	[185]
TiO$_2$ nano spindles	6M KOH	897 (0.21 A g^{-1})	75% retention (0.21 A g^{-1}) 5,000 cycles	–	[186]

(Continued)

TABLE 2.3
(Continued)

Electrode	Electrolyte	Gravimetric Capacitance (F g^{-1})	Cycling Stability	Flexibility	Ref.
NiCo$_2$O$_4$ NWARs/PPy	3M KOH	2244 (1 A g^{-1})	89.2% retention (1 A g^{-1}) 5,000 cycles	–	[187]
CoO-NiO-ZnO	3M KOH	2115.5 (1 A g^{-1})	87.9% retention (1 A g^{-1}) 5,000 cycles	–	[188]
ZIF/PPy	1M Na$_2$SO$_4$	554.4 (0.5 A g^{-1})	90.7% retention (0.5 A g^{-1}) 10,000 cycles	Little capacitance fading up to 180o bending	[189]
UiO-66/Poly pyrrole	3M KCl	90.5 (5 mV s^{-1})	96% retention (5 mV s^{-1}) 1,000 cycles	96% retention after 1,000 bending cycles	[190]
PANI/UiO-66	PVA/H$_2$SO$_4$	1015 (1 A g^{-1})	84% retention (1 A g^{-1}) 3,500 cycles	90% retention after 800 bending cycle	[191]

9   CONCLUSIONS

Reviewing the accurate gold properties of MXenes materials is critical in terms of improving performance and practical applications. Furthermore, fabricating finely structured films cheaply and reliably is challenging, and further research into the interactions and communication between guest materials and MXenes is considered an area requiring further review. In addition to material composition, an influential area for future research is the design of new advanced MXene-based SCs by improving electrochemical performance while maintaining the material's mechanical properties such as hardness, strength, and flexibility.

Key properties of MXenes, such as their pseudo-interaction capacitance, variable surface endpoints, density, metal-like conductivity, and 2D plate-like structure, make these good candidates. In this chapter, recent advances in MXene-based electrodes for energy storage applications have been described. This chapter reviews the research progress of various electrolytes in recent years in terms of basic knowledge of charge storage mechanisms, electrochemical characterization, chemical changes, structural design, and synthesis technology. And methods for the construction and testing of electrochemical performance of MXene-based hybrid supercapacitors and other devices. Since its discovery in 2011, MXenes have been prepared mainly by three synthesis methods: etching in fluorine-free electrolytes (e.g., molten salts of Lewis acids), etching in moderate fluorine-containing solutions (e.g., NH$_4$HF$_2$ or LiF/HCl), and HF etching and subsequent cation layering and embedding. Fluorine-free preparation is considered an ideal approach in terms of environmental friendliness, safety, and electrochemical performance. The subject of most studies reported in the literature is $Ti_3C_2T_X$, with little attention paid to alternative components. One of the key applications is energy storage, which could take advantage of the ability and diversity of generable MXene materials to change their surface chemistry. We need a fundamental research program to develop fundamental knowledge of the electrolyte/MXene interface and develop new ways to control surface chemistry. Although the 2D layered structure of MXene nanosheets improves ion accessibility. And the resulting electrochemical reaction rate is also improved, but it's also important to prevent restacking and agglomeration. To solve this problem, we have developed various electrode and structural architecture methods, such as controlling interlayer spacing and flake size and generating vertical alignment. Especially in terms of speed capabilities, reasonable structural design and appearance greatly enhance the display capabilities of MXene. Surface chemical modification is one of the most important technologies used to improve the electrochemical performance of MXenes. MXenes can be chemically transformed through surface modification and doping. When fabricating materials using fluorinated corrosive electrolytes, we can develop a post-processing procedure that removes the F-terminal and additional fluorine-free manufacturing methods, thereby reducing the number of active sites and disrupting payload transfer. Therefore, the choice of electrolyte is also a major factor affecting the electrochemical results of MXene electrodes, because improving the electrochemical results of MXene depends on the mastery of the nonaqueous electrolyte and the charge storage mechanism in the MXene/electrolyte interface. Furthermore, the MXene matrix can incorporate reinforcement materials without affecting the electrochemical results as well as flexibility of printable MXene-based supercapacitors to enhance their mechanical properties. The endless possibilities to generate new MXenes and tailor their functionality demonstrate that, despite many achievements and improvements, progress on SC MXenes is still in its early stages. The next steps are to create designs that support high-performance supercapacitors, manage surface chemistry, and understand capacitive energy storage mechanisms.

MXenes can provide many new potential applications due to their unique typical structural features and physicochemical properties. To date, in the field of energy storage, researchers have continued to report various methods to overcome MXene-related issues, and the design flow of MXene layers has continued to improve and realize further potential applications related to wearable SCs, sensors, and electromagnetic shielding. The continued rapid development of technologies and fundamental concepts related to MXenes holds great promise for the discovery of many more exciting objects.

Acknowledgments: This work was supported by the Technology Innovation Program (Development of Performance Improvement Technology for High-Power Capacitors for Demand Companies)) (00156073, Development of lithium-ion capacitors for the control of load variation of hydrogen vehicles, 2024) funded By the Ministry of Trade, Industry & Energy (MOTIE, Korea)

Author Contributions: Writing—original draft preparation, *Zuhong Ji*; Data collection, Kefaya Ullha; writing—review and editing and supervision, *Won-Chun Oh*. All authors have read and agreed to the published version of the manuscript.

Institutional Review Board Statement: Not applicable.

Informed Consent Statement: Not applicable.

Data Availability Statement: Data can be made available upon written request to the corresponding author and with a proper justification.

Conflicts of Interest: The authors declare no conflicts of interest.

REFERENCES

1. Z. Wen, M. H. Yeh, H. Guo, J. Wang, Y. Zi, W. Xu, J. Deng, L. Zhu, X. Wang and C. Hu, Self-powered textile for wearable electronics by hybridizing fiber-shaped nanogenerators, solar cells, and supercapacitors, *Sci. Adv.*, 2016, 2, e1600097.
2. J. M. Tarascon and M. Armand, Issues and challenges facing rechargeable lithium batteries, *Nature*, 2001, 414, 359–367.
3. F. Yi, J. Wang, X. F. Wang, S. M. Niu, S. M. Li, Q. L. Liao, Y. L. Xu, Z. You, Y. Zhang and Z. L. Wang, Stretchable and waterproof self-charging power system for harvesting energy from diverse deformation and powering wearable electronics, *ACS Nano,* 2016, 10, 6519–6525.
4. X. F. Wang, Y. J. Yin, F. Yi, K. R. Dai, S. M. Niu, Y. Z. Han, Y. Zhang and Z. You, Bioinspired stretchable triboelectric nanogenerator as energy-harvesting skin for self-powered electronics, *Nano Energy*, 2017, 39, 429–436.
5. B. Dunn, H. Kamath and J. M. Tarascon, Electrical energy storage for the grid: A battery of choices, *Science*, 2011, 334, 928–935.
6. J. M. Cao, J. Z. Li, L. Li, Y. Zhang, D. Cai, D. Chen and W. Han, Mn-Doped Ni/Co LDH nanosheets grown on the natural N-Dispersed PANI-derived porous carbon template for a flexible asymmetric supercapacitor, *ACS Sustain. Chem. Eng.*, 2019, 7, 10699–10707.
7. F. Yi, H. Y. Ren, K. R. Dai, X. F. Wang, Y. Z. Han, K. X. Wang, K. Li, B. L. Guan, J. Wang, M. Tang, J. Y. Shan, H. Yang, M. S. Zheng, Z. You, D. Wei and Z. F. Liu, Solar thermal-driven capacitance enhancement of supercapacitors, *Energy Environ. Sci.*, 2018, 11, 2016–2024.
8. J. M. Cao, J. Z. Li, L. Zhou, Y. L. Xi, X. Cao, Y. M. Zhang and W. Han, Tunable agglomeration of Co3O4 nanowires as the growing core for in-situ formation of Co2NiO4 assembled with polyaniline-derived carbonaceous fibers as the high-performance asymmetric supercapacitors, *J. Alloys Compd.*, 2021, 853, 157210.
9. N. S. Choi, Z. Chen, S. A. Freunberger, X. Ji, Y. K. Sun, K. Amine, G. Yushin, L. F. Nazar, J. Cho and P. G. Bruce, Challenges facing lithium batteries and electrical double-layer capacitors, *Angew. Chem., Int. Ed.*, 2012, 51, 9994–10024.
10. M. Winter and R. J. Brodd, What are batteries, fuel cells, and supercapacitors? *Chem. Rev.*, 2005, 105, 4245– 4269.
11. Y. Zhu, S. Murali, M. D. Stoller, K. J. Ganesh, W. Cai, P. J. Ferreira, A. Pirkle, R. M. Wallace, K. A. Cychosz and M. Thommes, Carbon-based supercapacitors produced by activation of graphene, *Science*, 2011, 332, 1537–1541.
12. G. Pandolfe and A. F. Hollenkamp, Carbon properties and their role in supercapacitors, *J. Power Sources*, 2006, 157, 11–27.
13. Z. Y. Yuan, L. L. Wang, J. M. Cao, L. J. Zhao and W. Han, Ultraviolet-assisted construction of Nitrogen-Rich Ag@Ti3C2Tx MXene for highly efficient hydrogen evolution electrocatalysis and supercapacitor, *Adv. Mater. Interfaces*, 2020, 7, 2001449.
14. P. Simon and Y. Gogotsi, Materials for electrochemical capacitors, *Nat. Mater.*, 2008, 7, 845–854.
15. Y. Wang, Y. Song and Y. Xia, Electrochemical capacitors: Mechanism, materials, systems, characterization and applications, *Chem. Soc. Rev.*, 2016, 45, 5925–5950.

16. F. Yi, H. Y. Ren, J. Y. Shan, X. Sun, D. Wei and Z. F. Liu, Wearable energy sources based on 2D materials, *Chem. Soc. Rev.*, 2018, 47, 3152–3188.

17. L. L. Zhang and X. S. Zhao, Carbon-based materials as supercapacitor electrodes, *Chem. Soc. Rev.*, 2009, 38, 2520– 2531.

18. R. Kotz and M. Carlen, Principles and applications of electrochemical capacitors, *Electrochim. Acta*, 2000, 45, 2483– 2498.

19. J. Cao, Z. Sun, J. Li, Y. Zhu, Z. Yuan, Y. Zhang, D. Li, L. Wang and W. Han, Microbe-assisted assembly of Ti3C2Tx MXene on fungi-derived nanoribbon heterostructures for ultrastable Sodium and Potassium Ion storage, *ACS Nano*, 2021, 15, 3423–3433.

20. J. Xiao, H. Li, H. Zhang, S. He, Q. Zhang, K. Liu, S. Jiang, G. Duan and K. Zhang, Nanocellulose and its derived composite electrodes toward supercapacitors: Fabrication, properties, and challenges, *J. Bioresour. Bioprod.*, 2022, 7 (4), 245–269.

21. L. Wei, W. Deng, S. Li, Z. Wu, J. Cai and J. Luo, Sandwich-like chitosan porous carbon Spheres/MXene composite with high specific capacitance and rate performance for supercapacitors, *J. Bioresour. Bioprod.*, 2022, 7 (1), 63–72.

22. S. Zheng, J. Zhang, H. Deng, Y. Du and X. Shi, Chitin derived nitrogen-doped porous carbons with ultrahigh specific surface area and tailored hierarchical porosity for high performance supercapacitors, *J. Bioresour. Bioprod.*, 2021, 6 (2), 142–151.

23. J. Zheng, B. Yan, Q. Zhang, C. Zhang, W. Yang, J. Han, S. Jiang and S. He, Potassium citrate assisted synthesis of hierarchical porous carbon materials for high performance supercapacitors, *Diam. Relat. Mater.*, 2022, 128, 109247.

24. B. Yan, L. Feng, J. Zheng, S. Jiang, C. Zhang, Y. Ding, J. Han, W. Chen and S. He, High performance supercapacitors based on wood-derived thick carbon electrodes synthesized via green activation process, *Inorg. Chem. Front.*, 2022, 9, 6108–6123.

25. Y. Shao, M. F. El-Kady, J. Sun, Y. Li, Q. Zhang, M. Zhu, H. Wang, B. Dunn and R. B. Kaner, Design and mechanisms of asymmetric supercapacitors, *Chem. Rev.*, 2018, 118, 9233–9280.

26. D. P. Dubal, K. Jayaramulu, J. Sunil, S. Kment, P. Gomez-Romero, C. Narayana, R. Zboril and R. A. Fischer, Metal-Organic Framework (MOF) derived electrodes with robust and fast lithium storage for Li-Ion hybrid capacitors, *Adv. Funct. Mater.*, 2019, 29, 1900532.

27. H. Yu, Y. Wang, Y. Jing, J. Ma, C. F. Du and Q. Yan, Surface modified MXene-based nanocomposites for electrochemical energy conversion and storage, *Small*, 2019, 15, 1901503.

28. D. Xiong, X. Li, Z. Bai and S. Lu, Recent advances in layered $Ti_3C_2T_x$ MXene for electrochemical energy storage, *Small*, 2018, 14, 1703419.

29. M. Q. Zhao, X. Xie, C. E. Ren, T. Makaryan, B. Anasori, G. Wang and Y. Gogotsi, Hollow MXene Spheres and 3D Macroporous MXene frameworks for Na-Ion storage, *Adv. Mater.*, 2017, 29, 1702410.

30. J. Liu, H.-B. Zhang, X. Xie, R. Yang, Z. Liu, Y. Liu and Z. Z. Yu, Multifunctional, superelastic, and lightweight MXene/Polyimide aerogels, *Small*, 2018, 14, 1802479.

31. X. Li, X. Yin, H. Xu, M. Han, M. Li, S. Liang, L. Cheng and L. Zhang, Ultralight MXene-coated, interconnected SiCnws three-dimensional lamellar foams for efficient microwave absorption in the X-Band, *ACS Appl. Mater. Interfaces*, 2018, 10, 34524–34533.

32. C. Xing, S. Chen, X. Liang, Q. Liu, M. Qu, Q. Zou, J. Li, H. Tan, L. Liu, D. Fan and H. Zhang, Two-dimensional MXene (Ti3C2)-integrated cellulose hydrogels: Toward smart three-dimensional network nanoplatforms exhibiting light-induced swelling and bimodal photothermal/chemotherapy anticancer activity, *ACS Appl. Mater. Interfaces*, 2018, 10, 27631–27643.

33. C. E. Ren, M. Q. Zhao, T. Makaryan, J. Halim, M. Boota, S. Kota, B. Anasori, M. W. Barsoum and Y. Gogotsi, Porous Two-dimensional transition metal carbide (MXene) flakes for high-performance Li-Ion storage, *ChemElectroChem*, 2016, 3, 689–693.
34. Y. T. Liu, P. Zhang, N. Sun, B. Anasori, Q.-Z. Zhu, H. Liu, Y. Gogotsi and B. Xu, Self-assembly of transition metal oxide nanostructures on MXene nanosheets for fast and stable lithium storage, *Adv. Mater.*, 2018, 30, 1707334.
35. J. Zhou, X. Zha, X. Zhou, F. Chen, G. Gao, S. Wang, C. Shen, T. Chen, C. Zhi, P. Eklund, S. Du, J. Xue, W. Shi, Z. Chai and Q. Huang, Synthesis and electrochemical properties of two-dimensional Hafnium Carbide, *ACS Nano*, 2017, 11, 3841–3850.
36. J. Halim, S. Kota, M. R. Lukatskaya, M. Naguib, M.-Q. Zhao, E. J. Moon, J. Pitock, J. Nanda, S. J. May, Y. Gogotsi and M. W. Barsoum, Synthesis and characterization of 2D Molybdenum Carbide (MXene), *Adv. Funct. Mater.*, 2016, 26, 3118–3127.
37. M. Ghidiu, M. Naguib, C. Shi, O. Mashtalir, L. M. Pan, B. Zhang, J. Yang, Y. Gogotsi, S. J. L. Billinge and M. W. Barsoum, Synthesis and characterization of two-dimensional Nb4C3 (MXene), *Chem. Commun.*, 2014, 50, 9517–9520.
38. S. Y. Pang, Y.T. Wong, S. Yuan, Y. Liu, M.-K. Tsang, Z. Yang, H. Huang, W. T. Wong, J. Hao, Universal strategy for HF-Free Facile and rapid synthesis of two-dimensional MXenes as multifunctional energy materials, *J. Am. Chem. Soc.*, 2019, 141, 9610–9616.
39. C. Xu, L. Wang, Z. Liu, L. Chen, J. Guo, N. Kang, X.-L. Ma, H.-M. Cheng and W. Ren, Large-area high-quality 2D ultrathin Mo2C superconducting crystals, *Nat. Mater.*, 2015, 14, 1135–1141.
40. Y. Gogotsi, Chemical vapour deposition: Transition metal carbides go 2D, *Nat. Mater.* 2015, 14, 1079–1080.
41. J. Hu, C. Liang, J. Li, C. Lin, Y. Liang, H. Wang, X. LI, Q. Wang and D. Dong, Lewis acidic molten salts etching route driven construction of double-layered MXene-Fe/carbon nanotube/silicone rubber composites for high-performance microwave absorption, *Carbon*, 2023, 204, 136–146.
42. P. Urbankowski, B. Anasori, T. Makaryan, D. Er, S. Kota, P. L. Walsh, M. Zhao, V. B. Shenoy, M. W. Barsoum and Y. Gogotsi, Synthesis of two-dimensional titanium nitride Ti4N3 (MXene), *Nanoscale,* 2016, 8, 11385–11391.
43. M. Li, J. Lu, K. Luo, Y. Li, K. Chang, K. Chen, J. Zhou, J. Rosen, L. Hultman, P. Eklund, P. O. Å. Persson, S. Du, Z. Chai, Z. Huang and Q. Huang, Element replacement approach by reaction with Lewis acidic molten salts to synthesize nanolaminated MAX phases and MXenes, *J. Am. Chem. Soc.*, 2019, 141, 4730–4737.
44. M. Naguib, M. Kurtoglu, V. Presser, J. Lu, J. Niu, M. Heon, L. Hultman, Y. Gogotsi, M. and W. Barsoum, Two-dimensional nanocrystals produced by exfoliation of Ti3AlC2, *Adv. Mater.*, 2011, 23, 4248–4253.
45. K. Arole, J. W. Blivin, S. Saha, D. E. Holta, X. Zhao, A. Sarmah, H. Cao, M. Radovic, J. L. Lutkenhaus and M. J. Green, Water-dispersible Ti3C2Tz MXene nanosheets by molten salt etching, *iScience*, 2021, 24 (12), 103403.
46. M. Naguib, R. R. Unocic, B. L. Armstrong and J. Nanda, Large-scale delamination of multi-layers transition metal carbides and carbonitrides "MXenes", *Dalton Trans.*, 2015, 44, 9353–9358.
47. O. Mashtalir, M. Naguib, V. N. Mochalin, Y. Dall'Agnese, M. Heon, M. W. Barsoum and Y. Gogotsi, Intercalation and delamination of layered carbides and carbonitrides, *Nat. Commun.*, 2013, 4, 1716.
48. M. Naguib, O. Mashtalir, J. Carle, V. Presser, J. Lu, L. Hultman, Y. Gogotsi and M. W. Barsoum, Two-dimensional transition metal carbides, *ACS Nano,* 2012, 6, 1322–1331.
49. M. R. Lukatskaya, O. Mashtalir, C. E. Ren, Y. Dall'Agnese, P. Rozier, P. L. Taberna, M. Naguib, P. Simon, M. W. Barsoum and Y. Gogotsi, Cation intercalation and high volumetric capacitance of two-dimensional Titanium Carbide, *Science*, 2013, 341, 1502–1505.

50. Y. Shao, M. F. El-Kady, J. Sun, Y. Li, Q. Zhang, M. Zhu, H. Wang, B. Dunn and B. Kaner, Design and mechanisms of asymmetric supercapacitors, *Chem. Rev.*, 2018, 118, 9233–9280.

51. B. Anasori, Y. Xie, M. Beidaghi, J. Lu, B. C. Hosler, L. Hultman, P. R. C. Kent, Y. Gogotsi and M. W. Barsoum, Two-dimensional, ordered, double transition metals carbides (MXenes), *ACS Nano*, 2015, 9, 9507–9516.

52. T. Li, L. Yao, Q. Liu, J. Gu, R. Luo, J. Li, X. Yan, W. Wang, P. Liu, B. Chen, W. Zhang, W. Abbas, R. Naz and D. Zhang, Fluorine-free synthesis of high-purity Ti3C2Tx (T=OH, O) via Alkali treatmentm, *Angew. Chem. Int. Ed. Engl.*, 2018, 57, 6115–6119.

53. M. Alhabeb, K. Maleski, T. S. Mathis, A. Sarycheva, C. B. Hatter, S. Uzun, A. Levitt and Y. Gogotsi, Selective etching of silicon from Ti3SiC2 (MAX) to obtain 2D Titanium Carbide (MXene), *Angew. Chem. Int. Ed. Engl.*, 2018, 57, 5444–5448.

54. G. S. Gund, J. H. Park, R. Harpalsinh, M. Kota, J. H. Shin, T. I. Kim, Y. Gogotsi and H. S. Park, MXene/polymer hybrid materials for flexible AC-filtering electrochemical capacitors, *Joule.*, 2019, 3, 164–176.

55. J. Li, A. Levitt, N. Kurra, K. Juan, N. Noriega, X. Xiao, X. Wang, H. Wang, H. N. Alshareef and Y. Gogotsi, MXene-conducting polymer electrochromic microsupercapacitors, *Energy Storage Mater.*, 2019, 20, 455–461.

56. T. Shang, Z. Lin, C. Qi, X. Liu, P. Li, Y. Tao, Z. Wu, D. Li, P. Simon and Q. Yang, 3D Macroscopic architectures from self-assembled MXene hydrogels, *Adv. Funct. Mater.*, 2019, 29, 1903960.

57. P. Urbankowski, B. Anasori, T. Makaryan, D. Er, S. Kota, P. L. Walsh, M. Zhao, V. B. Shenoy, M. W. Barsoum and Y. Gogotsi, Synthesis of two-dimensional titanium nitride Ti4N3 (MXene), *Nanoscale*, 2016, 8 (22), 11385–11391.

58. V. Kamysbayev, A. S. Filatov, H. C. Hu, X. Rui, F. Lagunas, D. Wang and R. F. Klie, Covalent surface modifications and superconductivity of two-dimensional metal carbide MXenes, *Science*, 2020, 369 (6506), 979–983.

59. C. Xu, L. Wang, Z. Liu, L. Chen, J. Guo, N. Kang, X. L. Ma, H. M. Cheng and W. Ren, Large-area high-quality 2D ultrathin Mo2C superconducting crystals, *Nat. Mater.*, 2015, 14 (11), 1135–1141.

60. D. Geng, X. Zhao, Z. Chen, W. Sun, W. Fu, J. Chen, W. Liu, W. Zhou and K. P. Loh, Direct synthesis of large-area 2D Mo2C on in situ grown graphene, *Adv. Mater.*, 2017, 29 (35), 1700072.

61. F. Turker, O. R. Caylan, N. Mehmood, T. S. Kasirga, C. Sevik and C. G. Buke, CVD synthesis and characterization of thin Mo2 C crystals, *J. Am. Ceramic Soc.*, 2020, 103 (10), 5586–5593.

62. Z. Sun, M. Yuan, L. Lin, H. Yang, C. Nan, H. Li, G. Sun and X. Yang, Selective lithiation–expansion–microexplosion synthesis of two-dimensional fluoride-free Mxene, *ACS Mater Lett.*, 2019, 1 (6), 628–632.

63. Z. Li, Y. Ren, L. Mo, C. Liu, K. Hsu, Y. Ding, X. Zhang, X. Li, L. Hu, D. Ji and G. Cao, Impacts of oxygen vacancies on zinc ion intercalation in VO2, *ACS Nano.*, 2020, 14 (5), 5581–5589.

64. N. Sun, Z. Guan, Q. Zhu, B. Anasori, Y. Gogotsi and B. Xu, Enhanced Ionic accessibility of flexible MXene electrodes produced by natural sedimentation, *Nano-Micro Lett.*, 2020, 12, 89.

65. Z. Ling, C. E. Ren, M.-Q. Zhao, J. Yang, J. M. Giammarco, J. Qiu, M. W. Barsoum and Y. Gogotsi, Flexible and conductive MXene films and nanocomposites with high capacitance, *Proc. Natl. Acad. Sci.*, 2014, 111, 16676–16681.

66. K. Hantanasirisakul, M.-Q. Zhao, P. Urbankowski, J. Halim, B. Anasori, S. Kota, C. E. Ren, M. W. Barsoum and Y. Gogotsi, Fabrication of Ti3C2Tx MXene transparent thin films with tunable optoelectronic properties, *Adv. Electron. Mater.*, 2016, 2, 1600050.

67. K. Maleski, V. N. Mochalin and Y. Gogotsi, Dispersions of two-dimensional Titanium Carbide MXene in organic solvents, *Chem. Mater.*, 2017, 29, 1632–1640.
68. K. Huang, Z. Li, J. Lin, G. Han and P. Huang, Correction: Two-dimensional transition metal carbides and nitrides (MXenes) for biomedical applications, *Chem. Soc. Rev.* 2018, 47, 5109–5124.
69. L. Verger, V. Natu, M. Carey and M. W. Barsoum, MXenes: An introduction of their synthesis, select properties, and applications, *Trends in Chem.*, 2019, 1, 656–669.
70. V. Kamysbayev, A. S. Filatov, H. Hu, X. Rui, F. Lagunas, D. Wang, R. F. Klie and D. V. Talapin, Covalent surface modifications and superconductivity of two-dimensional Metal Carbide MXenes, *Science*, 2020, 369, 979–983.
71. D. Dillon, M. J. Ghidiu, A. L. Krick, J. Griggs, S. J. May, Y. Gogotsi, M. W. Barsoum and A. T. Fafarman, Highly conductive optical quality solution-processed films of 2D Titanium Carbide, *Adv. Funct. Mater.*, 2016, 26, 4162–4168.
72. Anasori, M. R. Lukatskaya and Y. Gogotsi, 2D metal carbides and nitrides (MXenes) for energy storage, *Nat. Rev. Mater.*, 2017, 2, 16098.
73. H. Wang, Y. Wu, J. Zhang, G. Li, H. Huang, X. Zhang, Q. Jiang, Enhancement of the electrical properties of MXene Ti3C2 nanosheets by post-treatments of alkalization and calcination, *Mater. Lett.*, 2015, 160, 537–540.
74. J. Zhang, N. Kong, S. Uzun, A. Levitt, S. Seyedin, P. A. Lynch, S. Qin, M. Han, W. Yang, J. Liu, X. Wang, Y. Gogotsi and J. M. Razal, Scalable manufacturing of free-standing, strong Ti3C2 Tx MXene films with outstanding conductivity, *Adv. Mater.*, 2020, 32, 2001093.
75. M. Fatima, S. A. Zahra, S. A. Khan, D. Akinwande, J. Minár and S. Rizwan, Experimental and computational analysis of MnO2@V2C-MXene for enhanced energy storage, *Nanomaterials,* 2021, 11, 1707.
76. B. Mustafa, W. Lu, Z. Wang, F. Lian, A. Shen, B. Yang, J. Yuan, C. Wu, Y. Liu, W. Hu, L. Wang and G. Yu, Ultrahigh energy and power densities of d-MXene-based symmetric supercapacitors, *Nanomaterials,* 2022, 12, 3294.
77. S. Li, Y. Wang, Y. Li, J. Xu, T. Li and T. Zhang, In Situ growth of Ni-MOF nanorods array on Ti3C2Tx Nanosheets for supercapacitive electrodes, *Nanomaterials*, 2023, 13 (3), 610.
78. M. R. Lukatskaya, S. Kota, Z. Lin, M.-Q. Zhao, N. Shpigel, M. D. Levi, J. Halim, P.-L. Taberna, M. W. Barsoum, P. Simon and Y. Gogotsi, Ultra-high-rate pseudocapacitive energy storage in two-dimensional transition metal carbides, *Nat. Energy*, 2017, 2, 17105.
79. Y. Tao, X. Xie, W. Lv, D.-M. Tang, D. Kong, Z. Huang, H. Nishihara, T. Ishii, B. Li, D. Golberg, F. Kang, T. Kyotani and Q. H. Yang, Towards ultrahigh volumetric capacitance: Graphene derived highly dense but porous carbons for supercapacitors, *Sci. Rep.*, 2013, 3, 2975.
80. Z. Wang, P. Tammela, M. Strømme and L. Nyholm, Nanocellulose coupled flexible polypyrrole@graphene oxide composite paper electrodes with high volumetric capacitance, *Nanoscale*, 2015, 7, 3418–3423.
81. T. Li, L. Yao, Q. Liu, J. Gu, R. Luo, J. Li, X. Yan, W. Wang, P. Liu, B. Chen, W. Zhang, W. Abbas, R. Naz and D. Zhang, Fluorine-free synthesis of high-purity Ti3C2Tx (T=OH, O) via alkali treatment, *Angew. Chem.*, 2018, 130, 6223–6227.
82. X. Wang and L. Bannenbreg, Design and characterization of 2D MXene-based electrode with high-rate capability, *MRS Bulletin,* 2021, 46, 756–766.
83. L. Liu, H. Zschiesche, M. Antionietti, B. Daffos, N. V. Tarakina, M. Gibilaro, P. Chamelot, L. Massot, B. Duployer, P. L. Taberna and P. Simon, Tuning the surface chemistry of MXene to improve energy storage: Example of nitrification by salt melt, *Adv. Ener. Mater.*, 2023, 13 (2), 2202709.
84. X. Sang, D. Yilmaz, Y. Xie, M. Alhabeb, B. Anasori, X. Li, K. Xiao, P. R. C. Kent, A. van Duin, Y. Gogotsi and R. R. Unocic. Atomic defects and edge structure in single-layer Ti3C2Tx MXene, *Microsc. Microanal.*, 2017, 23 (1), 1704–1705.

85. J. Tang, T. S. Mathis, N. Kurra, A. Sarycheva, X. Xiao, M. N. Hedhili, Q. Jiang, H. N. Alshareef, B. Xu and F. Pan, Tuning the electrochemical performance of Titanium Carbide MXene by controllable in situ anodic oxidation, *Angew. Chem. Int. Ed.*, 2019, 58, 17849–17855.
86. X. Xiao, H. Wang, W. Bao, P. Urbankowski, L. Yang, Y. Yang, K. Maleski, L. Cui, S. J. Billinge and G. Wang, Two-dimensional arrays of transition metal nitride nanocrystals, *Adv. Mater.*, 2019, 31, 1902393.
87. Y. Tang, J. F. Zhu, C. H. Yang and F. Wang, Enhanced capacitive performance based on diverse layered structure of two-dimensional Ti3C2 MXene with long etching time, *J. Electrochem. Soc.*, 2016, 163, A1975–A1982.
88. K. Maleski, C. E. Ren, M.-Q. Zhao, B. Anasori and Y. Gogotsi, Size-dependent physical and electrochemical properties of two-dimensional MXene flakes, *ACS Appl. Mater. Interfaces*, 2018, 10, 24491–24498.
89. B. Mustafa, W. Lu, Z. Wang, F. Lian, A. Shen, B. Yang, J. Yuan, C. Wu, Y. Liu, W. Hu, L. Wang and G. Yu, Ultrahigh energy and power densities of d-MXene-based symmetric supercapacitors, *Nanomaterials,* 2022, 12 (19), 3294.
90. K. Zhu, Y. Jin, F. Du, S. Gao, Z. Gao, X. Meng, G. Chen, Y. Wei and Y. Gao, Synthesis of Ti2CTx MXene as electrode materials for symmetric supercapacitor with capable volumetric capacitance, *J. Energy Chem.*, 2019, 31, 11–18.
91. L. Li, F. Wang, J. Zhu, W. Wu, The facile synthesis of layered Ti2C MXene/carbon nanotube composite paper with enhanced electrochemical properties, *Dalton Trans.*, 2017, 46, 14880–14887.
92. B. Yao, M. Li, J. Zhang, L. Zhang, Y. Song, W. Xiao, A. Cruz, Y. Tong and Y. Li, TiN Paper for ultrafast-charging supercapacitors, *Nano-Micro Lett.*, 2020, 12, 3.
93. Djire, A. Bos, J. Liu, H. Zhang, E. M. Miller and N. R. Neale, Pseudocapacitive storage in nanolayered Ti2NTx MXene using Mg-Ion electrolyte, *ACS Appl. Nano Mater.*, 2019, 2, 2785–2795.
94. Q. Shan, X. Mu, M. Alhabeb, C. E. Shuck, D. Pang, X. Zhao, X.-F. Chu, Y. Wei, F. Du, G. Chen, Y. Gogotsi, Y. Gao and Y. Dall'Agnese, Two-dimensional vanadium carbide (V2C) MXene as electrode for supercapacitors with aqueous electrolytes, *Electrochem. Commun.*, 2018, 96, 103–107.
95. X. Wang, S. Lin, H. Tong, Y. Huang, P. Tong, B. Zhao, J. Dai, C. Liang, H. Wang, X. Zhu, Y. Sun and S. Dou, Two-dimensional V4C3 MXene as high performance electrodematerials for supercapacitors, *Electrochim. Acta,* 2019, 307, 414–421.
96. S. Zhao, C. Chen, X. Zhao, X. Chu, F. Du, G. Chen, Y. Gogotsi, Y. Gao and Y. Dall'Agnese, Flexible Nb4C3Tx film with large interlayer spacing for high-performance supercapacitors, *Adv. Funct. Mater.*, 2020, 30, 2000815.
97. R. Syamsai and A. N. Grace, Ta4C3 MXene as supercapacitor electrodes, *J. Alloy. Compd.*, 2019, 792, 1230–1238.
98. Y. Wen, T. E. Rufford, X. Chen, N. Li, M. Lyu, L. Dai and L. Wang, Nitrogen-doped $Ti_3C_2T_x$ MXene electrodes for high-performance supercapacitors, *Nano Energy*, 2017, 38, 368–376.
99. Y. Tang, J. Zhu, W. Wu, C. Yang, W. Lv and F. Wang, Synthesis of nitrogen-doped two-dimensional Ti3C2 with enhanced electrochemical performance, *J. Electrochem. Soc.*, 2017, 164, A923-A929.
100. L. Yang, W. Zheng, P. Zhang, J. Chen, W. Zhang, W. B. Tian and Z. M. Sun, Freestanding nitrogen-doped d-Ti3C2/reduced graphene oxide hybrid films for high performance supercapacitors, *Electrochim. Acta*, 2019, 300, 349–356.
101. C. Yang, Y. Tang, Y. Tian, Y. Luo, M. F. U. Din, X. Yin and W. Que, Flexible Nitrogen-doped 2D Titanium Carbides (MXene) films constructed by an ex situ solvothermal method with extraordinary volumetric capacitance, *Adv. Energy Mater.*, 2018, 8, 1802087.

102. C. Yang, W. Que, X. Yin, Y. Tian, Y. Yang and M. Que, Improved capacitance of nitrogen-doped delaminated two-dimensional titanium carbide by urea-assisted synthesis, *Electrochim. Acta*, 2017, 225, 416–424.

103. Y. Yoon, M. Lee, S. K. Kim, G. Bae, W. Song, S. Myung, J. Lim, S. S. Lee, T. Zyung and K.-S. An, A strategy for synthesis of carbon nitride induced chemically doped 2D MXene for high-performance supercapacitor electrodes, *Adv. Energy Mater.*, 2018, 8, 1703173.

104. C. Yang, W. Que, Y. Tang, Y. Tian and X. Yin, Nitrogen and sulfur Co-doped 2D titanium carbides for enhanced electrochemical performance, *J. Electrochem. Soc.*, 2017, 164, A1939–A1945.

105. P. Chakraborty, T. Das, D. Nafday, L. Boeri and T. Saha-Dasgupta, Manipulating the mechanical properties of Ti2C MXene: Effect of substitutional doping, *Phys. Rev. B.*, 2017, 95, 184106.

106. Z. W. Gao, W. Zheng and L. Y. S. Lee, Highly enhanced pseudocapacitive performance of vanadium-doped MXenes in neutral electrolytes, *Small*, 2019, 15, 1902649.

107. M. Fatima, J. Fatheema, N. B. Monir, A. H. Siddique, B. Khan, A. Islam, D. Akinwande and S. Rizwan, Nb-Doped MXene With enhanced energy storage capacity and stability, *Front. Chem.*, 2020, 8, 168.

108. J. L. Hart, K. Hantanasirisakul, A. C. Lang, B. Anasori, D. Pinto, Y. Pivak, T. v. J. Omme, S. J. May, Y. Gogotsi and M. L. Taheri, Control of MXenes' electronic properties through termination and intercalation, *Nat. Commun.*, 2019, 10, 522.

109. X. Zhang, Y. Liu, S. Dong, J. Yang and X. Liu, Surface modified MXene film as flexible electrode with ultrahigh volumetric capacitance, *Electrochim. Acta*, 2019, 294, 233–239.

110. M. Hu, T. Hu, Z. Li, Y. Yang, R. Cheng, J. Yang, C. Cui and X. Wang, Surface functional groups and interlayer water determine the electrochemical capacitance of $Ti_3C_2T_x$ MXene, *ACS Nano*, 2018, 12, 3578–3586.

111. Q. Tang, Z. Zhou and P. Shen, Are MXenes promising anode materials for Li ion batteries? Computational studies on electronic properties and Li storage capability of Ti_3C_2 and $Ti_3C_2X_2$ (X = F, OH) monolayer, *J. Am. Chem. Soc.*, 2012, 134, 16909–16916.

112. M. R. Lukatskaya, S.-M. Bak, X. Yu, X.-Q. Yang, M. W. Barsoum and Y. Gogotsi, Probing the mechanism of high capacitance in 2D Titanium Carbide using in situ x-ray absorption spectroscopy, *Adv. Energy Mater.*, 2015, 5, 1500589.

113. X. Zheng, W. Nie, Q. Hu, X. Wang, Z. Wang, L. Zou, X. Hong, H. Yang, J. Shen and C. Li, Multifunctional RGO/Ti3C2Tx MXene fabrics for electrochemical energy storage, electromagnetic interference shielding, electrothermal and human motion detection, *Materials and Design*, 2021, 200 (15), 109442.

114. M. Lu, W. Han, H. Li, H. Li, B. Zhang, W. Zhang and W. Zheng, Magazine-bending-inspired architecting Anti-T of MXene flakes with vertical Ion transport for high-performance supercapacitors, *Adv. Mater. Interfaces*, 2019, 6, 1900160.

115. O. Mashtalir, M. R. Lukatskaya, A. I. Kolesnikov, E. Raymundo-Piñero, M. Naguib, M. W. Barsoum and Y. Gogotsi, The effect of hydrazine intercalation on the structure and capacitance of 2D Titanium Carbide (MXene), *Nanoscale*, 2016, 8, 9128–9133.

116. Z. Fan, Y. Wang, Z. Xie, D. Wang, Y. Yuan, H. Kang, B. Su, Z. Cheng and Y. Liu, Modified MXene/Holey graphene films for advanced supercapacitor electrodes with superior energy storage, *Adv. Sci.*, 2018, 5, 1800750.

117. M. Q. Zhao, C. E. Ren, Z. Ling, M. R. Lukatskaya, C. Zhang, K. L. V. Aken, M. W. Barsoum and Y. Gogotsi, Flexible MXene/carbon nanotube composite paper with high volumetric capacitance, *Adv. Mater.*, 2015, 27, 339–345.

118. V. Mohammadi, M. Mojtabavi, N. M. Caffrey, M. Wanunu and M. Beidaghi, Assembling 2D MXenes into highly stable pseudocapacitive electrodes with high power and energy densities, *Adv. Mater.*, 2018, 31, 1806931.

119. Z. Lin, D. Barbara, P.-L. Taberna, K. L. V. Aken, B. Anasori, Y. Gogotsi and P. Simon, Capacitance of Ti3C2Tx MXene in ionic liquid electrolyte, *J. Power Sources*, 2016, 326, 575–579.

120. L. Yang, W. Zheng, P. Zhang, J. Chen, W. B. Tian, Y. M. Zhang and Z. M. Sun, J. In situ synthesis of CNTs@Ti3C2 hybrid structures by microwave irradiation for high-performance anodes in lithium ion batteries, *Electroanal. Chem.*, 2018, 6, 830–831.

121. L. Shen, X. Zhou, X. Zhang, Y. Zhang, Y. Liu, W. Wang, W. Si and X. Dong, Carbon-intercalated Ti3C2Tx MXene for high-performance electrochemical energy storage, *J. Mater. Chem. A*, 2018, 6, 23513–23520.

122. K. Wang, B. Zheng, M. Mackinder, N. Baule, H. Qiao, H. Jin, T. Schuelke and Q. H. Fan, Graphene wrapped MXene via plasma exfoliation for all-solid-state flexible supercapacitors, *Energy Storage Mater.*, 2019, 20, 299–306.

123. P. Zhang, Q. Zhu, R. A. Soomro, S. He, N. Sun, N. Qiao and B. Xu, In situ ice template approach to fabricate 3D flexible MXene film-based electrode for high performance supercapacitors, *Adv. Funct. Mater.*, 2020, 30, 2000922.

124. L. Li, M. Zhang, X. Zhang and Z. Zhang, New Ti3C2 aerogel as promising negative electrode materials for asymmetric supercapacitors, *J. Power Sources*, 2017, 364, 234–241.

125. T. Shang, Z. Lin, C. Qi, X. Liu, P. Li, Y. Tao, Z. Wu, D. Li, P. Simon and Q. H. Yang, 3D Macroscopic architectures from self-assembled MXene hydrogels, *Adv. Funct. Mater.*, 2019, 29, 1903960.

126. X. Zhang, Y. Liu, S. Dong, J. Yang and X. Liu, Self-assembled three-dimensional Ti3C2Tx nanosheets network on carbon cloth as flexible electrode for supercapacitors, *Appl. Surf. Sci.*, 2019, 485, 1–7.

127. S. Xu, G. Wei, J. Li, Y. Ji, N. Klyui, V. Izotov and W. Han, Binder-free Ti3C2Tx MXene electrode film for supercapacitor produced by electrophoretic deposition method, *Chem. Eng. J.*, 2017, 317, 1026–1036.

128. Z. Fan, J. Wang, H. Kang, Y. Wang, Z. Xie, Z. Cheng and Y. Liu, A compact MXene film with folded structure for advanced supercapacitor electrode material, *ACS Appl. Energy Mater.*, 2020, 3, 1811–1820.

129. M. Hu, T. Hu, R. Cheng, J. Yang, C. Cui, C. Zhang and X. Wang, MXene-coated silk-derived carbon cloth toward flexible electrode for supercapacitor application, *J. Energy Chem.*, 2018, 27, 161–166.

130. Halim, M. R. Lukatskaya, K. M. Cook, J. Lu, C. R. Smith, L.-Å. Näslund, S. J. May, L. Hultman, Y. Gogotsi, P. Eklund and M. W. Barsoum, Transparent conductive two-dimensional Titanium Carbide epitaxial thin films, *Chem. Mater.*, 2014, 26, 2374–2381.

131. O. Mashtalir, M. R. Lukatskaya, A. I. Kolesnikov, E. Raymundo-Piñero, M. Naguib, M. W. Barsoum and Y. Gogotsi, The effect of hydrazine intercalation on the structure and capacitance of 2D titanium carbide (MXene), *Nanoscale* 2016, 8, 9128–9133.

132. Y.Z. Fang, R. Hu, K. Zhu, K. Ye, J. Yan, G. Wang and D. Cao, Aggregation-resistant 3D Ti3C2Tx MXene with enhanced kinetics for potassium ion hybrid capacitors, *Adv. Funct. Mater.*, 2020, 30, 2005663.

133. W. Chen, J. Tang, P. Cheng, Y. Ai, Y. Xu and N. Ye, 3D porous MXene (Ti3C2Tx) prepared by Alkaline induced flocculation for supercapacitor electrodes, *Materials*, 2022, 15 (3), 925.

134. P. Zhang, Y. Oeng, Q. Zhu, R. A. Soomro, N. Sun and B. Xu, 3D form-based MXene architectures: Structural and electrolytic engineering for advanced potassium-ion storage, *Energy Environ. Mater.*, 2022, 9999, 1–10.

135. C. E. Ren, M.-Q. Zhao, T. Makaryan, J. Halim, M. Boota, S. Kota, B. Anasori, M. W. Barsoum and Y. Gogotsi, Porous two-dimensional transition Metal Carbide (MXene) flakes for high-performance Li-Ion storage, *ChemElectroChem.*, 2016, 3, 689–693.

136. M.Q. Zhao, X. Xie, C. E. Ren, T. Makaryan, B. Anasori, G. Wang and Y. Gogotsi, Hollow MXene spheres and 3D macroporous MXene frameworks for Na-Ion storage, *Adv. Mater.*, 2017, 29, 1702410.

137. J. Liu, H.-B. Zhang, R. Sun, Y. Liu, Z. Liu, A. Zhou and Z.-Z. Yu, Hydrophobic, flexible, and lightweight MXene foams for high-performance electromagnetic-interference shielding, *Adv. Mater.*, 2017, 29, 1702367.

138. P. Collini, S. Kota, A. D. Dillon, M. W. Barsoum and A. T. Fafarman, Electrophoretic deposition of two-dimensional Titanium Carbide (MXene) thick films, *J. Electrochem. Soc.*, 2017, 164, D573-D580.

139. L. Shi, S. Lin, L. Li, W. Wu, L. Wu, H. Gao and X. Zhang, Ti3C2Tx-foam as free-standing electrode for supercapacitor with improved electrochemical performance, *Ceram. Int.*, 2018, 44, 13901–13907.

140. Y. Yue, N. Liu, Y. Ma, S. Wang, W. Liu, C. Luo, H. Zhang, F. Cheng, J. Rao, X. Hu and J. Su, Y. Gao, Highly self-healable 3D microsupercapacitor with MXene-Graphene composite aerogel, *ACS Nano*, 2018, 12, 4224–4232.

141. X. Zhang, X. Liu, S. Dong, J. Yang and Y. Liu, Template-free synthesized 3D macroporous MXene with superior performance for supercapacitors, *Appl. Mater. Today*, 2019, 16, 315–321.

142. J. Guo, Y. Zhao, A. liu and T. Ma, Electrostatic self-assembly of 2D delaminated MXene (Ti3C2) onto Ni foam with superior electrochemical performance for supercapacitor, *Electrochim. Acta*, 2019, 305, 164–174.

143. M. Boota and Y. Gogotsi, MXene—conducting polymer asymmetric pseudocapacitors, *Adv. Energy Mater.*, 2019, 9 (7), 1802917.

144. L. Qin, Q. Tao, A. E. Ghazaly, J. Fernandez-Rodriguez, P. O. Å. Persson, J. Rosen and F. Zhang, High-performance ultrathin flexible solid-state supercapacitors based on solution processable Mo1.33C MXene and PEDOT:PSS, *Adv. Funct. Mater.*, 2018, 28, 1703808.

145. Z. Chen, Y. Han, T. Li, X. Zhang, T. Wang and Z. Zhang, Preparation and electrochemical performances of doped MXene/poly(3,4-ethylenedioxythiophene) composites, *Mater. Lett.*, 2018, 220, 305–308.

146. W. Wu, D. Wei, J. Zhu, D. Niu, F. Wang, L. Wang, L. Yang, P. Yang and C. Wang, Enhanced electrochemical performances of organ-like Ti3C2 MXenes/polypyrrole composites as supercapacitors electrode materials, *Ceram. Int.*, 2019, 45, 7328–7337.

147. H. Wang, L. Li, C. Zhu, S. Lin, J. Wen, Q. Jin and X. Zhang, In situ polymerized Ti3C2Tx/PDA electrode with superior areal capacitance for supercapacitors, *J. Alloy. Compd.*, 2019, 778, 858–865.

148. M. Boota, M. Pasini, F. Galeotti, W. Porzio, M.-Q. Zhao, J. Halim and Y. Gogotsi, Interaction of polar and nonpolar polyfluorenes with layers of two-dimensional Titanium Carbide (MXene): Intercalation and pseudocapacitance, *Chem. Mater.*, 2017, 29, 2731–2738.

149. X. Lu, J. Zhu, W. Wu and B. Zhang, Hierarchical architecture of PANI@TiO2/Ti3C2Tx ternary composite electrode for enhanced electrochemical performance, *Electrochim. Acta*, 2017, 228, 282–289.

150. Y. Ren, J. Zhu, L. Wang, H. Liu, Y. Liu, W. Wu and F. Wang, Synthesis of polyaniline nanoparticles deposited on two-dimensional titanium carbide for high-performance supercapacitors, *Mater. Lett.*, 2018, 214, 84–87.

151. Y. Tang, J. Zhu, C. Yang and F. Wang, Enhanced supercapacitive performance of manganese oxides doped two-dimensional titanium carbide nanocomposite in alkaline electrolyte, *J. Alloy. Compd.*, 2016, 685, 194–201.

152. J. Zhu, X. Lu and L. Wang, Synthesis of a MoO3/Ti3C2Tx composite with enhanced capacitive performance for supercapacitors, *RSC Adv.*, 2016, 6, 98506–98513.

153. W. Zheng, P. Zhang, W. Tian, Y. Wang, Y. Zhang, J. Chen and Z. M. Sun, Microwave-assisted synthesis of SnO2-Ti3C2 nanocomposite for enhanced supercapacitive performance, *Mater. Lett.,* 2017, 209, 122–125.

154. J. Zhu, Y. Tang, C. Yang, F. Wang and M. Cao, Composites of TiO2 Nanoparticles Deposited on Ti3C2 MXene nanosheets with enhanced electrochemical performance, *J. Electrochem. Soc.,* 2016, 163, A785–A791.

155. M. Cao, F. Wang, L. Wang, W. Wu, W. Lv and J. Zhu, Room temperature oxidation of Ti3C2 MXene for supercapacitor electrodes, *J. Electrochem. Soc.,* 2017, 164, A3933–A3942.

156. S. B. Ambade, R. B. Ambade, W. Eom, S. H. Noh, S. H. Kim and T. H. Han, 2D Ti3C2 MXene/WO3 Hybrid architectures for high-rate supercapacitors, *Adv. Mater. Interfaces,* 2018, 5, 1801361.

157. Y. Tian, C. Yang, W. Que, X. Liu, X. Yin and L. B. Kong, Flexible and free-standing 2D titanium carbide film decorated with manganese oxide nanoparticles as a high volumetric capacity electrode for supercapacitor, *J. Power Sources,* 2017, 359, 332–339.

158. H. Jiang, Z. Wang, Q. Yang, M. Hanif, Z. Wang, L. Dong and M. Dong, A novel MnO2/Ti3C2Tx MXene nanocomposite as high performance electrode materials for flexible supercapacitors, *Electrochim. Acta,* 2018, 290, 695–703.

159. H. Liu, R. Hu, J. Qi, Y. Sui, Y. He, Q. Meng, F. Wei, Y. Ren, Y. Zhao and W. Wei, One-step synthesis of nanostructured CoS2 Grown on Titanium Carbide MXene for high-performance asymmetrical supercapacitors, *Adv. Mater. Interfaces,* 2020, 7 (6), 1902659.

160. C. Choi, D. S. Ashby, D. M. Butts, R. H. DeBlock, Q. Wei, J. Lau and B. Dunn, Achieving high energy density and high power density with pseudocapacitive materials, *Nat. Rev. Mater.,* 2020, 5, 5–17.

161. M. Hu, Z. Li, T. Hu, S. Zhu, C. Zhang and X. Wang, High-capacitance mechanism for Ti3C2Tx MXene by in situ electrochemical Raman spectroscopy investigation, *ACS Nano,* 2016, 10, 11344–11350.

162. P. Simon and Y. Gogotsi, Materials for electrochemical capacitors, *Nat. Mater.,* 2008, 7, 845–853.

163. M. D. Levi, M. R. Lukatskaya, S. Sigalov, M. Beidaghi, N. Shpigel, L. Daikhin, D. Aurbach, M. W. Barsoum and Y. Gogotsi, Solving the capacitive paradox of 2D MXene using electrochemical quartz-crystal admittance and in situ electronic conductance measurements, *Adv. Energy Mater.,* 2015, 5, 1400815.

164. Y. Ando, M. Okubo, A. Yamada and M. Otani, Capacitive versus pseudocapacitive storage in MXene, *Adv. Funct. Mater.,* 2020, 30, 2000820.

165. M. Okubo, A. Sugahara, S. Kajiyama and A. Yamada, MXene as a charge storage host, *Acc. Chem. Res.,* 2018, 51, 591–599.

166. B. E. Francisco, C. M. Jones, S.-H. Lee and C. R. Stoldt, Nanostructured all-solid-state supercapacitor based on Li2S-P2S5 glass-ceramic electrolyte, *Appl. Phys. Lett.,* 2012, 100, 103902.

167. J. Duay, E. Gillette, R. Liu and S. B. Lee, Highly flexible pseudocapacitor based on freestanding heterogeneous MnO2/conductive polymernanowire arrays, *Phys. Chem. Chem. Phys.,* 2012, 14, 3329–3337.

168. C. W. Huang, C. A. Wu, S. S. Hou, P. L. Kuo, C. Te Hsieh and H. Teng, Gel electrolyte derived from poly(ethylene glycol) blending poly(acrylonitrile) applicable to roll-to-roll assembly of electric double layer capacitors, *Adv. Funct. Mater.,* 2012, 22, 4677–4685.

169. Y. N. Sudhakar, M. Selvakumar and D. K. Bhat, LiClO4-doped plasticized chitosan and poly(ethylene glycol) blend as biodegradable polymer electrolyte for supercapacitors, *Ionics,* 2013, 19, 277–285.

170. C. Ramasamy, J. Palma Del Vel and M. Anderson, An activated carbon supercapacitor analysis by using a gel electrolyte of sodium salt-polyethylene oxide in an organic mixture solvent, *J. Solid State Electrochem.*, 2014, 18, 2217–2223.

171. B. Pal, A. Yasin, R. Kunwar, S. Yang, M. M. Yusoff and R. Jose, Polymer versus cation of gel polymer electrolytes in the charge storage of asymmetric supercapacitors, *Ind. Eng. Chem. Res.*, 2019, 58, 654–664.

172. T. Li, L. Yao, Q. Liu, J. Gu, R. Luo, J. Li, X. Yan, W. Wang, P. Liu, B. Chen, W. Zhang, W. Abbas, R. Naz and D. Zhang, Fluorine-free synthesis of high-purity Ti3C2Tx (T=OH, O) via alkali treatment, *Angew. Chem., Int. Ed.*, 2018, 57, 6115–6119.

173. M. R. Lukatskaya, S. Kota, Z. F. Lin, M. Q. Zhao, N. Shpigel, M. D. Levi, J. Halim, P. L. Taberna, M. Barsoum, P. Simon and Y. Gogotsi, Ultra-high-rate pseudocapacitive energy storage in two-dimensional transition metal carbides, *Nat. Energy,* 2017, 2, 17105.

174. M. Heon, S. Loand, J. Applegate, R. Nolte, E. Cortes, J. D. Hettinger, P. L. Taberna, P. Simon, P. H. Huang, M. Brunet and Y. Gogotsi, Continuous carbide-derived carbon films with high volumetric capacitance, *Energy Environ. Sci.*, 2011, 4, 135–138.

175. J. Yan, C. E. Ren, K. Maleski, C. B. Hatter, B. Anasori, P. Urbankowski, A. Sarycheva and Y. Gogotsi, Flexible MXene/graphene films for ultrafast supercapacitors with outstanding volumetric capacitance, *Adv. Funct. Mater.*, 2017, 27, 1701264.

176. M. Q. Zhao, C. E. Ren, Z. Ling, M. R. Lukatskaya, C. F. Zhang, K. L. Van Aken, M. W. Barsoum and Y. Gogotsi, Flexible MXene/carbon nanotube composite paper with high volumetric capacitance, *Adv. Mater.,* 2015, 27, 339–345.

177. K. Li, X. H. Wang, S. Li, P. Urbankowski, J. M. Li, Y. X. Xu and Y. Gogotsi, An ultrafast conducting Polymer@MXene positive electrode with high volumetric capacitance for advanced asymmetric supercapacitors, *Small*, 2020, 16, 1906851.

178. H. Li, X. Chen, E. Zalnezhad, K. N. Hui, K. S. Hui and M. J. Ko, 3D hierarchical transition-metal sulfides deposited on MXene as binder-free electrode for high-performance supercapacitors, *J. Ind. Eng. Chem.*, 2020, 82, 309–316.

179. S. Kumar, M. A. Rehman, S. Lee, M. Kim, H. Hong, J. Y. Park, Y. Seo, Supercapacitors based on Ti3C2Tx MXene extracted from supernatant and current collectors passivated by CVD-graphene, *Sci. Rep.*, 2021, 11, 649.

180. G. Ge, Y. Z. Zhang, W. Zhang, W. Yuan, J. K. El-Demellawi, P. Zhang, E. D. Fabrizio, X. Dong and H. N. Alsshareef, Ti3C2Tx MXene-activated fast gelation of stretchable and self-healing hydrogels: A molecular approach, *ACS Nano,* 2021, 15 (2), 2698–2706.

181. Z. Yu and J. Thomas, Energy storing electrical cables: Integrating energy storage and electrical conduction, *Adv. Mater.,* 2014, 26, 4279–4285.

182. Y. Yan, P. Gu, S. Zheng, M. Zheng, H. Pang and H. Xue, Facile synthesis of an accordion-like Ni-MOF superstructure for high-performance flexible supercapacitors, *J. Mater. Chem. A,* 2016, 4, 19078–19085.

183. L. Wang, X. Feng, L. Ren, Q. Piao, J. Zhong, Y. Wang, H. Li, Y. Chen and B. Wang, Flexible solid-state supercapacitor based on a metal-organic framework interwoven by electrochemically-deposited PANI, *J. Am. Chem. Soc.*, 2015, 137 (15), 4920–4923.

184. Z. Zhou, Q. Zhang, J. Sun, B. He, J. Guo, Q. Li, C. Li, L. Xie and Y. Yao, Metal–organic framework derived spindle-like carbon incorporated α-Fe2O3 grown on carbon nanotube fiber as anodes for high-performance wearable asymmetric supercapacitors, *ACS Nano,* 2018, 12, 9333–9341.

185. Y. Liu, G. Li, Y. Guo, Y. Ying and X. Peng, Functionalized porous aromatic framework for efficient Uranium adsorption from aqueous solutions, *ACS Appl. Mater. Interfaces,* 2017, 9 (14), 12511–12517.

186. Y. Ding, W. Bai, J. Sun, Y. Wu, M.A. Memon, C. Wang, C. Liu, Y. Huang, J. Geng. Cellulose tailored anatase TiO2 nanospindles in three-dimensional graphene composites for high-performance supercapacitors, *ACS Appl. Mater. Interfaces*, 2016, 8 (19), 12165–12175.

187. D. Kong, W. Ren, C. Cheng, Y. Wang, Z. Huang and H.Y. Yang Three-dimensional NiCo2O4@ polypyrrole coaxial nanowire arrays on carbon textiles for high-performance flexible asymmetric solid-state supercapacitor, *ACS Appl. Mater. Interfaces*, 2016, 7 (38), 21334–21346.

188. M. Sun, J. Wang, M. Xu, Z. Fang, L. Jiang, Q. Han, J. Liu, M. Yan, Q. Wang and H. Bi, Hybrid supercapacitors based on interwoven CoO-NiO-ZnO nanowires and porous graphene hydrogel electrodes with safe aqueous electrolyte for high supercapacitance, *Adv. Electron. Mater.*, 2019, 5, 1900297.

189. X. Xu, J. Tang, H. Qian, S. Hou, Y. Bando, M. S. A. Hossain, L. Pan and Y. Yamauchi, Three-dimensional networked metal-organic frameworks with conductive polypyrrole tubes for flexible supercapacitors, *ACS Appl. Mater. Inter.*, 2017, 9, 38737–38744.

190. K. Qi, R. Hou, S. Zaman, Y. Qiu, B. Y. Xia and H. Duan, Construction of metal-organic framework/conductive polymer hybrid for all-solid-state fabric supercapacitor, *ACS Appl. Mater. Inter.*, 2018, 10, 18021–18028.

191. L. Shao, Q. Wang, Z. Ma, Z. Ji, X. Wang, D. Song, Y. Liu and N. Wang, A high-capacitance flexible solid-state supercapacitor based on polyaniline and Metal-Organic Framework (UiO-66) composites, *J. Power Sources*, 2018, 379, 350–361.

3 Carbon-Based Hybrid Materials as Advanced Electrodes for Structural Supercapacitors

Van Hoang Luan, Minh-Vien Le, Dat Ly and Won-Chun Oh

1 INTRODUCTION

1.1 IN ENERGY PRODUCTION

In recent years, the rapid development of aerospace, electric vehicles, and intelligent electronic equipment has led to increasing energy demands that cannot be sufficiently met by existing fossil fuels sources. Moreover, the combustion of fossil fuels releases carbon dioxide and other hazardous greenhouse gases, contributing to global warming and a cascade of environmental issues. As a result, the shift toward renewable energy sources, such as solar, wind, and ocean energy, presents a promising solution to mitigate energy deficits and reduce environmental pollution [1, 2]. However, these renewable energy sources are intermittent and unstable, as they depend on natural weather conditions and time of day. To address these challenges, rechargeable batteries and supercapacitors have become widely used energy storage devices. These devices are evaluated based on two key technical parameters: power density and energy density [3–5]. Power density refers to the rate at which energy is delivered, while energy density reflects the amount of energy storage in the device. The choice between supercapacitors and rechargeable batteries for energy storage depends on these parameters. Rechargeable batteries, though having a high energy density, suffer from a slow redox rate of electrolyte electrode materials during the charge-discharge process. On the other hand, supercapacitors offer high capacity, fast charge-discharge cycles, and long lifetime making them ideal for supplying power intermittently. Supercapacitors operate through electrochemical interactions between electrodes and electrolytes under biased conditions [6–8].

As depicted in the basic Ragone plot (Figure 3.1a), supercapacitors exhibit higher power density than batteries and greater energy density than traditional capacitors. Batteries and fuel cells provide high energy density (greater than 10 Wh.kg^{-1}) but suffer from lower power density (less than 100 W kg^{-1}). In contrast, capacitors, with exceptional power delivery capability (exceeding 10^5 W kg^{-1}), are limited by their

DOI: 10.1201/9781003561262-3

low energy density (less than 0.1 Wh.kg^{-1}). Therefore, batteries and fuel cells are more commonly used in applications requiring high energy storage, such as cell phones and electric vehicles [9–11].

Supercapacitors, on the other hand, can achieve power density above 10^4 W.kg^{-1} and energy density (less than 10 Wh.kg^{-1}) offering a more balanced performance. This contrasts with dielectric capacitors, which excel in power density, and batteries and fuel cells, which provide higher energy density. However, traditional supercapacitors, such as electric double-layer capacitance (EDLC) with activated carbon, are constrained by a low energy density of less than 20 Wh.kg^{-1}, significantly lower than that of lithium-ion batteries (LIBs) [12]. The primary challenge for the broader adoption of supercapacitors lies in improving energy density while maintaining high power density. Beyond their electrochemical energy storage capabilities, a notable innovation is the development of supercapacitors that can bear mechanical loads while simultaneously storing and releasing electrical energy. This led to the concept of structural energy storage device, which are designed to meet two critical requirements: high electrical capacity and mechanical endurance [13].

1.2 In Electric Vehicles (EVs)

As reported, the weight of the energy storage devices used in electric vehicles is approximately 150 kg for 12 KW system and 600 kg for 100 kW system [14, 15]. The energy consumption and the acceleration of electric vehicles increase proportionally with the vehicles. The power demand (P) and energy consumption (E) can be calculated using formulas (1) and (2) [16]:

$$P_{\text{demand}} = rV + sV^2 + tV^3 + mVA \qquad (1)$$

$$E = \frac{PS}{V} \qquad (2)$$

where m is the mass; V is the velocity; r, s, and t are coefficients to dependent on the electric vehicle; and S is the distance traveled.

Additionally, the vehicle's mass not only affects fuel consumption and speed, but also impacts the severity of injuries during accident. According to Newtonian mechanics, the change in velocity (Δv_1) of a vehicle at the point of impact can be determined using equation (3) [17]:

$$\Delta v_1 = \left(\frac{m_2}{m_1 + m_2} \right) (v_1 + v_2) \qquad (3)$$

where m_1 and m_2 are the masses of two vehicles; v_1 and v_2 are the velocities before the accident.

From these studies, integrating vehicle body components (such as roof, hood, …) with energy storage devices offers a promising solution to reduce the overall mass of electric vehicles by replacing traditional rechargeable battery packs [18]. Utilizing the vehicle's structure as an energy source reduces weight, frees up space in the engine compartment, and facilitates easier design and expansion of vehicle features.

1.3   KEY OF SUPERCAPACITORS

Supercapacitors, often used to enhance traditional capacitors, share a similar basic structure. Also known as electric double-layer capacitors (EDLCs) or ultra-capacitors, they are clean energy storage devices with high specific capacitance, long lifetime, and power density/energy density that fall between those batteries and capacitors, as shown in the Ragone plot from Srinivasan et al.'s work [9].

As illustrated in Figure 3.1b, supercapacitors consist of two collector electrodes (positive and negative), two working electrodes (typically carbon material electrodes), a separator, and an electrolyte [19]. The collector electrodes, usually made of copper or nickel, offer high electrical conductivity and anti-corrosion properties. The working electrodes, which are carbon-based materials with a high surface area or composite materials with excellent electrochemical performance, are coated on the collector electrodes. The separator plays a crucial role by providing electrical insulation while allowing ionic conduction between two working electrodes. The capacitance of supercapacitors is generated by the accumulating charges (ions) at the interface between the working electrode, and electrolyte, as shown in Figure 3.1b. The capacitance (C_e) of the EDLCs can be calculated using equation (4) [20]:

$$C_e = \varepsilon_0 \varepsilon_r \frac{S}{d} \tag{4}$$

where ε_0 (8.85–10^{-12} F m^{-1}) is the vacuum permittivity; ε_r is the relative dielectric constant of electrolyte; S is the surface area of the working electrode; and d is the distance between the ions and working electrodes.

As illustrated in Figure 3.1b, a supercapacitor device (cell) is essentially the combination of two capacitors in series, formed by the negative electrode and positive electrode. The capacitance of the cell (C_{cell}) is calculated using equation (5) [20]:

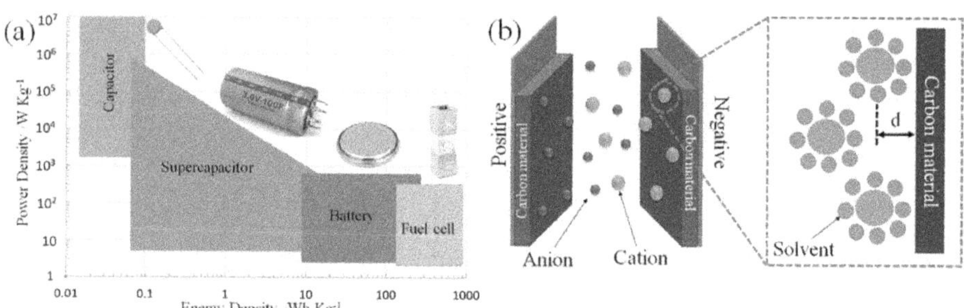

FIGURE 3.1   (a) The Ragone plot presents the relationship between power density and energy density of the different energy storage devices. (b) Schematic of EDLCs and electrical double-layer formation on a negative electrode with the solvation of a cation in a solvent.

$$C_{\text{cell}} = \frac{C_+ C_-}{C_+ + C_-} \tag{5}$$

where C_+ and C_- are the capacitance values of positive and negative electrodes.

As shown in the above discussion, energy and power density are two important parameters of supercapacitors. The energy density (E) and power density (P) are given by equations (6) and (7) [21]:

$$E = \frac{1}{2} C_{\text{cell}} \Delta V^2 \tag{6}$$

$$P = \frac{E}{t} \tag{7}$$

where ΔV is the operating voltage window and t is the discharge time.

To evaluate the specific capacitance (Cs) of supercapacitors, cyclic voltametric and galvanostatic charge-discharge techniques are commonly used, with calculations based on equations (8) and (9). Cyclic voltammetry investigates the relationship between current and potential at various scan rates, providing insights into the device's behavior. The galvanostatic charge-discharge curve is a widely used method for determining the specific capacitance at different load currents, primarily calculated from the discharge curve, which is based on the discharge time [22,23].

$$C_s = \frac{\int IdV}{m\vartheta\Delta V} \tag{8}$$

$$C_s = \frac{I}{m}\frac{dt}{dV} \tag{9}$$

where I is current; I/m is the current density; V is the operating potential; ϑ is the scan rate; and m is the total mass of the working electrode.

2 STRUCTURAL SUPERCAPACITORS

Multifunctional composite structural supercapacitors combine mechanical stability with electrochemical energy storage in a single device, enabling the storage of electrical charge while withstanding mechanical loads. The simplest structural supercapacitor, based on an EDLC configuration, consists of two carbon fiber fabrics serving as both current collectors and working electrodes, insulated by a separator layer with a SPE as the structural matrix. Carbon-based fiber fabric can be activated or surface-modified to meet two key requirements of active electrode materials: high electrical conductivity and large specific surface area [24, 25]. Additionally, carbon fiber's advantageous mechanical properties, such as high tensile strength and high Young's modulus, make it suitable as a load-bearing frame in structural devices. A SPE typically consists of a solid polymer (e.g., epoxy resin) and an ionic liquid (IL) such as Li-ion with an aqueous solvent. This combination forms the charge tunnel within the polymer matrix, facilitating charge transfer to the electrode during the electrochemical process. However, the interface between the working electrode and electrolyte is partially reduced due to the presence of a structural epoxy matrix. Charge mobility and electrical conductivity of SPEs are crucial factors in the development of structural supercapacitors.

With carbon-based fiber-reinforced polymers, structural supercapacitors exhibit mechanical properties with a modulus over 5–50 GPa and mechanical strength exceeding 85 MPa [26, 27]. Recent studies on the efficiency of the structural supercapacitor device focus not only on improving working electrode and electrolyte materials but also on the optimizing their design. EDLC-type structural supercapacitors are increasingly being replaced by carbon-based hybrid types, which integrate carbon-derived materials onto the carbon fiber surface (the current collector electrodes). New polymers and IL systems are continuously being developed to enhance the role of the electrolyte in practical applications. Especially, systems in particular, are gaining attention due to their low cost and excellent mechanical and thermal stability. Although structural supercapacitors exhibit long lifetimes and stable electrochemical performance, the charge-discharge process degrades over time due to the moisture ingress into the multifunctional electrolyte matrix. As a result, packing methods are being explored to protect these devices from environment exposure and prevent the depletion of IL-based SPEs [24].

2.1 Structural Supercapacitor Design

Currently, structural supercapacitors are classified according to designs as:

2.1.1 Integrated Structural Supercapacitors

The integrated structure of a single or multiple supercapacitor device is sandwiched between two layers of carbon fiber fabrics and encased in epoxy resin, as shown in Figure 3.2a. The carbon fiber fabric provides structural reinforcement and acts as both a protective layer and packaging for the supercapacitor, enhancing the mechanical stability of the device [28, 29]. However, integrated structural supercapacitors still face challenges, such as structural looseness and low volume fraction in carbon fiber-reinforced composite, which impact the multifunctional properties of the overall structure [30].

2.1.2 Lamination Structural Supercapacitors

Lamination structural supercapacitors are highly integrated systems that combine carbon fiber fabrics, glass/aramid fiber fabrics, and the polymer-based IL, as shown in Figure 3.2b. In this configuration, carbon fiber serves not only as a structural reinforcement for composite materials but also the working electrode in supercapacitors [31]. The carbon fiber fabric layers are interwoven with glass fiber fabric layers, and the entire structure is impregnated with a mixture of epoxy and electrolyte before being cured. In laminated structural supercapacitor, the liquid electrolyte is combined with the polymer to form a SPE with high tensile strength and Young's modulus. The SPE is formed by the dense structure of the cured polymer, which is modified to create continuous channels that serve as conductive paths. These channels allow ions from the electrolyte to reach the electrodes, facilitating efficient charge transfer.

2.2 CARBON FIBER-REINFORCED POLYMERS

A carbon fiber-reinforced polymer combines the strengths of two advanced materials (carbon fiber and polymer) making it highly suitable for aerospace and aircraft applications due to its excellent mechanical performance and low density. This composite is typically produced using vacuum-assisted resin transfer molding (VARTM) process, as shown in Figure 3.2c [32]. In this process, carbon fiber fabrics are on a stainless-steel plate (the mold). A peel ply layer and an infusion mesh are placed over the carbon fiber, adding to the peeling process after polymer curing. The infusion mesh facilitates uniform distribution of the liquid or gel polymer throughout the carbon fiber fabric under vacuum. The flow rate (Q) of the polymer injected into the carbon fiber fabric is controlled based on equation (10) , ensuring constant vacuum conditions during the process [33, 34].

$$Q = \frac{K\Delta p}{\eta x} \tag{10}$$

where K represents the permeability of the carbon fiber fabric; Δp is the pressure difference between the polymer infusion line and the vacuum port; η is the polymer viscosity, x is the flow position.

In carbon fiber-reinforced polymers, carbon fiber serves as the primary load-bearing component of the structure. The bulk of the polymer matrix, which holds the carbon fiber-together is formed by the adhesive or curing properties of various polymers, such as polyvinyl alcohol (PVA) or epoxy resin. The load-bearing capacity of carbon fiber is influenced by factors such as fiber length, diameter, density, matrix,

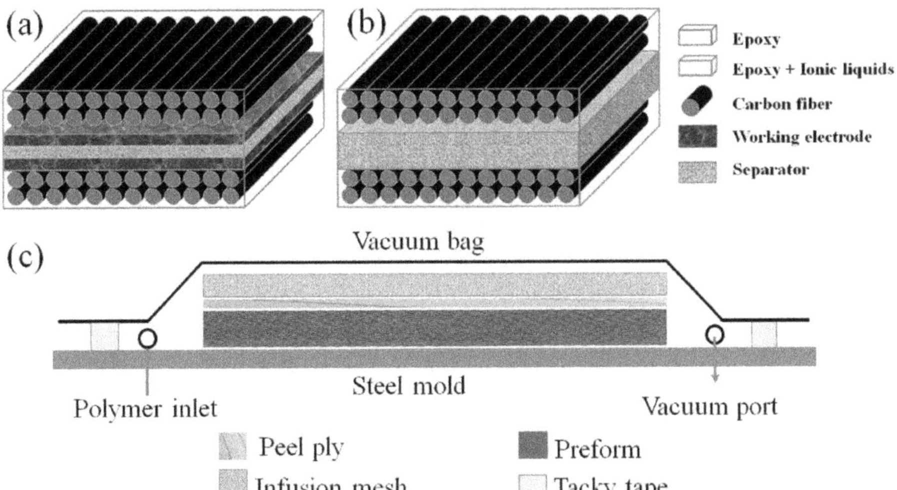

FIGURE 3.2 The different designs of (a) integrated structural supercapacitors and (b) lamination structural supercapacitors. (c) Schematic of the vacuum-assisted resin transfer mold (VARTM) process.

and orientation. As a result, mechanical properties of the composite are closely tied to and evaluated by the volume fraction (V_f) of carbon fibers. The carbon fiber volume fraction can be determined using equation (11) [35]:

$$V_f = \frac{m_f}{m_p} \cdot \frac{\rho_p}{\rho_f} \tag{11}$$

where m_f and m_p are the masses of the carbon fiber and polymer, respectively; ρ_f and ρ_p are the densities of the carbon fiber and polymer, respectively.

The carbon fiber volume fraction can be determined through various methods in combination with equation (11). In the structure, the value of m_f represents the mass of carbon fiber used, with ρ_f provided the manufacturer's specifications. The density ρ_p of the polymer after being cured is determined by simultaneously combining Archimedes's principle in water and a helium pycnometer. Consequently, the volume fraction is calculated by measuring the amount of polymer added to the composite structure. The methods used to determine the mass of the polymer include:

  i Wet-chemical analysis: This method uses concentrated sulfuric acid (H_2SO_4) at high temperatures to break down the polymer [35]. After the decomposition process, the remaining mass is the carbon fiber, allowing the polymer mass to be calculated by subtracting the carbon fiber mass (mf) from the total mass of the reinforced composite.

  ii Thermogravimetric analysis (TGA): In this method, the reinforced composite is analyzed under a N_2 gas atmosphere at high temperatures to fully decompose the polymer [35]. The carbon fiber volume fraction V_f is calculated by analyzing the first derivative of the degradation curve. The decomposition point, which marks the transition between the polymer and carbon fiber degradation, is used to determine the Vf value.

In structural energy storage devices, a carbon fiber volume fraction of approximately 60% is recommended. This allows enough space within the composite matrix for ions in the electrolyte to move efficiently, reducing the interface between the carbon fibers and electrolyte and enhancing electrical conductivity [36].

2.3 CARBON-BASED ELECTRODES

The capacitance of a structural supercapacitor device is directly influenced by the electrodes and their interfaces, as it depends on the amount of charge stored in the structure. During the electrochemical process, ions are separated from the electrolyte and move toward the surface of the opposite electrode. In EDLCs, charge is stored on the surface of electrodes through an electrostatic mechanism. Carbon-based materials are commonly used as working electrodes in EDLC devices due to high electric conductivity, strong redox activity, high mechanical strength, large surface area, and long cycling performance.

2.3.1 Carbon Fibers

The supercapacitor electrodes often endure significant mechanical loads, leading to a deterioration in their properties, consequently, a decline in device performance. Activated carbons, commonly used in conventional supercapacitors, have limited tensile strength (4.6–6.4 MPa) [37], and insufficient graphitization [38], making them unsuitable as electrodes for structural supercapacitors due to their exceptional mechanical strength (3.5-4 GPa) [39] and excellent electrical conductivity, which meet the stringent demands of these devices.

A basic structural supercapacitor consists of a separator layer (glass/aramid fiber fabric) sandwiched between two carbon fiber fabric working electrode layers, impregnated, and cured with a SPE composed of epoxy resin and IL-based electrolyte. However, compared to other carbon-based materials, carbon fiber structural electrodes exhibit a lower specific capacitance (11 mFg^{-1}), which is attributed to their relatively small specific surface area (0.2 m^{2}g^{-1}) [40]. Since the number of electrical charges stored in energy storage devices increases is proportional to the electrode's surface area, enhancing the electrochemical performance of carbon fiber electrodes by increasing their specific surface area while preserving their mechanical properties is a major challenge in structural supercapacitor research. Typically, improvements in the redox activity and surface area of carbon fiber are achieved through two methods: chemical activation of commercial carbon fibers and surface-deposited commercial carbon fibers (Figure 3.3).

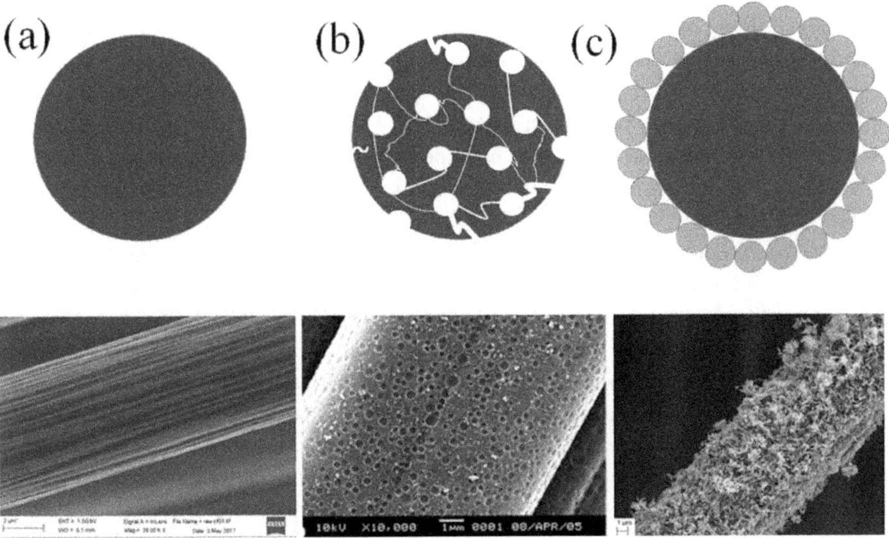

FIGURE 3.3 The cross-sectional and SEM images of (a) raw carbon fibers [41], (b) activated carbon fibers with porosity and channel [42], and (c) surface-deposited carbon fibers with functional materials coated on the surface [43].

2.3.1.1 Activated Carbon Fibers

Thermochemical activation of commercial carbon fibers is the process of modifying the structure using chemicals, such as gases (e.g., steam, CO_2, NH_3) or non-gases like inorganic acids, alkalis, salts at high temperatures (typically above 600°C) under an atmosphere or N_2 or Ar gas. This process enhances the performance of the carbon fiber by creating porosity and conductive channels, both on the surface and within the fiber, through reactions carbon and KOH as reaction (1) [44, 45, 46]. The result of thermochemical activation is a significant increase in specific surface area (up to100-fold) and electrochemical capacitance (up to 50-fold) with only a slight reduction in the strength of the fibers. However, precise control over the activation temperature and processing time is critical to avoid an excessive degradation of the fiber strength. During heat treatment, the initial phase involves the breakdown of cross-linking, which subsequently reduces tensile strength as the treatment progresses further [47].

$$6KOH + 2C \rightarrow 2K + 3H_2 + 2K_2CO_3 \tag{1}$$

$$H_2SO_4 + C \rightarrow SO_2 + CO_2 + H_2O \tag{2}$$

Wet-chemical activation of commercial carbon fibers involves treatment by immersing them in strong oxidants, acids or alkalis solutions or organic solvents like concentrated sulfuric acid, at high temperatures ranging from 100 to 200 °C, as per reaction (2) [48, 49]. This method is straightforward and does not require complex equipment. The carbon fiber is corroded or undergoes corrosion, forming interconnected pores or cave-like structures. Yu and coworkers observed that treatment with a mixture of $KMnO_4$, $NaNO_3$ H_2SO_4, and HI-acetic acid solution increased the specific surface area of carbon fibers by 15 times (from 6 to 92 m^2. g^{-1}) and improved specific capacitance by 23 times (from 0.6 to 14.2 $F.cm^{-3}$), with only a slight reduction in fiber strength [50]. Furthermore, the increased surface area following chemical activation enhances the adhesion between the electrodes and SPE, contributing to improved mechanical properties in structural supercapacitors [51].

Physical activation of commercial carbon fibers employs various techniques, including irradiation with ultraviolet (UV) light, plasma, gamma rays, or bombarded the surface by electron beams under gas atmospheres (e.g., air, O_2, N_2, mixed gases, organic gases...) [52]. During this process, a chemical reaction occurs between the carbon fiber surface and the activating gases, creating nano-pore structures and functional groups, that enhance both specific surface area and capacitance. However, the physical techniques typically operate under atmosphere conditions, necessitating the cleaning of the reaction chamber using vacuum methods before introducing gases. Consequently, this method requires modern and expensive equipment. Additionally, safety considerations must be considered particularly regarding the use of gamma-rays sources [53, 54]. Moreover, physical activation is often unsuitable for preserving the graphitic structure of carbon fibers, which negatively impacts their mechanical properties. Specifically, as noted by Qian et al [55], the tensile strength of carbon fibers subjected to physical activation significantly decreased due to severe pitting on the fiber surface.

Electrochemical activation of commercial carbon fibers is achieved by immersing carbon fibers in the aqueous electrolyte, typically strong acids and salts (e.g., sulfuric acid (H_2SO_4), ammonium sulfate ((NH_4)$_2SO_4$) solution, ...) [56, 57]. In this process, carbon fibers serve as the anode electrode, while platinum is commonly used as the cathode. During the electrochemical reaction, the carbon fiber electrode dissolved under applied potential conditions, allowing anion (e.g., SO_4^{2-}) from the electrolyte to intercalate between the layers on the carbon fiber surface and subsequently peeled off. The peeling rate is influenced by the electrolyte concentration and the voltage of electrodes. According to Bandar's report, the performance of the activated carbon fiber fabric improved by 205 times compared to untreated fibers due to this activation process [58]. However, the carbon fibers severely corroded during activation process, especially when high potentials, large anodic currents, or concentrated acid solutions are used [59]. Therefore, effectively activating carbon fibers while preserving their structural integrity remains a challenge, requiring precise control of the electrochemical activation conditions.

2.3.1.2 Surface-Deposited Carbon Fibers

Enhancing the surface area and functional groups of carbon fibers through activation has significantly improved the energy storage ability of structural supercapacitors. However, the treatment process can alter the mechanical properties of carbon fibers, potentially affecting the mechanical durability of the structural supercapacitors. To enhance the electrochemical storage performance, the surface of the carbon fiber is often coated with carbon- or inorganic-based materials that exhibit high surface area and strong redox activity. Carbon-based materials, in particular, have garnered considerable attention from researchers for modifying the surface of carbon fiber in structural supercapacitors, as they combine the advantages of two carbon-derived materials. This synergy at the interface between two materials enhances the stability of electrical conduction in working electrodes.

2.3.1.2.1 Carbon Aerogel-Modified Carbon Fiber Fabrics

Multifunctional carbon fiber fabrics with the porous monolith structure are formed through the integration of carbon aerogel on their surface. A layer of carbon aerogel is developed on the carbon fiber surface to increase the surface area while maintaining the mechanical stability of carbon fiber. In a study by Milo and coworkers, a structural supercapacitor was prepared using the carbon aerogel coated on the surface of the carbon fiber fabric [60]. The SPE used was a combination of 10% IL and polyethylene glycol diglycidyl ether (PEGDGE). The surface area of the working electrode increased approximately 100-fold due to the incorporation of carbon aerogel, which led to an enhancement in the specific capacitance, improving nearly tenfold from 6.52 (uncoated) to 602.5 mF g^{-1}. The mechanical performance, as measured by shear strength, also improved, with the uncoated sample achieving 5.36 MPa at 47.2% volume fraction, compared to 8.71 MPa at 42% volume fraction with a carbon aerogel-modified working electrode.

2.3.1.2.2 Graphene Coated on Carbon Fibers Graphene nanoplatelets are coated on the surface of carbon fiber fabric, serving as the working electrode in a structural supercapacitor, with PEGDGE-IL embedded in epoxy matrix acting as both the electrolyte and separator, as demonstrated in Xoan et al.'s work [61]. The specific capacitance of the structural supercapacitor, evaluated through cyclic voltammetry, was found to triple, increasing from 3 to 9 mF g^{-1} compared to uncoated carbon fiber fabric. Despite changes in the working electrode and the components of the epoxy

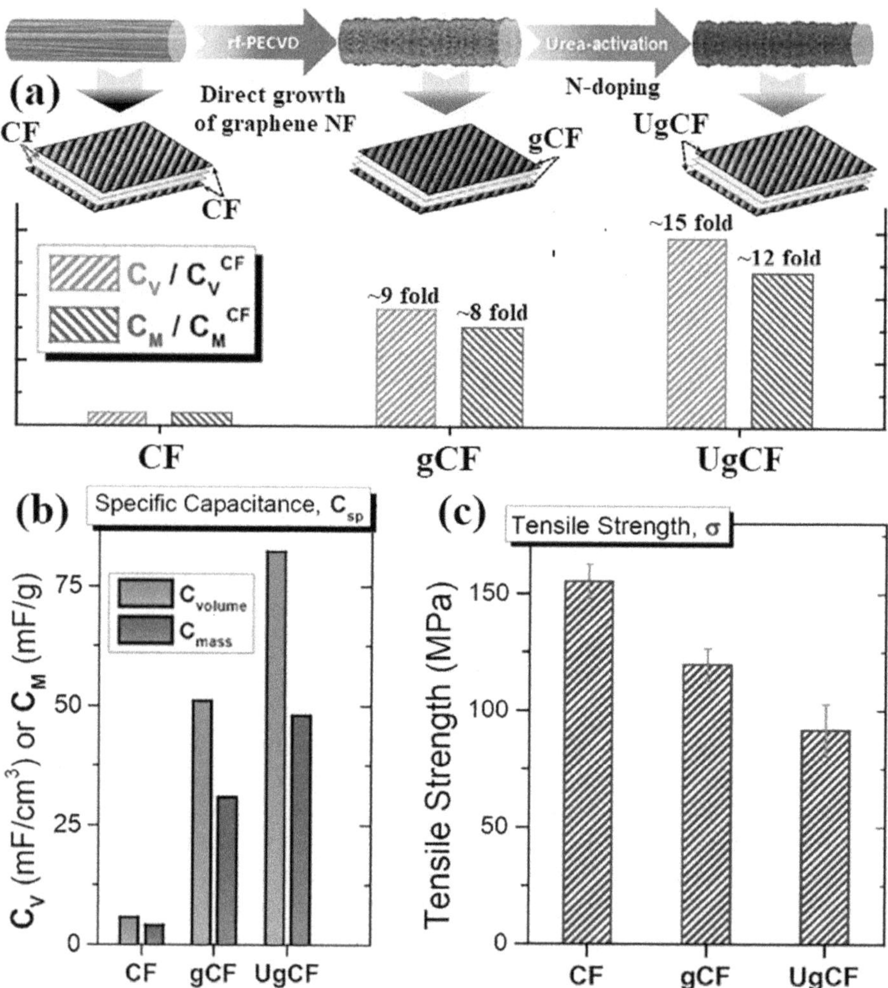

FIGURE 3.4 (a) Schematic of the preparation of three structural supercapacitor devices by carbon fiber (CF) fabric, graphene-deposited, and nitrogen-doped graphene-deposited carbon fibers. (b) and (c) The results of specific capacitance and tensile strength of CF, gCF, and UgCF devices. Reproduced from ref. [62].

https://doi.org/10.1016/j.electacta.2021.137746

resin mixture, the Young's modulus of all structural supercapacitors remained nearly constant.

However, the tensile strength showed a significant decrease—approximately by twofold due to the formation of channels in an epoxy matrix. Specifically, in the PEGDGE-IL epoxy, the tensile strength of graphene nanoplatelets-coated carbon fiber fabric was about 10% lower than that of the uncoated sample, primarily due to variations in volume fractions.

In a study by Pagona and coworkers, the specific capacitance of graphene-deposited carbon fibers (gCF ~30.8 mF g^{-1}) was enhanced approximately by ninefold, and increased by 14 times with nitrogen-doped graphene-deposited carbon fibers (UgCF ~40.8 mF.g^{-1}), compared to uncoated carbon fibers (CF ~4 mF.g^{-1}) [62]. However, this enhancement in electrochemical capacity also resulted a reduction of tensile strength because of temperature and chemicals treatments involved in the deposition process. The tensile strength of CF, gCF, and UCF structural devices decreased to 155, 120, and 90 MPa, respectively (Figure 3.4). Similarly, Alejandro et al. reported a decrease in tensile strength from 21 MPa to 9 MPa with graphene nanoparticle-deposited carbon fibers [63].

2.4 SOLID POLYMER ELECTROLYTES (SPEs)

The electrolyte is a crucial component of the supercapacitor structure, as it influences ionic size, electrical conductivity, operating potential window, and working temperatures range, all of which determining the supercapacitor's performance. In a structural supercapacitor, the transition from a liquid electrolyte to a solid polymer electrolyte is crucial for enhancing the mechanical stability. SPE materials have significantly advanced electrolyte performance due to their flexibility, mechanical strength, and suitability for various applications, including grid-scale energy systems, portable electronics, wearable electronics, micro-electronics, and especially flexible electronic devices that require design flexibility and safety. The shift from a liquid to a SPE simplifies cell design and enhance safety and durability. In some studies, a SPE also function as a separator between the two working electrodes.

The suitable electrolyte for Energy Storages (ESs) devices must exhibit a wide voltage window, high electrochemical stability, high ionic concentration, high mechanical stability, low resistivity, low volatility, low toxicity, low cost, and so on. Although SPEs provide high mechanical strength, their primary drawback includes relatively low ionic conductivity and poor contact surface area between the electrolyte and working electrode materials compared to traditional liquid electrolytes. This is often due to the distribution of the polymer matrix. SPEs typically consist of two core components: a polymer matrix and an IL. Core several polymers for SPE applications, such as poly(ethylene oxide) (PEO) [65], poly(vinyl alcohol) (PVA) [66], poly(acrylonitrile) (PAN) [67], polyvinylidene fluoride (PVdF) [68] and epoxy resins.

Besides the polymer matrix, ILs, also known as room-temperature molten salts, typically consist of a large asymmetric organic cation and an inorganic or an organic anion, giving them a low melting point (below 100 °C) [69]. When filled into a polymer matrix, ILs form small continuous channels that enhance ionic conduction. Commonly used ILs for ESs applications include imidazolium (EMIM$^+$) [70],

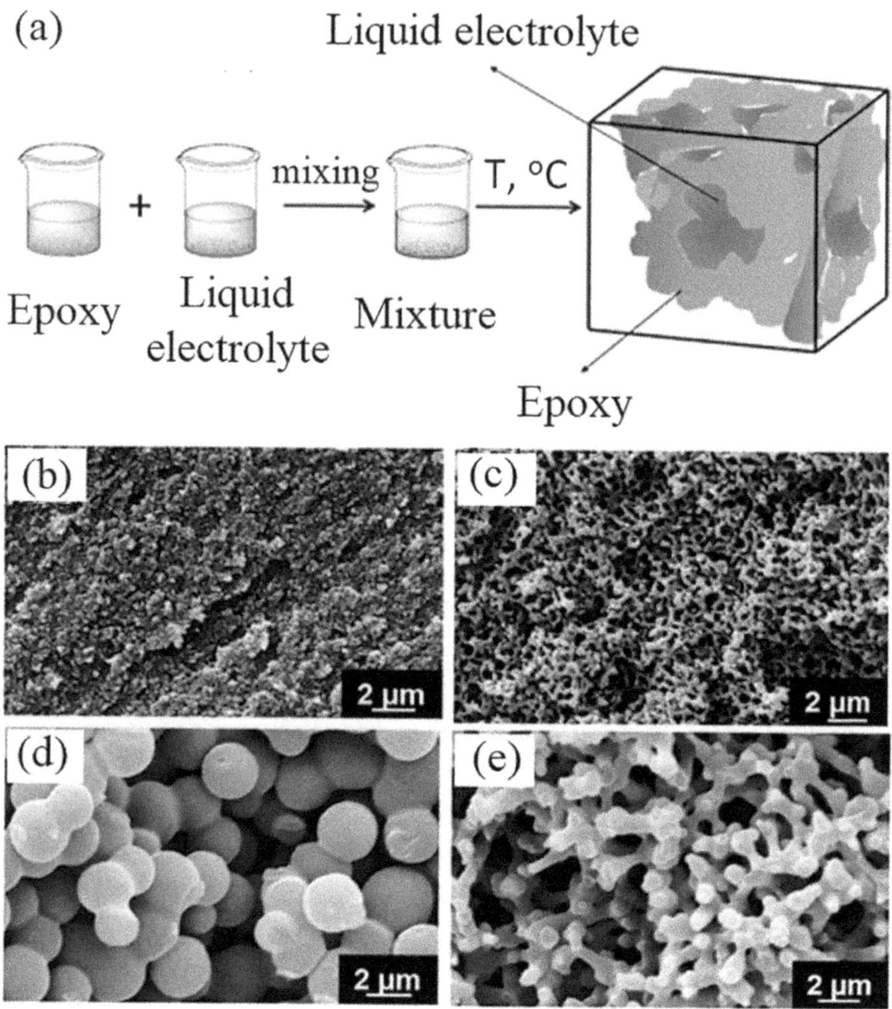

FIGURE 3.5 (a) Schematic of the internal structure of solid polymer electrolytes (SPE) from the combination of epoxy resins and ionic liquids (ILs). SEM images of epoxy resins/ILs (b) MVR444/50; (c) VTM266/50; (d) MTM57/50; and (e) MVR444/30 (all samples are measured after extraction of the electrolyte by ethanol). Reproduced from ref. [64].

pyrrolidinium (PYR_{14}^+) [71], ammonium ($DEME^+$) [72], sulfonium (Me_3S^+) [73], phosphonium (P_{2224}^+) [74] cations, while typical anions include tetra-fluoroborate (BF_4^-) [16], bis(trifluoromethanesulfonyl)imide ($TFSI^-$) [75], bis(fluorosulfonyl)imide (FSI^-) [76] and hexafluorophosphate (PF_6^-) [77]. The ILs are selected for their high ionic conductivity, broad electrochemical window, low vapor pressure and high thermal stability. The choices of polymer and ILs must ensure good solubility to create favorable conditions for the formation and uniform distribution of ionic channels within the matrix.

As discussed earlier, low ionic conductivity and poor interfacial properties are major drawbacks of SPEs, limiting their practical application. For instance, SPEs based on poly(ethylene oxide) (PEO) typically exhibit ionic conductivities in the range of 10^{-8}–10^{-4} S cm^{-1} at room temperature, which is significantly below the desired threshold of 10^{-3} S cm^{-1} for practical applications [78]. Recent studies on SPEs focus on addressing the critical challenge of enhancing ionic conductivity. Experimental evidence has shown that the ionic conductivity of SPEs is influenced by both the ionic size and the ability of the IL to migrate through the polymer matrix. Ionic transport primarily occurs within the amorphous regions of polymer and is governed by the segmental motion of the polymer chain [79]. Therefore, an amorphous phase in the SPE is crucial for achieving higher ionic conductivity.

One of the most effective methods for increasing the amorphous content and reducing crystallinity is the addition of low molecular weight plasticizers to the polymer electrolyte system [80]. While this increases the conductivity, it also comes with a trade-off, as plasticizers tend to compromise the mechanical properties of the electrolyte. Another promising approach involves incorporating nanosized fillers, such as CuO [81], SiO_2 [82], TiO_2 [83], or Al_2O_3 [84] into the host polymer matrix. This method improves both mechanical stability and ionic conductivity, making it an attractive option for enhancing SPEs. However, these traditional polymers and ILs, systems still face the challenge of a trade-off between conductivity and mechanical strength, as increased conductivity often results in a reduction of the modulus. This limits their applicability in structural supercapacitor systems, where both high conductivity and mechanical durability are essential.

The primary challenge in SPE design is achieving high modulus without compromising ionic conductivity. Recent research has have demonstrated that epoxy/IL-based electrolytes exhibit excellent mechanical properties alongside attractive ionic conductivities, ranging from 10^{-3} to 10^{-5} S cm^{-1} at room temperature [85–87]. Figure 3.5a illustrates the typical structure of epoxy-based solid electrolytes, showing the distribution of ionic channels within the epoxy matrix.

In another study, Kwon and coworkers demonstrated the performance of Al_2O_3 nanowires in enhancing both conductivity and mechanical properties of SPEs [88]. A SPE prepared using epoxy resin and BMIMTFSI/LiTFSI-IL exhibited a Young's modulus of 0.5 GPa and ionic conductivity of approximately 10^{-4} S.cm^{-1}. Notably, the addition of robust inorganic Al_2O_3 nanowires significantly improved the mechanical strength (to about 1 GPa) and boosted ionic conductivity ((to around 2.9×10^{-4} S.cm^{-1}) at 25°C. The structural supercapacitor using this SPE showed high electrochemical performance, with a specific power of 382 W kg^{-1} at the specific energy of 75 Wh.kg^{-1}, and 9.3 kW.kg^{-1} at the specific energy of 44 Wh.kg^{-1}. Additionally, a high specific capacitance of approximately 90 F.g^{-1} was achieved. To further enhance energy storage capacity and mechanical performance, Muralidharan N et al. [89] successfully developed a supercapacitor using carbon nanotube (CNT) reinforced electrodes, along with Kevlar or fiberglass, super-sap CCR (Clear Casting Resin) epoxy resin, lithium tetrafluoroborate (LiBF$_4$), and 1-butyl-3-methyl imidazolium tetrafluoroborate (BMIBF$_4$). The inclusion of CNTs and Kevlar significantly reinforced the supercapacitor, resulting in elastic modulus of over 5 GPa, mechanical strength greater than 85 MPa, and energy density up to 3 mWh.kg^{-1}.

Recently, Westover and coworkers [90] designed an IL-containing epoxy matrix suitable for use in structural supercapacitor capable of both storing and releasing energy under tensile stresses and extreme conditions, such as water immersion and high humidity (Figures 3.5a–e). The electrolytes used in this study consisted of bisphenol A/F, 1-butyl-3-methy-limidizolium-tetra-fluoroborate (BMIBF$_4$) and lithium tetra-fluoroborate (LiBF$_4$). These structural composites maintain their full energy storage capability (5–8 Wh.kg^{-1}) and power density close to 4 kW.kg^{-1} under tensile stresses of approximately 1.1 MPa. Notably, cycling tests revealed that the supercapacitors retain nearly 100% of their energy density over 4,000 cycles. These findings are promising and offer new opportunities for developing energy storage systems (ESs) that can operate effectively in extreme environments.

3 INTEGRATED STRUCTURAL SUPERCAPACITOR DEVELOPMENT

Three designs for structural supercapacitors labeled (1), (2), and (3), are illustrated in Figure 3.6. Each design is composed of one or more single cells, which include electrodes, separators, and electrolyte layers. Design (1) features a glass/ aramid nanofibers (ANFs) separator layer placed between two carbon fiber electrode layers all encapsulated in a SPE. The carbon fiber electrodes are connected to the negative and positive electrodes of an external circuit, as depicted in Figure 3.6a. This design represents a straightforward structural supercapacitor consisting of a single cell.

In design (2) (Figure 3.6b), the structure consists of alternating layers of carbon electrodes and separator layers, sandwiched between two carbon fiber layers on the upper and lower surfaces of the device. All the layers are securely fixed and filled with a single type of SPE. The upper and lower electrodes are connected to the positive and negative terminals of an external circuit. In this configuration, the equivalent capacitance (C_{eq}) of the device is calculated using equation (12) for the capacitor in series. Although the series design results in a lower capacitance compared to design (1) it allows for an enhanced operating voltage.

$$\frac{1}{C_{eq}} = \frac{1}{C_1} + \frac{1}{C_2} + \ldots + \frac{1}{C_{n-1}} \tag{12}$$

$$C_{eq} = C_1 + C_2 + \ldots + C_{n-1} \tag{13}$$

where C_1, C_2, ..., C_{n-1} are component capacitances; n is the number of working electrodes (number of carbon fiber layers).

In design (3) (Figure 3.6c), the device's preparation mirrors that of design (2) but the connection method of the external circuit differs. The carbon fiber fabric layers at odd positions (layer 1, 3, …) are connected to positive electrodes while even positions (layer 2, 4, …) connect to negative electrodes. The C_{eq} of this configuration is evaluated using equation (13) for capacitors in parallel. This parallel arrangement allows the total capacitance to equal the sum of the individual capacitances. In Yu's research, the voltage of two cells in series was doubled (from 1.8 V to 3.6 V) while

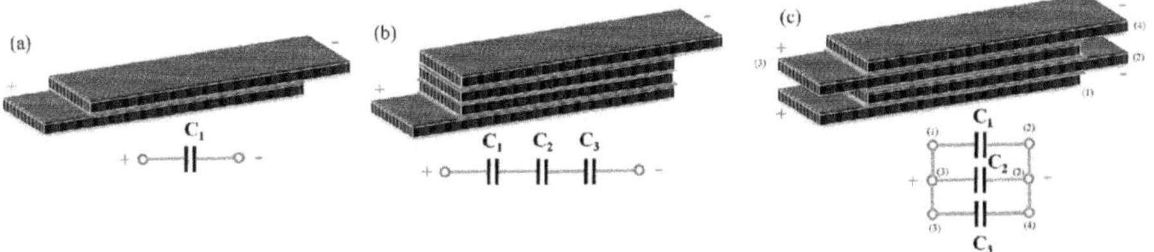

FIGURE 3.6 Three designs of a structural supercapacitor system with working electrodes, separators, and electrolyte layers. (a) Design (1) with a single cell consisting of two carbon fabric layers and one separator layer. (b) Design (2) with three cells in series consisting of four carbon fabric layers and three separator layers. (c) Design (3) with three cells in series consisting of four carbon fabric layers and three separator layers.

maintaining the same capacitance as single cell. The equivalent specific capacitance of two cells in parallel is enhanced approximately two times during charge-discharge conditions. Consequently, design (3) achieves the highest equivalent capacitance compared to designs (1) and (2). However, the need for carbon fiber fabric layers contacts with either the positive or negative electrode complicates the design. In the VARTM process, the electrical contacts are easily applied to the upper and lower surfaces in designs (1) and (2), while design (3) requires advanced technology [91].

4 CONCLUSIONS

The structural energy storage device is developed from the combination of electrochemical supercapacitors and the mechanical stability of the same device by carbon-derived materials. Carbon fibers with high mechanical properties, electrochemical performance, and high electrical conductivity are not only applied as working electrodes but also as bearing structures in structural supercapacitors. The surface modification of carbon fibers with carbon-derived materials such as carbon aerogel, and graphene is performed to improve the specific capacitance of the device. SPEs possess the mechanical properties and ionic conductivity for the electrochemical process of supercapacitors. The ratio of polymer and IL is an important factor in deciding the electrical and mechanical performances. Glass fiber fabric is normally used as the separator material to ensure the mechanical performance of the device. Although structural supercapacitor devices have low capacitance, they can be used in large applications such as electrical vehicles, aircrafts, and space fields.

5 ACKNOWLEDGMENTS

The authors would like to acknowledge Ho Chi Minh City University of Technology (HCMUT), VNU-HCM, for supporting this research.

REFERENCES

[1] Kamel AA, Rezk H, Abdelkareem MA. Enhancing the operation of fuel cell-photovoltaic-battery-supercapacitor renewable system through a hybrid energy management strategy. *International Journal of Hydrogen Energy,* 2021;46:6061–6075.

[2] Karmaker AK, Rahman MM, Hossain MA, Ahmed MR. Exploration and corrective measures of greenhouse gas emission from Fossil fuel power stations for Bangladesh. *Journal of Cleaner Production*, 2020;244:118645.

[3] Pereira T, Guo Z, Nieh S, Arias J, Hahn HT. Energy storage structural composites: A review. *Journal of Composite Materials*, 2009;43:549–560.

[4] Budde-Meiwes H, Drillkens J, Lunz B, Muennix J, Rothgang S, Kowal J, Sauer D. A review of current automotive battery technology and future prospects. *Proceedings of the Institution of Mechanical Engineers, Part D: Journal of Automobile Engineering*, 2013;227(5):761–776.

[5] Christen T, Carlen M. Theory of Ragone plots. *Journal of Power Sources*, 2000;91(2):210–216.

[6] Yu D, Goh K, Wang H, Wei L, Jiang W, Zhang Q, Dai L, Y Chen. Scalable synthesis of hierarchically structured carbon nanotube-graphene fibres for capacitive energy storage. *Nature Nanotechnology* 2014;9:555–562.

[7] Choi C, Kim JH, Sim HJ, Di J, Baughman RH, Kim SJ. Supercapacitors: Microscopically buckled and macroscopically coiled fibers for ultra-stretchable supercapacitors. *Advanced Energy Materials*, 2017;7:1602021.

[8] Fic K, Meller M, Menzel J, Frackowiak E. Around the thermodynamic limitations of supercapacitors operating in aqueous electrolytes. *Electrochimica Acta*, 2016;206:496–503.

[9] Aravindan V, Gnanaraj J, Lee YS, Madhavi S. Insertion-type electrodes for nonaqueous Li-Ion capacitors. *Chemical Reviews*, 2014;114:11619–11635.

[10] Deng D. Li-ion batteries: Basics, progress, and challenges. *Energy Science and Engineering*, 2015;3(5):385–418.

[11] Das HS, Tan CW, Yatim AHM. Fuel cell hybrid electric vehicles: A review on power conditioning units and topologies. *Renewable and Sustainable Energy Reviews*, 2017;76:268–291.

[12] Dong W, Xie M, Zhao S, Qin Q, Huang F. Materials design and preparation for high energy density and high power density electrochemical supercapacitors. *Materials Science and Engineering: R: Reports.Volume*, 2023;152:100713.

[13] Khan A, Saxena KK. A review on enhancement of mechanical properties of fiber reinforcement polymer composite under different loading rates. *Materials Today: Proceedings*, 2022;56:2316–2322.

[14] Lázaro VC, Lluc CC. Analysis of the future of mobility: The battery electric vehicle seems just a transitory alternative. *Energies*, 2022;15(23):9149.

[15] Berjoza D, Jurgena I. Influence of batteries weight on electric automobile performance. *In Proceedings of the 16th International Scientific Conference Engineering for Rural Development*, Jelgava, Latvia, 2017:24–26.

[16] Galvin R. Energy consumption effects of speed and acceleration in electric vehicles: Laboratory case studies and implications for drivers and policymakers. *Transportation Research Part D: Transport and Environment*, 2017;53:234–248.

[17] Tolouei R, Titheridge H. Vehicle mass as a determinant of fuel consumption and secondary safety performance. *Transportation Research Part D: Transport and Environment*, 2009;14:385–399.

[18] Javaid A, Khalid O, Shakeel A, Noreen S. Multifunctional structural supercapacitors based on polyaniline deposited carbon fiber reinforced epoxy composites. *Journal of Energy Storage*, 2021;33:102168.

[19] Conway BE. *Electrochemical Capacitors: Scientific Fundamentals and Technological Applications*. Kluwer Academic/Plenum, New York; 1999.

[20] Shao H, Wu YC, Lin Z, Tabernaab PL, Simon P. Nanoporous carbon for electrochemical capacitive energy storage. *Chemical Society Reviews*, 2020;49(10):3005–3039.

[21] Xu Y, Pei S, Yan Y, Wang L, Xu G, Yarlagadda S, Chou T. High-performance structural supercapacitors based on aligned discontinuous carbon fiber electrodes and solid polymer electrolytes. *ACS Applied Materials & Interfaces*, 2021;13 (10):11774–11782.

[22] Mathis TS, Kurra N, Wang X, Pinto D, Simon P, Gogotsi Y. Energy storage data reporting in perspective-guidelines for interpreting the performance of electrochemical energy storage systems. *Advanced Energy Materials*, 2019; 9 (39):1902007.

[23] Zhang S, Pan N. Supercapacitors performance evaluation. *Advanced Energy Materials*, 2015; 5 (6):1401401.

[24] Evgeny S, Yunfu O, Juan JT, Federico S, Carlos G, Rebeca M, Juan JV. *Scientific Reports* 2018;8:3407.

[25] Javaid A, Ho KKC, Bismarck A, Shaffer MSP, Steinke JHG, Greenhalgh ES. Multifunctional structural supercapacitors for electrical energy storage applications. *Journal of Composite Materials,* 2014;48(12):1409–1416.

[26] Muralidharan N, Teblum E, Westover AS, Schauben D, Itzhak A, Muallem M, Nessim GD, Pint CL. *Scientific Reports,* 2018;8:17662.

[27] Natasha S, Hui Q, Milo SPS, Joachim HGS, Emile SG, Paul C, Anthony K, Alexnder B. Structural composite supercapacitors. *Composites Part A: Applied Science and Manufacturing.* 2013;46:96.

[28] Zhou H, Li H, Li L, Liu T, Chen G, Zhu Y, Zhou L, Huang H. Structural composite energy storage devices—A review. *Material Today Energy,* 2022;24:100924.

[29] Senokos E, Ou Y, Torres JJ, Sket F, González C, Marcilla R, Vilatela JJ. Energy storage in structural composites by introducing CNT Fiber/Polymer electrolyte interleaves. *Scientific Reports,* 2018;8:3407.

[30] Reece R, Lekakou C, Smith PA. A high-performance structural supercapacitor. *ACS Applied Materials & Interfaces,* 2020;12:25683–25692.

[31] Drechsler K, Heine M, Mitschang P. 12.1 Carbon fiber reinforced polymers. In: Jäger H, Frohs W, editors. *Industrial Carbon and Graphite Materials.* 1st ed., Volume I. Wiley, Hoboken, NJ; 2021, pp. 697–739.

[32] Abe K, Ohya Y. An investigation of flow fields around flanged diffusers using CFD. *Journal of Wind Engineering and Industrial Aerodynamics,* 2004;92(3–4):315–330.

[33] Dominik B, Jens S, Dirk H. Flow rate control during vacuum-assisted resin transfer molding (VARTM) processing. *Composites Science and Technology,* 2006;66:2265–2271.

[34] Johnson RJ, Pitchumani R. Active flow control in a VARTM Process using localized induction heating. *In: Proceedings of the FPMC7,* 2004:247–252.

[35] Dominik G, Manuel O, Iman T. Determination of fiber volume fraction of carbon fiber-reinforced polymer using thermogravimetric methods. *Polymer Testing,* 2019;75:358–366.

[36] Chen X, Wen K, Wang C, Cheng S, Wang S, Ma H, Tian H, Zhang J, Li X, Shao J. Enhancing mechanical strength of carbon fiber-epoxy interface through electrowetting of fiber surface. *Composites Part B: Engineering,* 2022;234:109751.

[37] Kumar N, Kim SB, Lee SY, Park SJ. Recent Advanced Supercapacitor: A Review of Storage Mechanisms, Electrode Materials, Modification, and Perspectives. *Nanomaterials,* 2022;12:3708.

[38] Zhao Y, Zhang X. In situ activation graphitization to fabricate hierarchical porous graphitic carbon for supercapacitor. *Scientific Reports,* 2021;11:6825

[39] Qi G, Wu J, Yao B, Cui Q, Ren L, Zhang B, Du S. High modulus carbon fiber based composite structural supercapacitors towards reducing internal resistance and improving multifunctional performance. *Composites Science and Technology,* 2024;254:110670.

[40] Wu KJ, Young WB, Young C. Structural supercapacitors: A mini-review of their fabrication, mechanical & electrochemical properties. *Journal of Energy Storage,* 2023;72:108358.

[41] Qiu J, Li J, Yuan Z, Zeng H, Chen X. Surface modification of carbon fibres for interface improvement in textile composites. *Applied Composite Materials*, 2018;25:853–860.

[42] Chen Y, Wu Q, Pan N, Gong J, Pan D. Rayon-based activated carbon fibers treated with both alkali metal salt and Lewis acid. *Microporous and Mesoporous Materials*, 2008;109(1–3):138–146.

[43] Xu Y, Lu W, Xu G, Chou TW. Structural supercapacitor composites: A review. *Composites Science and Technology*, 2021;204:108636.

[44] Chen S, Qiu L, Cheng HM. Carbon-based fibers for advanced electrochemical energy storage devices. *Chemical Reviews*, 2020;120:2811–2878.

[45] Sankar S, Lee H, Jung H, Kim A, Ahmed ATA, Inamdar AI, Kim H, Sejoon Lee, Im H, Kim DY. Ultrathin graphene nanosheets derived from rice husks for sustainable supercapacitor electrodes. *New Journal of Chemical*, 2017;41:13792.

[46] Qian H, Diao H, Shirshova N, Greenhalgh ES, Steinke JG, Shaffer MS, Bismarck A. Activation of structural carbon fibres for potential applications in multifunctional structural supercapacitors. *Journal of Colloid Interface and Science*, 2013;395:241–248.

[47] Xiao H, Lu Y, ZhaoW, Qin X. The effect of heat treatment temperature and time on the microstructure and mechanical properties of PAN-based carbon fibers. *Journal of Materials Science*, 2014;49:794–804.

[48] Chen S, Ma W, Cheng Y, Weng Z, Sun B, Wang L, Chen W, Li F, Zhu M, Cheng HM. Scalable non-liquid-crystal spinning of locally aligned graphene fibers for high-performance wearable supercapacitors. *Nano Energy*, 2015;15:642–653.

[49] Sharma M, Gao S, Mäder E, Sharma H, Wei LY, Bijwe J. Carbon Fiber surfaces and composite interphases. *Composites Science and Technology*, 2014;102:35–50.

[50] Yu DS, Zhai SL, Jiang WC, Goh KL, Wei L, Chen XD, Jiang RR, Chen Y. Transforming Pristine carbon fiber tows into high performance solid-state fiber supercapacitors. *Advanced Materials*, 2015;27:4895–4901.

[51] Yao L, Zheng K, Koripally N, Eedugurala N, Azoulay JD, Zhang X, Nga Ng T, Structural pseudocapacitors with reinforced interfaces to increase multifunctional efficiency. *Science Advances*, 2023;9:eadh0069.

[52] Mädler J, Richter B, Wolz DSJ, Behnisch T, Böhm R, Jäger H, Gude M, Urbas L. Hybride semi-parametrische modellierung Der Thermooxidativen stabilisierung von PAN-Precursorfasern. *Chemie Ingenieur Technik*, 2022;94:889–896.

[53] Okajima K, Ohta K, Sudoh M. Capacitance behavior of activated carbon fibers with oxygen-plasma treatment. *Electrochimica Acta*, 2005;50:2227–2231.

[54] Xiao H, Lu Y, Wang M, Qin X, Zhao W, Luan J. Effect of Gamma-irradiation on the mechanical properties of polyacrylonitrile-based carbon fiber. *Carbon*, 2013;52:427–439.

[55] Jang D, Lee ME, Choi J, Cho SY, Lee S. Strategies for the production of PAN-Based carbon fibers with high tensile strength. *Carbon*, 2022;186:644–677.

[56] Momma T, Liu X, Osaka T, Ushio Y, Sawada Y. Electrochemical Modification of active carbon fiber electrode and its application to double-layer capacitor. *Journal of Power Sources*, 1996;60:249–253.

[57] Qin T, Peng S, Hao J, Wen Y, Wang Z, Wang X, He D, Zhang J, Hou J, Cao G. Flexible and wearable all-solid-state supercapacitors with ultrahigh energy density based on a carbon fiber fabric electrode. *Advanced Energy Materials*, 2017;7:1700409.

[58] AlBatool AA, Basheer AA, Ghzzai NA, Feraih SA, Mohammed FA, Asma MA, Abdullah GA, Thamer SA, Bandar MA. Enhancing the performance of a metal-free self-supported carbon felt-based supercapacitor with facile two-step electrochemical activation. *Nanomaterials*, 2022;12(3):427.

[59] Zhu W, Zhang X, Yin Y, Qin Y, Zhang J, Wang Q. In-situ electrochemical activation of carbon fiber paper for the highly efficient electroreduction of concentrated nitric acid. *Electrochimica Acta*, 2018;291:328–334.

[60] Hui Q, Anthony RK, Emile SG, Alexander B, Milo SPS. Multifunctional structural supercapacitor composites based on carbon aerogel modified high performance carbon fiber fabric. *ACS Applied Materials & Interfaces*, 2013;5(13):6113–6122.

[61] Xoan FSR, Antonio DB, Joaquín AA, Bianca KM, María S, Alejandro U. A proof of concept of a structural supercapacitor made of graphene coated woven carbon fibers: EIS study and mechanical performance. *Electrochimica Acta*, 2021;370:137746.

[62] Abhijit G, Anastasios K, John B, Shahzad H, Pagona P. Multifunctional structural supercapacitor based on urea-activated graphene nanoflakes directly grown on carbon fiber electrodes. *ACS Applied Materials & Interfaces*, 2020;3(5):4245–4254.

[63] Joaquín AA, Xoan FSR, María S, Alejandro U. Effect of electrode surface treatment on carbon fiber based structural supercapacitors: Electrochemical analysis, mechanical performance and proof-of-concept. *Journal of Energy Storage*, 2023;59:106599.

[64] Shirshova N, Bismarck A, Carreyette S, Fontana QPV, Greenhalgh ES, Jacobsson P, Johansson P, Marczewski MJ, Kalinka G, Kucernak ARJ, Scheers J, Shaffer MSP, Steinke JHG, Wienrich M. Structural supercapacitor electrolytes based on bicontinuous ionic liquid–epoxy resin systems. *Journal of Materials Chemistry A*, 2013;1:15300–15309.

[65] Singh NK, Verma ML, Minakshi M. PEO nanocomposite polymer electrolyte for solid state symmetric capacitors. *Bulletin of Materials Science*, 2015;38(6):1577–1588.

[66] Zhang X, Wang L, Peng J, Cao P, Cai X, Li J, Zhai M. A flexible Ionic liquid Gelled PVA-Li2SO4 polymer electrolyte for semi-solid-state supercapacitors. *Advanced Materials Interfaces*, 2015;2(15):1500267.

[67] Tamilarasan P, Ramaprabhu S. Graphene based all-solid-state supercapacitors with ionic liquid incorporated polyacrylonitrile electrolyte. *Energy*, 2013;51:374–3781.

[68] Pandey GP, Liu T, Hancock C, Li Y, Sun XS, Li J. Thermostable gel polymer electrolyte based on succinonitrile and ionic liquid for high-performance solid-state supercapacitors. *Journal of Power Sources*, 2016;328:510–519.

[69] Rogers RD, Voth GA. Ionic liquid. *Account of Chemical: Research*, 2007;40(11):1077–1078.

[70] Bhise SC, Awale DV, Vadiyar MM, Patil SK, Kokare BN, Kolekar SS. Facile synthesis of CuO nanosheets as electrode for supercapacitor with long cyclic stability in novel methyl imidazole-based ionic liquid electrolyte. *Journal of Solid State Electrochemistry*, 2017;21(9):2585–2591.

[71] Tiruye GA, Munoz-Torrero D, Palma J, Anderson M, Marcilla R. All-solid state supercapacitors operating at 3.5 V by using ionic liquid based polymer electrolytes. *Journal of Power Sources*, 2015;279:472–480.

[72] Noor NA, Isa MI. Investigation on transport and thermal studies of solid polymer electrolyte based on carboxymethyl cellulose doped ammonium thiocyanate for potential application in electrochemical devices. *International Journal of Hydrogen Energy*, 2019;44(16):8298–8306.

[73] Pereira ND, Trigueiro JP, Monteiro ID, Montoro LA, Silva GG. Graphene oxide–ionic liquid composite electrolytes for safe and high-performance supercapacitors. *Electrochimica Acta*, 2018;259:783–792.

[74] Pilathottathil S, Kottummal TK, Thayyil MS, Perumal PM, Purakakath JA. Inorganic salt grafted ionic liquid gel electrolytes for efficient solid state supercapacitors: Electrochemical and dielectric studies. *Journal of Molecular Liquids*, 2018;264:72–79.

[75] Liao KS, Sutto TE, Andreoli E, Ajayan P, McGrady KA, Curran SA. Nano-sponge ionic liquid–polymer composite electrolytes for solid-state lithium power sources. *Journal of Power Sources*, 2010;195(3):867–871.

[76] Alexandre SA, Silva GG, Santamaría R, Trigueiro JP, Lavall RL. A highly adhesive PIL/IL gel polymer electrolyte for use in flexible solid state supercapacitors. *Electrochimica Acta*, 2019;299:789–799.

[77] Luo Q, Wei P, Huang Q, Gurkan B, Pentzer EB. Carbon capsules of ionic liquid for enhanced performance of electrochemical double-layer capacitors. *ACS Applied Materials & Interfaces*, 2018;10(19):16707–16714.

[78] Wang Y, Zhong WH. (2015). Development of Electrolytes towards achieving safe and high-performance energy-storage devices: A review. *ChemElectroChem*, 2015:2(1): 22–36.

[79] Wieczorek W, Raducha D, Zalewska A, Stevens JR. Effect of salt concentration on the conductivity of PEO-based composite polymeric electrolytes. *Journal of Physical Chemistry B*, 1998;102:8725–8731.

[80] Das S, Ghosh A. Effect of plasticizers on ionic conductivity and dielectric relaxation of PEO-LiClO4 polymer electrolyte. *Electrochimica Acta*, 2015;171:59–65.

[81] Johan MR, Fen LB. Combined effect of CuO nanofillers and DBP plasticizer on ionic conductivity enhancement in the solid polymer electrolyte PEO–LiCF3SO3. *Ionic*, 2010;16(4):335–338.

[82] Qi W, George C, Abdulsalam SA. Influence of nanofillers on electrical characteristics of epoxy resins insulation. In: *2010 10th IEEE International Conference on Solid Dielectrics. IEEE*, 2010:1–4.

[83] Ketabi S, Lian K. The effects of SiO2 and TiO2 nanofillers on structural and electrochemical properties of poly (ethylene oxide)–EMIHSO4 electrolytes. *Electrochimica Acta*, 2015;154:404–412.

[84] Florin C, Alexandru H, Mihail-Gabriel H. Influence of temperature on dielectric performance of epoxy nanocomposites with inorganic nanofillers. *UPB Scientific Bulletin, Series A: Applied Mathematics and Physics*, 2013;75(3):159–168.
Matsumoto K, Endo T. Confinement of ionic liquid by networked polymers based on multifunctional epoxy resins. *Macromolecules*, 2008;41(19):6981–6986.

[85] Matsumoto K, Endo T. Synthesis of ion conductive networked polymers based on an ionic liquid epoxide having a quaternary ammonium salt structure. *Macromolecules*, 2009;42(13):4580–4584.

[86] Ly Nguyen TK, Obadia MM, Serghei A, Livi S, Duchet-Rumeau J, Drockenmuller E. 1, 2, 3-Triazolium-based Epoxy–Amine networks: Ion-conducting polymer electrolytes. *Macromolecular Rapid Communications*, 2016;37(14):1168–1174.

[87] Soares BG, Silva AA, Livi S, Duchet-Rumeau J, Gerard JF. New epoxy/Jeffamine networks modified with ionic liquids. *Journal of Applied Polymer Science*, 2014;131(3):39834.

[88] Kwon SJ, Kim T, Jung BM, Lee SB, Choi UH. Multifunctional Epoxy-based solid polymer electrolytes for solid-state supercapacitors. *ACS Applied Materials & Interfaces*, 2018;10(41):35108–35117.

[89] Muralidharan N, Teblum E, Westover AS, Schauben D, Itzhak A, Muallem M, Nessim GD, Pint CL. Carbon nanotube reinforced structural composite supercapacitor. *Scientific Reports*, 2018;8(1):17662.

[90] Westover AS, Baer B, Bello BH, Sun H, Oakes L, Bellan LM, Pint CL. Multifunctional high strength and high energy epoxy composite structural supercapacitors with wet-dry operational stability. *Journal of Materials Chemistry A*, 2015;3(40):20097–20102.

[91] Lim L, Liu Y, Liu W, Tjandra R, Rasenthiram L, Chen Z, Yu A. All-in-one graphene based composite fiber: Toward wearable supercapacitor. *ACS Applied Materials & Interfaces,* 2017;9(45):39576–39583.

4 Carbon Derivatives-Based Silicon as Anode Materials for Lithium-Ion Batteries

Asadullah Zaffar, Safia Ali, Khushnuma Gul, Won-Chun Oh and Kefayat Ullah

1 INTRODUCTION

The need to safeguard the environment and the day-by-day increase in energy consumption concerns the fossil fuels dilemma and calls/pushes for alternative energy sources. New research shows that the best and most notable renewable energy source can be lithium-ion batteries (LIBs) [1]. A rechargeable battery, also known as a secondary battery, is an electrochemical device that can be charged and discharged several times and can efficiently, and with almost no gaseous emissions, transform chemical energy into electrical energy, for example, Lithium-ion (Li-ion) [2], Nickel-cadmium (Ni-Cd) [3] batteries. LIBs are a fantastic technological and commercial choice for electric vehicles, backup power systems, and portable electronics (such as laptops, mobile phones, and electronic watches) [4, 5]; that is why we will only discuss LIBs in this chapter. LIBs' exceptional power density, energy density, accessibility, and safety have made it the most successful and desired recyclable energy source [6]. LIBs are an essential permitting technology for hybrid electric automobiles, so Sony's researchers created the first marketable LIBs in 1980 [7–10]. Due to safety concerns, alloys, oxides, chalcogenides, and carbonaceous materials have replaced lithium as anode materials [11].

Silicon, due to its high hypothetical limit, legitimate working voltage, bountiful accessibility, and ecological goodwill, is the best other option to serve as anodes for LIBs because of its tremendous volume change, unsteady robust solid electrolyte interface (SEI), and low electrical as well as ionic conductivity, which leads to the crushing of silicon and extreme lessening of limit. Among various arrangements, compounding with carbon is a promising, compelling one that draws increasingly more consideration. In the silicon/carbon (Si/C) half-breed anodes, Si goes about as the dynamic material, which gives a high limit, and carbon works on the conductivity and mitigates the extension of Si [12]. The "Miracle substance," graphene, in an sp^2-hybridized state, is a 2D sheet of carbon atoms arranged into a periodic structure of lattice like a honeycomb. Graphene, exfoliated from graphite, has shown remarkable potential in materials sciences, physics, chemistry, and other fields [13].

DOI: 10.1201/9781003561262-4

Following the introduction to carbon's nanostructure types, such as fullerene (created through wrapping the graphene layer into a 0D molecule) and carbon nanotubes (created through cylindrically rolling the graphene layer), the development of graphene has staged various fascinating new disciplines in science and industry, including graphite [14, 15]. At environmental temperatures, graphene also shows the rare half-integral quantum Hall effect in case of both electrons and holes [13, 16]. The recent development of economical, accessible, and expandable production processes for mono-layer graphene has made this one of the notable alluring nanomaterials [17].

In spite of these excellent characteristics, graphene has several downsides; the volumetric capacity of graphene sheets is significantly limited due to their low density. If graphene sheets are restacked, their unique properties may be lost. The production of a large surface area and notably conductive graphene is complicated, and regulating the formation of contaminants/flaws on graphene sheets is complicated [18]. Li-ion transportation ways are lengthy due to the slender proportion of graphene nanosheets [19, 20]. *Ex situ* X-ray Photoelectron Spectroscopy (XPS) and X-ray Diffraction (XRD) studies show that the transformation process within the composite electrodes is reversible [21]. Porous structure graphite is created via elevated heat graphitization of semi-coke, trailed by a catalytic pore-formation method. The obtained permeable graphite elements remained enveloped by reduced graphene oxide (rGO) layers, which were exfoliated using graphitized semi-coke to create three-dimensional (3D) hierarchical permeable graphite/rGO composites. This suggests the potential to convert affordable coal products into high-performance anode materials for lithium-ion batteries and other energy storage systems, offering a more economical and versatile approach to energy solutions [22–24]. Polymeric Imide (PI) Nano flakes were grown on carbon nanotubes (CNTs) to form a cable-like structure (PI@CNTs) using an *in situ* polymerization method. The distributed PI Nano flakes reduce the distance between lithium ions, and the CNTs substrate improves electron transport, resulting in an enhanced electrochemical performance [25]. Carbon nanotubes (CNTs) acts as a crucial function in the design to boost the electrochemical performances for P3BT (a one-dimensional structure-design for a poly(thiophene) derivative (i.e., poly 3-butyl-thiophene, abbreviated as P3BT)), according to the experimental investigation and Density Functional Theory (DFT) calculation. The firm-interaction among P3BT and CNT calms the charged-discharged conditions of P3BT to improve cycling constancy. Still, it also reimburses for the unfavorable outcome of butyl side chains on the energy band and electron affinity [26]. A 3D carbon-coated stable silicon/graphene/CNT (C@Si/GN/CNT/Polydopamine-derive Carbon (PDA-C) derivative was developed to address the volumetric expansion of silicon during charge-discharge cycles [27]. In this chapter, we will discuss the vital act of carbon derivatives-based silicon as anode materials, which are used to upgrade or boost the performance of LIBs.

2 METHODS AND MATERIALS

In this section, several synthesis methods of carbon derivatives-based silicon are discussed.

2.1 Graphite-based Silicon

Graphite is the marketable anode material for LIBs due to its large specific capacity and short charge and discharge cycle. Silicon (Si) is a potential alternative for graphite, and it interacts with lithium to generate the $Li_{4.4}Si$ alloy [28, 29]; their semi-conductive charge transport impedes electrode redox procedures and electron diffusion [30]. These strategies for improving the electrochemical presentation of Si-based anode constituents have been explored. One is constructed on nanostructures of Si materials' nanoparticles [31], 1D nanowires [32, 33], or nanotubes [34], hollow or yolk-shell structures [35–37], and 3D nanoporous frameworks [38]. The other also combines silicon/carbon nanocomposites with several conductive carbon constituents like carbon black [39], graphite [40], carbon nanotubes [41], nanofibers [42], and Titanium-Carbide coated Carbon (TiC/C) nanofibers [43, 44].

2.2 Graphene-based Silicon

Numerous investigations have explored the incorporation of graphene with silicon nanoparticles through a process involving freeze-drying [45], followed by sonication and vacuum-filtration [46, 47] or motorized mixing [48]. The manual elasticity of graphene ascribed to the enhanced volume retention of the Si-graphene anode related to Si-based anodes. The electrochemical impedance measurement validates the Si-graphene electrode's improved electrical conductivity and solid electrolyte boundary [49]. An organic vapor deposition method is employed in fabricating anode materials, combining graphene with silicon, for lithium-ion batteries. Specifically, well-organized crystal-like silicon elements are generated atop graphene layers, utilizing fluid dichlorosilane as the silicon source. During charge-discharge cycles, the silicon-graphene composite demonstrates high silicon utilization [50]. An openly placed aluminum oxide (Al_2O_3) layer is applied on the silicon-graphene electrode. This achievement is realized through the integration of graphene with silicon as the anode and a lithium additional layer structured compound as the cathode, specifically $Li_{1.2}Ni_{0.2}Mn_{0.6}O_2$ [50].

2.3 Graphene Oxide-Based Silicon

The composite material comprising graphene and silicon nanoparticles was synthesized by incorporating silicon into graphene oxide (GO) and reduced graphene oxide (rGO) as an anode for LIBs. Electrochemical properties were examined, including charge-discharge capacity and retention rate [51]. Wang et al. created a complex film of graphene-silicon using the *in situ* filtration method [52]. Chang et al. produced a hybrid material of complex Si/reduced graphene oxide (Si/rGO) through a dip-coating process [53]. Wu et al. synthesized a binder-free multilayered composite of Si-embedded porous carbon/graphene (Si–C/G) using the electrospray deposition technique [54].

2.4 Carbon Nanotube-Based Silicon

The $Li_{22}Si_5$ composite exhibits a highly lithiated phase, which is ten times greater than graphite. Nevertheless, the lithiated phase's development is linked to an enormous volume expansion that eventually causes the active material to be ground up, which causes a vast capacity to fade in tandem [55]. Using techniques that manage the arrangement, morphology, and unit size may significantly increase capacity retention and succession life. Active-inactive media [56], nanostructured and amorphous silicon procedures, including silicon nanoparticles [57], nanowires [58], nanotubes [59], and other nanoscale structures are often used in these techniques [60]. Until now, every binder-less method that has been investigated has needed a current gatherer to act as the substrate for silicon development.

2.5 Fullerene-based Silicon

Fullerenes-based silicon nanowires (F-Si NWs) with thicknesses of 3, 4, and 5 nm are randomly embedded with different kinds of fullerenes (C60, C180, and C320) with varying weight ratios (0.5%, 0.75%, and 1.5%), with Young's modulus, tensile strength, and tensile strain, up to 23.13%, 140.74%, and 70.59%, respectively [61]. Due to its availability, stability, lack of toxicity, and ease of production, silicon is a material that is widely used in the electronics industry [62] because of its remarkable and encouraging optical, electrical, and thermoelectric capabilities, and mechanical, 1D silicon nanowires (Si NWs) have been more popular among conservative silicon-based nanostructures in recent years [63, 64]. Silicon carbide (Si/C) nanowires are gaining a lot of attention as one of the essential elements for these nanocomposites because of their unique qualities, which include low heat expansion quantity, large strength, small density, and large hardness [65]. Fullerenes have enhanced electrical [66], mechanical [67], and optical [68] properties; these are a group of carbon-based nanostructures with an enclosed morphology, have gathered a lot of attention, especially in the fields of optoelectronics [69], electronics [70], hydrogen storage technology [71], and fuel cells [72]. Numerous experimental and computational studies have observed how fullerenes affect the mechanical behavior of nanocomposites based on metals or polymers [73–84].

3 DISCUSSION

3.1 LIBs

M.S. Whittingham, along with his advisor, proposed the first idea for LIBs in the years between 1970 and 1980 at Binghamton University [85]. Whittingham focused on the intercalation of various ions into complex metal oxides and their potential application as a new electrode [86]. In the initial era of commercial LIBs, Akira Yoshino discovered that a carbonaceous material with a specific crystalline structure could be a suitable anode material [87]. In 1991, Sony, led by Yoshio Nishi and a team at Sawka sei, successfully developed and introduced the first ever commercially available LIBs [88]. Because of the invention of LIBs that led to a rechargeable planet, three researchers, M. Stanley Whittingham, Akira Yoshino, and John B. Goodenough, shared the Nobel Prize in 2019 [89].

LIBs are electrochemical apparatus designed to pack electrical power in the custom of chemical energy throughout charging and convert chemical energy back into electric energy (as electric power) throughout discharging [85, 90, 91].

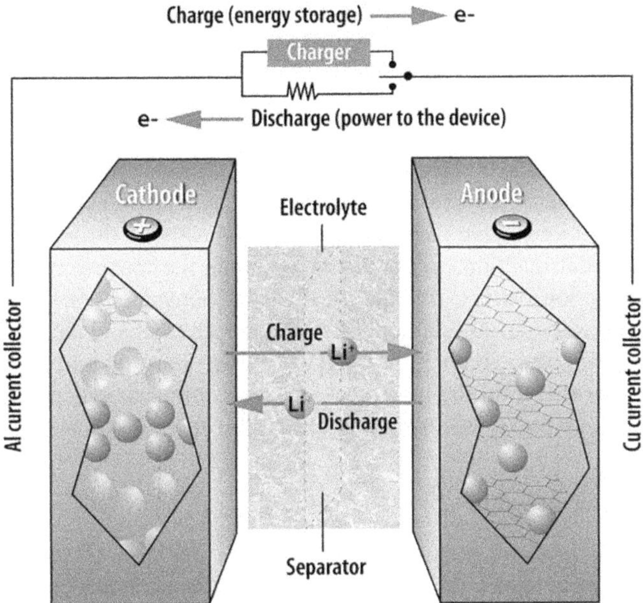

FIGURE 4.1 Charging and discharging of LIBs with essential elements; (diagram by Argonne National Laboratory, licensed under CC BY-NC-SA 2.0) from an open access journal [92].

Like any other battery, LIBs also contain three essential elements: electrodes (anode and cathode), separator, and electrolyte [93] as shown in Figure 4.1 [92]. The cathode typically employs metal oxides, while the anode commonly utilizes porous carbon [93]. The anode, where oxidation occurs during a battery's discharge, is where electrons are produced, flowing out of the battery and generating an electric current. In the case of rechargeable batteries, the anode transforms into the site of reduction during the charging process, accepting electrons. Conversely, the cathode, functioning as the electrode where reduction occurs during a battery's discharge, is where electrons are received from the external circuit, completing the electrical circuit.

When a rechargeable battery is charged, the cathode becomes the oxidation site, releasing electrons into the external circuit. While discharging, electrons are generated at the anode, traversing the external circuit to reach the cathode and creating an electric current. Concurrently, positive ions travel through the electrolyte from the anode to the cathode, preserving overall electrical neutrality in the system [90, 94, 95]. Notably, the materials specific to the anode and cathode can vary, depending on the battery type. For instance, in a lithium-ion battery, the anode typically consists of a material capable of intercalating or alloying with lithium ions. At the same time, the cathode is composed of a material proficient in reversibly hosting lithium ions. The electrochemical reactions at the anode and cathode are foundational to battery operation and play a central part in the storage and release of electrical energy across diverse battery types [90, 96]. The electrons released in the anode's chemical reaction travel through the external circuit, thereby generating an electric current.

Concurrently, positively charged ions traverse from the anode to the cathode through the electrolyte in the battery's electrolyte. At the cathode, another chemical reaction occurs, often involving the reduction of a material—such as the acceptance of lithium ions by the cathode. Consequently, the energy from the external power source is stored in the battery as chemical energy. During the discharging phase, the stored chemical energy is released. Electrons move from the anode through the external circuit, creating an electric current that powers the connected device. Simultaneously, positively charged ions move from the anode to the cathode through the electrolyte. A reverse chemical reaction occurs at the anode as the material undergoes reduction. In a lithium-ion battery, this manifests as the movement of lithium ions from the anode to the cathode. The cathode then accepts electrons in a chemical reaction, thereby completing the circuit [94, 95]. This cyclic charging and discharging process can be reiterated until the battery's capacity diminishes due to various factors, including chemical reactions, material degradation, and stresses.

3.2   SILICON PRODUCTION

Silicon (Si) is a promising candidate for electric devices, electric vehicles, and energy storage applications, and it can be acquired through various methods. In one approach, Si particles were deposited at approximately 200 mbar pressure using a $SiHCl_3$ breakdown process in a hydrogen (H_2) atmosphere at 1000°C. The resulting Si elements exhibited a distinct polygonal shape with well-defined borders, maintaining a consistent size of around 600 nm. These Si particles were observed to adhere to the flat surfaces of the graphene layers, as evidenced by scanning electron microscopy (SEM) [50]. Through the adsorption of gas molecules on an apparent stable nucleus, Si growth can take place [97]. A highly organized crystalline structure is observed in transmission electron microscope (TEM) pictures with a (111) interplanar spacing of 3.12 Å of the implanted Si particles. Diverging from other studies where SiH_4 was employed as the silicon source, resulting in spherical or shapeless Si nanoparticles, the distinct sharp-bordered polygonal shape and highly organized crystal-like configuration observed in this research can be attributed to the elevated synthesis temperature [35, 98].

3.3   ANODE MATERIALS

Through an examination of mass loss during graphene combustion and complete silicon oxidation to SiO_2 in the composite, thermogravimetric analysis (TGA) results indicate an initial composition of 14% silicon and 86% graphene in the Si-graphene compound. The Raman spectra of graphene exhibit characteristic features, including a disorder-induced D band at approximately 1350 cm^{-1}, a graphitic G band at around 1580 cm^{-1}, and a second-order 2D band at about 2700 cm^{-1}. Significantly, quantifying graphene sheets can be achieved using the full-width at half-maximum (FWHM) of the 2D band, which is a well-established quantitative method [99]. The produced graphene, with an FWHM of around 80 cm^{-1}, indicates more than five layers. Despite graphene's typical limitations, graphene reported in this study, with a small sheet count and ample surface area, presents a unique opportunity. Using an organic vapor deposition method, the Si-graphene composite in LIBs exhibits high silicon utilization, achieving a 90% capacity retention after 500 cycles and an average Coulombic efficiency exceeding 99.5%. An openly deposited Al_2O_3 coating on the Si-graphene electrode suppresses unwanted reactions, enhancing primary Coulombic efficiency and adjustable capacity. Combining a Si-graphene anode with a designed $Li_{1.2}Ni_{0.2}Mn_{0.6}O_2$ cathode forms a fully functional 3.6 V cell device, showing promise for high-energy-density LIBs [50]. Silicon deposition on graphene reduces the compound's surface area to approximately 8 m^2 g^{-1}. Despite this, Raman spectra show minimal changes in the 2D band FWHM or ratio between Disorder-Band (D Band) and Graphene-Band (G Band) (D/G band intensity ratio), highlighting graphene's stability during silicon deposition at 1000°C. The emergence of a distinct Raman band at 520 cm^{-1} confirms successful crystal-like silicon incorporation [50].

Graphene nanosheets (GNSs) display flexibility, thanks to their open, porous structure without rigid connections between neighboring nanosheets. This permeable and elastic framework can be utilized as a reinforcing structure with significant protective capabilities, thereby effectively reducing electrode dissolution or degradation [100]. This unique property of graphene nanosheets (GNSs) is harnessed by employing a distinctive elastic embedded structure. This structure alternates layers of graphene nanosheets (GNSs) and silicon (Si) anode material, thereby effectively bonding them. The aim is to reduce Si anode dissolution, enhance cyclability, and minimize GNS restacking during cycling. Embedded GNSs play a crucial role in facilitating electron and lithium-ion transport pathways. This, in turn, improves the electrode's electrical conductivity and facilitates a higher rate of lithium-ion diffusion [101]. Numerous studies have explored the incorporation of graphene with silicon nanoparticles through a process involving freeze-drying, followed by sonication, and subsequent vacuum-filtration [46, 47] or machine-driven blending [48]. Si-graphene composites fell short of expectations, especially in long-term cyclability. Challenges in achieving uniform scattering of silicon nanoparticles on the graphene substrate were encountered, as they tended to form agglomerates due to high shared potential [102]. Such clusters would degrade and eventually fail the electrode due to volume expansion mechanisms triggered by lithiation reactions. Recent research has shown that vapor deposition of an endless Si film made of nanoparticles onto a graphene layer can result in much higher performance [98]. At 500.8°C, graphene was covered with Si through SiH_4 breakdown; (it is generally known that SiH_4 is a hazardous and explosive chemical molecule) [103]. The documentation on the deposition of crystal-like silicon elements from dichlorosilane onto graphene layers is yet to be published. In our approach to synthesis, trichlorosilane vapor and H_2 gas are concurrently introduced into a heated tube furnace containing graphene samples. This method has been designed to facilitate the integration of trichlorosilane vapor and H_2 gas in a simultaneous manner, thereby promoting the effective synthesis of graphene samples within the heated environment of the tube furnace. Resultantly, both sides of the graphene layers are coated with highly crystalline silicon elements, characterized by uniform sizes of approximately 600 nm. The initial Si-graphene composite exhibits a high silicon utilization rate, with nearly complete activation of the silicon component during the lithiation process. It demonstrates consistent cycling performance, sustaining over 500 cycles in half-cells, as the counter electrode, with lithium foil.

The obtained binding energy values were then meticulously compared to provide insights into the atomic composition and chemical states present in the sample [45, 104–106]. The composite material comprising graphene and silicon nanoparticles, both possessing substantial theoretical capacities, was synthesized by incorporating silicon into GO and rGO. The material's physicochemical attributes were scrutinized when employed in LIBs as an anode [51]. The design aimed to surpass the theoretical total of carbon-based materials, addressing the issue of the short half-life of lithium secondary batteries by leveraging graphene's buffering effect [51]. Wang et al. generated a compound film of graphene-silicon using the *in situ* filtration method [52]. Chang et al. produced a hybrid material of complex Si/reduced graphene oxide (Si/rGO) through a dip-coating process [53]. Wu et al. synthesized a binder-free multilayered composite of Si-embedded porous carbon/graphene (Si–C/G) using the electrospray deposition technique [54]. The Ministry of Education. Science and Technology at South Korea (MEST) MEST and The National Research Foundation (NRF) NRF in Korea et al. crafted rGO/Si composites through a straightforward dip-coating method devoid of a binder, serving as anodes for LIBs [51]. Graphite is the marketable anode material for LIBs due to its large specific capacity and short charge-discharge potential. Silicon (Si) is a potential alternative for graphite, and it interacts with lithium to generate the $Li_{4.4}Si$ alloy [28, 29]. Another practical constraint of Si-based anodes, with extraordinary performance, is their semi-conductive charge transport, which impedes electrode redox procedures and electron diffusion [30]. These strategies for improving the electrochemical presentation of Si anode constituents have been explored. One is constructed on nanostructures of Si materials' nanoparticles [31], 1D nanowires [32, 33] or nanotubes [34], hollow or yolk-shell structures [35–37], and 3D nanoporous frameworks [38] because nanostructures are more operative at releasing strain, such nanostructured Si materials practice less essential degradation and reduction during volume change processes than bulk Si. Another approach is to combine silicon-carbon nanocomposites with several conductive carbon constituents like carbon black [39], graphite [40], carbon nanotubes [41], carbon nanofibers [42], and TiC/C nanofibers [43] To protect against the volumetric changes in silicon (Si) and enhance electrical conductivity, the incorporation of a carbon element is crucial [44, 107].

The exploration of interface stability between silicon-based anode materials and electrolytes has been approached through various investigations. A particularly effective method entails incorporating a carbon coating onto the surface of a nanostructured silicon electrode, strategically engineered with void spaces. Atomic Layer Deposition (ALD) has recently emerged as a promising coating process for the customization of cathodes ($LiCoO_2$ [108]), anodes (natural graphite [109]), separators (polypropylene [110]), and Reactive-ion Etching (RIE)RIE-patterned Si [111]). Atomic Layer Deposition (ALD) has proven to be effective in enhancing the safety, rate performance, and longevity of lithium-ion batteries (LIBs). ALD relies on successive, self-limiting surface reactions and provides the capability to precisely adjust coating thickness at the atomic level, down to the Angstrom level. In our study, a straightforward ALD procedure was applied directly to the Si-graphene electrode immediately after fabrication, using trimethylaluminum (TMA) and water as precursors. This resulted in a conformal Al_2O_3 coating that successfully inhibited side reactions, leading to a significantly improved initial Coulombic efficiency and a larger reversible capacity [50]. Graphene-based silicon sheets are synthesized in two phases. First, graphite powders undergo oxidation using a modified Hummers process. The resulting graphite oxides are then rapidly exfoliated at 1050°C in a nitrogen environment, thereby yielding Si-graphene sheets [112]. The Si coating on graphene layers was performed in a horizontal cylinder oven heated to 1000°C, at a pressure of around 200 mbar filled with pure H_2 gas; $SiHCl_3$ vapors were introduced at flow rates of 5 cm and 40 cm, respectively. After removal from the oven below 50°C, materials were characterized using SEM and TEM [50]. Featuring a composite structure with an additional lithium layer, $Li_{1.2}Ni_{0.2}Mn_{0.6}O_2$ cathode, and Si/carbon anode, a 3.6 V complete cell performed well. The new method shows great promise for obtaining energy densities greater than commercial $LiCoO_2$-graphite-linked batteries already in the market [50].

Silicon nanowires (Si NWs) with thicknesses of 3, 4, and 5 nm are randomly embedded with different kinds of fullerenes (C60, C180, and C320) with varying weight ratios (0.5%, 0.75%, and 1.5%). Fullerene-embedded silicon nanowires (F-Si NWs) exhibit improved Young's modulus, tensile strength, and tensile strain, up to 23.13%, 140.74%, and 70.59%, respectively. The notable benefits of fullerenes are struggle to adjacent contraction and prevention of fracture circulation; this improves the mechanical properties of silicon nanowires (Si NWs) while minimally affecting the underlying bulk silicon's cubic diamond lattice structure [61]. Due to its availability, stability, lack of toxicity, and ease of production, silicon is a material that is widely used in the electronics industry [62]. Because of their remarkable and encouraging optical, electrical, thermoelectric, and mechanical properties, 1D silicon nanowires (Si NWs) have been more popular among conservative silicon-based nanostructures in recent years [63, 64]. Apart from their inherent exceptional qualities, silicon nanostructures perform better in applications due to their ability to build hybrid structures [113]. Due to these beneficial characteristics, Si NWs are much superior to their bulk and thin-film silicon counterparts [114, 115]. In addition to the sectors indicated above, Si NWs have also been demonstrated to be potential candidates for biological and chemical sensors used in environmental monitoring applications [63], medical diagnostics [64], and thermoelectric devices [116]. Ho et al. [117] suggest that fullerene C84-embedded Si nanostructure is an alternative material composition. The investigation of the mechanical, electrical, and magnetic properties of the fullerene C84-embedded Si substrate in that work led to the conclusion that this nanocomposite may be used instead of SiC as an efficient abrasive material [117]. Fullerene-like structures are created when carbon atoms are gradually added to Si_{13} nanoclusters [118]. Because of their superior electrical [66], mechanical [67], and optical [68] properties, fullerenes, a class of carbon-based nanostructures with a cage-like morphology, have garnered a lot of attention, especially in the fields of optoelectronics [69], electronics [70], hydrogen storage technology [71], and fuel cells [72]. Numerous experimental and computational studies have observed how fullerenes affect the mechanical behavior of nanocomposites based on metals or polymers [73–84]. Dreyer et al. show that adding fullerene to polymer reinforcements can increase the stiffness of the polymers by as much as 400% [75]. Similarly, adding fullerenes to the ionic polymer membrane raises tensile strength and Young's modulus [76]. Unal et al.'s computational study on the mechanical properties of graphene nanoribbon-fullerene (GNRF) hybrid nanostructures suggests that the incorporation of fullerenes can enhance the tensile and compressive properties of GNRs [79]. A brand new CNT-fullerene hybrid nanostructure known as NanoBud was created by Nasibulin et al. and is prepared using fullerenes covalently bound to the surface of CNTs. Characterization studies show that these hybrid nanostructures exhibit better mechanical, electrical, and optical capabilities than pure CNTs and fullerenes [81]. Tayfun et al. studied thermoplastic polyurethane (TPU) composites at various fullerene concentrations ranging from 0.5 to 2.5 wt%. Adding fullerene molecules increased the composites' tensile strength, Young's modulus, and elongation percent [83]. The study by Ogasawara et al. that examined the mechanical properties of fullerene-dispersed carbon-fiber reinforced epoxy matrix composites (CFRPs) found that fullerene dispersion of 0.1– 0.5 wt% can improve the nanocomposite's tensile, compression, and compression after impact (CAI) strengths as well as their interlaminar fracture toughness [84]. Ogasawara et al. observe that fullerene addition can lead to a 60% improvement in interlaminar fracture toughness [84]. Additionally, fullerenes are used in a type of nanomaterial known as a nanopeapod, which is a hybrid structure made up of molecules, such as fullerenes, enclosed within carbon or non-carbon nanotubes [61, 119, 120]. Gao et al. point out that the improved coating capacity of ductile Si NWs may be used for the LIBs anodes to achieve one of the advantageous properties of flexibility for applications in the semiconductor industry [121]. The second step randomly distributes the fullerenes across the Si nanowire's volumetric region. The nanowire specimens' crystal orientation is selected as [109] because nanowires with sizes less than 10 nm typically grow along the [109] direction when created using the widely used CVD technique [122].

4   LIMITATIONS AND SUGGESTIONS

The manufacturing, functionalization, and usage of carbon-based silicon and its nanocomposites have significantly developed in the modern era. Worldwide attention in electric automobiles and green energy storage opens the pathway for utilizing them as anode materials in upcoming lithium-ion batteries (LIBs). Inappropriately, numerous problems must be resolved before truly marketizing. Silicon undergoes significant volume expansion during lithiation, which can lead to mechanical stress and the formation of cracks in the anode material. It exhibits reduced cycling stability, as charging and discharging cycles lead to the extension and shrinkage of the silicon material, which results in the degradation of the anode structure and, ultimately, reduced battery performance. As capacity fades over multiple charge and discharge cycles, this is often associated with forming a solid electrolyte interface (SEI) on the surface of the anode, which can limit the accessibility of lithium-ions. The production of silicon-based anode materials can be more expensive compared to traditional graphite anodes. Achieving cost-effectiveness and scalability in the manufacturing processes is a significant challenge for widespread adoption. Precise control over the synthesis of silicon/carbon composites is crucial for achieving optimal electrochemical performance. Variability in synthesis techniques and parameters can result in inconsistent material properties, affecting the overall performance of LIBs. Our suggestions are: explore advanced nanostructuring techniques to design silicon-based anode materials at the nanoscale, which can help mitigate volume expansion issues and enhance cycling stability; investigate the use of coatings and additives to improve the strength of the solid electrolyte interface (SEI); reduce capacity fading and volume changes throughout cycling; use advanced characterization techniques, such as *in situ* and operando methods, to gain a deeper understanding of the electrochemical processes at the silicon-based anode interface, which can guide the development of more effective anode materials; develop sustainable and scalable methods for the synthesis and production of silicon-based anode materials. This includes exploring environmentally friendly processes and optimizing production techniques to reduce costs and increase efficiency.

5 CONCLUSION

In this modern era, attention to electric vehicles, electric devices, and other charge packing technologies are the foremost factors demanding the improvement of LIBs. The field of battery material research is ever-changing, with ongoing studies bringing new insights. Choosing the best option depends on application requirements; costs; and trade-offs in capacity, cycling stability, and manufacturing feasibility. Researchers are working to enhance silicon-based anodes by integrating other materials or modifying their structures to overcome challenges. Practical applications vary, and the choice of anode material depends on the specific design and performance goals of the LIBs. The exploration of carbon derivatives-based silicon as anode materials for LIBs shows a promise for advancing energy storage systems. Silicon's high theoretical capacity makes it an attractive candidate to replace or complement traditional graphite anodes. However, several challenges need addressing to fully harness the potential of silicon-based anodes in practical applications. We recommend considering graphene-based silicon due to its exceptional electrical conductivity and mechanical strength, potentially mitigating volume expansion issues associated with silicon-based anodes. In essence, while carbon-based silicon anodes present challenges, they also offer exciting opportunities to advance energy storage technology. Ongoing research in this field is crucial for unlocking the full potential of silicon-based anode materials and enhancing the development of high-performance, durable, sustainable LIBs for the future.

REFERENCES

1. Khan, B.M., W.C. Oh, P. Nuengmatch, and K. Ullah, Role of Graphene-Based Nanocomposites as Anode Material for Lithium-Ion Batteries. *Materials Science and Engineering: B*, 2023, 287: p. 116141.
2. Obeid, M.M., and Q. Sun, Recent Advances in Topological Quantum Anode Materials for Metal-Ion Batteries. *Journal of Power Sources*, 2022, 540: p. 231655.
3. Koehler, U., General Overview of Non-Lithium Battery Systems and their Safety Issues. *Electrochemical Power Sources: Fundamentals, Systems, and Applications*, 2019: pp. 21–46. https://doi.org/10.1016/b978-0-444-63777-2.00002-5.

4. Hou, J., S. Qu, M. Yang, and J. Zhang, Materials and Electrode Engineering of High Capacity Anodes in Lithium Ion Batteries. *Journal of Power Sources*, 2020, 450: p. 227697.

5. Fang, Z., J. Wang, H. Wu, Q. Li, S. Fan, and J. Wang, Progress and Challenges of Flexible Lithium Ion Batteries. *Journal of Power Sources*, 2020, 454: p. 227932.

6. Yang, C., X. Zhang, J. Li, J. Ma, L. Xu, J. Yang, S. Liu, S. Fang, Y. Li, X. Sun, and X. Yang, Holey Graphite: A Promising Anode Material with Ultrahigh Storage for Lithium-Ion Battery. *Electrochimica Acta*, 2020, 346: p. 136244.

7. Nishi, Y., The Development of Lithium Ion Secondary Batteries. *The Chemical Record*, 2001, 1(5): pp. 406–413.

8. Li, H., Z. Wang, L. Chen, and X. Huang, Research on Advanced Materials for Li-Ion Batteries. *Advanced Materials*, 2009, 21(45): pp. 4593–4607.

9. Nagaura, T., Lithium Ion Rechargeable Battery. *Progress in Batteries & Solar Cells*, 1990, 9: p. 209.

10. Arico, A.S., P. Bruce, B. Scrosati, J.M. Tarascon, and W. Van Schalkwijk, Nanostructured Materials for Advanced Energy Conversion and Storage Devices. *Nature Materials*, 2005, 4(5): pp. 366–377.

11. Armand, M., "Intercalation Electrode" in Materials for Advanced Batteries. In *Nato Conference Series, Series VI: Material Science,* 1980. Plenum Press.

12. Li, X., M. Zhang, S. Yuan, and C. Lu, Research Progress of Silicon/Carbon Anode Materials for Lithium-Ion Batteries: Structure Design and Synthesis Method. *ChemElectroChem*, 2020, 7(21): pp. 4289–4302.

13. Novoselov, K.S., A.K. Geim, S.V. Morozov, D.E. Jiang, Y. Zhang, S.V. Dubonos, I.V. Grigorieva, and A.A. Firsov, Electric Field Effect in Atomically Thin Carbon Films. *Science*, 2004, 306(5696): pp. 666–669.

14. Castro Neto, A. H., F. Guinea, N.M. Peres, K.S. Novoselov, and A.K. Geim, The Electronic Properties of Graphene. *Reviews of Modern Physics*, 2009, 81(1): p. 109.

15. Teobaldi, G., F. Al Ma'Mari, M. Rogers, S. Alghamdi, T. Moorsom, S. Lee, T. Prokscha, H. Luetkens, M. Valvidares, M. Flokstra, and R. Stewart, Proceedings of the National Academy of Sciences. *Proceedings of the National Academy of Sciences of USA*, 2017, 114(22): pp. 5583–5588.

16. Zhang, Y., Y.W. Tan, H.L. Stormer, and P. Kim, Experimental Observation of the Quantum Hall Effect and Berry's Phase in Graphene. *Nature*, 2005, 438(7065): pp. 201–204.

17. Choi, W., I. Lahiri, R. Seelaboyina, and Y.S. Kang, Solid State Mater. *Science*, 2010, 35: pp. 52–71.

18. Qi, W., J.G. Shapter, Q. Wu, T. Yin, G. Gao, and D. Cui, Nanostructured Anode Materials for Lithium-Ion Batteries: Principle, Recent Progress and Future Perspectives. *Journal of Materials Chemistry A*, 2017, 5(37): pp. 19521–19540.

19. Srivastava, M., J. Singh, T. Kuila, R.K. Layek, N.H. Kim, and J.H. Lee, Recent Advances in Graphene and Its Metal-Oxide Hybrid Nanostructures for Lithium-Ion Batteries. *Nanoscale*, 2015, 7(11): pp. 4820–4868.

20. Rosaiah, P., T. Niyitanga, S. Sambasivam, and H. Kim, Graphene Based Magnetite Carbon Nanofiber Composites as Anodes for High-Performance Li-Ion Batteries. *New Journal of Chemistry*, 2023, 47(1): pp. 482–490.

21. Singhbabu, Y.N., P.N. Didwal, K. Jang, J. Jang, C.J. Park, and M.H. Ham, Green Synthesis of a Reduced-Graphene-Oxide Wrapped Nickel Oxide Nano-Composite as an Anode for High-Performance Lithium-Ion Batteries. *ChemistrySelect*, 2022, 7(17): p. e202200676.

22. Zhong, M., J. Yan, L. Wang, Y. Huang, L. Li, S. Gao, Y. Tian, W. Shen, J. Zhang, and S. Guo, Hierarchic Porous Graphite/Reduced Graphene Oxide Composites Generated from Semi-Coke as High-Performance Anodes for Lithium-Ion Batteries. *Sustainable Materials and Technologies*, 2022, 33: p. e00476.

23. Xu, Z., J. Yang, S. Hou, H. Lin, S. Chen, Q. Wang, H. Wei, J. Zhou, and S. Zhuo, Thiophene-Diketopyrrolopyrrole-Based Polymer Derivatives/Reduced Graphene Oxide Composite Materials as Organic Anode Materials for Lithium-Ion Batteries. *Chemical Engineering Journal*, 2022, 438: p. 135540.

24. Yang, X., C. Lin, D. Han, G. Li, C. Huang, J. Liu, X. Wu, L. Zhai, and L. Mi, In Situ Construction of Redox-Active Covalent Organic Frameworks/Carbon Nanotube Composites as Anodes for Lithium-Ion Batteries. *Journal of Materials Chemistry A*, 2022, 10(8): pp. 3989–3995.

25. Liu, B., K. Jiang, K. Zhu, X. Liu, K. Ye, J. Yan, G. Wang, and D. Cao, Cable-Like Polyimide@ Carbon Nanotubes Composite as a Capable Anode for Lithium Ion Batteries. *Chemical Engineering Journal*, 2022, 446: p. 137208.

26. Li, T., L. Wang, and J. Li, Carbon Nanotube Enables High-Performance Thiophene-Containing Organic Anodes for Lithium Ion Batteries. *Electrochimica Acta*, 2022, 408: p. 139947.

27. Wang, F., S. Lin, X. Lu, R. Hong, and H. Liu, Poly-Dopamine Carbon-Coated Stable Silicon/Graphene/Cnt Composite as Anode for Lithium Ion Batteries. *Electrochimica Acta*, 2022, 404: p. 139708.

28. Boukamp, B., G. Lesh, and R. Huggins, All-Solid Lithium Electrodes with Mixed-Conductor Matrix. *Journal of the Electrochemical Society*, 1981, 128(4): p. 725.

29. Zhang, H., and P.V. Braun, Three-Dimensional Metal Scaffold Supported Bicontinuous Silicon Battery Anodes. *Nano Letters*, 2012, 12(6): pp. 2778–2783.

30. Wang, X.-L., and W.-Q. Han, Graphene Enhances Li Storage Capacity of Porous Single-Crystalline Silicon Nanowires. *ACS Applied Materials & Interfaces*, 2010, 2(12): pp. 3709–3713.

31. Zhuo, Y., H. Sun, M.H. Uddin, M.K. Barr, D. Wisser, P. Roßmann, J.D. Esper, S. Tymek, D. Döhler, W. Peukert, and M. Hartmann, An Additive-Free Silicon Anode in Nanotube Morphology as a Model Lithium Ion Battery Material. *Electrochimica Acta*, 2021, 388: p. 138522.

32. Chan, C.K., H. Peng, G. Liu, K. McIlwrath, X.F. Zhang, R.A. Huggins, and Y. Cui, High-Performance Lithium Battery Anodes Using Silicon Nanowires. *Nature Nanotechnology*, 2008, 3(1): pp. 31–35.

33. Peng, K., J. Jie, W. Zhang, and S.T. Lee, Silicon Nanowires for Rechargeable Lithium-Ion Battery Anodes. *Applied Physics Letters*, 2008, 93(3): p. 33105.

34. Wu, H., G. Chan, J.W. Choi, I. Ryu, Y. Yao, M.T. McDowell, S.W. Lee, A. Jackson, Y. Yang, L. Hu, and Y. Cui, Stable Cycling of Double-Walled Silicon Nanotube Battery Anodes through Solid–Electrolyte Interphase Control. *Nature Nanotechnology*, 2012, 7(5): pp. 310–315.

35. Yao, Y., M.T. McDowell, I. Ryu, H. Wu, N. Liu, L. Hu, W.D. Nix, and Y. Cui, Interconnected Silicon Hollow Nanospheres for Lithium-Ion Battery Anodes with Long Cycle Life. *Nano Letters*, 2011, 11(7): pp. 2949–2954.

36. Wu, H., G. Zheng, N. Liu, T.J. Carney, Y. Yang, and Y. Cui, Engineering Empty Space between Si Nanoparticles for Lithium-Ion Battery Anodes. *Nano Letters*, 2012, 12(2): pp. 904–909.

37. Liu, N., H. Wu, M.T. McDowell, Y. Yao, C. Wang, and Y. Cui, A Yolk-Shell Design for Stabilized and Scalable Li-Ion Battery Alloy Anodes. *Nano Letters*, 2012, 12(6): pp. 3315–3321.

38. Bang, B.M., J.I. Lee, H. Kim, J. Cho, and S. Park, High-Performance Macroporous Bulk Silicon Anodes Synthesized by Template-Free Chemical Etching. *Advanced Energy Materials*, 2012, 2(7): pp. 878–883.

39. Magasinski, A., P. Dixon, B. Hertzberg, A. Kvit, J. Ayala, and Yushin, G. J. N. M., High-Performance Lithium-Ion Anodes Using a Hierarchical Bottom-up Approach. *Nature Materials*, 2010, 9(4): pp. 353–358.

40. Fuchsbichler, B., C. Stangl, H. Kren, F. Uhlig, and S. Koller, High Capacity Graphite–Silicon Composite Anode Material for Lithium-Ion Batteries. *Journal of Power Sources*, 2011, 196(5): pp. 2889–2892.

41. Gohier, A., B. Laïk, K.H. Kim, J.L. Maurice, J. P. Pereira-Ramos, C.S. Cojocaru, and P.T. Van, High-Rate Capability Silicon Decorated Vertically Aligned Carbon Nanotubes for Li-Ion Batteries. *Advanced Materials*, 2012, 24(19): pp. 2592–2597.

42. Chen, P.-C., J. Xu, H. Chen, and C. Zhou, Hybrid Silicon-Carbon Nanostructured Composites as Superior Anodes for Lithium Ion Batteries. *Nano Research*, 2011, 4: pp. 290–296.

43. Yang, Y., H. Zhang, J. Chen, S. Lee, T.C. Hou, and Z.L. Wang, Simultaneously Harvesting Mechanical and Chemical Energies by a Hybrid Cell for Self-Powered Biosensors and Personal Electronics. *Energy & Environmental Science,* 2013, 6(6): pp. 1744–1749.

44. Bridel, J.-S., T. Azais, M. Morcrette, J.M. Tarascon, and D. Larcher, Key Parameters Governing the Reversibility of Si/Carbon/Cmc Electrodes for Li-Ion Batteries. *Chemistry of Materials*, 2010, 22(3): pp. 1229–1241.

45. Zhou, M., T. Cai, F. Pu, H. Chen, Z. Wang, H. Zhang, and S. Guan, Graphene/Carbon-Coated Si Nanoparticle Hybrids as High-Performance Anode Materials for Li-Ion Batteries. *ACS Applied Materials & Interfaces*, 2013, 5(8): pp. 3449–3455.

46. Zhao, X., C.M. Hayner, M.C. Kung, and H.H. Kung, In-Plane Vacancy-Enabled High-Power Si–Graphene Composite Electrode for Lithium-Ion Batteries. *Advanced Energy Materials*, 2011, 1(6): pp. 1079–1084.

47. Lee, J.K., K.B. Smith, C.M. Hayner, and H.H. Kung, Silicon Nanoparticles–Graphene Paper Composites for Li Ion Battery Anodes. *Chemical Communications*, 2010, 46(12): pp. 2025–2027.

48. Xiang, H., K. Zhang, G. Ji, J.Y. Lee, C. Zou, X. Chen, and J. Wu, Graphene/Nanosized Silicon Composites for Lithium Battery Anodes with Improved Cycling Stability. *Carbon*, 2011, 49(5): pp. 1787–1796.

49. Lou, D., S. Chen, S. Langrud, A.A. Razzaq, M. Mao, H. Younes, W. Xing, T. Lin, and H. Hong, Scalable Fabrication of Si-Graphene Composite as Anode for Li-Ion Batteries. *Applied Sciences*, 2022, 12(21): p. 10926.

50. Ren, J.G., Q.H. Wu, G. Hong, W.J. Zhang, H. Wu, K. Amine, J. Yang, and S.T. Lee, Silicon–Graphene Composite Anodes for High-Energy Lithium Batteries. *Energy Technology*, 2013, 1(1): pp. 77–84.

51. Lee, S.-H., Y.J. Kim, Y.S. Nam, S.H. Park, H. Lee, Y. Hyun, and C.S. Lee, Synthesis and Characterization of Silicon/Reduced Graphene Oxide Composites as Anodes for Lithium Secondary Batteries. *Journal of Nanoscience and Nanotechnology*, 2018, 18(7): pp. 5026–5032.

52. Wang, J.-Z., C. Zhong, S.L. Chou, and H.K. Liu, Flexible Free-Standing Graphene-Silicon Composite Film for Lithium-Ion Batteries. *Electrochemistry Communications*, 2010, 12(11): pp. 1467–1470.

53. Chang, J., X. Huang, G. Zhou, S. Cui, P.B. Hallac, J. Jiang, P.T. Hurley, and J. Chen, Multilayered Si Nanoparticle/Reduced Graphene Oxide Hybrid as a High-Performance Lithium-Ion Battery Anode. *Advanced Materials (Deerfield Beach, Fla.)*, 2013, 26(5): pp. 758–764.

54. Wu, J., X. Qin, H. Zhang, Y.B. He, B. Li, L. Ke, W. Lv, H. Du, Q.H. Yang, and F. Kang, Multilayered Silicon Embedded Porous Carbon/Graphene Hybrid Film as a High Performance Anode. *Carbon*, 2015, 84: pp. 434–443.

55. Beaulieu, L., K.W. Eberman, R.L. Turner, L.J. Krause, and J.R. Dahn, Colossal Reversible Volume Changes in Lithium Alloys. *Electrochemical and Solid-State Letters*, 2001, 4(9): p. A137.

56. Kim, I.-s., G. Blomgren, and P. Kumta, Nanostructured Si/Tib2 Composite Anodes for Li-Ion Batteries. *Electrochemical and Solid-State Letters*, 2003, 6(8): p. A157.

57. Liu, X.H., L. Zhong, S. Huang, S.X. Mao, T. Zhu, and J.Y. Huang, Size-Dependent Fracture of Silicon Nanoparticles during Lithiation. *ACS Nano*, 2012, 6(2): pp. 1522–1531.

58. Chan, C., H.L. Peng, G. Liu, G.K. Mcilwrath, X.F. Zhang, Ra Huggins, and Y. Cui. High-Performance Lithium Battery Anodes Using Silicon Nanowires. *Nature Nanotechnology*, 2008, 3: pp. 31–35.

59. Ryu, I., J.W. Choi, Y. Cui, and W.D. Nix, Size-Dependent Fracture of Si Nanowire Battery Anodes. *Journal of the Mechanics and Physics of Solids*, 2011, 59(9): pp. 1717–1730.

60. Epur, R., M.K. Datta, and P.N. Kumta, Nanoscale Engineered Electrochemically Active Silicon–Cnt Heterostructures-Novel Anodes for Li-Ion Application. *Electrochimica Acta*, 2012, 85: pp. 680–684.

61. Erbas, B., S. Yardim, and M. Kirca, Mechanical Properties of Fullerene Embedded Silicon Nanowires. *Archive of Applied Mechanics*, 2023, 93(1): pp. 355–367.

62. Peng, K.-Q., X. Wang, L. Li, Y. Hu, and S.T. Lee, Silicon Nanowires for Advanced Energy Conversion and Storage. *Nano Today*, 2013, 8(1): pp. 75–97.

63. Rashid, J.I.A., J. Abdullah, N.A. Yusof, and R. Hajian, The Development of Silicon Nanowire as Sensing Material and Its Applications. *Journal of Nanomaterials*, 2013, 2013: p. 328093.

64. Shi, L., D. Yao, G. Zhang, and B. Li, Size Dependent Thermoelectric Properties of Silicon Nanowires. *Applied Physics Letters*, 2009, 95(6): p. 063102.

65. Zhang, H., W. Ding, and D.K. Aidun, Mechanical Properties of Crystalline Silicon Carbide Nanowires. *Journal of Nanoscience and Nanotechnology*, 2015, 15(2): pp. 1660–1668.

66. Jishi, R., M. Dresselhaus, and G. Dresselhaus, Electron-Phonon Coupling and the Electrical Conductivity of Fullerene Nanotubules. *Physical Review B*, 1993, 48(15): p. 11385.

67. Taherpour, A., Quantitative Relationship Study of Mechanical Structure Properties of Empty Fullerenes. *Fullerenes, Nanotubes, and Carbon Nonstructures*, 2008, 16(3): pp. 196–205.

68. Kataura, H., Y. Maniwa, M. Abe, A. Fujiwara, T. Kodama, K. Kikuchi, H. Imahori, Y. Misaki, S. Suzuki, and Y. Achiba, Optical Properties of Fullerene and Non-Fullerene Peapods. *Applied Physics A*, 2002, 74: pp. 349–354.

69. Lv, Z., Z. Deng, D. Xu, X. Li, and Y. Jia, Efficient Organic Light-Emitting Diodes with C60 Buffer Layer. *Displays*, 2009, 30(1): pp. 23–26.

70. Yu, D., K. Park, M. Durstock, and L. Dai, Fullerene-Grafted Graphene for Efficient Bulk Heterojunction Polymer Photovoltaic Devices. *The Journal of Physical Chemistry Letters*, 2011, 2(10): pp. 1113–1118.

71. Orbzd, A., and A. Mashhadzadeh, Density Functional Theory Based Molecular Dynamics Study on Hydrogen Storage Capacity of C-24, B12n12, Al-12 N-12, Be12o12, Mg12o12, and Zn12o12 Nanocages. *International Journal of Hydrogen Energy*, 2020, 45(11): pp. 6745–6756.

72. Coro, J., M. Suarez, L.S. Silva, K.I. Eguiluz, and G.R. Salazar-Banda, Fullerene Applications in Fuel Cells: A Review. *International Journal of Hydrogen Energy*, 2016, 41(40): pp. 17944–17959.

73. Giannopoulos, G.I., S.K. Georgantzinos, and N.K. Anifantis, Thermomechanical Response of Fullerene-Reinforced Polymers by Coupling Md and Fem. *Materials*, 2020, 13(18): p. 4132.

74. Giannopoulos, G.I., Linking Md and Fem to Predict the Mechanical Behaviour of Fullerene Reinforced Nylon-12. *Composites Part B: Engineering*, 2019, 161: pp. 455–463.

75. Dreyer, D.R., K.A. Jarvis, P.J. Ferreira, and C.W. Bielawski, Graphite Oxide as a Carbocatalyst for the Preparation of Fullerene-Reinforced Polyester and Polyamide Nanocomposites. *Polymer Chemistry*, 2012, 3(3): pp. 757–766.

76. Jung, J.-H., S. Vadahanambi, and I.-K. Oh, Electro-Active Nano-Composite Actuator Based on Fullerene-Reinforced Nafion. *Composites Science and Technology*, 2010, 70(4): pp. 584–592.

77. Barrera, E., J. Sims, and D. Callahan, Development of Fullerene-Reinforced Aluminum. *Journal of Materials Research*, 1995, 10(2): pp. 366–371.

78. Barrera, E.V., J. Sims, and D.L. Callahan, Development of Fullerene-Reinforced Aluminum. *Journal of Materials Research*, 1995, 10(2): pp. 366–371.

79. Degirmenci, U., and M. Kirca, Design and Mechanical Characterization of a Novel Carbon-Based Hybrid Foam: A Molecular Dynamics Study. *Computational Materials Science*, 2018, 154: pp. 122–131.

80. Gulmez, D.E., Y.O. Yildiz, and M. Kirca, Nanoporous Gold Reinforced with Carbon Based Nanomaterials: A Molecular Dynamics Study. *Composites Part B: Engineering*, 2018, 151: pp. 62–70.

81. Nasibulin, A.G., P.V. Pikhitsa, H. Jiang, D.P. Brown, A.V. Krasheninnikov, A.S. Anisimov, P. Queipo, A. Moisala, D. Gonzalez, G. Lientschnig, and A. Hassanien, A Novel Hybrid Carbon Material. *Nature Nanotechnology*, 2007, 2(3): pp. 156–161.

82. Pu, J., Y. Mo, S. Wan, and L. Wang, Fabrication of Novel Graphene–Fullerene Hybrid Lubricating Films Based on Self-Assembly for Mems Applications. *Chemical Communications*, 2014, 50(4): pp. 469–471.

83. Tayfun, U., Y. Kanbur, U. Abaci, H.Y. Guney, and E. Bayramli, Mechanical, Flow and Electrical Properties of Thermoplastic Polyurethane/Fullerene Composites: Effect of Surface Modification of Fullerene. *Composites Part B: Engineering*, 2015, 80: pp. 101–107.

84. Ogasawara, T., Y. Ishida, and T. Kasai, Mechanical Properties of Carbon Fiber/ Fullerene-Dispersed Epoxy Composites. *Composites Science and Technology*, 2009, 69(11–12): pp. 2002–2007.

85. Qiao, H., and Q. Wei., Functional Nanofibers in Lithium-Ion Batteries. In *Functional Nanofibers and Their Applications*. Elsevier, Woodhead Publishing, 2012, pp. 197–208. https://doi.org/10.1533/9780857095640.2.197

86. Ramanan, A., Nobel Prize in Chemistry 2019. *Resonance*, 2019, 24(12): pp. 1381–1395.

87. Ogawa, S., and P.V. Kamat, Lithium-Ion Batteries and Beyond: Celebrating the 2019 Nobel Prize in Chemistry–A Virtual Issue. *ACS Energy Letters* 2019. 4(11): pp. 2754–2756.

88. Xie, C., Y. Gao, and Z. Zheng, Textile composite electrodes for flexible batteries and supercapacitors: opportunities and challenges. *Advanced Energy Materials*, 2021, 11(3): p. 2002838.

89. Kamat, P.V., Lithium-Ion Batteries and Beyond: Celebrating the 2019 Nobel Prize in Chemistry–A Virtual Issue. *ACS Energy Letters*, 2019, 4(11): pp. 2757–2759.

90. Divakaran, A.M., M. Minakshi, P.A. Bahri, S. Paul, P. Kumari, A.M. Divakaran, and K.N. Manjunatha, Rational Design on Materials for Developing Next Generation Lithium-Ion Secondary Battery. *Progress in Solid State Chemistry*, 2021, 62: p. 100298.

91. Deng, D., Li-Ion Batteries: Basics, Progress, and Challenges. *Energy Science & Engineering*, 2015, 3(5): pp. 385–418.

92. Iwuoha, E., P.U. Nzereogu, A.D. Omah, F.I. Ezema, and A.C. Nwanya, Anode Materials for Lithium-Ion Batteries: A Review. *Applied Surface Science Advances*, 2022, 9: p. 100233.

93. Mekonnen, Y., A. Sundararajan, and A.I. Sarwat, A Review of Cathode and Anode Materials for Lithium-Ion Batteries. *SoutheastCon*, 2016, 2016: pp. 1–6.

94. Nguyen, M.Y., D.H. Nguyen, and Y.T. Yoon, A New Battery Energy Storage Charging/ Discharging Scheme for Wind Power Producers in Real-Time Markets. *Energies*, 2012, 5(12): pp. 5439–5452.

95. Li, Y., D.I. Stroe, Y. Cheng, H. Sheng, X. Sui, and R. Teodorescu, On the Feature Selection for Battery State of Health Estimation Based on Charging–Discharging Profiles. *Journal of Energy Storage*, 2021, 33: p. 102122.

96. Deng, D., M.G. Kim, J.Y. Lee, and J. Cho, Green Energy Storage Materials: Nanostructured Tio2 and Sn-Based Anodes for Lithium-Ion Batteries. *Energy & Environmental Science*, 2009, 2(8): pp. 818–837.

97. Dai, L., D.W. Chang, J.B. Baek, and W. Lu, Carbon Nanomaterials for Advanced Energy Conversion and Storage. *Small*, 2012, 8(8): pp. 1130–1166.

98. Evanoff, K., A. Magasinski, J. Yang, and G. Yushin, Nanosilicon-Coated Graphene Granules as Anodes for Li-Ion Batteries. *Advanced Energy Materials*, 2011, 1(4): pp. 495–498.

99. Hao, Y., Y. Wang, L. Wang, Z. Ni, Z. Wang, R. Wang, C.K. Koo, Z. Shen, and J.T. Thong, Probing Layer Number and Stacking Order of Few-Layer Graphene by Raman Spectroscopy. *Small*, 2010, 6(2): pp. 195–200.

100. Zhou, G., D.W. Wang, F. Li, L. Zhang, N. Li, Z.S. Wu, L. Wen, G.Q. Lu, and H.M. Cheng, Graphene-Wrapped Fe3o4 Anode Material with Improved Reversible Capacity and Cyclic Stability for Lithium Ion Batteries. *Chemistry of Materials*, 2010, 22(18): pp. 5306–5313.

101. Sun, Y., Q. Wu, and G. Shi, Graphene Based New Energy Materials. *Energy & Environmental Science*, 2011, 4(4): pp. 1113–1132.

102. Boal, A.K., F. Ilhan, J.E. DeRouchey, T. Thurn-Albrecht, T.P. Russell, and V.M. Rotello, Self-Assembly of Nanoparticles into Structured Spherical and Network Aggregates. *Nature*, 2000, 404(6779): pp. 746–748.

103. Boettcher, S.W., J.M. Spurgeon, M.C. Putnam, E.L. Warren, D.B. Turner-Evans, M.D. Kelzenberg, J.R. Maiolo, H.A. Atwater, and N.S. Lewis, Energy-Conversion Properties of Vapor-Liquid-Solid–Grown Silicon Wire-Array Photocathodes. *Science*, 2010, 327(5962): pp. 185–187.

104. Chang, J., X. Huang, G. Zhou, S. Cui, S. Mao, and J. Chen, Three-Dimensional Carbon-Coated Si/Rgo Nanostructures Anchored by Nickel Foam with Carbon Nanotubes for Li-Ion Battery Applications. *Nano Energy*, 2015, 15: pp. 679–687.

105. Lee, J.W., A.S. Hall, J.D. Kim, and T.E. Mallouk, A Facile and Template-Free Hydrothermal Synthesis of Mn3o4 Nanorods on Graphene Sheets for Supercapacitor Electrodes with Long Cycle Stability. *Chemistry of Materials*, 2012, 24(6): pp. 1158–1164.

106. Kim, H., S.W. Kim, J. Hong, Y.U. Park, and K. Kang, Electrochemical and Ex-Situ Analysis on Manganese Oxide/Graphene Hybrid Anode for Lithium Rechargeable Batteries. *Journal of Materials Research*, 2011, 26(20): pp. 2665–2671.

107. Jia, H., P. Gao, J. Yang, J. Wang, Y. Nuli, and Z. Yang, Novel Three-Dimensional Mesoporous Silicon for High Power Lithium-Ion Battery Anode Material. *Advanced Energy Materials*, 2011, 1(6): pp. 1036–1039.

108. Jung, Y.S., A.S. Cavanagh, A.C. Dillon, M.D. Groner, S.M. George, and S.H. Lee, Enhanced Stability of Licoo2 Cathodes in Lithium-Ion Batteries Using Surface Modification by Atomic Layer Deposition. *Journal of the Electrochemical Society*, 2009, 157(1): p. A75.

109. Jung, Y.S., A.S. Cavanagh, L.A. Riley, S.H. Kang, A.C. Dillon, M.D. Groner, S.M. George, and S.H. Lee, Ultrathin Direct Atomic Layer Deposition on Composite Electrodes for Highly Durable and Safe Li-Ion Batteries. *Advanced Materials*, 2010, 22(19): pp. 2172–2176.

110. Jung, Y.S., A.S. Cavanagh, L. Gedvilas, N.E. Widjonarko, I.D. Scott, S.H. Lee, G.H. Kim, S.M. George, and A.C. Dillon, Improved Functionality of Lithium-Ion Batteries Enabled by Atomic Layer Deposition on the Porous Microstructure of Polymer Separators and Coating Electrodes. *Advanced Energy Materials*, 2012, 2(8): pp. 1022–1027.

111. He, Y., X. Yu, Y. Wang, H. Li, and X. Huang, Alumina-Coated Patterned Amorphous Silicon as the Anode for a Lithium-Ion Battery with High Coulombic Efficiency. *Advanced Materials*, 2011, 23(42): pp. 4938–4941.

112. Gu, W., W. Zhang, X. Li, H. Zhu, J. Wei, Z. Li, Q. Shu, C. Wang, K. Wang, W. Shen, and F. Kang, Graphene Sheets from Worm-Like Exfoliated Graphite. *Journal of Materials Chemistry*, 2009, 19(21): pp. 3367–3369.

113. Mikolajick, T., A. Heinzig, J. Trommer, S. Pregl, M. Grube, G. Cuniberti, and W.M. Weber, Silicon Nanowires–A Versatile Technology Platform. *Physica Status Solidi (RRL)–Rapid Research Letters*, 2013, 7(10): pp. 793–799.

114. Kelzenberg, M.D., S.W. Boettcher, J.A. Petykiewicz, D.B. Turner-Evans, M.C. Putnam, E.L. Warren, J.M. Spurgeon, R.M. Briggs, N.S. Lewis, and H.A. Atwater, Enhanced Absorption and Carrier Collection in Si Wire Arrays for Photovoltaic Applications. *Nature Materials*, 2010, 9(3): pp. 239–244.

115. Baek, E., S. Pregl, M. Shaygan, L. Römhildt, W.M. Weber, T. Mikolajick, D.A. Ryndyk, L. Baraban, and G. Cuniberti, Optoelectronic Switching of Nanowire-Based Hybrid Organic/Oxide/Semiconductor Field-Effect Transistors. *Nano Research*, 2015, 8: pp. 1229–1240.

116. Boukai, A.I., Y. Bunimovich, J. Tahir-Kheli, J.K. Yu, W.A. Goddard, III, and J.R. Heath, Silicon Nanowires as Efficient Thermoelectric Materials. *Nature*, 2008, 451(7175): pp. 168–171.

117. Ho, M.-S., C.P. Huang, J.H. Tsai, C.F. Chou, and W.J. Lee, Probing C84-Embedded Si Substrate Using Scanning Probe Microscopy and Molecular Dynamics. *JoVE (Journal of Visualized Experiments)*, 2016, 115: p. e54235. https://doi.org/10.3791/54235.

118. Kartel, M., V.S. Kuts, A.G. Grebenyuk, and Y.A. Tarasenko, Comparative Quantum Chemical Examination of Lithiation/Delithiation Processes in Sin Nanoclusters and Cmsin Nanocomposites. *Хімія, фізика та технологія поверхні*, 2015, 6(1): pp. 32–41.

119. Rostamiyan, Y., V. Mohammadi, and A. Hamed Mashhadzadeh, Mechanical, Electronic and Stability Properties of Multi-Walled Beryllium Oxide Nanotubes and Nanopeapods: A Density Functional Theory Study. *Journal of Molecular Modeling*, 2020, 26(4): p. 76.

120. Albooyeh, A., A. Dadrasi, A. Hamed Mashhadzadeh, and M.R. Saeb, Theory for Designing Mechanically Stable Single-and Double-Walled Sige Nanopeapods. *Journal of Molecular Modeling*, 2021, 27(7): p. 214.

121. Gao, A., S. Mukherjee, I. Srivastava, M. Daly, and C.V. Singh, Atomistic Origins of Ductility Enhancement in Metal Oxide Coated Silicon Nanowires for Li-Ion Battery Anodes. *Advanced Materials Interfaces*, 2017, 4(23): p. 1700920.

122. Wu, Y., Y. Cui, L. Huynh, C.J. Barrelet, D.C. Bell, and C.M. Lieber, Controlled Growth and Structures of Molecular-Scale Silicon Nanowires. *Nano Letters*, 2004, 4(3): pp. 433–436.

5 Carbon-Based Composites for Energy Storage Applications

Ganjar Fadiilah, Rahmat Hidayat and Is Fatimah

1 INTRODUCTION

The development of advanced materials plays a crucial role in pursuing sustainable energy solutions, particularly in energy storage. Hybrid nanostructures based on carbon have emerged as a promising avenue for enhancing the performance and efficiency of energy storage devices, such as batteries and supercapacitors. Carbon-based hybrid nanostructures offer exceptional properties and versatility in energy storage applications, such as catalysts and electrodes. Carbon-based nanomaterials, such as carbon nanotubes (CNTs), graphene, and fullerenes, have attracted much attention as materials that may be useful in the next phase of energy conversion and storage. These materials have unique physicochemical properties, including exceptional porosity, strong and flexible structures, good cyclic stability, and outstanding electrical conductivity [1, 2]. These qualities make them the perfect subjects for research within these ever-expanding fields of science. Improvements in both the atomic and electronic domains are critical for the development of durable and improved electronics [3]. Adding metal nanoparticles (NPs), metal oxides, polymers, and LDHs to carbon nanomaterials is one of the most promising approaches to creating effective electrocatalysts for energy storage applications. Due to their high surface area and rapid charge/discharge kinetics, carbon-based materials are valued for their use as electrodes in energy storage devices [4, 5].

Additionally, hybrid nanostructures based on carbon have emerged as promising catalysts for electrochemical reactions central to energy conversion and storage processes. The tunable properties of carbon-based hybrid catalysts enable precise control over reaction kinetics, selectivity, and long-term stability, thereby offering significant advantages for improving the efficiency and durability of fuel cell systems. On the one hand, these carbon-based nanohybrid materials' large pore structure and specific area, notable conductivity, and superior stability significantly benefit their enhanced electrochemical performance, as shown in Figure 5.1a. On the other hand, carbon-based materials' structures can be changed to create nanomaterials in a variety of dimensions, including one-dimensional (1D) materials in the form of nanofibers or nanorods, two-dimensional (2D) materials in the form of nanosheets or nanofilms, and three-dimensional (3D) materials in complex forms (Figure 5.1b). The reactivity

DOI: 10.1201/9781003561262-5

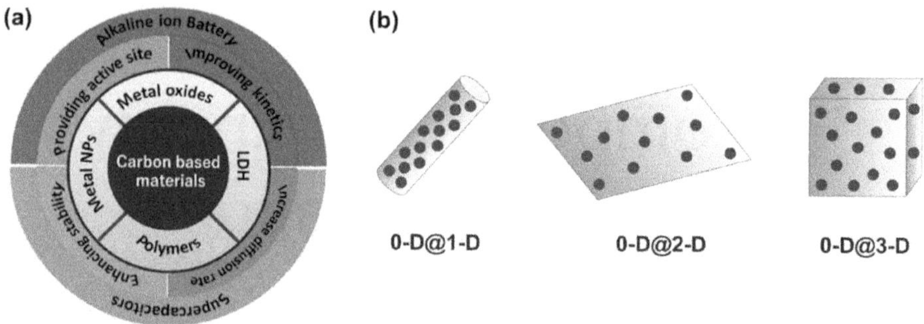

FIGURE 5.1 (a) Different carbon-based nanohybrid materials for energy storage applications and (b) an illustration of incorporation of 0D structures into other morphologies of carbon-based materials.

and specific surface area of the materials are increased; more active sites are provided; and the absolute volume change is weakened by the impact of the small size of zero-dimensional (0D) carbon-based nanomaterials. 1D carbon-based nanomaterials have their own linear electron and ion transport pathways, which may encourage the movement of ions and quicken the electrode-to-electrolyte charge transfer. Because of their vast specific surface area, flexibility of the nanosheets, and changeable layer spacing, 2D carbon-based nanomaterials have the potential to be good candidates for electrodes. Excellent structural and mechanical stability and high connectivity and conductivity are some critical components of 3D carbon-based nanomaterials' superior electrochemical performance. In addition to enhancing the electrode materials' electrochemical performance, the design of these structural dimensions can offer appropriate options for various application scenarios and settings [6–8].

Hybrid nanostructures based on carbon hold promise for many other energy storage applications, including batteries, thermoelectric devices, and supercapacitors. The unique properties of carbon and the collaborative efforts of researchers and scientists in the field are pushing the boundaries of energy storage and conversion technologies, thereby striving toward arriving at more efficient, sustainable, and scalable solutions for the growing global demand for clean energy. Despite the progress made, significant efforts are required to address the evolving energy storage and conversion challenges. This chapter discusses the practical uses and optimization benefits of carbon-based nanohybrid materials in various energy storage devices such as supercapacitors and alkaline ion batteries (AIBs). It also suggests the main obstacles and prospects for advancing carbon-based nanohybrid materials. This chapter aims to help readers understand the interactions between various dimensional material structures and offer insightful advice for synthesizing high-performance carbon-based nanohybrid energy storage systems.

2 CARBON-BASED MATERIALS AS CATALYSTS AND ELECTRODES

Carbonaceous materials serve as versatile substances in catalysis and electrode applications. These carbon-based materials are categorized by their dimensional structures: 0D (nanodots and nanospheres), 1D (nanorods, nanowires, nanofibers, and nanotubes), 2D (graphene), and 3D (graphite and carbon network) [9]. Additionally, graphene stands out for its unique ability to exhibit 0D properties in the form of fullerenes, 1D properties as nanotubes, and 3D properties as graphite. Carbon-based materials serve as catalysts in various applications due to their unique properties and versatility. Their role as catalysts includes both traditional chemical reactions and advanced electrochemical processes [10, 11].

In the realm of traditional catalysis, carbon-based materials excel at boosting reaction rates and selectivity while remaining intact throughout the process. Their large surface area, adjustable electronic characteristics, and chemical robustness equip them for a broad spectrum of chemical reactions [12]. For example, substances like activated carbon and CNTs find utility in heterogeneous catalysis for tasks such as organic synthesis, environmental pollutant degradation, and air purification [13]. In the field of electrocatalysis, carbon-based materials are pivotal for various electrochemical processes found in fuel cells [14], batteries [15], and electrolyzers [16]. They adeptly convert electrical energy to chemical forms and back again, and they do so with remarkable efficiency. Graphene, CNTs, and metal NPs supported by carbon stand out as popular choices for electrocatalytic applications. Their customized surface properties, exceptional electrical conductivity, and resilience promote swift charge transfer and optimal reaction rates, thereby positioning them as prime candidates for energy storage and conversion tasks. Modifying carbon-based materials further enhances their catalytic performance [17]. Introducing specific functional groups or doping them with heteroatoms can fine-tune their catalytic efficiency, selectivity, and endurance for specific roles. Moreover, integrating carbon-based materials with different catalysts or supporting structures can yield synergistic benefits, thus improving their overall effectiveness [18]. The primary strength of carbon-based electrodes lies in their adaptability. Their characteristics can be refined using techniques such as surface modification, doping, and nanostructuring to boost their electrochemical capabilities [19]. For instance, graphene doped with nitrogen demonstrates an enhanced electrocatalytic performance in oxygen reduction reactions (ORRs) compared to its pristine counterpart [20]. Carbon-based materials find extensive use as electrodes across a range of electrochemical applications, thanks to their distinct attributes like high electrical conductivity, stability, and large surface area. Their applicability as electrodes extends to energy storage devices [21], sensors [22], electrochemical reactors [23], etc. In energy storage applications, carbon-based materials like graphite, CNTs, and graphene serve as electrodes in batteries, supercapacitors, and fuel cells. For instance, graphite is commonly used as an anode material in lithium-ion batteries due to its ability to intercalate lithium ions efficiently [24]. CNTs and graphene exhibit a high surface area and enhanced electrical conductivity, making them ideal candidates for supercapacitor electrodes, which store energy through electrostatic charge accumulation.

Carbon-based electrodes are also extensively utilized in sensor technology by modifying their surfaces to selectively capture specific analytes, enabling the detection of a wide range of chemical substances. Adapted carbon electrodes play a vital role in electrochemical sensors by exhibiting high sensitivity and selectivity in detecting pollutants, biomolecules, and other target substances. In electrochemical reactors, these electrodes enable a variety of redox reactions by serving as a conductive platform for electron transfer. They find applications in areas such as wastewater treatment, electroorganic synthesis, and chemical sensing. By selecting an appropriate carbon material and modifying its surface, the material can be tailored to enhance the electrode's performance and selectivity for specific reaction types. Thus, carbon-based electrodes offer versatile solutions across different electrochemical applications – from environmental remediation to advanced sensing technologies.

3 INTEGRATION OF CARBON WITH OTHER MATERIALS

The integration of carbon with other materials has not only emerged but also surged as a promising approach to significantly enhance the performance and functionality of energy storage devices such as batteries and supercapacitors. In order to utilize carbon electrodes to their fullest potential in energy storage applications, it is necessary to optimize their synthesis and fabrication. Previous studies have reported that they used various methods to modify the composition, structure, and properties of carbon compounds, with the aim of improving their energy storage capabilities. There are several practical methods for producing graphene as an electrode material, such as chemical vapor deposition (CVD), arc discharge approach, micromechanical exfoliation, electrochemical epitaxial growth, chemical approaches, and inclusion in graphite. In order to enhance its effectiveness as an energy storage tool, the structure of the carbon catalyst material is usually altered through surface modification and functionalization with other materials. Various materials have been developed to act as modifiers, including metal NPs, metal oxides/metal sulfides, conductive polymers, and LDHs. One of the significant benefits of integrating carbon with other materials is its ability to improve conductivity and charge transfer kinetics within energy storage devices. In a study by Song et al. (2018), metal NPs grafted onto CNTs were used for H_2 storage applications [25]. Researchers investigated and synthesized Ag NPs and Cu NPs using electrostatic adsorption and *in situ* reduction methods. They found that the different metal NPs had varying storage capacities, which was attributed to their distinct characteristics. The thermal conductivity of the material, influenced by the characteristics of metal NPs, can affect the reaction time and reaction equilibrium during the catalysis process. Scientists are developing a material called Transition Metal Oxide (TMO) for electrochemical applications. TMO has fast and reversible surface redox reactions, making it an excellent choice for applications requiring high specific capacitance and energy density. Various metal oxides, such as ruthenium oxide (RuO_2), manganese dioxide (MnO_2), iron hydroxide oxide ($Fe(OH)_2$), cobalt oxide (Co_3O_4), nickel oxide (NiO), molybdenum trioxide (MoO_3), tin oxide (SnO_2), vanadium oxide (V_2O_5), and tungsten oxide (WO_3) have been studied to determine their suitability as electrode materials for pseudocapacitors. Different methods can be used to produce porous transition metal-based materials, including electrospinning, etching, and deposition. However, metal oxide electrodes have some drawbacks, such as self-discharge and a tiny operational potential window with low electrical conductivity and decreased cycling capability [26]. To overcome these issues, scientists have studied the development of binary transition metal oxides (BTMOs). BTMOs consist of at least one transition metal ion and one electrochemically active/inactive ion, providing superior advantages such as structural stability and better electronic conductivity compared to a single TMO[27].

4 ROLE AND CATALYTIC PROPERTIES OF HYBRID NANOSTRUCTURES

Hybrid nanostructures have become increasingly popular due to their unique properties and wide range of applications. They play an essential role in catalysis by exhibiting enhanced catalytic activity, selectivity, and stability compared to their individual components. One of the main advantages of hybrid nanostructures in catalysis is their ability to synergistically combine the intrinsic properties of various nanomaterials, such as metals, metal oxides, and carbon-based materials, to achieve superior catalytic performance [28]. Noble metals, such as Pt NPs, have been traditionally considered as highly effective catalysts for the ORR. However, electrodes based on Pt suffer from several challenges, including low durability, inadequate power sources, methanol crossover, and high expenses [29, 30]. Consequently, the development of hybrid materials by combining noble metal NPs with other materials has been explored to enhance their performance. For example, when noble metals such as Pt or Pd are integrated with metal oxides such as TiO_2 or CeO_2, the catalytic activity for hydrogenation, oxidation, and fuel cell reactions can increase significantly [31, 32]. The synergy between the metal and metal oxide components allows for increased active site dispersion, improved charge transfer kinetics, and optimized adsorption-desorption properties, thus resulting in improved catalytic efficiency.

Furthermore, hybrid nanostructures offer tunable catalytic properties through precise control over their composition, morphology, and nanostructure. The catalytic behavior of hybrid nanomaterials can be fine-tuned by adjusting parameters such as particle size, shape, and surface chemistry to meet specific reaction requirements and optimize performance metrics such as reaction rate, conversion, and selectivity. Maity et al. (2022) reported the CuO-stabilized oxy-functionalized boron nitride-CNT nanohybrid (CuO/mK-BN/CNT) material for supercapacitor electrodes [33]. The developed material showed high stability after 10,000 cycles. By functionalizing CuO with BN, it is possible to create a material with a high surface area and additional electronic energy states that can affect its electrochemical activity. Additionally, CuO functionalization as a nanohybrid material results in the partial overlap of the valence and conduction bands, thereby making BN more suitable for charge transport through a variable range hopping mechanism. In another study, introducing multilayer ZnO with a high piezoelectric constant into wood (around 9.93 pm V^{-1}) leads to a further increase in performance as a hybrid material with a six-fold increase in the resulting current. This showed that the benefits of deep ZnO decoration are huge [34]. The uniform distribution of ZnO, the crystal structure of Wurtzite ZnO, and its good mechanical properties strongly support its application as a piezoelectric device. Besides that, this tunability allows the tailoring of catalyst designs for various catalytic applications, from environmental remediation to energy conversion and storage. In addition to their enhanced catalytic activity, hybrid nanostructures show remarkable stability and durability under harsh reaction conditions, making them suitable for long-term catalytic applications. The synergistic interaction between various components in hybrid nanostructures can reduce catalyst poisoning, surface oxidation, and sintering, which are common degradation mechanisms in catalysis. The high surface area and unique nanostructure of hybrid nanomaterials facilitate efficient mass transport and diffusion of reactants and products, thereby minimizing mass transfer limitations and improving catalytic performance [35].

The catalytic properties of hybrid nanostructures are governed by various factors; the properties of the constituent materials, their chemical composition, and the architecture of the hybrid nanostructures. For example, heterointerfaces between different components in $SrAl_2O_4/rGO$ hybrid nanostructures can create active sites with unique electronic and chemical properties, thereby enhancing electrocatalytic reactivity [36]. When two materials are mixed to form a hybrid, it can result in some irregularities and defects in the structure. These defects create more electrochemical active sites and reduce the length of pathways for ion and electron migration. This ultimately increases the redox activity, thus allowing for faster charging and polarity and making it an ideal choice for electrochemical energy storage devices. Additionally, the presence of defects, vacancies, and surface functional groups in hybrid nanostructures can modulate their catalytic behavior by influencing the adsorption energy, reaction pathway, and reaction kinetics. Moreover, the catalytic properties of hybrid nanostructures can be tuned through advanced synthetic techniques and engineering approaches. Bottom-up synthesis methods such as sol-gel synthesis, CVD, and colloidal synthesis enable precise control over hybrid nanostructure size, composition, and surface properties. In addition, surface modification techniques such as doping, surface functionalization, and ligand exchange enable further tuning of the catalytic properties, thus improving catalytic activity, selectivity, and stability. In conclusion, the unique properties and tunable catalytic behavior of hybrid nanostructures make them attractive candidates for catalytic applications. Carbon-based materials can be modified with conductive polymer materials, just like metal NPs and metal oxides. The choice of polymer material and synthesis route is crucial in this process. Modification with conductive polymers is promising as it makes it suitable for various industrial applications. The mechanical strength of the modified material is an essential property with broad significance. Additionally, properties, such as electrical conductivity, can be tailored for applications in specific electronics, energy storage, and automotive sectors. Modified carbon-based materials exhibit exceptional mechanical strength, corrosion resistance, tunable properties and enable lightweight construction, which differentiate them from traditional carbon materials like graphite, graphene, CNTs, and carbon black [37]. Although these carbon materials exhibit high electrical conductivity and surface area, they lack the properties required for structural energy storage systems, such as mechanical integrity, corrosion resistance and enabling lightweight construction and scalable manufacturability.

5 APPLICATIONS

5.1 BATTERIES

Nanohybrid structures have emerged as a promising avenue for improving the performance of batteries. They offer a unique blend of properties that leverage the strengths of individual components while mitigating their weaknesses. Nanohybrid structures can achieve synergistic effects by combining different materials at the nanoscale, which can go beyond what is achievable with conventional materials alone. This makes them a compelling solution for energy storage, where efficiency, stability, and capacity are essential. One of the key advantages of nanohybrid structures in batteries is their ability to enhance the kinetics of electrochemical reactions. Narayanasamy et al. (2021) reported the nanohybrid structure of rGO-VSe$_2$ as a cathode material in zinc-ion battery (ZIB) [38]. Combining conductive rGO with another material, VSe$_2$, creates a 3D pathway for electrons to move more efficiently during chemical reactions. This hybrid structure was used as a cathode, and it displayed a reversible potential of 221.5 mAh g^{-1} at 0.5 A g^{-1}, along with excellent conversion efficiency and cycling stability. After 150 cycles, the cathode retained about 91.6% of its initial performance. In addition to conventional battery materials, there have been recent developments in the field of battery technology, such as the In$_2$S$_3$ electrode for alkaline ion batteries (AIBs) [39]. This hybrid material has been noted for its capacity for increased energy storage, cycle capability, and rate capability, while also being environmentally friendly. In$_2$S$_3$ has a unique crystal structure that enables the rapid diffusion of alkali ions (Na$^+$, Li$^+$, or K$^+$) within the lattice during the charging process, thereby providing an excellent theoretical specific capacity. In practical terms, this means that In$_2$S$_3$ can store large amounts of electrical energy and exhibit fast charge and discharge rates.

Materials at the nanoscale exhibit a higher surface area-to-volume ratio, thus providing more active sites for electrochemical reactions. This increased interfacial area facilitates faster ion and electron transport within the battery, thereby improving charge/discharge rates and overall efficiency. Moreover, nanohybrids can alleviate issues related to electrode swelling and volume expansion during cycling, which are common challenges in high-capacity battery systems. The mechanism of how the graphene/In$_2$S$_3$ composite works as an electrode in a battery is illustrated in Figure 5.2b. The 3D structure of the composite provides a large surface area which increases the active sites for Na+ adsorption, thereby resulting in improved reaction kinetics for sodium storage. Additionally, the porous layered structure of the composite allows the electrolyte to penetrate the nanostructure of the electrode material, which lowers the Na+ diffusion barrier and improves kinetics. Moreover, the homogeneous carbon coating on In$_2$S$_3$ enhances electronic conductivity; and when combined with the other factors, the sodium storage capacity is improved. Lastly, the reversible capacity of the material is influenced by the type of nanohybrid composite developed (Figure 5.2a, data sources: [40–46]).

Through careful design and engineering, nanohybrid architectures can offer enhanced mechanical stability and structural integrity, thereby prolonging the cycling life of the battery. Nanohybrids also enable precise control over the composition and morphology of battery materials, hence allowing for tailored properties to meet specific performance requirements. For instance, nanohybrids can enhance electronic conductivity and promote more uniform electrode-electrolyte interactions by incorporating conductive nanomaterials such as CNTs or graphene into the electrode structure. This can mitigate electrode degradation and impedance growth, thus improving cycling stability and capacity retention. Additionally, introducing nanostructured catalysts within the electrode or electrolyte can catalyze critical

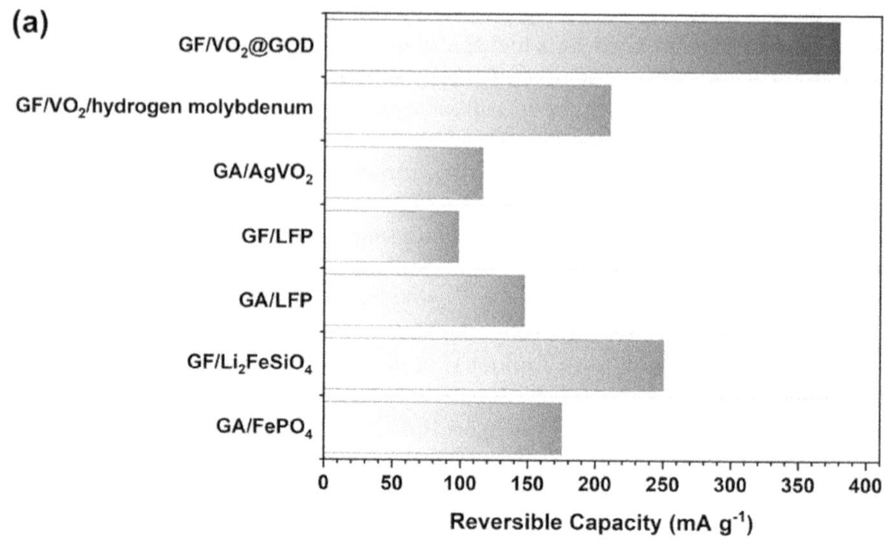

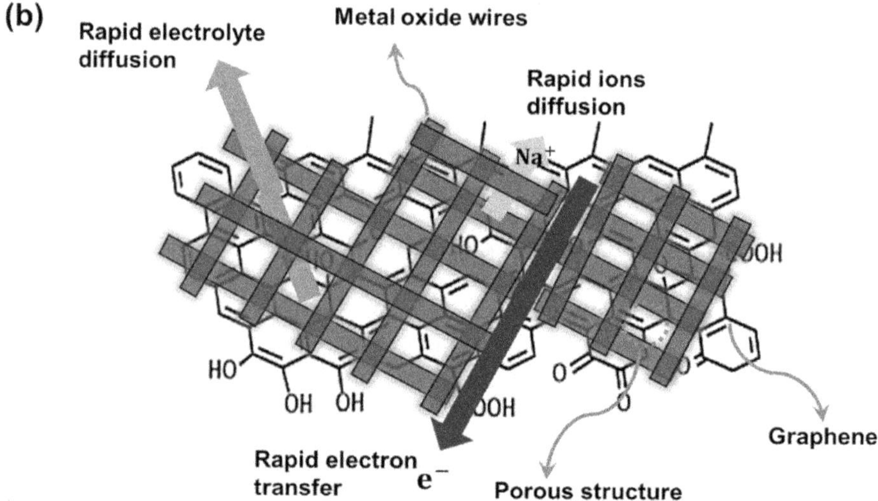

FIGURE 5.2 (a) The performance of the cathode based on graphene nanohybrid materials for an AIB and (b) the mechanism of alkaline transport in graphene hybrid materials.

electrochemical reactions, thereby further enhancing the overall energy conversion efficiency of the battery system. Apart from performance enhancements, nanohybrid structures hold promise for addressing critical challenges related to battery safety and environmental sustainability. By carefully selecting and designing nanomaterial components, researchers can minimize the use of toxic or environmentally harmful substances commonly found in traditional battery chemistries. Furthermore, nano-hybrid batteries' improved efficiency and longevity can contribute to reducing over-all energy consumption and greenhouse gas emissions, thereby supporting efforts toward a more sustainable energy future.

However, there are some challenges that need to be overcome before widespread commercialization can be achieved. These include scalability issues associated with the synthesis and fabrication of nanomaterials, as well as concerns regarding cost-effectiveness and long-term stability. Additionally, the potential impacts of nanomaterials on human health and the environment must be carefully evaluated to ensure the responsible development and deployment of nanohybrid battery technologies. In conclusion, nanohybrid structures offer a pathway toward enhanced performance, safety, and sustainability. Through the strategic integration of nanoscale materials and innovative design strategies, nanohybrid batteries have the potential to transform energy storage and power a wide range of applications, from portable electronics to electric vehicles and grid-scale energy storage systems. However, further research and development are crucial to unlock the full potential of nanohybrid structures and accelerate the transition toward a cleaner, more efficient energy landscape.

5.2 SUPERCAPACITORS

Supercapacitors are energy storage devices that store electrical energy via the reversible adsorption of ions at the electrode-electrolyte interface. While traditional batteries are limited in their ability to deliver high power densities and exhibit rapid charge/discharge rates, supercapacitors can do so; this makes them ideal for applications that require bursts of energy or frequent cycling. Nanohybrid structures, with their ability to enhance the performance metrics of supercapacitors, have emerged as a promising approach in energy storage. Nanohybrid structures offer numerous benefits over traditional supercapacitors: increasing the electrodes' effective surface area and enhancing ion adsorption and charge storage capacity. The electrochemical performance of these materials is excellent, with good power delivery, operational safety, and a long service life. These are easy to synthesize and manufacture on a large scale; these must be binder-free to ensure high electron conductivity.

Nanomaterials such as graphene, CNTs, layered hydroxide (LDH), and metal oxides are ideal for supercapacitor electrode materials due to their high specific surface areas and excellent electrical conductivity [47]. By integrating these nanomaterials into the electrode matrix, nanohybrid structures can significantly enhance the electrochemical performance of supercapacitors. Furthermore, nanohybrid structures offer avenues for precise control over the morphology and composition of supercapacitor electrodes, allowing for tailored properties to meet specific application requirements. LDH-based electrodes have demonstrated promising electrochemical energy storage capabilities due to their faster ion exchange and large surface area. The abundant oxygen vacancies in LDHs create intermediate gaps in electronic states, which enhance the electrochemical charge transfer at the electrode and make them suitable for application in energy storage. However, despite their high electrochemical activity, LDHs exhibit low conductivity and instability after several cycles [48]. Therefore, to address this issue, the core-shell structure has been found to be the most efficient for increasing the electrical conductivity and specific surface area of the developed hybrid electrode materials. In line with this, a hybrid supercapacitor device has been fabricated using $NiCo_2S_4$@NiV-LDH/NF and biochar as a battery-type electrode and a capacitive-type electrode, respectively, resulting in an improved high specific capacity of 1778.8 Cg $^{-1}$ (3557.6 Fg^{-1}) [49]. A recent research project examined the development of an electrode composed of CuO/NiCo-LDHs on a highly conductive Cu foam substrate. This core-shell assembly was subjected to an electrochemical activation process that modified its electronic structure to enhance its energy storage capabilities. As a result, the interfacial structure of the CuO and NiCo-LDHs was significantly altered; this led to improved ion diffusion and electron conduction properties [50].

Researchers can optimize the nanohybrid architecture through innovative design and engineering approaches to maximize energy storage capacity, minimize internal resistance, and improve long-term cycling stability. In addition to improving super-capacitors' performance metrics, nanohybrid structures also promise to expand their range of applications in various fields, including portable electronics, renewable energy systems, and electric vehicles. Hybrid materials such as metal sulfides and mesoporous materials like LDH and graphene are widely used in various applications. To enhance the flexibility and conductivity of the electrode, researchers have replaced oxygen with sulfur. A recent study by Cheng et al. (2022) utilized NiCo$_2$S$_4$ NPs that have richer redox active sites in the electrode composition [52]. However, due to the small surface area, chemical instability, and weak mechanical properties, researchers had to incorporate graphene and LDH materials. The use of Ni-Mo LDH with NPs led to abundant active sites, while GO increased conductivity and electro-chemical activity. Researchers employed flower-like 3D hierarchical structures in asymmetric supercapacitors, which exhibited a specific capacitance of 1346.0 Fg^{-1} after 10,000 cycles. Related studies have also proven that the hybrid formation with CuCo$_2$S$_4$ can maintain the capacitance retention rate reduction value of 92.1% after 5,000 cycles at a current density of 5 A g^{-1}, as shown in Figure 5.3.

Nanohybrid supercapacitors offer significant performance advantages; these enable the development of advanced energy storage solutions for emerging technologies such as wearable electronics, IoT devices, and hybrid electric vehicles. Furthermore, utilizing nanohybrid structures in supercapacitors can contribute to advancing sustainable energy technologies by improving energy efficiency and reducing environmental impact. By enhancing the energy storage capacity and cycling

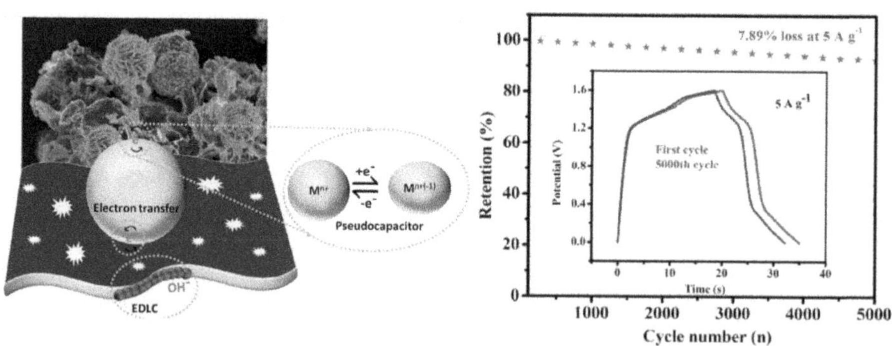

FIGURE 5.3 The purposed mechanism and cycling performance of the supercapacitor based on metal sulfides and mesoporous materials [51].

stability of supercapacitors, nanohybrid structures can enable the development of more efficient energy storage systems for renewable energy integration, grid stabilization, and energy harvesting applications. While nanohybrid structures have demonstrated significant advancements in supercapacitors, there are several challenges in realizing their full potential. These challenges include scalability issues associated with the synthesis and fabrication of nanomaterials and concerns regarding cost-effectiveness and long-term stability. Additionally, integrating nanohybrid structures into commercial supercapacitor devices requires careful optimization of manufacturing processes and electrode architectures to ensure consistent performance and reliability. In conclusion, applying nanohybrid structures in supercapacitors represents a promising avenue for advancing energy storage technology. By leveraging the unique properties of nanomaterials and innovative design approaches, nanohybrid supercapacitors offer enhanced performance characteristics, improved energy efficiency, and expanded functionality compared to traditional energy storage devices. The potential applications of nanohybrid supercapacitors in emerging technologies underscore their significance in the future of energy storage solutions.

6 CONCLUSION AND FUTURE PERSPECTIVE

Incorporating carbon-based hybrid nanostructures as catalysts and electrodes exhibit great potential to advance energy storage technology. Hybrid nanostructures synergistically combine conducting polymers, LDHs, metal NPs, and oxides with carbonaceous materials to provide improved stability, scalability, and electrochemical performance. These hybrid designs solve significant issues related to charge/discharge rates, cycle stability, and capacity retention by taking advantage of carbon's unique qualities, including high conductivity, surface area, and chemical stability.

Future research efforts in this sector should focus on refining the design and synthesis of hybrid nanostructures to achieve even higher performance metrics. Techniques such as surface functionalization, interface engineering, and exact control over nanostructure morphology will be essential for modifying the characteristics of hybrid materials to meet the demands of various energy storage systems. Investigating cutting-edge hybrid designs and novel carbon-based materials may present fresh opportunities to push the limits of energy storage performance and enable game-changing uses in grid-scale storage, electric cars, and renewable energy. Further research and development into carbon-based hybrid nanostructures can completely change the field of energy storage and accelerate the shift to sustainable energy sources.

REFERENCES

[1] F.I. Fajarwati, R. Hidayat, G. Fadillah, Synthesis and transformation of graphene-like structures from bamboo waste for photoelectrochemical devices, *Carbon Trends*, 15 (2024), 100351.

[2] K. Wakabayashi, Graphene, electronic properties and topological properties, in: T. Chakraborty (Ed.) *Encyclopedia of Condensed Matter Physics* (Second Edition), Academic Press, Oxford, 2024, pp. 273–287.

[3] H.-c. Liu, S. Zhu, Y.-z. Chang, W.-j. Hou, G.-y. Han, Pitch-based carbon materials: A review of their structural design, preparation and applications in energy storage, *New Carbon Materials*, 38 (2023), 459–473.

[4] T.-W. Chen, S.-M. Chen, G. Anushya, R. Kannan, P. Veerakumar, A.G. Al-Sehemi, V. Mariyappan, S. Alargarsamy, M.M. Alam, T.C. Mahesh, R. Ramachandran, P. Kalimuthu, Electrochemical energy storage applications of functionalized carbon-based nanomaterials: An overview, *International Journal of Electrochemical Science*, 19 (2024), 100548.

[5] T. Lalire, C. Longuet, A. Taguet, Electrical properties of graphene/multiphase polymer nanocomposites: A review, *Carbon*, 225 (2024), 119055.

[6] C.-y. Zhu, Y.-w. Ye, X. Guo, F. Cheng, Design and synthesis of carbon-based nanomaterials for electrochemical energy storage, *New Carbon Materials*, 37 (2022), 59–92.

[7] M. Li, B. Mu, Effect of different dimensional carbon materials on the properties and application of phase change materials: A review, *Applied Energy*, 242 (2019), 695–715.

[8] S. Taioli, Enabling materials by dimensionality: From 0D to 3D carbon-based nanostructures, in: T. Onishi (Ed.) *Theoretical Chemistry for Advanced Nanomaterials: Functional Analysis by Computation and Experiment*, Springer Singapore, Singapore, 2020, pp. 135–200.

[9] Y. Wang, L. Zhang, H. Hou, W. Xu, G. Duan, S. He, K. Liu, S. Jiang, Recent progress in carbon-based materials for supercapacitor electrodes: A review, *Journal of Materials Science*, 56 (2021), 173–200.

[10] D. Xue, H. Xia, W. Yan, J. Zhang, S. Mu, Defect engineering on carbon-based catalysts for electrocatalytic CO_2 reduction, *Nano-Micro Letters*, 13 (2021), 1–23.

[11] Y. Chen, J. Wei, M.S. Duyar, V.V. Ordomsky, A.Y. Khodakov, J. Liu, Carbon-based catalysts for Fischer–Tropsch synthesis, *Chemical Society Review*, 50 (2021), 2337–2366.

[12] L.H. Zhang, Y. Shi, Y. Wang, N.R. Shiju, Nanocarbon catalysts: Recent understanding regarding the active sites, *Advanced Science*, 7 (2020), 1902126.

[13] M. He, S. Zhang, Q. Wu, H. Xue, B. Xin, D. Wang, J. Zhang, Designing catalysts for chirality-selective synthesis of single-walled carbon nanotubes: Past success and future opportunity, *Advanced Materials*, 31 (2019), 1800805.

[14] S. Zaman, M. Wang, H. Liu, F. Sun, Y. Yu, J. Shui, M. Chen, H. Wang, Carbon-based catalyst supports for oxygen reduction in proton-exchange membrane fuel cells, *Trends in Chemistry*, 4 (2022), 886–906.

[15] T. Zhang, F. Ran, Design strategies of 3D carbon-based electrodes for charge/ion transport in lithium ion battery and sodium ion battery, *Advanced Functional Materials*, 31 (2021), 2010041.

[16] J. Wang, H. Kong, J. Zhang, Y. Hao, Z. Shao, F. Ciucci, Carbon-based electrocatalysts for sustainable energy applications, *Progress in Materials Science*, 116 (2021), 100717.

[17] H. Xu, J. Yang, R. Ge, J. Zhang, Y. Li, M. Zhu, L. Dai, S. Li, W. Li, Carbon-based bifunctional electrocatalysts for oxygen reduction and oxygen evolution reactions: Optimization strategies and mechanistic analysis, *Journal of Energy Chemistry*, 71 (2022), 234–265.

[18] J. Wu, B. Liu, X. Fan, J. Ding, X. Han, Y. Deng, W. Hu, C. Zhong, Carbon-based cathode materials for rechargeable zinc-air batteries: From current collectors to bifunctional integrated air electrodes, *Carbon Energy*, 2 (2020), 370–386.

[19] G. Kothandam, G. Singh, X. Guan, J.M. Lee, K. Ramadass, S. Joseph, M. Benzigar, A. Karakoti, J. Yi, P. Kumar, Recent advances in carbon-based electrodes for energy storage and conversion, *Advanced Science*, 10 (2023), 2301045.

[20] J.H. Dumont, U. Martinez, K. Artyushkova, G.M. Purdy, A.M. Dattelbaum, P. Zelenay, A. Mohite, P. Atanassov, G. Gupta, Nitrogen-doped graphene oxide electrocatalysts for the oxygen reduction reaction, *ACS Applied Nano Materials*, 2 (2019), 1675–1682.

[21] S. Jha, B. Akula, P. Boddu, M. Novak, H. Enyioma, R. Cherradi, H. Liang, Roles of molecular structure of carbon-based materials in energy storage, *Materials Today Sustainability*, 22 (2023), 100375.

[22] G. Fadillah, R. Hidayat, I. Yanti, I. Fatimah, T.A. Saleh, Electrochemical-assisted synthesis of molecularly imprinted graphene oxide/magnetite for highly selective enantiomer separation, *Microchemical Journal*, 200 (2024), 110354.

[23] F.I. Fajarwati, R. Hidayat, G. Fadillah, Synthesis and transformation of graphene-like structures from bamboo waste for photoelectrochemical devices, *Carbon* Trends, 15 (2024), 100351.

[24] H. Zhang, Y. Yang, D. Ren, L. Wang, X. He, Graphite as anode materials: Fundamental mechanism, recent progress and advances, *Energy Storage Materials*, 36 (2021), 147–170.

[25] Y.-M. Song, F. Wang, G. Guo, S.-J. Luo, R.-B. Guo, Energy-efficient storage of methane in the formed hydrates with metal nanoparticles-grafted carbon nanotubes as promoter, *Applied Energy*, 224 (2018), 175–183.

[26] S. Mandal, J. Hu, S.Q. Shi, A comprehensive review of hybrid supercapacitor from transition metal and industrial crop based activated carbon for energy storage applications, *Materials Today Communications*, 34 (2023), 105207.

[27] M.V. Reddy, G.V. Subba Rao, B.V.R. Chowdari, Metal Oxides and oxysalts as anode materials for Li Ion batteries, *Chemical Reviews*, 113 (2013), 5364–5457.

[28] D. Ma, Chapter 1- Hybrid nanoparticles: An introduction, in: S. Mohapatra, T.A. Nguyen, P. Nguyen-Tri (Eds.) *Noble Metal-Metal Oxide Hybrid Nanoparticles*, Cambridge, Cambridgeshire, UK: Woodhead Publishing, 2019, pp. 3–6.

[29] R. Ma, G. Lin, Y. Zhou, Q. Liu, T. Zhang, G. Shan, M. Yang, J. Wang, A review of oxygen reduction mechanisms for metal-free carbon-based electrocatalysts, *NPJ Computational Materials*, 5 (2019), 78.

[30] N. Hassan, Catalytic performance of nanostructured materials recently used for developing fuel cells'electrodes, *International Journal of Hydrogen Energy*, 46 (2021), 39315–39368.

[31] U. Caudillo-Flores, I. Barba-Nieto, M.J. Muñoz-Batista, D. Motta Meira, M. Fernández-García, A. Kubacka, Thermo-photo production of hydrogen using ternary Pt-CeO2-TiO2 catalysts: A spectroscopic and mechanistic study, *Chemical Engineering Journal*, 425 (2021), 130641.

[32] D.-C. Tsai, B.-H. Kuo, H.-P. Chen, E.-C. Chen, F.-S. Shieu, Enhanced performance of proton exchange membrane fuel cells by Pt/carbon/antimony-doped tin dioxide triple-junction catalyst, *Scientific Reports*, 13 (2023), 23076.

[33] C.K. Maity, S. De, S. Acharya, S.H. Siddiki, S. Sahoo, K. Verma, V.K. Thakur, G.C. Nayak, Copper oxide stabilized oxy-functionalized boron nitride-carbon nanotube nanohybrid: An ultra-stable electrode for flexible asymmetric supercapacitor device in ionic electrolyte, *Journal of Energy Storage*, 56 (2022), 105928.

[34] Y. Gao, F. Ram, B. Chen, J. Garemark, L. Berglund, H. Dai, Y. Li, Scalable hierarchical wood/ZnO nanohybrids for efficient mechanical energy conversion, *Materials & Design*, 226 (2023), 111665.

[35] S. Pal, P.S. Das, M.K. Naskar, S. Ghosh, Metal oxide nanocrystals embedded polypyrrole nanohybrid: Exploring role of interface in photocatalytic hydrogen generation, *Materials Today Sustainability*, 25 (2024), 100610.

[36] S. Khan, M. Usman, K. Jabbour, M. Abdullah, P. John, F. Asghar, M.S. Waheed, A.M. Karami, M.F. Ehsan, M.N. Ashiq, Facile fabrication of SrAl2O4/rGO nanohybrid via hydrothermal route for electrochemical energy storage devices, *Journal of Energy Storage*, 83 (2024), 110721.

[37] K.B.M. Ismail, M.A. Kumar, S. Mahalingam, B. Raj, J. Kim, Carbon fiber-reinforced polymers for energy storage applications, *Journal of Energy Storage*, 84 (2024), 110931.

[38] M. Narayanasamy, L. Hu, B. Kirubasankar, Z. Liu, S. Angaiah, C. Yan, Nanohybrid engineering of the vertically confined marigold structure of rGO-VSe2 as an advanced cathode material for aqueous zinc-ion battery, *Journal of Alloys and Compounds*, 882 (2021), 160704.

[39] S.R. Mishra, V. Gadore, G. Yadav, M. Ahmaruzzaman, Next-generation energy storage: In2S3-based materials as high-performance electrodes for alkali-ion batteries, *Next Energy*, 2 (2024), 100071.

[40] D. Chao, C. Zhu, X. Xia, J. Liu, X. Zhang, J. Wang, P. Liang, J. Lin, H. Zhang, Z.X. Shen, H.J. Fan, Graphene quantum dots coated VO2 arrays for highly durable electrodes for Li and Na ion batteries, *Nano Letters*, 15 (2015), 565–573.

[41] X. Xia, D. Chao, C.F. Ng, J. Lin, Z. Fan, H. Zhang, Z.X. Shen, H.J. Fan, VO2 nano-flake arrays for supercapacitor and Li-ion battery electrodes: Performance enhancement by hydrogen molybdenum bronze as an efficient shell material, *Materials Horizons*, 2 (2015), 237–244.

[42] L. Liang, Y. Xu, Y. Lei, H. Liu, 1-Dimensional AgVO3 nanowires hybrid with 2-dimensional graphene nanosheets to create 3-dimensional composite aerogels and their improved electrochemical properties, *Nanoscale*, 6 (2014), 3536–3539.

[43] N. Li, Z. Chen, W. Ren, F. Li, H.-M. Cheng, Flexible graphene-based lithium ion batteries with ultrafast charge and discharge rates, *Proceedings of the National Academy of Sciences of the United States of America*, 109 (2012), 17360–17365.

[44] J. Yang, J. Wang, D. Wang, X. Li, D. Geng, G. Liang, M. Gauthier, R. Li, X. Sun, 3D porous LiFePO4/graphene hybrid cathodes with enhanced performance for Li-ion batteries, *Journal of Power Sources*, 208 (2012), 340–344.

[45] H. Zhu, X. Wu, L. Zan, Y. Zhang, Three-dimensional macroporous Graphene–Li2FeSiO4 composite as cathode material for Lithium-Ion batteries with superior electrochemical performances, *ACS Applied Materials & Interfaces*, 6 (2014), 11724–11733.

[46] Y. Huang, D. Wu, J. Jiang, Y. Mai, F. Zhang, H. Pan, X. Feng, Highly oriented macroporous graphene hybrid monoliths for lithium ion battery electrodes with ultrahigh capacity and rate capability, *Nano Energy*, 12 (2015), 287–295.

[47] M. Jafari, F. Ganjali, R. Eivazzadeh-Keihan, A. Maleki, S. Geranmayeh, Recent advances in applications of graphene-layered double hydroxide nanocomposites in supercapacitors and batteries, *FlatChem*, 45 (2024), 100658.

[48] X. Wang, C. Yan, A. Sumboja, J. Yan, P.S. Lee, Achieving high rate performance in layered hydroxide supercapacitor electrodes, *Advanced Energy Materials*, 4 (2014) 1301240.

[49] J. Pan, S. Li, F. Li, W. Zhang, D. Guo, L. Zhang, D. Zhang, H. Pan, Y. Zhang, Y. Ruan, Design and construction of core-shell heterostructure of Ni-V layered double hydroxide composite electrode materials for high-performance hybrid supercapacitor and L-Tryptophan sensor, *Journal of Alloys and Compounds*, 890 (2022), 161781.

[50] G. Zhang, X. Wu, Y. Xu, Y. Cao, Y. Luo, H. Yang, J. Wang, W. Li, X. Li, Modulating electronic and interface structure of CuO/CoNi-LDH via electrochemical activation for high areal-performance supercapacitors, *Journal of Energy Storage*, 83 (2024), 110426.

[51] F. Heidari Gourji, D. Velauthapillai, M. Keykhaei, Ultra- ordered array of CuCo2S4 microspheres on co-doped nitrogen, sulfur-porous graphene sheets with superior electrochemical performance for supercapacitor application, *Energy Reports*, 8 (2022), 7712–7723.

[52] C. Cheng, Y. Zou, F. Xu, C. Xiang, Q. Sui, J. Zhang, L. Sun, Z. Chen, Ultrathin graphene@NiCo2S4@Ni-Mo layered double hydroxide with a 3D hierarchical flowers structure as a high performance positive electrode for hybrid supercapacitor, *Journal of Energy Storage*, 52 (2022), 105049.

6 Nanomaterials for the Development of Electrodes

*Karna Wijaya, Amalia Kurnia Amin, Latifah Hauli,
Wahyu Dita Saputri, Dewi Yuanita Lestari,
Aldino Javier Saviola, Riska Astin Fitria,
Adyatma Bhagaskara, Iqmal Tahir, Won-Chun Oh,
Aneu Aneu and Remi Ayu Pratika*

1 INTRODUCTION

Nanomaterials are a class of materials engineered and manipulated at the nanoscale, typically with dimensions ranging from 1 to 100 nm [1]. The critical aspects of nanomaterials are their size and dimensionality. Nanomaterials are characterized primarily by their small size and high surface area-to-volume ratio. This characteristic is often associated with quantum effects, which can significantly impact their properties. At this scale, materials often exhibit unique and enhanced properties that differ from those observed in bulk materials. The field of nanomaterials has opened up new possibilities across various scientific, industrial, and technological domains. Nanomaterials represent a diverse and exciting class of materials with exceptional properties and a wide range of applications [2]. Their unique characteristics have the potential to drive innovations in numerous industries and technologies, while also requiring careful consideration of safety and ethical concerns.

Energy storage materials, including batteries, supercapacitors (SCs), and fuel cells, are attracting increasing attention as primary agents of energy storage. This growing interest reflects the increasing demand in various sectors, ranging from small-scale batteries to extensive electric transportation systems. Therefore, there is an urgent need for more effective, reliable, responsive, accessible, and cost-effective energy storage solutions in the future to meet this growing demand [3]. Carbon-based nanomaterials, including activated carbon, graphene, fullerenes, metal oxides (MOs), and transition metals, have garnered significant interest over the past decade for improving energy storage devices. Their hierarchical structure, excellent porosity, improved mechanical and electrical properties, and large specific surface area make them ideal electrode materials for enhancing the electrochemical performance and achieving long-term stability of energy storage devices, which are crucial for modern energy storage devices reliant on efficient sources with high energy and power densities[4].

DOI: 10.1201/9781003561262-6

1.1 Types of Nanomaterials

Nanoparticles (such as metal nanoparticles and quantum dots), nanotubes (such as carbon nanotubes (CNTs)), nanowires, nanosheets (such as graphene), nanocomposites (combinations of various nanomaterials), and more are among the many diverse types of nanomaterials. Every kind has unique characteristics and uses [5,6].

1.2 Unusual Properties of Nanomaterials

Nanomaterials can have extraordinary qualities like high mechanical strength, remarkable electrical and thermal conductivity, increased chemical reactivity, enormous surface area, and quantum confinement effects in nanoparticles. Nanomaterials are useful for a variety of applications because of these characteristics [7,8]. The physicochemical properties of base materials of electrodes are listed in Table 6.1.

1.3 Synthesis Methods of Nanomaterials

There are various methods for synthesizing nanomaterials, which are mainly divided into two approaches, bottom-up and top-down processes (Figure 6.1). In top-down approaches, bulk materials are crushed and ground to produce nanostructure materials. Top-down methods include mechanical ball milling, etching, sputtering, laser ablation, and electro-explosion. Meanwhile, in bottom-up approaches, small building blocks are used to build the nanomaterial's structure. Bottom-up methods include chemical vapor deposition (CVD), metal organic decomposition (MOD), laser pyrolysis, molecular beam epitaxy, sol-gel method, wet synthesis, and self-assembly processes. The choice of method depends on the desired material and its intended application [6,21,22].

Some methods for synthesizing nanomaterials are discussed in Sects. 1.3.1 and 1.3.2. The choice of the method adopted depends on the desired material properties and application.

1.3.1 Top-Down Methods for Nanomaterial Synthesis

- **Mechanical Ball Milling**: Mechanical ball milling reduces the material's particle size to nanoscale through high-speed rotation or vibration in a closed chamber with grinding media. This generates friction for crushing particles until desired sizes are attained. Widely used in pharmaceuticals, construction, and nanotechnology, it exhibits precise size control and uniform distribution. Cooling may be necessary to manage heat generation. The optimization of parameters, including media selection, is essential for achieving efficient outcomes [23].
- **Etching:** This method uses chemical or physical processes to selectively remove a material from a surface. In the field of nanomaterial synthesis for electrode development, this technique plays an important role. Initially, a mask or template is applied to designate regions for material removal. Next, an etching process is employed to meticulously eradicate the material from the exposed area, leaving behind a carefully crafted nanostructured

TABLE 6.1

Physicochemical Properties of Base Materials of Electrodes

Material	Surface Area	Electrical Conductivity	Chemical Stability	Advantages	Disadvantages	Ref
Activated carbon	High	Moderate	High	High capacitance, low cost	Low density, can swell in electrolytes	[9], [10]
Graphene	Very high	High	High	Excellent conductivity, high mechanical strength	Difficult to process. Can restack	[11], [12]
Fullerene	High	Moderate	High	Unique, spherical structure, good electron acceptor	Limited scalability, high cost	[13], [14]
Activated carbon composites	Depends on composites	Variable	Depends on composites	Tailored properties, improved performance over a single material	Complexity of fabrication	[15], [16]
Material oxides	Variable	Variable	High	High redox activity, diverse options	Can be brittle, exhibit lower conductivity compared to metals	[17], [18]
Transition metals	High	High	Variable	Excellent conductivity	Can be expensive and may corrode	[19], [20]

Notes:

- *Surface area is a key factor for capacitance, with higher surface area enabling greater ion storage.*
- *Electrical conductivity is crucial for efficient electron transport within the electrode.*
- *Chemical stability ensures material integrity during charge/discharge cycles.*

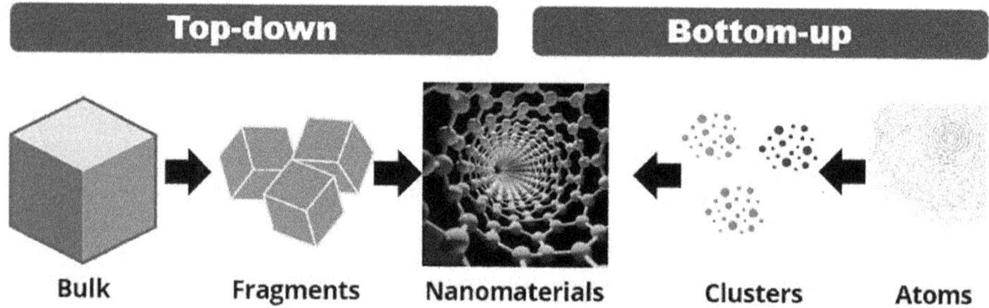

FIGURE 6.1   Illustration of bottom-up and top-down approaches.

pattern. This method allows precise control over the electrode's morphology and surface characteristics, which are important factors that affect its performance in various applications such as energy storage and conversion devices. By adjusting the nanostructured pattern through etching, researchers can improve electrode properties such as surface area, porosity, and electrochemical activity, thus advancing the efficiency and durability of electrode materials for next-generation energy technologies [24].

- **Sputtering:** Sputtering, a prominent method in nanomaterial synthesis, involves bombarding a target material with high energy ions, typically from a plasma discharge source. This process ejects atoms from the target surface, which then condense onto a nearby substrate and form a thin film. Its precise control over parameters like ion energy and pressure allows for tailored film thickness and nanostructure, facilitating the creation of nanomaterials with diverse properties for applications in electronics, optics, and catalysis [25].

- **Laser Ablation:** It employs a high energy laser beam to vaporize a target material, which subsequently condenses onto a substrate and yields nanoparticles. The laser's intense heat can also induce the formation of distinctive nanostructures directly on the target material's surface. This method allows precise control over the size, shape, and composition of nanoparticles, exhibiting versatility in synthesizing a wide range of nanomaterials with tailored properties. By adjusting parameters such as laser power, wavelength, and irradiation time, researchers can fine-tune the characteristics of the produced nanoparticles to meet a specific application's requirements. Laser ablation finds applications in various fields, including nanoelectronics, biomedicine, and catalysis, owing to its capability to produce nanoparticles with high purity and uniformity. Additionally, the direct fabrication of nanostructures on target surfaces expands the potential for creating novel materials with enhanced functionalities and performance [26].

- **Electro-explosion:** Electrical explosion of metal wires is a promising technique for synthesizing highly active nanoparticles. Subjecting a metal wire to a high-density current pulse causes rapid heating and exploding and subsequently forming nanoparticles when exposed to a gas atmosphere.

This method provides numerous advantages over alternative nanopowder synthesis techniques: high energy efficiency, ease of modification of process parameters and nanopowder characteristics, minimal variation in nanoparticle size distribution, stability of nanopowder properties, and enhanced activity in chemical processes. It also enables the production of nanopowders from various metals, alloys, oxides, and nitrides [27].

1.3.2 Bottom-Up Methods for Nanomaterial Synthesis

- **Chemical vapor deposition (CVD):** Chemical vapor deposition (CVD) is a process in which gases undergo chemical reactions at high temperatures to form thin solid layers on surfaces. By carefully selecting and controlling the precursor gas, this method enables the precise manipulation of film thickness and chemical composition. Applications such as electronics and coatings find widespread use of this method due to its ability to produce high-quality nanomaterials [28].
- **Metal organic decomposition (MOD):** This method employs organometallic precursors, which are complex molecules containing a metal atom bonded to organic (carbon-based) groups. When subjected to high temperatures, these precursors break down, resulting in the formation of metal nanoparticles. This technique, known as metal organic deposition (MOD), enables the precise regulation of particle size and shape, thereby rendering it valuable for applications such as catalysts and sensors [29].
- **Laser pyrolysis:** Laser pyrolysis is an advanced nanoparticle synthesis technique that utilizes the energy of a high-powered laser beam to vaporize a solid target material, which leads to the formation of nanoparticles through nucleation and condensation processes. This method demonstrates precise control over the size, shape, and composition of the nanoparticles by modifying the laser's parameters and the composition of the target material. Laser pyrolysis finds widespread applications in areas such as catalysis, sensing, electronics, and biomedical sciences. Its properties such as scalability and speed, and its environmentally friendly nature make it an attractive option for nanoparticle synthesis, providing customized materials for research and technological advancement in various industries [30].
- **Molecular beam epitaxy (MBE):** Molecular beam epitaxy (MBE) is a method based on the precise deposition of atoms or molecules onto a substrate surface in a high-vacuum environment, following the principles of epitaxial growth. This technique ensures meticulous control over composition and doping at the nanoscale, thus leading to the creation of advanced structures. It enables the formation of 2D structures that are characterized by remarkably smooth interfaces and 3D nanostructures that effectively trap carriers. By comprehending kinetic growth mechanisms, MBE enables the design and production of structures with custom properties, instilling assurance in their functionality. This capability has facilitated the development of diverse innovative semiconductor structures and devices essential for nanoscale photonics and electronics [31].

- **Sol-gel method:** The sol-gel method, a wet chemical process, is widely utilized in synthesizing nanomaterials, especially high-quality metal-oxide-based nanomaterials. Its name stems from the conversion of liquid precursors into sols and progression to gel formation. MO nanoparticles are produced by dissolving a molecular precursor, usually a metal alkoxide, in water or alcohol, followed by inducing gel formation through hydrolysis or alcoholysis via heating and stirring. Various drying techniques, such as alcohol evaporation, are employed before powdering and calcination. Sol-gel technique enables precise control over particle size and is versatile for creating ceramics and coatings. It's cost-effective and favors the synthesis of MO thin films that can be used in optics, electronics, energy, and pharmaceutical applications [32].

- **Self-assembly processes:** Self-assembly in nanostructure formation is the spontaneous arrangement of atoms, molecules, or nanoscale components into ordered structures without human intervention. This highly viable, cost-effective method offers promising prospects for nanofabrication, with high throughput. Utilizing molecular interactions, self-assembly processes guide the formation of nanoscale architectures based on principles like molecular recognition and entropy. Some examples include template-directed assembly and supramolecular chemistry. This approach provides simplicity, scalability, and precise control over nanostructure properties [33].

1.4 APPLICATIONS OF NANOMATERIALS

Nanomaterials are used in transistors, sensors, and displays to improve performance and efficiency. By increasing energy density and charge/discharge rates, nanomaterials improve the energy storage and conversion efficiency of batteries, SCs, solar cells, and fuel cells. In chemical reactions, nanomaterials work as efficient catalysts to speed up and improve the selectivity of the reactions. Because of their biocompatibility and targeting capabilities, nanomaterials are utilized in medical imaging, medication delivery systems, and diagnostics applications [34,35].

Nanomaterials can purge air and water from impurities. Nanomaterials enhance the mechanical characteristics of materials such as composites and polymers. Their unique qualities also spark concerns over nanoparticles' possible environmental and human health effects. Many studies have been focused on comprehending and reducing these risks. Due to the evolving nature of nanomaterials, governments and international bodies have developed regulations and standards to ensure their safe use and handling. Nanomaterials' research is highly interdisciplinary, involving physics, chemistry, materials science, biology, and engineering [36,37].

2 NANOMATERIALS-BASED ELECTRODES

In modern materials science and engineering, nanomaterials have emerged as a revolutionary class of substances with transformative potential across many applications.

Nanomaterials-Based Electrodes
energy storage and conversion system

Batteries

Supercapacitors

Fuel cells

FIGURE 6.2 Nanomaterial-based electrodes as energy storage and conversion systems.

Among their myriad uses, none is more promising than their role in developing advanced electrodes in electrochemical systems. Electrodes, which are the fundamental components of numerous electrochemical and electronic devices, have experienced a profound transformation, thanks to the unique properties offered by nanomaterials [38]. These materials, engineered at the nanoscale, exhibit extraordinary electrical, thermal, mechanical, and chemical attributes that enhance the electrodes' performance and enable entirely new functionalities [39]. In this context, integrating nanomaterials into electrode design represents a pivotal advancement with far-reaching implications for applications in energy storage, sensing, electrocatalysis, and beyond.

This introductory exploration delves into the realm of nanomaterials and their pivotal role in shaping the future of electrode technology. We will elucidate the exceptional characteristics of nanomaterials that make them ideal candidates for electrode applications and explore the diverse domains in which their incorporation has catalyzed innovation and progress. From SCs and lithium-ion batteries to sensors, fuel cells, and neural interfaces, the influence of nanomaterials on electrode development is profound. As we navigate this transformative landscape, we will witness how nanomaterials continue to revolutionize how we harness and manipulate electrical energy, thereby driving progress toward a sustainable, efficient, and technologically advanced future.

Nanomaterials have emerged as prominent candidates for the development of electrodes in various energy storage and conversion systems, including batteries, SCs, and fuel cells (Figure 6.2). Their unique properties, derived from their small size and high surface area-to-volume ratio, enable significant performance enhancements compared to their bulk counterparts [35]. Here is an overview of some critical nanomaterials used for electrode development.

3 COMPUTATIONAL CHEMISTRY PARAMETERS INFLUENCING THE FABRICATION OF ELECTRODE MATERIALS

Researchers aiming to create new nanomaterials for applications in electrodes have primarily concentrated on understanding the relationships between synthesis, structure, and properties within various nanomaterial families. This approach, while informative, could be more efficient due to the parameters of computational chemistry involved. Scientists aspire to produce materials with ideal properties and this is an achievable goal, if the optimal atomic environments and corresponding processing conditions are known in advance. However, comprehending atomic environments is challenging, especially in complex structures.

Various experimental techniques, such as X-ray, neutron, and electron diffraction (XRD, ND, and ED), nuclear magnetic resonance (NMR), and X-ray absorption fine structure spectroscopy (XAFS), can probe atomic arrangements in complex structures. Yet, interpreting data on an atomic scale often relies on hypotheses and speculation. Modern computational approaches offer valuable insights into the optimal material (phase) for specific applications, guiding experimental design [40].

Molecular modeling, established over the past decades, provides fundamental insights into atomistic-level problems. Force-field-based methods focus on dynamic processes, while *ab initio* methods address reactivity and electronic effects. Molecular dynamics simulations have been employed to model porous nanomaterials like activated carbon in electrodes. Despite their usefulness, these simulations often neglect electronic effects, which can significantly influence molecular properties. *Ab initio* simulations can address this gap by exploring interactions between carbon electrodes and electrolyte ions and studying the impact of surface functionalization [41,42].

Due to the computational demands of *ab initio* calculations, smaller model systems are commonly used to mimic essential characteristics. By gradually increasing the system size, researchers assess the influence of a larger environment. Smaller models also facilitate efficient screening of functionalization effects and explore hypothetical or impractical states. Computational methods separate interfering effects, providing insights that are not always feasible experimentally. This interplay between simulation and experimentation accelerates product development, thereby saving time and costs by optimizing properties and performance based on critical information obtained from both approaches [43].

In the realm of energy storage and conversion devices, such as batteries, SCs, and fuel cells, electrodes play a critical role in determining the overall performance and efficiency of the system. Computational chemistry offers valuable insights into the design, optimization, and understanding of electrode materials due to its ability to simulate and predict the properties and behavior of materials at the atomic and molecular levels [44,45]. In this article, we explore the computational chemistry considerations on electrode materials, including the role of simulations in material discovery, characterization, and performance prediction.

3.1 Material Discovery and Screening

Computational chemistry techniques like density functional theory (DFT), molecular dynamics (MD), and Monte Carlo simulations enable researchers to explore vast chemical spaces and identify promising electrode materials with desired properties. By calculating electronic structures, energetics, and stability of materials, computational simulations can predict the feasibility of various synthesis routes and guide experimental efforts toward synthesizing novel electrode materials with optimized performance [46,47].

3.2 Structural and Morphological Optimization

Understanding electrode materials' atomic scale structure and morphology is crucial for enhancing their electrochemical properties. Computational methods, including DFT calculations and molecular modeling, can elucidate the effects of crystal structure, surface morphology, and defects on the electrochemical behavior of materials. By exploring different structural configurations and surface terminations, researchers can optimize electrode materials for specific applications, such as maximizing surface area for enhanced charge storage or minimizing diffusion barriers for improved ion transport [47,48].

Computational chemistry uses theory-based molecular modeling to estimate chemical events for overcoming challenges in solving exact equations for systems smaller than a few microns. The most promising of these techniques relies on complex combinations of two general strategies: (i) the adequate stochastic sampling of many dimensions and (ii) the use of compact models that capture the essential many-particle effects by construction. These techniques come in various forms, which comprise the computational chemistry toolbox (Figure 6.3).

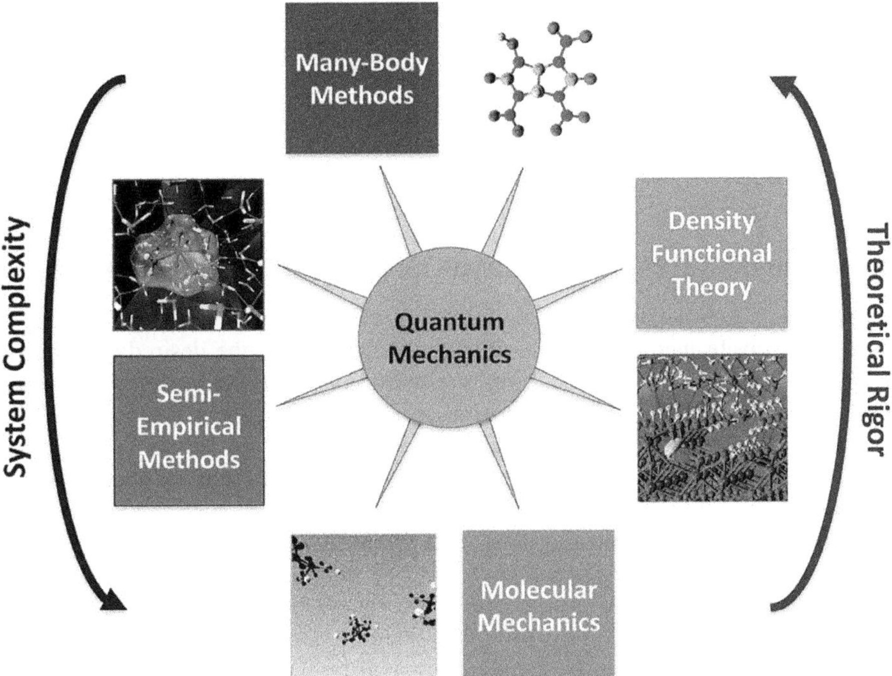

FIGURE 6.3 Summary of computational methods [49] Royal Society of Chemistry 2017.

3.3 ELECTROCHEMICAL PROPERTIES AND MECHANISMS

Computational chemistry is vital in elucidating the underlying electrochemical mechanisms and processes at electrode interfaces. Combining quantum mechanical calculations and electrochemical models can provide insights into charge transfer kinetics, reaction pathways, and redox reactions during charge/discharge cycles. Understanding these fundamental processes is essential for designing electrodes with high efficiency, stability, and cycling performance [50].

3.4 DEFECT ENGINEERING AND SURFACE FUNCTIONALIZATION

Defects and surface-functional groups significantly influence the electrochemical behavior of electrode materials. Computational chemistry techniques can predict the formation, energetics, and effects of defects, such as vacancies, dopants, and grain boundaries, on material properties and performance. Additionally, simulations can guide the design of surface modifications, such as using functional groups or coatings, to improve electrode materials' stability, conductivity, and catalytic activity [51,52].

3.5 RATIONAL DESIGN OF NEW MATERIALS AND ALLOYS

Researchers can rationally design new electrode materials and alloys with tailored properties for specific energy storage and conversion applications by combining computational predictions with experimental validation. High-throughput computational screening methods enable the exploration of large material databases, accelerating the discovery of promising candidates with superior electrochemical performance. Furthermore, computational modeling facilitates the design of multi-component materials and composites with synergistic effects, enhancing functionality and performance [46,47,49,50,53,54].

4 CARBON-BASED NANOCOMPOSITES FOR ENERGY STORAGE

Composites are promising materials for use in energy storage systems. The properties of composite materials can be modified according to the desired application. Many studies have shown that composite materials exhibit better properties than their components [55]. Using composites in energy storage systems enhances the contact surface area between the electrolyte and the electrode. Carbon nanomaterials are suitable materials for use as composite materials for electrode applications in various electrochemical systems because they exhibit a large surface area and high electronic conductivity and improved inert properties that support energy storage system applications [56].

However, the ideal carbon-based electrode should be robust, chemically stable, cost-effective, conductive and exhibit high surface area and suitable hydrophilicity [57]. These features can be achieved by modifying thermal [58] and chemical [59] treatments. One of the applications for carbon electrodes is aqueous organic redox flow batteries (AORFBs). This type of battery is a promising technology to store energy for grid applications, since it is a safe and scalable technology [60]. In AORFBs, carbon-based electrodes are primarily used, including carbon paper electrodes (CPEs) and carbon cloth electrodes (CCEs). The difference between CPEs and CCEs is the use of binders as the enhancer of the mechanical stability of electrodes [61,62]. The CCEs' unique structure facilitates diffusion-dominated mass transport between the bundle weaves [63]. The difference in their structure plays a pivotal role in influencing the electrochemical characteristics and performance of these two types of carbon-based electrodes in AORFBs.

4.1 Composite of Biobased Activated Carbon

Activated carbon material is highly suitable for electrodes in SCs due to its abundance, low cost, environmentally friendly nature, and excellent performance in energy storage systems. Additionally, activated carbon has a high surface area, enabling it to store charge effectively. Activated carbon also exhibits high conductivity, making it well-suited for electrode materials. However, supercapacitor components utilizing activated carbon from biomass typically demonstrate low energy density. This drawback is attributed to poor dispersion, low theoretical specific capacitance values, inefficient rechargeability, and multi-layer fragility [64].

The limitations of activated carbon and MOs can be overcome by creating a composite comprising activated carbon and MOs. Carbon provides electrical conductivity by creating channels for electron transport obtained from MOs during redox reactions. Consequently, the composite matrix will alter porosity, and pore volume will increase, enabling the storage of a considerable amount of charge. The exceptional electrochemical properties are obtained from the combination of activated carbon's cycling ability and charge storage capability with the increased specific capacitance obtained by Faradaic redox reactions in MOs. The surface of the MOs becomes the site of redox reactions, and then electrolyte ions are stored on the carbon matrix. Many ions are produced from the composite electrode due to the presence of MOs and the carbon matrix. This facilitates electron conduction and reduces the overall resistance of the composite electrode [65].

Rami can be used as a biomass precursor to produce activated carbon. This activated carbon and MnO can be utilized as a composite for electrode fabrication. The activated carbon composite electrode from rami-MnO (RAC/MnO) exhibits superior electrochemical properties compared to carbon electrodes. The RAC/MnO composite electrode has a working voltage value of 1.4 V. In contrast, the carbon electrode only has a working voltage value of 1.0 V, which implies that the working voltage value of the RAC/MnO composite electrode is 40% higher than that of the carbon electrode. The specific capacitance value for the RAC/MnO composite electrode is 720 F g^{-1} at a current density of 0.5 A g^{-1} [66].

A composite made from activated carbon material derived from bamboo can be produced using bamboo leaves and copper oxide. This composite is synthesized by dispersing copper oxide onto the surface of the activated carbon derived from bamboo leaves. Electrodes made from this composite have a specific capacitance value of 147 F g^{-1}. The capacitance retention of these electrodes is reported to be very good (93%) after 10,000 cycles [67].

4.2 Carbon Dot (CD)-Based Nanocomposite

Carbon dots (CDs) are members of carbon-based nanomaterials. This material is less than 20 nm in size. As seen in Figure 6.4, CDs can be divided into Graphene Quantum Dots (GQDs), Carbon Quantum Dots (CQDs), Carbon Nanodots (CNDs), and Carbonized Polymer Dots (CPDs). Generally, CDs have a carbon core in their structure that can appear as a carbon lattice, in a semi-crystalline or amorphous form, or polymer chains and functional groups on the surface. GQDs demonstrate a disc-like arrangement of 2D nanoparticles or small fragments of graphene sheets, with single or multiple layers. GQDs are generally anisotropic and have lateral sizes not exceeding 20 nm with a height of less than five graphene layers. CQDs are multilayered and are quasi-spherical nanoparticles with a crystalline graphite core. Quantum here means nanoparticles that have a delocalized band structure due to contributions from molecular orbitals and surface states. CNDs have an amorphous quasi-spherical nanoparticle shape. The structure of CPDs comprises a polymer/carbon hybrid with many polymer chains and functional groups grafted on the carbon core and surface [68].

One CD-based composite exhibiting exceptional electrochemical properties is the CQD/RuO_2 composite. This composite is synthesized using the standard impregnation method. It can be used as an active material for supercapacitor electrodes. This composite electrode demonstrates outstanding performance, with a capacitance value of 594 F g^{-1} at 1 A g^{-1} in 1 M H_2SO_4. This value exhibits a capacitance retention

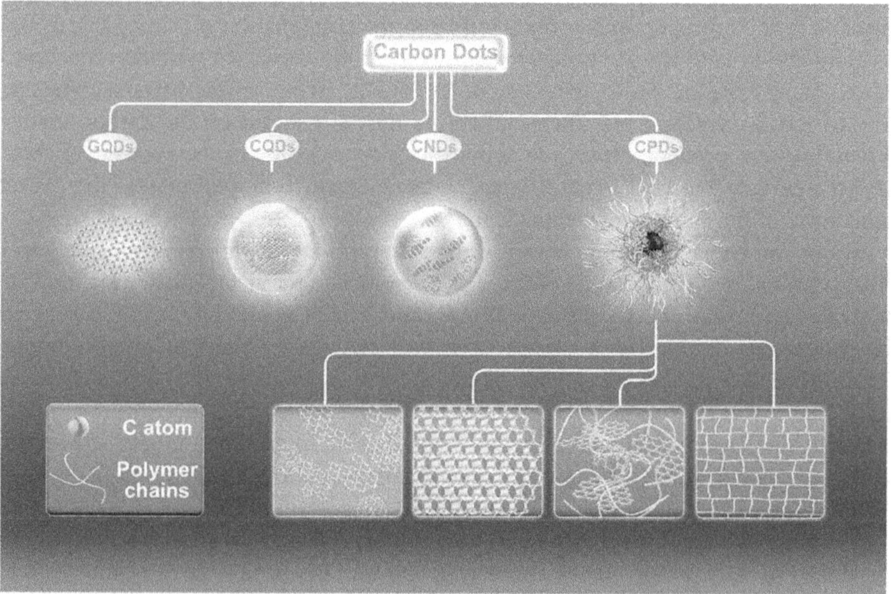

FIGURE 6.4 Classification of CDs: graphene quantum dots (GQDs), carbon quantum dots (CQDs), carbon nanodots (CNDs), and carbonized polymer dots (CPDs); and the possible structures of the carbon core of CPDs. [70] Copyright 2019 Wiley-VCH.

of 77.4% at a current density of 50 A g^{-1}. The highest capacitive activity is obtained with a Ru loading weight of 41.9%. This composite electrode exhibits an excellent cycle stability with a retention value of almost 96.9% after 5,000 charge/discharge cycles at 5 A g^{-1}. This value is significantly better than the capacitance retention value for RuO_2 under the same conditions, which only reaches 73.9%. The superior electrochemical properties of this composite are attributed to the excellent dispersion of RuO_2 on CQD, which prevents RuO_2 agglomeration. The unique CQD network is crucial in exhibiting fast charge transfer and ionic penetration during redox reactions [69].

Biomass from cauliflower leaves can be used to synthesize N-doped CQDs, which are subsequently utilized to synthesize hybrid materials with reduced graphene oxide (rGO). This hybrid material is synthesized using the hydrothermal method. The rGO/CDs hybrid material yields a specific capacitance of 278 F g^{-1} at 0.2 A g^{-1} in a 1 M H_2SO_4 electrolyte solution. The capacitance value remains high at high current densities, reaching 165 F g^{-1} at 100 A g^{-1}, or in other words, exhibiting a capacitance retention of 59.9%. The rGO/CDs electrode demonstrates good durability with a capacitance retention of 96.1% over 10,000 charge/discharge cycles at a current density of 20 A g^{-1}. This exceptional electrochemical performance is attributed to the ability of CQDs to intercalate rGO layers, preventing the stacking of rGO sheets. This is crucial for increasing the surface area (415 m^2 g^{-1}) and enhancing the pore volume (2.94 cm^3 g^{-1}) [71].

4.3 FUNCTIONALIZED GRAPHENE AND THEIR DERIVATIVES

Generally, renewable energy sources are coupled with energy storage devices, viz., batteries, SCs, and hydrogen storage technologies for maintaining continuous power supply. Recent research indicates that the efficiency of those electrochemical energy storage applications depends on various factors [72]. Batteries store energy through various faradaic reactions on the electrodes; these reactions typically result in relatively low power densities but deliver high energy density [73,74]. SCs store energy through fascinating mechanisms that distinguish them from traditional capacitors and batteries. SCs store energy in two ways: (1) Electrical double-layer capacitance (electric double-layer capacitors, EDLCs), which involves the adsorption of accumulated ions from the electrolyte on the electrode surface without chemical reactions, thus creating a double layer of charge separation (electrostatic charge storage (Helmholtz Layer)); and (2) pseudo-capacitance/Faraday quasi-capacitance (pseudocapacitors), which involves rapid surface redox electrochemical reactions and/or intercalation on the electrode surface [74,75,76,77]. By applying two dissimilar electrodes, combining these two energy storage capacitance lines creates a hybrid supercapacitor system that balances high energy density and rapid charge/discharge [77].

The overall performance of batteries, SCs, and hydrogen storage technology systems is highly dependent on the structure and properties of the materials used. Therefore, the development of electrode design and engineering is crucial for the advancement of energy storage applications. In particular, 2D van der Waals layered carbon allotrope materials of graphene, graphene oxide (GO), reduced graphene oxide (rGO), and its derivatives that are without or with functionalization groups are attractive solid-state candidates to meet the needs of next-generation energy and hydrogen storage technologies [78]. High-porosity powders with large specific surface area per unit mass, high electrical conductivity, excellent electron mobility and flexibility, and exceptional mechanical strength can be employed in energy storage applications [78,79,80,81].

Hydrogen ions could be stored in porous graphene-based electrodes and recovered back, in which a higher internal pore surface area enables increased hydrogen storage in the ionic form [81]. These characteristics also contribute to graphene oxide, which is an ideal material for SC electrodes, enabling adequate charge storage and rapid electrical energy discharge. Specific capacitance measures the charge an SC can hold using its mass or surface area, and graphene oxide-based SCs benefit significantly from their elevated specific capacitance [80]. Graphene-based materials have high intrinsic capacitance. Electrodes for applications in SCs, hydrogen storage, and advanced batteries can be fabricated from various graphitic materials, ranging from zero-dimensional (0D) to one-dimensional (1D), two-dimensional (2D), and three-dimensional (3D) materials. These might profoundly impact the development of portable wearable electronics and electric vehicles [81].

4.3.1   Zero-Dimensional (0D) Structures: Graphene Quantum Dots (GQDs)

One of the outstanding types of 0D structures that exists within a 2D graphene derivative owing to its remarkable quantum confinement and unique edge effect is a 2D quantum dot, namely graphene quantum dots (GQDs), which are graphene sheets that are less than 100 nm in their lateral lattice dimension (nanosheets) and having less than ten layers of π-π stacked graphene nanosheets [82]. GQDs exhibit excellent graphene properties, but their typical quantum confinement and edge effects benefit from their lateral dimension of less than 10 nm and thickness of less than 2 nm [83,84]. The GQD material's advantage for energy storage stems from its decreased dimensionality, resulting in a high charging energy that safeguards the quantized charged state of the dot, enabling SET (Single-Electron Transistor) operation at elevated temperatures [81].

The quantum confinement effect (QCE) occurs when electrons in a material become confined to a limited area, enabling their energy levels to be quantized. As a result, discrete electronic transitions arise, leading to a distinct array of optical and electrical characteristics that can be potentially modified by altering the quantum dots' size, shape, and composition [60]. The QCE of GQDs results from significant changes in their electronic behavior that are exhibited due to the GQD's lateral dimensions approaching that of an electron's wavelength. Discrete energy levels of GQDs exhibit quantized energy levels due to confinement in the plane, making GQDs ideal for energy-related applications. The localized electronic state of confined electrons in the structure of GQDs leads to unique electronic properties. In contrast, the tunable bandgap and charge storage capacity of GQDs are altered by controlling the size (size-dependent properties). The size and shape of GQDs can be modified by controlling the conditions of synthesis [83,84].

The surface structure of quantum dots, known as edge effects, significantly boosts their quantum yield [85]. Augmenting the quantity of edge sites on the surface of GQDs can result in elevated quantum yields. The specific types of edge (zigzag, armchair, or mixed) of GQDs play a critical role in impacting the performance of SCs. Zigzag-edged GQDs enhance the SC's performance as zigzag edges provide an additional surface area for charge storage; the effect of its small size, the presence of more chemically reactive zigzag edges due to dangling bonds that improve charge transfer, and the functionalization of zigzag edges can tailor their behavior for specific applications. Armchair-edged GQDs contribute to the overall stability of SCs because armchair edges provide stability to the GQD structure, and the reduced reactivity of armchair edges (less reactive than zigzag edges) ensures long-term performance. Armchair edges maintain a balance between reactivity and stability. Mixed-edge GQDs are a flexible way to work with SCs because they enable the adjustment of SC characteristics by controlling zigzag and armchair edge quantities. This allows for the tailoring of properties and the development of optimal reactivity, as the presence of mixed edges achieves a balance between reactivity and stability. Mixed edges can be customized for specific SC requirements.

GQDs as electrode materials for SCs generally fall under the electrical double-layer category. Functionalized GQDs, such as GQDs doped with N, S, and transition metals or MOs, are pseudocapacitive. The original GQDs can also be used as an auxiliary (conductive agent) in SCs to improve the properties of the composite electrode in pursuit of higher capacity [83,84]. The properties of GQDs, such as tunable bandgap and increased charge transport, have been successfully utilized to modify the electrode surfaces. The graphene structure inside GQDs boosts the electrical conductivity, electrolyte diffusion, and charge transportation kinetics between the electrode and analyte and decreases the system impedance, thus leading to improved electrochemical performance [79,84]. The electrochemical energy storage phenomenon can be significantly influenced by the presence of crystallinity, defect states, doped heteroatoms, surface groups, and the conductivity of the electrode material [84].

4.3.2 One-Dimensional (1D) Structures: Graphene Fibers (GFs) and Graphene Nanoribbons (GNRs)

Graphene fibers (GFs) are macroscopically assembled by aligning 2D graphene sheets and their derivatives regularly along the fiber axis [86,87]. According to Zhai and Chen [88], the predominant obstacle impeding the development of wearable electronics is the lack of dependable energy storage systems capable of being worn directly without compromising the devices' compact size, user-friendliness, and flexibility [74]. The GF flexible electrode has been widely regarded as a suitable option for applications in smart wearable energy storage devices in the form of graphene fiber supercapacitors (GFSCs), photovoltaic cells, and cable-shaped batteries [73,74,86,88,89]. This is primarily due to its distinctive characteristics, including a high specific surface area, high mechanical flexibility, exceptional electrical conductivity, and electrochemical properties; and its lightweight nature and potential for functionalization [74,86]. The architected pores of GFs play a crucial role in facilitating the diffusion and storage of electrolyte ions. These pores serve as reservoirs for ion buffering (macropores), channels for ion transport (mesopores), and sites for charge accommodation (micropores). Furthermore, to achieve perfect capacitive performance of GFs, it is essential that the sizes of the pores match those of the electrolyte ions [86].

Graphene nanoribbons (GNRs) represent another type of 1D nanostructures originating from repeating hexagonal graphene carbon cells. Graphene can be divided into quasi-one-dimensional graphene nanoribbon (GNR) strips. This allows the bandgap to be manipulated by controlling charge carriers' quantum confinement [90]. Bandgap energy plays a crucial role in shaping the properties of electrons, which can impact the electrical conductivity of materials. The electronic properties of GNRs are susceptible to their specific width and edge termination, which depend upon the synthesis method used (lithography, chemical synthesis, the unzipping of CNTs, or epitaxial growth on silicon carbide sidewalls). There are four prevalent types of GNRs: armchair (AGNRs), zigzag (ZGNRs), chiral (chGNRs), and chevron (cGNRs). AGNRS and ZGNRs can be distinguished only based on their width. In the case of AGNRs, the width measurement is commonly denoted by the number of dimer lines, while in ZGNRs, the measurement of width is determined by the number of zigzag lines.

In contrast, identifying chGNRs requires the consideration of two parameters: width and edge orientation. The edge orientation of chGNRs is determined by the number of graphene unit cell vectors, ma1 and na2, along the edge. Meanwhile, cGNRs with meandering topology incorporate armchair and zigzag edge terminations [90].

4.3.3   Two-Dimensional (2D) Materials: Graphene Oxide (GO), Reduced Graphene Oxide (rGO), Graphene Papers, Graphene Sheets, Graphene Films

Graphene has an impressive ability to withstand mechanical deformation, allowing it to be folded without any risk of breaking. Thanks to its sp^2 hybridization, pristine single-layer graphene (SLG) possesses remarkable strength, measuring at 48,000 kN·m·kg^{-111}. The SLG nanosheet exhibits two distinct structural regions at the molecular level: (1) the basal plane, which is built up of 2D conjugated sp^2 carbon atoms, and (2) the edge, which is composed of a one-atom-thick defective graphitic line of carbon atoms with dangling bonds and different capping moieties like hydrogen, hydroxyl, carbonyl, and carboxyl groups [91]. The specific capacitance of the graphene edge was significantly higher, with a much faster electron transfer rate and more robust electrocatalytic activity compared to the graphene basal plane. An intriguing phenomenon was observed when the electrochemical reactions were accelerated at the edge of the SLG, which was only a few atoms thick. With its exceptional conductivity, the graphene edge is considered an excellent electrode material for electrocatalysis and a potential candidate for the storage of capacitive charges [91].

Nevertheless, obtaining individual graphene nanosheets or single-layer graphene (SLG) remains a significant challenge. The 2D graphene nanosheets tend to restack or aggregate into non-porous nanostructures during the synthesis process, which significantly reduces the surface area and pore volume. To prevent graphene nanosheets from restacking and to increase the energy storage capacity, a process that can be adopted is molecular functionalization, which involves introducing functional species like polymer electrolytes, MXenes, LDHs (layered double hydroxides), MOs, metal sulfides, and/or other redox-active molecules into graphene through covalent, non-covalent, and π-π* interactions [72,82, 92, 93].

4.3.4   Three-Dimensional (3D) Structures: Graphene Networks (Foams, Sponges, and Aerogels)

Graphite is a typical 3D carbon material [93]. On the other side, there were graphene networks, including graphene foams (GFs), graphene sponges (GSs), and graphene aerogels (GAs). The initial synthesis of GFs involved the utilization of nickel foam as the template. Consequently, the GFs have acquired the macropores' configuration of the nickel foam, characterized by a robust and interconnected 3D graphene network. The porous structure of GSs resembles that of GFs; however, the graphene sheets exhibit partial orientation or alignment that is closely parallel to one another, forming an anisotropic lamellar structure [94]. Whereas, GAs are a 3D metamaterial with a highly open porous (micro, meso, and macro) structure assembled by an interconnected network of 2D graphene-based nanosheets. The assembly of micro (less than 2 nm) and mesopores (2–50 nm) contributes to the high specific surface area, while the macropores (greater than 50 nm) provide accessibility to the active surfaces. Each pore size contributes uniquely to the efficacy of a supercapacitor. Macropores store electrolyte ions, whereas mesopores facilitate the transportation of electrolyte ions. Charge accommodation processes are enabled by micropores [95].

These 3D networks avoid restacking of individual graphene sheets, generating ultralight density graphene aerogels with ultrahigh porosity, outstanding mass transfer, and ultralow thermal conductivity, which is a crucial macroscopic form to harness the extraordinary surface properties of nanomaterials and has exhibited great potential in energy storage applications [94,96]. The enhanced surface area of 3D graphene networks and/or hybrid composite electrodes with 3D structures contributes to their more significant potential in electrochemical performance. As dimensionality increases, electrolytes come in contact with an increased number of active sites. The utilization of 3D architectures facilitates uninterrupted pathways to enhance electrolyte contact while concurrently accelerating charge transfer through the reduction of diffusion paths, as this structure combines the intrinsic high conductivity of graphene with the synergistic effect of 3D porous materials [93].

4.4   Fullerene-Based Nanocomposite

Buckminster fullerene (C60) is a carbon allotrope with a ball-like structure [97]. Fullerene has a large surface area, facilitating the diffusion of ions within and without. Additionally, fullerene possesses excellent structural and mechanical strength, thus supporting long-term cycle performance. The electron cloud on fullerenes dramatically aids in the charging/discharging process and electrolyte binding phenomenon on its surface [98].

Ramadan et al. (2020) [99] conducted an intercalation of the fullerene derivative Phenyl-C60-butyric acid methyl ester (PCBM) into a polyaniline (PANI) matrix. The morphology of the composite is that of irregular sponges with a porous structure. BET data indicate the presence of meso and macropores as well as a few micropores. Electrochemical measurement results show that the specific capacitance value of the PANI/PCBM5 electrode increased more than twice compared to the pure PANI electrode. The particular capacitance value for the PANI/PCBM5 electrode is 2201 F g^{-1} at a current density of 2 A g^{-1} with a reasonable rate capability of around 73% at 10

A g^{-1}. The PANI/PCBM5 composite exhibits an energy density of 61.9 Wh kg^{-1}, which is about twice as high as pure PANI's (31.2 Wh kg^{-1}). The PANI/PCBM5 electrode shows an excellent capacitance retention of 96% even after 1,000 cycles.

Ma et al. (2015) [100] successfully synthesized a C60/graphene composite. The C60/graphene composite was initially prepared using a simple solution method from graphene oxide and C60 particles functionalized by Li. Electrochemical measurement results show that the specific capacitance value of the C60/graphene composite is higher than that of graphene. The particular capacitance value for the C60/graphene composite is 135.36 F g^{-1} at a current density of 1 A g^{-1}, while the specific capacitance value of graphene is 101.88 F g^{-1}. This composite material exhibits an outstanding retention rate of 92.35% after 1,000 charge/discharge cycles. These results indicate that the C60/graphene composite is a promising material for use as a supercapacitor electrode.

5 MO NANOMATERIALS AS ELECTRODES

With their nanostructured morphology and inherent electrochemical properties, MO nanomatcrials havc garncrcd significant attention as electrodes in energy storage and conversion devices. These materials provide advantages such as high specific surface area, increased number of active sites, and facile charge transfer kinetics, making them ideal candidates for electrochemical applications. MOs have been exhaustively studied and examined as an effective material for supercapacitor (SC) storage applications due to their superior electrochemical stability and energy density, accessibility, affordability, and environmental friendliness.

MOs have faster redox reactions with the electrolyte because of their distinct oxidation states. Regarding selecting MOs for supercapacitor applications, several factors must be considered: The oxides must exhibit electrical conductivity, the variable oxidation state serves as an advantage for rapidly reversible redox processes, and protons may easily be incorporated into the layered structure of oxides during reduction and oxidation processes, facilitating the conversion of O$_2$ and OH$^-$ in the electrolyte [101,102,103]. The search results show several electrodes based on MOs for energy storage.

This technology can function as a fundamental electrochemical storage device, which is a renewable energy source, by effectively mitigating the disparity between batteries and conventional capacitors [104]. However, it is essential to know that these materials have restrictions, including the hindered transport of electrons and ions and the subpar electronic conductivity of their material. Those limitations can have a pivotal effect on the overall efficacy of SCs. The development of electrode materials is of dominant significance in attaining elevated energy density, enhanced specific power, and accelerated charging/discharging rates. Those factors will improve the overall efficacy of SCs from MO electrodes.

5.1 Transition Metal Oxide (TMO) Materials

Transition metal oxides (TMOs) are appropriate materials for pseudocapacitors. Pseudo-capacitance increases the charge storage capacity through faradaic redox

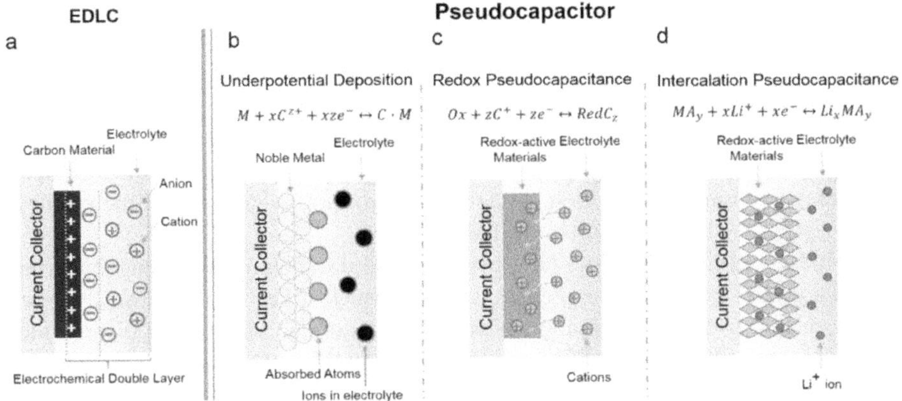

FIGURE 6.5 Illustration of the charge storage mechanism for EDLCs and (b-d) different types of pseudocapacitive electrodes: (b) deposition under potential, (c) redox pseudocapacitors, and (d) ion intercalation pseudocapacitors. [105]. Copyright 2018 American Chemical Society.

processes. Figure 6.5 illustrates the charge storage mechanism. There are two ways that pseudo-capacitance works: surface redox and intercalation pseudo-capacitances. Surface redox pseudo-capacitance is a charge storage phenomenon that occurs at or near the surface of the electrode material. The adsorption or desorption of electrolyte cations and anions causes it. Materials primarily exhibiting surface redox pseudo-capacitance exhibit a behaviour similar to EDLC-based materials, characterized by linear and triangle galvanic charge/discharge curves and practically rectangular cyclic voltammetry curves. In the intercalation pseudo-capacitance process, electrolyte cations (e.g. Na^+, K^+, Li^+, H^+, etc.) are reversibly intercalated and de-intercalated into the crystal structure of electrode materials to store energy. Additionally, the intercalation pseudo-capacitance process does not involve any phase transition of the material.

Conductive polymers and TMOs are among the electrode materials intensively researched as potential candidates for surface redox pseudo-capacitance applications. Conductive polymers experience alterations in their material structures due to volume expansion that transpires during the charge and discharge process. This results in a lack of cyclic stability, thereby constraining the practical applications of these polymers in SCs. TMO materials extensively researched for surface redox pseudo-capacitance include RuO_2, Mn_3O_4, MnO_2, $MnOOH$, $FeOOH$, Fe_3O_4, Fe_2O_3, etc. Ruthenium oxides (RuO_2) have exceptional properties, leading to an excellent theoretical capacitance of 1360 F g^{-1}, making them preferred materials and an optimal pseudocapacitive electrode material. However, their elevated cost is the primary practical limitation [81]. Hence, it is imperative to utilize alternative, inexpensive electroactive materials based on transition metals. Some TOM electrode materials that can experience intercalation pseudo-capacitance are V_2O_5, WO_3, MoO_3, Nb_2O_5, TiO_2, and so on [106].

TMO materials possess electroactive sites, chemical stability, and high thermal stability, ensuring excellent pseudocapacitive performance and cycle stability. This condition could occur if the TMOs are fabricated as MO nanostructures. Typically, novel nanostructured MOs have a high surface specific area and are porous, thereby facilitating the infiltration of electrodes, complete contact between active materials and electrolytes, ion transport in the electrolytes, and high charging and discharging rates of the active materials. Compared to bulk MOs, their practical applications have been impeded by the slow diffusion of ions in the bulk phase and uncontrollably expanding volumes [107,108,109]. As a result, it is important to consider the synthesis and design procedures for nanoscale TMOs.

Various synthetic techniques have been employed to control TMO nanomaterials' size, shape, and structure. The physical characteristics of the TMO nanostructure are closely related to their morphology. Nanostructures often exhibit three distinct morphologies: zero-dimensional (0D), one-dimensional (1D), and two-dimensional (2D) nanostructures. Zero-dimensional (0D) nanostructures frequently refer to particles with dimensions measuring a few nanometers, affording distinct exposed surfaces. Several preparation techniques have been used to synthesize nanoparticles with various morphologies, including sphere, cube, octahedron, dodecahedron, etc. Hydrothermal and electrochemical deposition are the most widely used methods for preparing this morphological dimension [110,111]. Hydrothermal synthesis has been regarded as one of the most accessible and most practical methods. It requires no complex preparation conditions or specialized equipment. Reaction parameters, including temperature, reactant concentration, and reaction duration, are controlled to determine product chemical composition, particle size, and morphology. This process involves heating precursors such as metal salts, metal complexes in water (hydrothermal), or solvents (solvothermal) at a high temperature and pressure inside a specific closed reaction vessel or autoclave. The synthesis process typically lasts 6–48 hours, resulting in the dissolution, nucleation, and crystallization of materials with low solubility or insolubility under standard conditions [112,113]. Electrochemical deposition of TMO nanomaterials is a straightforward, quick, and cost-efficient method that typically requires setting up a two- or three-electrode cell system. Transition metal and MO nanoparticles might be deposited on an electrode surface by adjusting the voltage or current of an electrochemical cell. Various transition metals and MO nanomaterials can be produced by electrochemical methods such as cyclic voltammetry (CV), chronoamperometry (CA), chronopotentiometry (CP), chronocoulometry (CC), and pulse electrodeposition at constant or varying potential [89].

Several types of one-dimensional (1D) nanostructures exist, including nanotubes, nanofibers, nanowires, and nanobelts [114,115,116,117]. Nanostructures can be synthesized in this dimension using either vapor-based or solution-based methods. The vapor-based technique imposes strict standards regarding reaction temperature and experimental apparatus. In contrast, the solution-based technique facilitates cost reduction and simplifies experimental procedures at moderate temperatures. However, compared to solution-based procedures, vapor-based methods have several advantages. Controlling the dimension, content, structure, and position of one-dimensional nanomaterials can be accomplished by modifying growth sites, catalysts, or precursors by vapor deposition techniques.

Additionally, highly crystalline metallic oxide nanowires can be produced under deposition conditions characterized by high vacuum and temperature [118].

Vapor-based preparation can be achieved via physical vapor deposition (PVD) and chemical vapor deposition (CVD). PVD involves the sublimation of surface atoms from bulk TMOs into TMO vapor. The TMO vapor then condenses onto substrates to form 1D TMO nanostructures. CVD methods are extensively utilized in the semiconductor industry due to their superior performance characteristics compared to PVD methods. The formation of resultant compounds on the substrates is facilitated by critical chemical reactions and dissociations of volatile precursors and reactive gases that occur through the CVD process at various pressures and flow rates. The CVD process, one of the conventional methods for producing 1D TMO nanocrystals, primarily employs metal catalyst-assisted vapor-liquid-solid (VLS) growth [110].

2D nanostructures can be classified into three categories based on their structural characteristics: layered, lamellar, and non-layered TMOs. Despite recent successful preparations, developing a scalable, simple, effective, and inexpensive strategy for synthesizing the many components of 2D MO nanoparticles remains a crucial problem. In recent years, significant advancements have been achieved in developing 2D nanostructures using top-down exfoliation, bottom-up synthesis, and liquid metal methods. The top-down exfoliation process involves layering the primary crystal precursors of MOs (frequently layered oxides) into elemental layers. The bottom-up synthesis involves acquiring the necessary metal atoms/molecules and oxygen and constraining their development in the 2D direction under defined conditions. The liquid metal technique involves synthesizing ultrathin 2D MOs on the surface of liquid metals or alloys by self-limiting reaction [119].

5.2 Binary Transition MOs (BTMOs)

BTMOs consist of a transition metal ion and one or more electrochemically active or inactive ions. BTMOs can enhance capacitive performance by combining the advantages of individual components, resulting in a more significant potential window, better conductivity, more active sites, and increased stability [120]. Several methods used to synthesize BTMOs have been widely reported, such as hydrothermal/solvothermal, microwave-assisted, electrodeposition, template-assisted, sol-gel, and co-precipitation methods [121]. Each method can exhibit variations in morphology, surface area, structure, and stability based on the desired result. Various BTMO crystal forms, including spinel, scheelite, and $CaFe2O_4$-type, have been investigated extensively as potential supercapacitor materials. Spinel cobaltite (MCo_2O_4, M = Ni, Cu, Zn, Mn, etc.) and metal molybdate ($MMoO4$, M = Ni, Mn, Co, etc.) recently attracted great interest in research due to their affordability, improved electrochemical activity, and natural availability. Among the several types of spinel cobaltites, the most well-known pseudo-capacitance material is $NiCo_2O_4$. $NiCo_2O_4$ has a comparatively high electrical conductivity; it exhibits an electrical conductivity at least two orders of magnitude greater than the electrical conductivities of NiO and Co_3O_4, which are single MOs, due to the electron transport between cations facilitated by the multivalences of transition metal cations that requires a comparatively low activation energy [122]. Furthermore, the presence of Ni and Co enhances the electrochemical activity of the $NiCo_2O_4$ electrode. Compared to single metal compounds, the most advantageous characteristic of binary metal compounds is that the electrocapacitive material conducts more faradic redox reactions.

5.3 Ternary Transition MOs (TTMOs)

TTMOs have shown superior supercapacitive performance compared to single MOs and are considered the most advantageous and cost-effective material group for pseudocapacitors. Huang et al. (2018) [123] even reported that TTMO ($Ni_xCo_yMo_zO$) shows higher specific capacitance values than BTMOs (Ni_xMo_yO, Co_xMo_yO, and Ni_xCo_yO). This is due to the increased electrocapacitive capability and electrical conductivity of the ternary MOs [123]. The hydrothermal method is the most widely used method to synthesize TTMOs. This method has shown to enhance the specific capacitance of TTMOs compared to single or binary MO states [124,125]. The hydrothermal method is a straightforward and efficient technique for producing water-insoluble ternary MO nanostructures with high purities and adjustable shapes. This process involves employing water-soluble metal compounds as precursors under high pressure and moderate temperature conditions. This process is advantageous since it does not require dangerous catalysts, seeds, harmful surfactants, or templates, making it environmentally friendly. Alternative approaches involving BTMO synthesis can be utilized for TTMO synthesis to achieve specific morphological characteristics.

5.4 MO-Based Composite

The investigation of MO-based composites has garnered significant attention due to their potential to enhance the electrocapacitive capabilities of SCs. However, TMOs in single, binary, or ternary states have also proven to perform well in SC applications. MO-based composites widely studied include TMO/TMO, TMO/conductive polymers, and TMO/carbon-based composites. TMO/TMO composites have been extensively synthesized using hydrothermal and sol-gel techniques. The TMO and TMO composite have exhibited superior performance and stability in addition to their high specific capacitance values. Conductive polymers have been shown to have poor stability in SC applications, as previously discussed. Combining TMOs with conductive polymers is a solution to overcome that problem. It has been proven to increase specific capacitance value and good electrical conductivity. TMO/carbon-based composites have been widely studied and developed. Carbon-based materials (such as CNTs, graphene, activated carbon, etc.) have been extensively employed in energy storage applications due to their excellent surface area and abundant availability in the natural environment. A composite material of TMO and carbon-based components signifies excellent potential as an electrode material. The synthesis process for these composite materials can also be determined or adjusted to obtain the desired morphology and characteristics [126].

5.5 Synthesis Methods

MO nanomaterials can be synthesized using various techniques, including sol-gel method, hydrothermal synthesis, chemical vapor deposition, and template-assisted synthesis. These methods enable precise control over the size, shape, and composition of the nanomaterials, leading to tailored electrochemical properties for specific applications.

5.6 ELECTROCHEMICAL PROPERTIES

MO nanomaterials exhibit unique electrochemical properties that make them well-suited for electrode applications. MO nanomaterials offer high specific capacitance, making them ideal for energy storage applications in electrochemical capacitors (SCs). Their large surface area facilitates rapid ion adsorption and desorption, resulting in high charge storage capacity and fast charge/discharge kinetics. MO nanomaterials demonstrate excellent rate capability, enabling efficient charge transfer at high current densities. This property is crucial for applications requiring rapid energy storage and delivery, such as electric vehicles and grid-scale energy storage systems. MO nanomaterials exhibit good cycling stability, with minimal degradation in electrochemical performance over multiple charge/discharge cycles. This durability is essential for energy storage devices' long-term reliability and longevity.

5.7 APPLICATIONS IN ENERGY STORAGE AND CONVERSION

MO nanomaterials find diverse applications in energy storage and conversion technologies. MO nanomaterials serve as electrodes in lithium-ion batteries, offering high energy density and long cycle life. Their ability to intercalate lithium ions into the crystal lattice enables efficient energy storage and delivery for portable electronics, electric vehicles, and renewable energy storage systems. The performance of Li-ion batteries depends on each component, such as the cathode, anode, electrolyte, and separator. Among these, cathode materials are those that require development to increase the energy density and reduce the production price of Li-ion batteries. Cathodes and anodes are also closely related to battery capacity and stability by improving the ionic conductivity and electrochemical properties by using methods such as doping (anion, cation, and co-doping) and coating [127].

MO nanomaterials are utilized as electrode materials in electrochemical capacitors (SCs) for rapid energy storage and release. Their high specific capacitance and excellent rate capability make them ideal candidates for high-power density and fast-charging applications, such as hybrid vehicles and regenerative braking systems. 2D MXene (M = transition metal, X = carbides, nitrides, and carbonitrides) is the newest frontier in the development of supercapacitor electrodes. This material was discovered by Gogotsi et al. in 2011. MXene has excellent flexibility and diverse surface-functional groups, because it can be synthesized using various methods [128].

MO nanomaterials are employed as photoelectrodes in photoelectrochemical water-splitting cells for hydrogen production. Their light-absorbing properties and efficient charge separation enable the conversion of solar energy into chemical energy, paving the way for sustainable hydrogen fuel production. In recent times, the application of TMNs (Transition Metal Nitrides) in water-splitting, including Hydrogen Evolution Reaction (HER) and Oxygen Evolution Reaction (OER), has gained increased attention for the enhancement of electronic, catalytic, chemical, and physical properties [129].

Despite their promising potential, several challenges remain in implementing MO nanomaterials as electrodes. Improving the stability and durability of MO nanomaterials under prolonged cycling conditions is crucial for their widespread adoption in commercial energy storage devices. Addressing cost-related issues associated with the synthesis and scalability of MO nanomaterials is essential for large-scale production and commercialization. Future research should study the synergistic effects and interfaces between MO nanomaterials and other electrode materials to enhance overall device performance and efficiency.

6 CONCLUSION

Nanomaterials for developing electrodes explore the utilization of various nanomaterials in advancing electrode technologies. Carbon-based materials such as graphene, fullerene, and MOs are investigated for their potential applications in electrode development. These nanomaterials offer unique properties such as high surface area, exceptional conductivity, and chemical stability, making them promising candidates for enhancing the performance of electrodes in various electrochemical applications. By harnessing the remarkable properties of these nanomaterials, researchers aim to overcome existing limitations and pave the way for developing more efficient and sustainable electrode systems.

Integrating these nanomaterials into electrode structures holds great potential for achieving high-performance electrochemical devices, contributing to advancements in energy storage and conversion technologies. Continued research efforts in synthesizing, characterizing, and optimizing these nanomaterials will be crucial for furthering their practical applications and addressing the challenges in sustainable energy systems.

ACKNOWLEDGMENTS

The authors would like to express their gratitude to Universitas Gadjah Mada for funding this research through Riset Kolaborasi Indonesia (RKI): (contract number: 1557/UN1/DITLIT/DIT-LIT/PT.01.03/2022).

REFERENCES

[1] B. Mekuye, B. Abera, Nanomaterials: An overview of synthesis, classification, characterization, and applications, *Nano. Select.*, 4 (2023) 486–501. https://doi.org/10.1002/nano.202300038.

[2] S. Ravi, S. Vadukumpully, Sustainable carbon nanomaterials: Recent advances and its applications in energy and environmental remediation, *J. Environ. Chem. Eng.*, 4 (2016) 835–856.

[3] T. W. Chen, S. M. Chen, G. Anushya, R. Kannan, P. Veerakumar, A. G. Al-Sehemi, V. Mariyappan, S. Alargarsamy, M. M. Alam, T. C. Mahesh, R. Ramachandran, P. Kalimuthu, Electrochemical energy storage applications of functionalized carbon-based nanomaterials: An overview, *Int. J. Electrochem. Sci.*, 19 (2024) 100548. https://doi.org/10.1016/j.ijoes.2024.100548.

[4] N. Waris, M. S. Chaudhary, A. H. Anwer, S. Sultana, P. P. Ingole, S. A. A. Nami, M. Z. Khan, A review on development of carbon-based nanomaterials for energy storage devices: Opportunities and challenges, *Energy and Fuels,* 37 (2023) 19433–19460. https://doi.org/10.1021/acs.energyfuels.3c03213.

[5] V. Harish, D. Tewari, M. Gaur, A. B. Yadav, S. Swaroop, M. Bechelany, A. Barhoum, Review on nanoparticles and nanostructured materials: Bioimaging, biosensing, drug delivery, tissue engineering, antimicrobial, and agro-food applications, *Nanomaterials (Basel)*, 12 (2022) 456. https://doi.org/10.3390/nano12030457.

[6] N. Baig, I. Kammakakam, W. Falath, Nanomaterials: A review of synthesis methods, properties, recent progress, and challenges, *Mater. Adv.*, 2 (2021) 1821–1871. https://doi.org/10.1039/D0MA00807A.

[7] I. Khan, K. Saeed, I. Khan, Nanoparticles: Properties, applications and toxicities, *Arabian J. Chem.*, 12 (2019) 908–931. https://doi.org/10.1016/j.arabjc.2017.05.011.

[8] N. Joudeh, D. Linke, Nanoparticle classification, physicochemical properties, characterization, and applications: A comprehensive review for biologist, *J. Nanobiotechnology*, 20 (2022) 262. https://doi.org/10.1186/s12951-022-01477-8.

[9] E. Frackowiak, F. Béguin, Carbon materials for the electrochemical storage of energy in capacitors, *Carbon N. Y.*, 39 (2001) 937–950. https://doi.org/10.1016/S0008-6223(00)00183-4.

[10] G. Pandolfo, A. F. Hollenkamp, Carbon properties and their role in supercapacitors, *J. Power Sources*, 157 (2006) 11–27. https://doi.org/10.1016/j.jpowsour.2006.02.065.

[11] Y. Zhu, S. Murali, W. Cai, X. Li, J. W. Suk, J. R. Potts, and R. S. Ruoff, Graphene and graphene oxide: Synthesis, properties, and applications, *Adv. Mater.*, 22 (2010) 3906–3924. https://doi.org/10.1002/adma.201001068.

[12] M. F. El-Kady, V. Strong, S. Dubin, R. B. Kaner, Laser Scribing of high-performance and flexible graphene-based electrochemical capacitors, *Science,* 335 (2012) 1326–1330. https://doi.org/10.1126/science.1216744.

[13] F. Wudl, Fullerene materials, *J. Mater. Chem.*, 12 (2002) 1959–1963. https://doi.org/10.1039/B201196D.

[14] L. Duclaux, Review of the doping of carbon nanotubes (multiwalled and single-walled), *Carbon N. Y.* 40 (2002) 1751–1764. https://doi.org/10.1016/S0008-6223(02)00043-X.

[15] L. Zhang, X. S. Zhao, Carbon-based materials as supercapacitor electrodes, *Chem. Soc. Rev.*, 38 (2009) 2520–2531. https://doi.org/10.1039/B813846J.

[16] D. Qu, H. Shi, Studies of activated carbons used in double-layer capacitors, *J. Power Sources*, 74 (1998) 99–107. https://doi.org/10.1016/S0378-7753(98)00038-X.

[17] H. Y. Lee, J. B. Goodenough, Supercapacitor behavior with KCl electrolyte, *J. Solid State Chem.*, 144 (1999) 220–223. https://doi.org/10.1006/jssc.1998.8128.

[18] J. P. Zheng, P. J. Cygan, T. R. Jow, Hydrous ruthenium oxide as an electrode material for electrochemical capacitors, *J. Electrochem. Soc.*, 142 (1995) 2699. https://iopscience.iop.org/article/10.1149/1.2050077.

[19] M. Jayalakshmi, K. Balasubramanian, Simple capacitors to supercapacitors - An overview, *Int. J. Electrochem.*, 3 (2008) 1196–1217. https://doi.org/10.1016/S1452-3981(23)15517-9.

[20] R. Kötz, M. Carlen, Principles and applications of electrochemical capacitors, *Electrochim. Acta*, 45 (2000) 2483–2498. https://doi.org/10.1016/S0013-4686(00)00354-6.

[21] A. M. El-Khawaga, A. Zidan, A. I. A. A. El-Mageed, Preparation methods of different nanomaterials for various potential applications: A review, *J. Mol. Struct.*, 1281 (2023) 135148. https://doi.org/10.1016/j.molstruc.2023.135148.

[22] N. Abid, A. M. Khan, S. Shujait, K. Chaudhary, M. Ikram, M. Imran, J. Haider, M. Khan, Q. Khan, M. Maqbool, Synthesis of nanomaterials using various top-down and bottom-up approaches, influencing factors, advantages, and disadvantages: A review, *Adv. Colloid Interface Sci.*, 300 (2022) 102597. https://doi.org/10.1016/j.cis.2021.102597.

[23] M. S. El-Eskandarany, A. Al-Hazza, L. A. Al-Hajji, N. Ali, A. A. Al-Duweesh, M. Banyan, F. Al-Ajmi, Mechanical milling: A superior nanotechnological tool for fabrication of nanocrystalline and nanocomposite materials, *Nanomaterials,* 11 (2021) 2484. https://doi.org/10.3390/nano11102484.

[24] L. Chen, B. Kishore, E. Kendrick, Nanostructured materials for sodium-ion batteries in nanomaterials for electrochemical energy storage, in R. Raccichini and U. B. T.-F. of N. Ulissi, Eds. Amsterdam, Netherlands: Elsevier (2021) vol. 19, pp. 165–197. https://doi.org/10.1016/B978-0-12-821434-3.00009-0.

[25] N. Baig, I. Kammakakam, W. Falath, Nanomaterials: A review of synthesis methods, properties, recent progress, and challenges, *Mater. Adv.,* 2 (2021) 1821–1871. https://doi.org/10.1039/D0MA00807A.

[26] H. Zeng, X. W. Du, S. C. Singh, S. A. Kulinich, S. Yang, J. He, W. Cai, Nanomaterials via laser ablation/irradiation in liquid: A review, *Adv. Funct. Mater.,* 22 (2012) 1333–1353. https://doi.org/10.1002/adfm.201102295.

[27] D. S. Kryzhevich, K. P. Zolnikov, A. V. Korchuganov, S. G. Psakhie, Nanopowder synthesis based on electric explosion technology, *AIP Conf. Proc.,* 1893 (2017) 030125-1-4. https://doi.org/10.1063/1.5007583.

[28] J.-O. Carlsson, P. M. Martin, *Chapter 7- Chemical Vapor Deposition,* P. M. B. T.-H. of D. T. for F. and C. (Third E. Martin, Ed.). Boston: William Andrew Publishing (2010) pp. 314–363. https://doi.org/10.1016/B978-0-8155-2031-3.00007-7.

[29] B. Kalanyan, W. A. Kimes, R. Beams, S. J. Stranick, E. Garratt, I. Kalish, A. V. Davydov, R. K. Kanjolia, J. E. Maslar, Rapid wafer-scale growth of polycrystalline 2H-MoS2 by pulsed metal–organic chemical vapor deposition, *Chem. Mater.,* 29 (2017) 6279–6288. https://doi.org/10.1021/acs.chemmater.7b01367.

[30] M. Malekzadeh, M. T. Swihart, Vapor-phase production of nanomaterials, *Chem. Soc. Rev.,* 50 (2021) 7132–7249. https://doi.org/10.1039/D0CS01212B.

[31] P. Frigeri, L. Seravalli, G. Trevisi, S. Franchi, Eds. Molecular beam Epitaxy: An overview, in *Reference Module in Materials Science and Materials Engineering* (2016) Elsevier. https://doi.org/10.1016/B978-0-12-803581-8.00796-7.

[32] D. Bokov, A. Turki Jalil, S. Chuprad, W. Suksatan, M. Javed Ansari, I. H. Shewael, G. H. Valiev, E. Kianfar, Nanomaterial by sol-gel method: Synthesis and application, *Adv. Mater. Sci. Eng.,* 1 (2021) 1–21. https://doi.org/10.1155/2021/5102014.

[33] W. Lu, *Self-Assembly of Nanostructures BT - Encyclopedia of Nanotechnology,* B. Bhushan, Ed. Dordrecht: Springer Netherlands (2012) pp. 2371–2382. https://doi.org/10.1007/978-90-481-9751-4_274.

[34] M. Z. Al Mahmud, A concise review of nanoparticles utilized energy storage and conservation, *J. Nanomaterials,* 2023 (2023) 5432099. https://doi.org/10.1155/2023/5432099.

[35] A. Nayak, B. Bhushan, S. Kotnala, N. Kukretee, P. Chaudhary, A. R. Tripathy, K. Ghai, S. L. Mudliar, Nanomaterials for supercapacitors as energy storage application: Focus on its characteristics and limitations, *Mater. Today Proceed.,* 73 (2023) 227–232. https://doi.org/10.1016/j.matpr.2022.11.259.

[36] E. A. Kumah, R. D. Fopa, S. Harati, P. Boadu, F. V. Zohoori, T. Pak, Human and environmental impacts of nanoparticles: A scoping review of the current literature, *BMC Public Health,* 23 (2023) 1059. https://doi.org/10.1186/s12889-023-15958-4.

[37] S. Singh, Chapter 9 – Adverse effects of nanoparticles on human health and the environment, in *Antiviral and Antimicrobial Coatings Based on Functionalized Nanomaterials,* Amsterdam, Netherlands: Elsevier (2023) pp. 305–330. https://doi.org/10.1016/B978-0-323-91783-4.00016-4.

[38] K. Wijaya, R. A. Pratika, A. Nadia, F. Rahmawati, R. Zainul, W. C. Oh, W. D. Saputri, Recent trends and applications of nanomaterial based on carbon paste electrodes: A short review, *Evergreen,* 10 (2023) 1374–1387. https://doi.org/10.5109/7151686.

[39] Y. Li, K. Zhang, L. Zhang, M. Nie, Q. Wang, Highly conductive and durable electro-thermal electrode though the steric confinement effect of graphene on helically intersected carbon fiber network, *Compos. Sci. Technol.,* 213 (2021) 108900. https://doi.org/10.1016/j.compscitech.2021.108900.

[40] P. Aich, C. Meneghini, L. Tortora, Advances in structural and morphological characterization of thin magnetic films: A review, *Materials (Basel)* 16 (2023) 7331. https://doi.org/10.3390/ma16237331.

[41] S. Y. Guan, Z. H. Zhang, Wu, X. K. Gu, C. Y. Zhao, Boiling on nano-porous structures: theoretical analysis and molecular dynamics simulations, *Int. J. Heat Mass. Transfer,* 191 (2022) 122848. https://doi.org/10.1016/j.ijheatmasstransfer.2022.122848.

[42] V. V. Chaban, N. A. Andreeva, Aqueous electrolytes at the charged graphene surface: electrode-electrolyte coupling, *J. Mol. Liq.,* 387 (2023) 122724. https://doi.org/10.1016/j.molliq.2023.122724.

[43] Y. R. Jain, F. Callaway, T. L. Griffiths, P. Dayan, R. He, P. M. Krueger, F. Lieder, A computational process-tracing method for measuring people's planning strategies and how they change over time, *Behav. Res. Methods,* 55 (2023) 2037–2079. https://doi.org/10.3758/s13428-022-01789-5.

[44] E. Jonsson, Ionic liquids as electrolytes for energy storage applications – A modelling perspective, *Energ. Storage Mat.,* 25 (2020) 827–835. https://doi.org/10.1016/j.ensm.2019.08.030.

[45] K. fan, Y. H. Tsang, H. Huang, Computational design of promising 2D electrode materials for Li-ion and Li-S battery applications, *Mater. Reports: Energy,* 3 (2023) 100213. https://doi.org/10.1016/j.matre.2023.100213.

[46] D. Adekoya, S. Qian, X. Gu, W. Wen, D. Li, J. Ma, S. Zhang, *DFT-Guided Design and Fabrication of Carbon-Nitride-Based Materials for Energy Storage Devices: A Review.* Singapore: Springer Nature, 2021. https://doi.org/10.1007/s40820-020-00522-1.

[47] H. Ma, J. F. Chen, H. F. Wang, P. J. Hu, W. Ma, Y. T. Long, Exploring dynamic interactions of single nanoparticles at interfaces for surface-confined electrochemical behavior and size measurement, *Nat. Commun.,* 11 (2020) 1–9. https://doi.org/10.1038/s41467-020-16149-0.

[48] Y. Yang, S. Bremner, C. Menictas, M. Kay, Battery energy storage system size determination in renewable energy systems: A review, *Renew. Sustain. Energy Rev.,* 91 (2018) 109–125. https://doi.org/10.1016/j.rser.2018.03.047.

[49] P. G. Tratnyek, E. J. Bylaska, E. J. Weber, In silico environmental chemical science: properties and processes from statistical and computational modelling, *Environ. Sci. Process. Impacts,* 19 (2017) 188–202. https://doi: 10.1039/c7em00053g.

[50] S. Roy, S. Singh, M. Khan, E. Chamanehpour, S. Sain, T. Goswami, S. S. Roy, Y. K. Mishra, A. Mathur, Electrochemistry at 2D and 3D nanoelectrodes: The interplay between interface kinetics and surface density of states, *Electrochim. Acta,* 477 (2024) 143762. https://doi.org/10.1016/j.electacta.2024.143762.

[51] C. Cesari, B. Berti, T. Funaioli, C. Femoni, M. C. Iapalucci, D. Pontiroli, G. Magnani, M. Riccò, M. Bortoluzzi, F. M. Vivaldi, S. Zacchini, Atomically precise platinum carbonyl nanoclusters: Synthesis, total structure, and electrochemical investigation of [Pt27(CO)31]4-Displaying a defective structure, *Inorg. Chem.,* 61 (2022) 12534–12544. https://doi.org/10.1021/acs.inorgchem.2c00965.

[52] A. García-Miranda Ferrari, D. A. C. Brownson, A. S. Abo Dena, C. W. Foster, S. J. Rowley-Neale, C. E. Banks, Tailoring the electrochemical properties of 2D-hBN: Via physical linear defects: Physicochemical, computational and electrochemical characterisation, *Nanoscale Adv.,* 2 (2020) 264–273. https://doi.org/10.1039/c9na00530g.

[53] S. Ghasemi, K. Moth-Poulsen, Single molecule electronic devices with carbon-based materials: Status and opportunity, *Nanoscale,* 13 (2021) 659–671. https://doi.org/10.1039/d0nr07844a.

[54] S. Schweizer, J. Landwehr, B. J. M. Etzold, R. H. Meißner, M. Amkreutz, P. Schiffels, J. R. Hill, Combined computational and experimental study on the influence of surface chemistry of carbon-based electrodes on electrode-electrolyte interactions in supercapacitors, *J. Phys Chem. C,* 123 (2019) 2716–2727. https://doi.org/10.1021/acs.jpcc.8b07617.

[55] L. S. Anthony, M. Vasudevan, V. Perumal, M. Ovinis, P. B. Raja, T. N. J. I. Edison, Bioresource-derived polymer composites for energy storage applications: Brief review. *J. Environ. Chem. Eng.*, 9 (5) (2021) 105832. https://doi.org/10.1016/j.jece.2021.105832.

[56] S. Yasami, S. Mazinani, M. Abdouss, Developed composites materials for flexible supercapacitors electrode: "Recent progress & future aspects," *Journal of Energy Storage*, 72 (2023) 108807. https://doi.org/10.1016/j.est.2023.108807.

[57] J. Winsberg, T. Hagemann, T. Janoschka, M. D. Hager, U. S. Schubert, Redox-flow batteries: From metals to organic redox-active materials, *Angew. Chem. Int. Ed.*, 56 (2017) 686–711. https://doi.org/10.1002/anie.201604925.

[58] Z. Yang, L. Tong, D. P. Tabor, E. S. Beh, M. A. Goulet, D. De Porcellinis, A. AspuruGuzik, R. G. Gordon, M. J. Aziz, Alkaline benzoquinone aqueous flow battery for large-scale storage of electrical energy, *Adv. Energy Mater.*, 8 (2018) 1702056. https://doi.org/10.1002/aenm.201702056.

[59] B. Huskinson, M. P. Marshak, C. Suh, S. Er, M. R. Gerhardt, C. J. Galvin, X. Chen, A. Aspuru-Guzik, R. G. Gordon, M. J. Aziz, A metal-free organic–inorganic aqueous flow battery, *Nature*, 505 (2014) 195–198. https://doi.org/10.1038/nature12909.

[60] Sánchez-Díez, E. Ventosa, M. Guarnieri, A. Trovo, C. Flox, R. Marcilla, F. Soavi, P. Mazur, E. Aranzabe, R. Ferret, Redox flow batteries: Status and perspective towards sustainable stationary energy storage, *J. Power Sources*, 481 (2021) 228804. https://doi.org/10.1016/j.jpowsour.2020.228804.

[61] H. J. Ahn, J. H. Lee, Y. Jeong, J. H. Lee, C. S. Chi, H. J. Oh, Nanostructured carbon cloth electrode for desalination from aqueous solutions, *Mater. Sci. Eng. A*, 449–451 (2007) 841–845. https://doi.org/10.1016/j.msea.2006.02.448.

[62] A. Forner-Cuenca, E. E. Penn, A. M. Oliveira, F. R. Brushett, Exploring the role of electrode microstructure on the performance of non-aqueous redox flow batteries, *J. Electrochem. Soc.*, 166 (2019) A2230–A2241. https://doi.org/10.1149/ 2.0611910jes.

[63] A. A. Wong, M. J. Aziz, Method for comparing porous carbon electrode performance in redox flow batteries, *J. Electrochem. Soc.*, 167 (2020) 110542. https://doi.org/ 10.1149/1945–7111/aba54d.

[64] S. Mandal, J. Hu, S. Q. Shi, A comprehensive review of hybrid supercapacitor from transition metal and industrial crop based activated carbon for energy storage applications, *Mater. Today Commun.*, 34 (2023) 105207. https://doi.org/10.1016/j. mtcomm.2022.105207.

[65] L. Guan, L. Yu, G. Z. Chen, Capacitive and non-capacitive faradaic charge storage, *Electrochimica Acta*, 206 (2016) 464–478. https://doi.org/10.1016/j.electacta.2016.01.213.

[66] X. Zhou, S. Cao, H. Li, H. Guo, Y. Chen, Wide voltage-window biomass carbon-based MnO electrodes for supercapacitors, *J. Nanoparticle Res.*, 23(4) (2021) 110. https://doi. org/10.1007/s11051-021-05210-8.

[67] Q. Wang, Y. Zhang, J. Xiao, H. Jiang, T. Hu, C. Meng, Copper oxide/cuprous oxide/hierarchical porous biomass-derived carbon hybrid composites for high-performance supercapacitor electrode, *J. Alloys Compd.*, 782 (2019) 1103–1113. https://doi.org/10.1016/j. jallcom.2018.12.235.

[68] P. P. Falara, A. Zourou, K. V. Kordatos, Recent advances in carbon dots/2-D hybrid materials, *Carbon*, 195 (2022) 219–245. https://doi.org/10.1016/j.carbon.2022.04.029.

[69] V. C. Hoang, K. Dave, V. G. Gomes, Carbon quantum dot-based composites for energy storage and electrocatalysis: Mechanism, applications and future prospects, *Nano Energy*, 66 (2019) 104093. https://doi.org/10.1016/j.nanoen.2019.104093.

[70] C. Xia, S. Zhu, T. Feng, M. Yang, B. Yang, Evolution and synthesis of carbon dots: From carbon dots to carbonized polymer dots, *Adv. Sci.*, 23 (2019) 1901316. https://doi: 10.1002/advs.201901316.

[71] H. Jindal, A. S. Oberoi, I. S. Sandhu, M. Chitkara, B. Singh, Graphene for hydrogen energy storage - a comparative study on GO and RGO employed in a modified reversible PEM fuel cell, *Int. J. Energy Res.*, 45 (2021) 5815–5826. https://doi.org/10.1002/er.6202.

[72] E. Boateng, A. R. Thiruppathi, C. K. Hung, D. Chow, D. Sridhar, A. Chen, Functionalization of graphene-based nanomaterials for energy and hydrogen storage, *Electrochim. Acta,* 452 (2023) 142340. https://doi.org/10.1016/j.electacta.2023.142340.

[73] S. Zhai, L. Wei, H. E. Karahan, X. Chen, C. Wang, X. Zhang, J. Chen, X. Wang, Y. Chen, 2D Materials for 1D electrochemical energy storage devices, *Energy Storage Mater.,* 19 (2019) 102–123. https://doi.org/10.1016/j.ensm.2019.02.020.

[74] S. Zhai, Y. Chen, Graphene-based fiber supercapacitors, *Acc. Mater. Res.,* 3 (2022) 922–934. https://doi.org/10.1021/accountsmr.2c00087.

[75] H. Ji, X. Zhao, Z. Qiao, J. Jung, Y. Zhu, Y. Lu, Y. Zhang, Y. Macdonald, A. H. Ruoff, Capacitance of carbon-based electrical double-layer capacitors, *Nat. Commun.,* 5 (2014) 3317. https://doi.org/10.1038/ncomms4317.

[76] Z. Sadat, R. E. Keihan, V. D. Esfahlan, S. Dalvand, Fabrication of a novel porous nanostructure based on embedded with graphene oxide/Poly (p - Phenylenediamine) to construct an efficient supercapacitor, *Sci. Rep.,* 0123456789 (2024) 1–11. https://doi. org/10.1038/s41598-024-53241-7.

[77] K. O. Oyedotun, J. O. Ighalo, J. F. Amaku, C. Olisah, A. O. Adeola, K. O. Iwuozor, K. G. Akpomie, J. Conradie, K. A. Adegoke, Advances in supercapacitor development: Materials, processes, and applications, *J. Electron. Mater.,* 52 (2023) 96–129. https:// doi.org/10.1007/s11664-022-09987-9.

[78] S. H. Choi, S. J. Yun, Y. S. Won, C. S. Oh, S. M. Kim, K. K. Kim, Y. H. Lee, Large-scale synthesis of graphene and other 2D materials towards industrialization, *Nat. Commun.,* 13 (2022) 1–5. https://doi.org/10.1038/s41467-022-29182-y.

[79] S. A. Carminati, I. Rodríguez-Gutiérrez, A. De Morais, B. L. Da Silva, M. A. Melo, F. L. Souza, A. F. Nogueira, Challenges and prospects about the graphene role in the design of photoelectrodes for sunlight-driven water splitting, *RSC Adv.,* 11 (2021) 14374–14398. https://doi.org/10.1039/d0ra10176a.

[80] S. M. Mousavi, S. A. Hashemi, M. Y. Kalashgrani, A. Gholami, M. Binazadeh, W. H. Chiang, M. Rahman, Recent advances in energy storage with graphene oxide for supercapacitor technology, *Sustain. Energy Fuels,* 7 (2023) 5176–5197. https://doi. org/10.1039/d3se00867c.

[81] A. C. Ferrari, F. Bonaccorso, V. Fal'ko, K. S. Novoselov, S. Roche, P. Bøggild, S. Borini, F. H. Koppens, V. Palermo, N. Pugno, J. A. Garrido, Science and technology road-map for graphene, related two-dimensional crystals, and hybrid systems, *Nanoscale,* 7 (2015) 4598–4810. https://doi.org/10.1039/c4nr01600a.

[82] A. T. Smith, A. M. LaChance, S. Zeng, B. Liu, L. Sun, Synthesis, properties, and applications of graphene oxide/reduced graphene oxide and their nanocomposites, *Nano Mater. Sci.,* 1 (2019) 31–47. https://doi.org/10.1016/j.nanoms.2019.02.004.

[83] Q. Liu, J. Sun, K. Gao, N. Chen, X. Sun, D. Ti, C. Bai, R. Cui, L. Qu, Graphene quantum dots for energy storage and conversion: From fabrication to applications, *Mater. Chem. Front.,* 4 (2020), 421–436. https://doi.org/10.1039/c9qm00553f.

[84] S. A. Ansari, Graphene quantum dots: Novel properties and their applications for energy storage devices, *Nanomaterials* 12 (2022) 3814. https://doi.org/10.3390/nano12213814.

[85] Y. A. Kumar, G. Koyyada, T. Ramachandran, J. H. Kim, H. H. Hegazy, S. Singh, Recent advancement in quantum dot-based materials for energy storage applications: A review, *Dalt. Trans.,* 52 (2023) 8580–8600. https://doi.org/10.1039/d3dt00325f.

[86] H. Cheng, Q. Li, L. Zhu, S. Chen, Graphene fiber-based wearable supercapacitors: Recent advances in design, construction, and application, *Small Methods,* 5 (2021) 2100502. https://doi.org/10.1002/smtd.202100502.

[87] B. Fang, D. Chang, Z. Xu, C. Gao, A Review on graphene fibers: Expectations, advances, and prospects, *Adv. Mater.,* 32 (2019) 1902664. https://doi.org/10.1002/adma.201902664.

[88] X. Zhao, B. Zheng, T. Huang, C. Gao, Graphene-based single fiber supercapacitor with a coaxial structure, *Nanoscale,* 7 (2015) 9399–9404. https://doi.org/10.1039/ C5NR01737H.

[89] C. Dai, G. Sun, L. Hu, Y. Xiao, Z. Zhang, L. Qu, Recent progress in graphene-based electrodes for flexible batteries, *InfoMat.*, 2 (2020) 509–526. https://doi.org/10.1002/inf2.12039.

[90] R. S. K. Houtsma, J. Rie, Atomically precise graphene nanoribbons: Interplay of structural and electronic properties, *Chem. Soc. Rev.*, 50 (2021) 6541–6568. https://doi.org/10.1039/d0cs01541e.

[91] W. Yuan, Y. Zhou, Y. Li, C. Li, H. Peng, J. Zhang, Z. Liu, L. Dai, The edge- and basal-plane-specific electrochemistry of a single-layer graphene sheet, *Sci. Rep.*, 3 (2013) 1–7. https://doi.org/10.1038/srep02248.

[92] X. Li, L. Zhi, Graphene hybridization for energy storage applications, *Chem. Soc. Rev.*, 47 (2018) 3189–3216. https://doi.org/10.1039/c7cs00871f.

[93] N. Kumar, S. Ghosh, D. Thakur, C. P. Lee, P. K. Sahoo, Recent advancements in zero- to three-dimensional carbon networks with a two-dimensional electrode material for high-performance supercapacitors, *Nanoscale Adv.*, 5 (2023) 3146–3176. https://doi.org/10.1039/d3na00094j.

[94] Y. Ma, Y. Chen, Three-dimensional graphene networks: Synthesis, properties and applications, *Natl. Sci. Rev.*, 2 (2015) 40–53. https://doi.org/10.1093/nsr/nwu072.

[95] Y. Beeran, H. Reddy, Z. Ahmad, Graphene based aerogels: Fundamentals and applications as supercapacitors, *J. Energy Storage*, 30 (2020) 101549. https://doi.org/10.1016/j.est.2020.101549.

[96] K. Pang, X. Song, Z. Xu, X. Liu, Y. Liu, L. Zhong, Y. Peng, J. Wang, J. Zhou, F. Meng, J. Wang, C. Gao, Hydroplastic foaming of graphene aerogels and artificially intelligent tactile sensors, *Sci. Adv.*, 6 (2020) 1–8. https://doi.org/10.1126/sciadv.abd4045.

[97] Z. Jiang, Y. Zhao, X. Lu, J. Xie, Fullerenes for rechargeable battery applications: Recent developments and future perspectives, *J. Energ. Chem.*, 55 (2021) 70–79. https://doi.org/10.1016/j.jechem.2020.06.065.

[98] R. Mohanty, G. Swain, K. Parida, K. Parida, Enhanced electrochemical performance of flexible asymmetric supercapacitor based on novel nanostructured activated fullerene anchored zinc cobaltite, *J. Alloys Compound.*, 919 (2022) 165753. https://doi.org/10.1016/j.jallcom.2022.165753.

[99] A. Ramadan, M. Anas, S. Ebrahim, M. Soliman, A. I. Abou-Aly, Polyaniline/fullerene derivative nanocomposite for highly efficient supercapacitor electrode, *Int. J. Hydrogen Energy*, 45 (2020) 16254–16265. https://doi.org/10.1016/j.ijhydene.2020.04.093.

[100] J. Ma, Q. Guo, H. L. Gao, X. Qin, Synthesis of C60/Graphene composite as electrode in supercapacitors, *Fullerenes Nanotubes Carbon Nanostructures*, 23 (2015) 477–482. https://doi.org/10.1080/1536383X.2013.865604.

[101] I. Shaheen, Recent advancements in metal oxides for energy storage materials: Design, classification, and electrodes configuration of supercapacitor, *J. Energy Storage*, 72 (2023) 108719. https://doi.org/10.1016/j.est.2023.108719.

[102] M. Buldu-Akturk, M. Toufani, A. Tufani, E. Erdem, ZnO and reduced graphene oxide electrodes for all-in-one supercapacitor devices, *Nanoscale*, 14 (2022) 3269–3278. https://doi.org/10.1039/d2nr00018k.

[103] L. Liu, Large-scale mechanical preparation of graphene containing nickel, nitrogen and oxygen dopants as supercapacitor electrode material, *Chem. Eng. J.*, 430 (2022) 132815. https://doi.org/10.1016/j.cej.2021.132815.

[104] F. Ahmad, A. Shahzad, M. Danish, M. Fatima, M. Adnan, S. Atiq, M. Asim, M. Ahmed Khan, Q. Ul Ain, R. Perveen, Recent developments in transition metal oxide-based electrode composites for supercapacitor applications, *J. Energy Storage*, 81 (2024) 110430. https://doi.org/10.1016/j.est.2024.110430.

[105] Y. Shao, M. F. El-Kady, J. Sun, Y. Li, Q. Zhang, M. Zhu, H. Wang, B. Dunn, and R. B. Kaner, Design and mechanisms of asymmetric supercapacitors, *Chem. Rev.*, 118 (2018) 9233–9280. https://doi: 10.1021/acs.chemrev.8b00252.

[106] T. N. Jebakumar Immanuel Edison, R. Atchudan, Y. R. Lee, Facile synthesis of carbon encapsulated RuO$_2$ nanorods for supercapacitor and electrocatalytic hydrogen evolution reaction, *Int. J. Hydrogen Energy,* 44 (2019) 2323–2329. https://doi.org/10.1016/j.ijhydene.2018.02.018.

[107] V. Augustyn, P. Simon, B. Dunn, Pseudocapacitive oxide materials for high-rate electrochemical energy storage, *Energy Environ. Sci.,* 7 (2014) 1597–1614. https://doi.org/10.1039/c3ee44164d.

[108] F. Wang, Latest advances in supercapacitors: From new electrode materials to novel device designs, *Chem. Soc. Rev.,* 46 (2017) 6816–6854. https://doi.org/10.1039/c7cs00205j.

[109] R. R. Salunkhe, Y. V. Kaneti, Y. Yamauchi, Metal-organic framework-derived nanoporous metal oxides toward supercapacitor applications: Progress and prospects, *ACS Nano,* 11 (2017) 5293–5308. https://doi.org/10.1021/acsnano.7b02796.

[110] F. Diao, Y. Wang, Transition metal oxide nanostructures: premeditated fabrication and applications in electronic and photonic devices, *J. Mater. Sci.,* 53 (2018) 4334–4359. https://doi.org/10.1007/s10853-017-1862-3.

[111] T. Guo, M. S. Yao, Y. H. Lin, C. W. Nan, A comprehensive review on synthesis methods for transition-metal oxide nanostructures, *Cryst. Eng. Comm.,* 17 (2015) 3551–3585. https://doi.org/10.1039/c5ce00034c.

[112] A. Nandagudi, Hydrothermal synthesis of transition metal oxides, transition metal oxide/carbonaceous material nanocomposites for supercapacitor applications, *Mater. Today Sustain.,* 19 (2022) 100214. https://doi.org/10.1016/j.mtsust.2022.100214.

[113] G. Yang, S. J. Park, Conventional and microwave hydrothermal synthesis and application of functional materials: A review, *Materials* (Basel), 12 (2019) 1177. https://doi.org/10.3390/ma12071177.

[114] T. Close, G. Tulsyan, C. A. Diaz, S. J. Weinstein, C. Richter, Reversible oxygen scavenging at room temperature using electrochemically reduced titanium oxide nanotubes, *Nat. Nanotechnol.,* 10 (2015) 418–422. https://doi.org/10.1038/nnano.2015.51.

[115] G. Yanalak, A. Aljabour, E. Aslan, F. Ozel, I. H. Patir, A systematic comparative study of the efficient co-catalyst-free photocatalytic hydrogen evolution by transition metal oxide nanofibers, *Int. J. Hydrogen Energy,* 43 (2018) 17185–17194. https://doi.org/10.1016/j.ijhydene.2018.07.113.

[116] S. Rackauskas, In situ study of noncatalytic metal oxide nanowire growth, *Nano Lett.,* 14 (2014) 5810–5813. https://doi.org/10.1021/nl502687s.

[117] H. J. Park, J. M. Lee, M. Nasir, S. J. Yoo, C. J. Choi, K. Lee, Synthesis of ultra-thin nanobelt-like vanadium-oxide and its abnormal optical-electrical properties, *Ceram. Int.,* 49 (2023) 6419–6428. https://doi.org/10.1016/j.ceramint.2022.10.086.

[118] G. Zhang, X. Xiao, B. Li, P. Gu, H. Xue, H. Pang, Transition metal oxides with one-dimensional/one-dimensional-analogue nanostructures for advanced supercapacitors, *J. Mater. Chem. A,* 5 (2017) 8155–8186. https://doi.org/10.1039/c7ta02454a.

[119] H. Xie, Recent advances in the fabrication of 2D metal oxides, *iScience,* 25 (2022) 1–30. https://doi.org/10.1016/j.isci.2021.103598.

[120] M. V. Reddy, G. V. Subba Rao, B. V. R. Chowdari, Metal oxides and oxysalts as anode materials for Li ion batteries, *Chem. Rev.,* 113 (2013) 5364–5457. https://doi.org/10.1021/cr3001884.

[121] Y. Zhang, L. Li, H. Su, W. Huang, X. Dong, Binary metal oxide: Advanced energy storage materials in supercapacitors, *J. Mater. Chem. A,* 3 (2015) 43–59. https://doi.org/10.1039/c4ta04996a.

[122] H. Wang, Q. Gao, L. Jiang, Facile approach to prepare nickel cobaltite nanowire materials for supercapacitors, *Small,* 7 (2011) 2454–2459. https://doi.org/10.1002/smll.201100534.

[123] Y. Y. Huang, L. Y. Lin, Synthesis of ternary metal oxides for battery-supercapacitor hybrid devices: influences of metal species on redox reaction and electrical conductivity, *ACS Appl. Energy Mater,* 1 (2018) 2979–2990. https://doi.org/10.1021/acsaem.8b00781.

[124] S. Sahoo, T. T. Nguyen, J. J. Shim, Mesoporous Fe–Ni–Co ternary oxide nanoflake arrays on Ni foam for high-performance supercapacitor applications, *J. Ind. Eng. Chem.*, 63 (2018) 181–190. https://doi.org/10.1016/j.jiec.2018.02.014.

[125] M. Usman, Facile synthesis of iron–nickel–cobalt ternary oxide (FNCO) mesoporous nanowires as electrode material for supercapacitor application, *J. Mater.*, 8 (2022) 221–228. https://doi.org/10.1016/j.jmat.2021.03.012.

[126] N. S. George, L. M. Jose, A. Aravind, *Review on transition metal oxides and their composites for energy storage application*, Intechopen, pp. 1–15, 2022 [Online].

[127] H. Cho, J. Kim, M. Kim, H. An, K. Min, K. Park, A review of problems and solutions in Ni-rich cathode-based Li-ion batteries from two research aspects: Experimental studies and computational insights, *J. Power Sources*, 597 (2024) 234132. https://doi.org/10.1016/j.jpowsour.2024.234132.

[128] B. D. Sankar, S. Sekar, S. Sathish, S. Dhanasekaran, R. Nirmala, D. Y. Kim, Y. Lee, S. Lee, R. Navamathavan, Recent advancements in M\Xene with two-dimensional transition metal chalcogenides/oxides nanocomposites for supercapacitor application – A topical review, *J. Alloys Compd.*, 978 (2024) 173481. https://doi.org/10.1016/j.jallcom.2024.173481.

[129] S. A. Kadam, L. M. Jose, N. S. George, S. Sreehari, D. A. Nayana, D. V. Pham, K. P. Kadam, A. Aravind, Y. Ma, Recent progress in transition metal nitride electrodes for supercapacitor, water splitting, and battery applications, *J. Alloys Compd.*, 976 (2024) 173083. https://doi.org/10.1016/j.jallcom.2023.173083.

7 Application of Carbon Nanomaterials in Supercapacitors

Jing Wang

1 INTRODUCTION

The discovery of carbon nanotubes (CNTs) has revolutionized the field of nanotechnology since the early 1990s, marking the birth of nanoscale tubular carbon materials [1]. These tubular structures, consisting of one or more graphene (GR) layers, possess a series of unique properties, such as one-dimensional structure, ordered pore arrangement, high specific surface area, excellent electrical conductivity, and stable chemical properties, which make them particularly important for SC development. CNTs are ideal for SC electrode materials due to their special hollow structure, high electrical conductivity, large specific surface area, and pore structure suitable for electrolyte ion migration (typically larger than 2 nm), and these are particularly suitable for high-power applications [2–5]. Although alternating currents provide a larger contact surface, CNTs perform similarly in terms of capacitive performance, thanks to their maximized surface utilization and optimized charge distribution [6]. Its mesoporous structure also induces the diffusion of the electrolyte, which reduces series resistance and improves energy output efficiency; and as a result, CNTs have attracted a wide range of research interests.

GR has also been widely studied as a promising electrode material for SCs[7]. However, the strong van der Waals forces and π-π interactions between GR sheets lead to easy lamination, which limits the efficient utilization of its surface area. To break through this limitation, GR's electrical conductivity has been enhanced by modifying it, which in turn has been applied in SCs. For example, Yuan's team prepared SCs exhibiting high energy density and good cycling stability by designing porous GR frameworks. And Bhattacharya et al. [8–11]. successfully prepared GR SCs with excellent performance by the laser-induced technique.

In addition, decompression and cutting of CNTs can transform it into GR flakes, which not only overcomes the stacking problem of GR, but also combines the one-dimensional advantage of CNTs with the two-dimensional (2D) characteristics of GR, thus significantly improving the electrochemical performance of the material. For example [12], Wang's team successfully prepared curved GR nanosheets exhibiting high specific capacitance and excellent cycling stability by utilizing the Hummers method for multi-walled CNTs and reducing them with $NaBH_4$.

DOI: 10.1201/9781003561262-7

TABLE 7.1
Selected Carbon Nanomaterials

Material	Preparation Method	Specific Capacitance	Cycling Life	Energy Density (Power Density)
CNTs	*In situ* synthesis	272 mFcm^{-2} at 0.5 mA cm^{-2}	100% after 5,000 cycles	–
GR	Buried-protection KOH activation technology	316.8 F g^{-1} at 1 A g^{-1}	92.5% after 2,000 cycles	–
AC	Hard template method	428.8 F g^{-1} at 0.25 A g^{-1}	98.5% after 20,000 cycles	11.2 Wh kg^{-1}
CNCs	*In situ* MgO template method	313 F g^{-1} at 1 A g^{-1}	100% after 20,000 cycles	10.90 Wh kg^{-1} (22.22 kW kg^{-1})
NiCoMn(OH)6/ CNT	Simple hydrothermal method	2136.2 F g^{-1} at 1 A g^{-1}	77% after 2,000 cycles	–
NiCo-LDH/RGO	*In situ* growth and hydrothermal method	1675 F g^{-1} at 1 A g^{-1}	62.2% after 1,000 cycles	49.9 Wh kg^{-1} (3747.9 W kg^{-1})
NiSe2@CNT	Rapid microwave process in 2 min	980.5 F g^{-1} at 1 A g^{-1}	82% after 9,000 cycles	25.61 Wh kg^{-1} (810 W kg^{-1})
Ni-CO LDH/GNR	Solvent thermal method	1765 F g^{-1}	80% after 2,000 cycles	749 W kg^{-1} (25.4 W kg^{-1})
NPMOF	Direct discourse	220 F g^{-1}	99.1% after 10,000 cycles	–
PANI-SRPG	*In situ* oxidation polymerization	939.96 F g^{-1} at 1 A g^{-1}	98.46% after 5,000 cycles	145 Wh kg^{-1} (17578.13 W kg^{-1})
DMC/PANI	*In situ* polymerization	985 F g^{-1} at 0.5 A g^{-1}	82% after 5,000 cycles	32 Wh kg^{-1} (125 W kg^{-1})
PRGO/CNTs	The vacuum filter and solvothermal methods	236 F g^{-1} at 0.5 A g^{-1}	–	–
CMK-3/CNTs	Ultrafast microwave synthesis	315.6 F g^{-1} at 1 A g^{-1}	99.32% at 10,000 cycles	–
LEC	Liquefaction technology	288 F g^{-1} at 0.2 A g^{-1}	97% after 3,000 cycles	93.73 Wh kg^{-1} (850 W kg^{-1})
CC@T-Nb2O5@ MnO2	Two-step hydrothermal method	319 F g^{-1} at 0.2 A g^{-1}	85.7% after 2,000 cycles	31.76 Wh kg^{-1} (2250 W kg^{-1})
MnO2/AC/GR	The self-limiting growth method	813.0 F g^{-1} at 1 A g^{-1}	98.4% after 1,000 cycles	33.9 Wh kg^{-1} (319.3 W kg^{-1})

Three-dimensional (3D) GR has become a key material for enhancing the performance of SCs due to its high specific surface area, unique porous structure, and fast electron and ion transport properties. GR with this 3D structure can be used directly as an electrode material without the use of adhesives, showing great potential for application [13–16].

Carbon quantum dots (CQDs), an emerging type of carbon-based nanoparticles, consist of a conjugated structure of sp^2/sp^3 carbon atoms, and their diameters usually do not exceed 10 nm. They have attracted attention due to their tiny size, uniform dispersion, excellent electron transfer capability, unique structure and composition, fluorescent properties [17], and obvious quantum edge effects. In 2004, Xu et al. discovered novel fluorescent carbon materials with diameters less than 18 nm while analyzing carbon soot produced by electric arcs. This discovery has stimulated an extensive research interest in such carbon nanomaterials with unique optical properties [18] and has led to the exploration of carbon dot synthesis methods, which are broadly categorized into two routes: "top-down" and "bottom-up" [19] (Table 7.1).

2   CNTS IN SCS

The synthesis of polypyrrole (PPy, a conductive polymer) usually requires the use of specific chemicals, such as ferric chloride as an oxidizing agent [20]. The process is relatively simple, but the properties of the final material obtained depend heavily on the synthesis conditions and the specific ratios of the reactants. In this process, carbon nanotubes (CNTs) are mixed with pyrrole (a monomer), an oxidizing agent (e.g., ferric chloride), and other additives, resulting in the formation of a composite material called PPy/CNTs [21–23]. This approach is not only simple and cost-effective, but can also be easily scaled up to achieve commercial production. However, the dispersion of carbon nanotubes in water can be a challenge.

To improve the dispersibility of CNTs in water, researchers have attempted to modify the surface of CNTs by adding hydrophilic groups to aid in their uniform dispersion in water [24]. This surface modification not only helped in the uniform dispersion of CNTs in the reaction mixture, but also enhanced the interaction between PPy and CNTs.

A common method employed is to treat CNTs in an acidic environment in order to introduce oxygen-containing groups such as hydroxyl and carboxylic acids on its surface. These groups not only help in stabilizing the dispersion of CNTs, but also in promoting the formation of polymers on the CNT surface. Although some studies have shown that acid treatment can increase the electrical conductivity of CNTs [25], most of them concluded that this method may damage the structure of CNTs and reduce their electrical conductivity.

Another method of surface modification is by air plasma treatment, which introduces oxygen- and nitrogen-containing groups on the CNT surface. This treatment is believed to improve the electrical conductivity of CNTs and GR by adding impure elements such as nitrogen, boron, or sulfur, which in turn improves the electrochemical performance of PPy/CNT composites [26]. It was also shown that air plasma-treated CNTs could improve the thermal stability of the composites. Through plasma treatment, special nitrogen-containing groups can also be introduced on the surface of CNTs, which can form a conjugated structure with the PPy backbone and

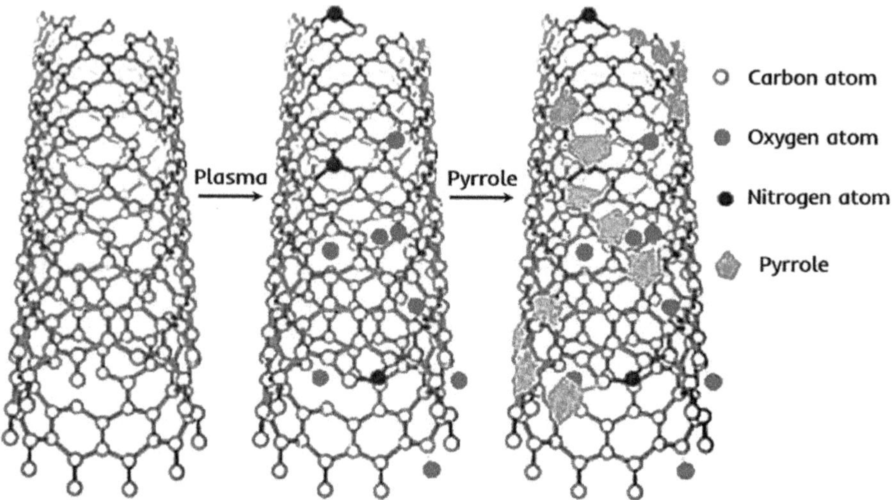

FIGURE 7.1   Schematic diagram of air plasma activation.

also enhance the interaction between PPy and CNTs through noncovalent interactions [27] (Figure 7.1).

The treatment of carbon nanotubes (CNTs) to improve their coordination with polypyrrole (PPy) and enhance the performance of PPy/CNT-based SCs is an effective approach [28]. The use of concentrated acid and air plasma treatment can significantly improve the interactions between CNTs and PPy, which in turn enhances the supercapacitive performance of the composite electrodes. For example, Fu et al. successfully prepared PPy/CNT-based SCs by chemical oxidative polymerization of pyrrole in a mixture of acid-treated CNTs [29–32], cetyltrimethylammonium bromide (a surfactant), and ammonium persulfate (an oxidant). They observed that the prepared capacitor exhibited a high specific capacitance value of 233.5 farad g^{-1} at a current density of 0.5 A g^{-1} and 1.0 M KCl solution. Similarly, Yang et al. employed ammonium persulfate as an oxidizing agent to polymerize pyrrole on untreated (pristine) and air plasma-treated CNTs. The results showed that PPy composite capacitors prepared using air plasma-treated CNTs displayed significantly higher specific capacitance values (264 farad g^{-1}) compared to PPy composite capacitors prepared using untreated CNTs (210 faradg^{-1}) in 1.0 M sulfuric acid solution [33].

These studies indicate that the interfacial compatibility between CNTs and PPy can be effectively enhanced by appropriate surface treatment techniques, thereby enhancing the performance of SCs [34].

3 SURFACE MODIFICATIONS AND ANALYTICAL APPLICATIONS OF GRAPHENE OXIDE (GO)

3.1 INTRODUCTION

It is still a challenge and an urgent task for researchers to design electrode materials with high energy density and high rate capability for SC applications. In order to achieve rapid energy storage at high charge/discharge rates, it is very essential to design and fabricate electrode materials with stable chemical and electrical properties [35–42]. GR with porous structure and high aspect ratio has been proved to greatly shorten the ion transport distance [43–50]. However, because of the strong π-π interaction between GR sheets, GR flakes tend to restack to form graphite-like powders when they are processed into electrode materials, which generally decrease the specific surface area and lead to the inefficient use of GR layers for electrochemical energy storage. As an alternative, multilayered graphene oxide (GO) seems to be a promising candidate for supercapacitors due to its advantages of high surface area, facile mass production, easy dispersion, and readily functional design. Moreover, rGO with remaining oxygen functional groups can exhibit not only electric double-layer capacitance (EDLC) but also pseudocapacitance properties for SCs [51–58]. However, the poor conductivity of functionalized GO with rich and interactive oxygen-containing groups inhibited the practical use in SCs. Promoting the redox reactions of functional groups and enhancing the pseudocapacity requires a conductive network (sp2 carbon) and electron transfer channel in GO, especially at high scan rates [59–66].

The structures of surface-modified GO are illustrated in Figure 7.2. GR comprises either a single layer of crystalline carbon atoms packed in a 2D honeycomb (hexagonal) lattice or several coupled layers of crystalline carbon atoms arranged to form a honeycomb structure [67–77]. Figure 7.2 presents the structures of surface-modified GO with different functional groups (epoxy bridges, hydroxyl groups, and pairwise carboxyl groups) and organic ligands/nanocomposites on the surface. The presence of such functional groups on its surface facilitates the covalent modification, noncovalent modification, and elemental doping of GO. In turn, the modification of GO provides synergistic qualities and remarkable properties, which originate from both GR and the modifiers, thereby resolving the issues related to its manufacture, storage, and handling, and their applications in the applied sciences [78].

3.1.1 Modification of GO Via Covalent Interactions

The covalent bonding of functional groups to the surface of GO leads to performance optimization [79]. It is easier to achieve covalent functionalization on GO than on GR, owing to the presence of O-containing groups on the surface of the former. Indeed, numerous hydroxyl, epoxy, and carboxyl groups exist on the surface of GO [80–81], which facilitate typical chemical reactions such as carboxylic acylation, epoxy ring opening, diazotization, and addition [82]. Modification via covalent bonding enables the introduction of fresh functionalities and induces an increase in processability. The covalent functionalization of GO can be achieved via multiple techniques [83].

3.1.2 Modification of GO Via Noncovalent Interactions

Noncovalent modification is a facile technique that allows GO to retain approximately the entirety of its initial stability. It was noticed that the feasibility of the noncovalent modification of GO is higher than that of covalent modification [84]. The noncovalent functionalization of GO results in the synthesis of composites with special functions via the generation of H-bonding and electrostatic interactive forces between GO and the functional molecules. The surface modification of GO via noncovalent interactions ensures both the preservation of the bulk structure and quality of the products and optimal dispersibility [85]. Noncovalent functionalization can

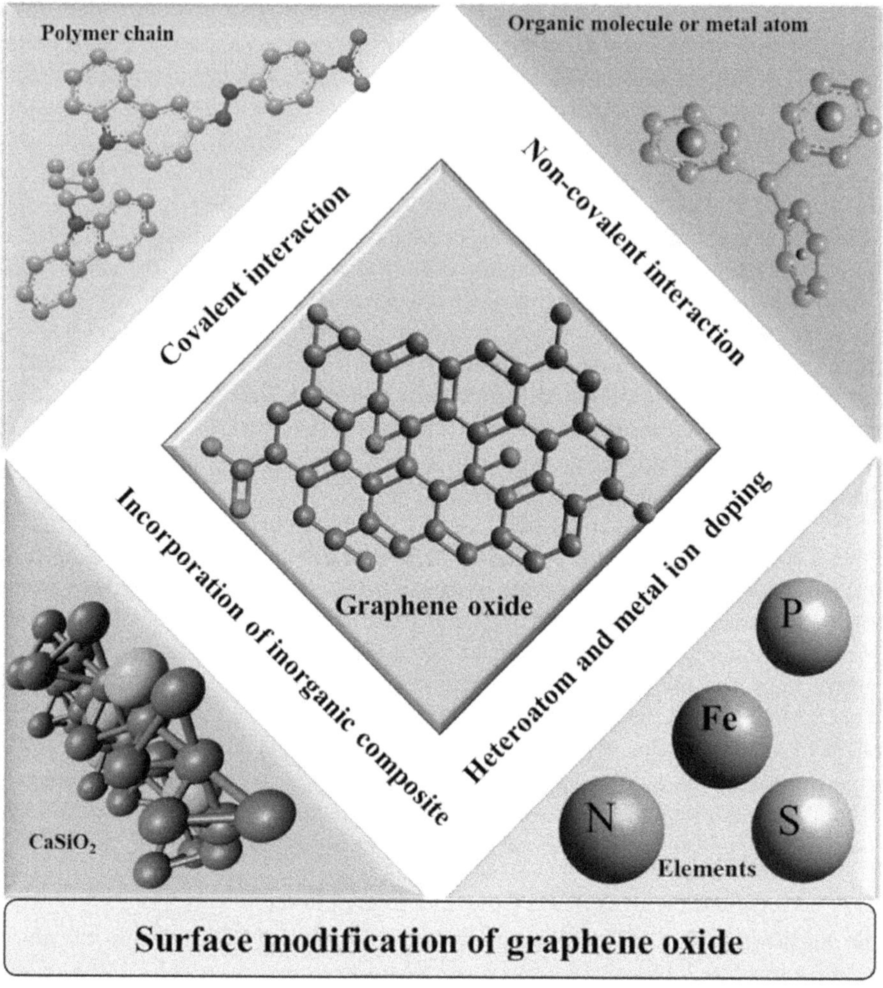

FIGURE 7.2 Structures of GO and its modification with covalent, noncovalent, heteroatoms/metals, and nanocomposites.

be realized via pep interactions, hydrogen bonding, ionic bonding, and electrostatic interaction modification [86]. Notably, the noncovalent modification of GO is a facile process under mild conditions, and it ensures the preservation of the structure and properties of GO [87].

3.2 EXPERIMENTAL

3.2.1 Synthesis of GO with Covalent Modification

The organoamine-modified GO was simply fabricated by the solvothermal reaction. Firstly, 0.12 g GO was dispersed to form a suspension in 60 mL DMF (N,N-dimethylformamide) under ultrasonication. Afterward, the dark homogeneous suspension was transferred into a 100 mL Teflon-lined stainless steel autoclave and heated at 140, 160, or 180°C for 12 h. The products were washed with deionized water several times and freeze-dried at −60°C to obtain organoamine-modified GO. The samples obtained at different temperatures were denoted as DMFrGO140, DMFrGO160, and DMFrGO180, respectively, while DMFrGO is the general name given for DMFrGO140, DMFrGO160, and DMFrGO180.

3.2.2 Material Characterization

The morphologies of the as-synthesized products were observed by a field emission scanning electron microscope (SEM, Hitachi S–4800) and transmission electron microscope (TEM, JEM-2100F) with an energy dispersive X-ray spectrometer (EDX).

3.2.3 Electrochemical Measurements

The electrodes were prepared by using DMFrGO materials as the active materials. A homogeneous slurry was prepared by mixing 80 wt% active materials, 10 wt% acetylene black, and 10 wt% polytetrafluoroethylene (PTFE) under ultrasonication. The working electrodes were fabricated by homogeneously coating the slurry on nickel foam and drying at 60°C for 12 h. Without additional explanation, each working electrode contained about 1 mg cm^{-2} of electroactive materials. The electrochemical performances were measured both in the three-electrode system in 6 M KOH solution and on a symmetric quasi-solid SC device by using PVA/KOH gel as the electrolyte.

3.3 Results and Discussion

DMF often pyrolyzes into formic acid and dimethylamine under the assistance of water molecules during the solvothermal process. Dimethylamine molecules possibly react with carboxylic groups on the GO sheet. Covalently modified GO with dimethylamine is readily produced. The chemical process is schematically illustrated in Figure 7.3. Such unique nanoarchitecture gives rise to a seamless Ohmic contact between the functional groups and the GO substrate to promote the ion transportation and enhance the redox reactions of the functional groups, which will greatly improve the supercapacitive performances [88–94].

The solvothermal modification of GO at various temperatures produces DMFrGO samples that are made of randomly aggregated wrinkle nanosheets with spitball-like morphology (Figure 7.4a and Figure S1). Due to the similar morphologies of all DMFrGO samples, DMFrGO180 is selected for further characterization. TEM observation (Figure 7.4b) reveals that DMFrGO180 is constructed by stacked and wrinkled nanosheets with smooth surface. Moreover, the HRTEM image (Figure 7.4c) clearly demonstrates the presence of sp3 carbon domains that are disorderly distributed within the continuous sp2 carbon network. These sp3 carbon domains possibly bond to heteroatom-containing functional groups. Elemental mapping analyses (Figure 7.4d) confirm the existence of prevailed C composition and small amount of N and O elements. The occurrence of N species indicates the successfully grafted dimethylamine groups onto GO nanosheets. The evenly distributed O and N

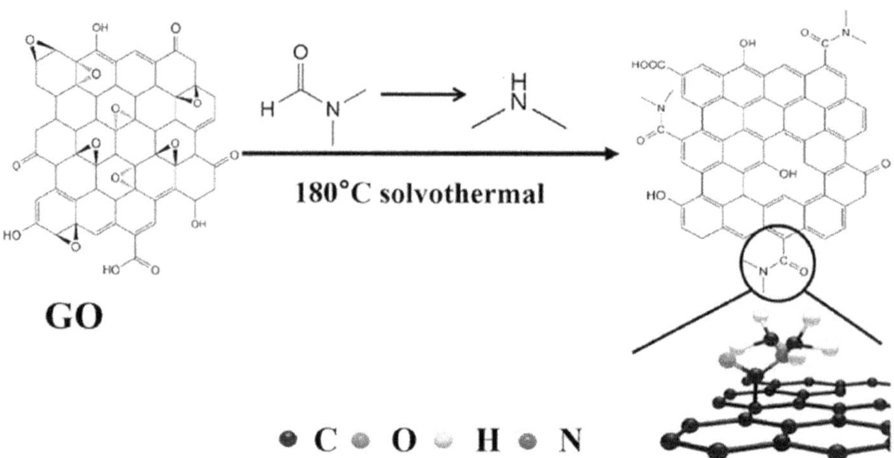

FIGURE 7.3 Covalent modification of GO with dimethylamine.

elements also imply that those functional groups on GO nanosheets are mixtures of oxygen-containing and nitrogen-containing groups.

To evaluate the electrochemical properties of DMFrGO materials, the CV measurements are firstly performed in a three-electrode system with 6 M KOH as the electrolyte. Figure 7.5a shows the CV curves of raw GO and DMFrGO materials at 50 mV s^{-1}. All DMFrGO materials exhibit disordered rectangular shape with broad redox peaks, suggesting the coexistence of EDLC and pseudocapacitance-friendly properties. The broad redox peaks are slightly intensified with the elevated solvothermal temperature, indicating that the reaction temperature plays an important role in driving redox reactions for providing improved pseudocapacitance. The capacitances of the samples are calculated by formula 2 (shown in Supporting Information) with CV data at a scan rate of 50 mV s^{-1}. The capacitances of raw GO, DMFrGO140, DMFrGO160, and DMFrGO180 are 8.8, 180, 197, and 207 F g^{-1}, respectively.

The galvanostatic charge/discharge (GCD) measurements were carried out from −0.9 to 0.1 V to further evaluate the capacitive behaviors of raw GO and DMFrGO materials. Raw GO and DMFrGO materials display symmetrical and linear shapes at a current density of 1 A g^{-1}, which suggested the excellent electrochemical reversibility and high coulombic efficiency (Figure 7.5b). The capacitances of the samples are calculated using formula 1 (shown in Supporting Information) with GCD data. The DMFrGO180 electrode delivered a specific capacitance of 287 F g^{-1} at 1 A g^{-1}, which is higher than those of raw GO (30.2 F g^{-1}), DMFrGO140 (226 F g^{-1}), and DMFrGO160 (252 F g^{-1}) electrodes. In comparison with the data at 1 A g^{-1}, the capacitance retention of DMFrGO140, DMFrGO160, and DMFrGO180 are 65%, 48.8%, and 67.9% at a current density of 10 0 A g^{-1}, respectively (Figure 7.5c and Figure S5). The high capacity retention of DMFrGO180 implies its excellent rate performance.

EIS analyses of DMFrGO materials confirm decreased charge transfer resistances (Rct) with increased processing temperatures (Figure 7.5d). The Rs, Rct, CPE, and Wo in an equivalent circuit (inset Figure 7.5d) represent intrinsic Ohmic resistance,

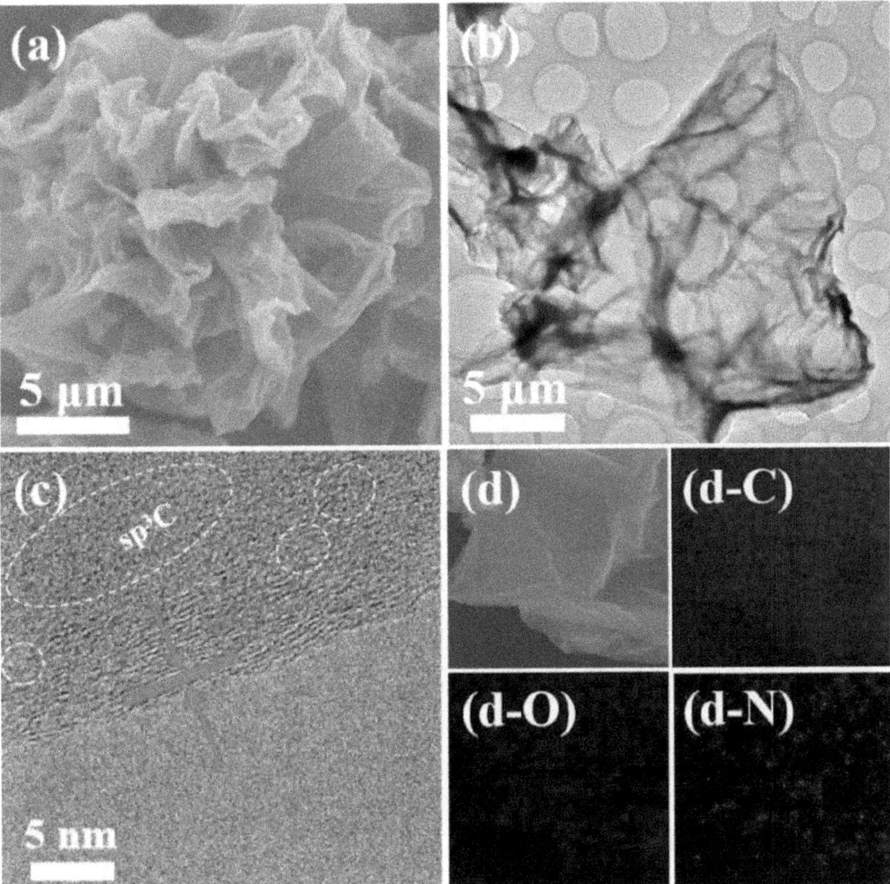

FIGURE 7.4 (a) SEM and (b) TEM images of DMFrGO180. (c) HRTEM image and (d) elemental mapping images of DMFrGO180.

charge transfer resistance, electronic double-layer capacitor, and Warburg resistance, respectively. Fitting the EIS data was performed by ZView 2 software. The Rct of the DMFrGO materials are 0.28, 0.17, and 0.09, respectively (with error%<10%). During the solvothermal process, GO is reduced. Simultaneously, DMF often pyrolyzes into formic acid and dimethylamine under the assistance of water molecules. We can find that there is no obvious covalent bond at 140°C, and a weak linkage is produced between dimethylamine cation and carboxylic anion over 160°C. With an increase in the solvothermal temperature to 180°C, the ratio of the covalent bond appears to increase. These results demonstrate that the covalent bonding of dimethylamine units to GR generates a seamless Ohmic contact for electron transfer [80,81]. The presence of numerous electrically conductive paths in covalent modified DMFrGO electrodes is significantly beneficial for the contribution of both electric double-layer capacitance and pseudocapacitance. The pseudocapacitance depended not only on the content of O-containing groups but also the redox reversibility of these groups.

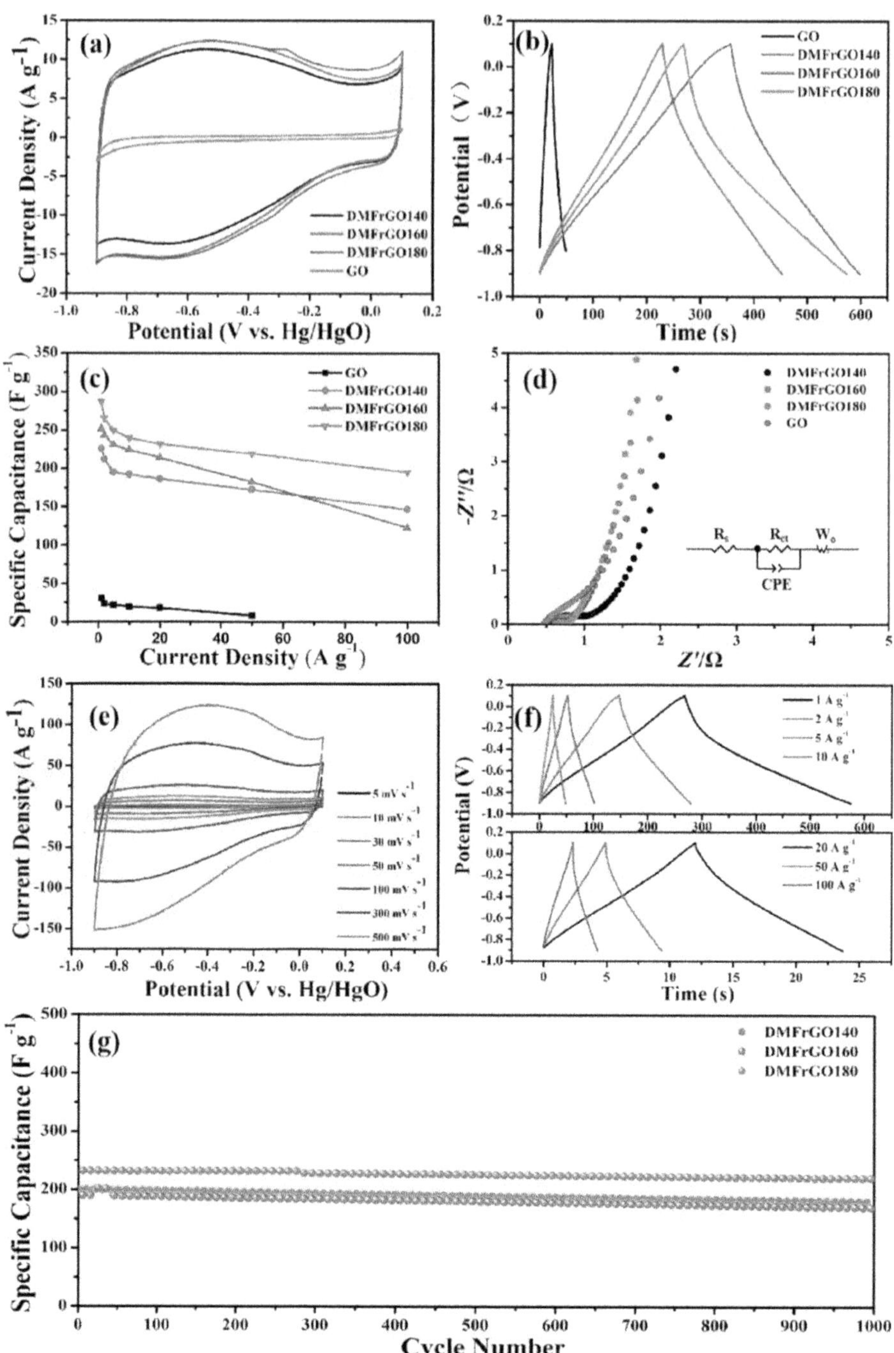

FIGURE 7.5 Supercapacitive performances of raw GO and DMFrGO materials. (a) CV curve of raw GO and DMFrGO materials at 50 mV s^{-1}. (b) GCD curves of raw GO and DMFrGO materials at the current densities of 1 A g^{-1}. (c) Specific capacitance versus current density of raw GO and DMFrGO materials. (d) Nyquist plot of raw GO and DMFrGO materials. Inset shows the equivalent circuit model. (e) CV curves of DMFrGO180 at scan rates of in the range of 5–500 mV s^{-1}. (f) GCD curves of DMFrGO180 at different current densities in the range of 1–500 A g^{-1} (g) Cycling performance of DMFrGO materials at a current density of 20 A g^{-1}

The redox reversibility related to the electron migration between the functional groups and the carbon network. High temperature treatment enhanced the conductivity of DMFrGO180. Also, the O-containing group content of DMFrGO180 is only slightly less than that of DMFrGO140 or DMFrGO160. The best electrical conductivity rendered DMFrGO180 the highest redox reversibility to exhibit increased pseudocapacitance-friendly properties.

Figure 7.5e shows the CV curves of DMFrGO180 at various scan rates from 5 to 500 mV s^{-1}. Even at a very high scan rate of 500 mV s^{-1}, the CV curves of the DMFrGO180 remain in a quasi-rectangular shape, indicating superior rate capability and excellent frequency response. The galvanostatic charge/discharge curves of DMFrGO180 (Figure 7.5f) show a nearly triangular shape at current densities from 1 to 10 0 A g^{-1}, without an obvious iR drop even at 10 0 A g^{-1}, suggesting fast kinetics for charge-storage processes. The specific capacitance values of DMFrGO180 are 287.1, 265.4, 249.5, 240, 232, 219.5, 195 F g^{-1} at 1, 2, 5, 10, 20, 50, and 10 0 A g^{-1}, respectively. A high capacitance of 195 F g^{-1} is sustained even at an ultrahigh current density of 10 0 A g^{-1}, which is 67.9% of the value at 1 A g^{-1}. These data reveal the excellent rate performance of DMFrGO180. While elevating the mass loading to 5 mg cm^{-2}, DMFrGO180 electrode still exhibits a high specific capacitance of 233 F g^{-1} at 1 A g^{-1} and 129.5 F g^{-1} at 50 A g^{-1}, indicating excellent electrochemical performances for its practical application in SCs (Figure S6). DMFrGO180 also displays the best cycling stability. DMFrGO180 exhibits 94% capacitance retention at 20 A g^{-1} over 1,000 cycles, which is better than 90% of DMFrGO140 and 89% of DMFrGO160 (Figure 7.5g).

3.4 Conclusions

High reaction temperatures produce covalent acid-amide linkage between GO nanosheets and dimethylamine. DMFrGO180 with the most covalent acid-amide bonding exhibit the best conductivity and the highest capacity. Electrochemical measurements in the three-electrode system demonstrate the good SC performances of DMFrGO180 with 287 F g^{-1} capacitance at 1 A g^{-1} and superior rate capability. The improved SC performances stem from the synergistic effect of the unique structures. The covalently bonded dimethylamine groups prevent the aggregation of GR sheets, increase the number of available active sites, and generate seamless Ohmic contact between functional redox groups and the GO substrate to offer more electronic transport paths and improve conductivity. The seamless Ohmic contact and improved conductivity render the oxygen-containing functional groups enhanced redox reversibility and accelerated redox reaction rate.

4 POLYMER COMPOSITES WITH QUANTUM DOTS AS POTENTIAL ELECTRODE MATERIALS FOR SC APPLICATIONS

4.1 Introduction

Amalgamation of new-age nanotechnology and utilization of favorable nanomaterials for SCs have created enormously encouraging scope for development, which raised promising future prospects for electronic technologies. These approaches will hopefully shape the next generation electronics to be more compact and economical as well. Mainly, the SCs are comprised of two working electrodes with a separator positioned in between them to isolate them and along with that a fitting electrolyte has been employed to make a complete working setup. Though in recent studies, restrictions have been applied on the utilization of materials which have a toxic impact on the environment [95]. Recent research has also concentrated on the utilization of everyday life generated waste materials or abandoned wastages to be employed for energy application purposes. These waste-derived materials are used for different components in the SCs. Through these approaches, not only the scientific fields are getting adequate advantages but this is also is a prominent way to reduce growing issues in handling waste materials. Regarding this, biomass-derived materials have exhibited great opportunities and even utilizing plastic materials for energy storage devices is taken into consideration in recent research [96,97]. Owing to the favorable characteristics like higher surface-to-volume ratio, enriched surface area, superior concentration of edge atoms, quantum confinement effect, easy solubility in non-aqueous and aqueous solvents, functionalization and easy doping conditions, Quantum Dots (QDs) have been considered for a various range of applications like photovoltaics, bio-sensing, light-emitting diodes, bio-imaging, batteries, fuel cells, etc. [98,99]. Moreover, research findings have also shown their viability in coalescing with other nanomaterials to develop better workable composites for application purposes, commonly QDs.

In recent studies, the utilization of polymers, especially conductive polymers, has attracted much attention. Along with delivering excellent specific energy, they have exhibited optimum pseudocapacitive properties, which resulted in faster redox reactions and hence superior electrochemical activity [100,101]. Researchers have utilized conductive polymers of different sizes or dimensions and different morphologies (nanoparticles, nanosheets, nanofibers) in energy storage and conversion applications [102]. However, limitations like low mechanical strength, inadequate solubility, shrinkage, and swelling during charge/discharge cycles have restricted their capacitive behavior. In order to achieve further improvements, studies have recommended compositing carbonaceous compounds with these polymer materials. This composite preparation approach has resulted in improved conductivity, thermal stability, and intermolecular chain reaction, and hence significant changes like better rate capability and enhanced capacitance were observed in SC device performance [103].

4.2   SYNTHESIS OF QUANTUM DOTS, POLYMERS, AND NANOCOMPOSITES

4.2.1   Electrochemical Process

Apart from hydrothermal and microwave-irradiated QD preparation, the electrochemical process is another well-known and widely studied technique. This process is one of the major categories of the top-down synthesis approach. Unlike other synthesis methods, this process has been reported to be examined at reaction times ranging from several minutes to several days. However, few studies have mentioned its swift reaction period and lesser temperature requirement which can easily lead to scaling up of material production [104]. For an instance, Li et al. prepared 4 nm, uniform, and monodispersed CQDs with strong and stable photoluminescence (PL) via employing graphite rods as both cathode and anode in the NaOH/EtOH electrolyte where the author mentioned that production rate of CQDs was around 10 mg hour^{-1} for each processing setup [105]. Similar to CQDs, GQDs also can be synthesized under facile conditions. Li et al. utilized GR sheets as an electrode with phosphate buffer solution to prepare water-soluble GQDs of 3–5 nm size [106]. So, this process has appropriately exhibited electrochemical oxidation of the carbon precursors to synthesize CQDs as well as appropriate pathways for GQDs or doped GQDs [107,108]. By employing the single-stage facile alternating voltage electrochemical technique, the preparation of NiO QDs/GR composite was completed where both Ni flakes and graphite rods were exfoliated using an alternating voltage of 5 V and NaOH (2 M) as the electrolyte [109]. With Ni flakes and graphite rods as electrodes, the as-prepared NiO quantum dots were uniformly dispersed on the surface of GR and an enhanced electrochemical activity was achieved using the prepared composite electrode material. Preparation of WS$_2$ and MoS$_2$ via electrochemical technique is also explored by many researchers [110]. Valappil et al. prepared WS$_2$ QDs by the synergistic effect of lithium perchlorate intercalation and propylene carbonate used as an electrolyte [111]. A potential of 2 V was applied with the described electrochemical setup. The as-prepared WS2 QDs have an average size of 3 nm with few layers and exhibited PL emission (photoluminescence quantum yield (PLQY) = 5%). Similarly, MoS2 QDs (size of 2.5–6 nm) were prepared via the one step electrochemical method from its bulk material along with using diluted aqueous ionic liquid solutions of 1-butyl-3-methylimidazolium chloride and bis-trifluoromethylsulphonylimide [112]. As-prepared MoS$_2$ QDs exhibited excitation-dependent luminescence, which could be further enhanced via surface passivation. So, this process has successfully shown

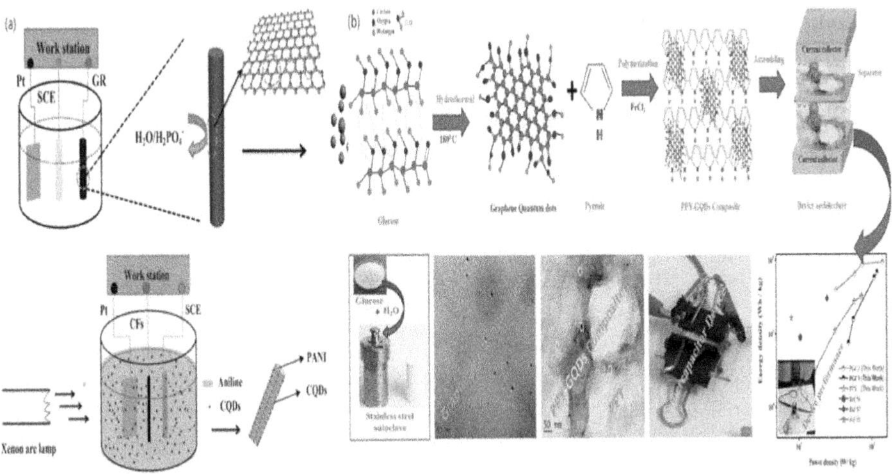

FIGURE 7.6 (a and b) Schematic illustrations of electrochemical fabrication of CQDs and CQDs/PANI/CF(LI). A schematic diagram represents the synthesis mechanism of GQDs and PPy-GQDs composite fabricated as the SC device, (c) schematic illustration showing the preparation of electrochemically reduced GQDs (ERGQDs)/PPy-2, and (d) schematic diagram of the preparation procedure of CQD@PANI nanoparticles.

advantages like enhanced production yield, flexible usage of electrolytes, superior stability without any additive binders during synthesis, and efficient exfoliation via regulating the potential or current [113,114]. More importantly, size controllability is another crucial aspect which can be achieved with this process by altering the temperature of the electrolyte [115–117]. However, the electrochemical process involves graphite or expensive carbon precursors for synthesis, which raises overall production costs. Also, unnecessarily attached residual additives on the surface of QDs have restricted the efficiency of this method. Zhao et al. [118] reported CQDs/polyaniline (PANI) hybrid material; PANI was grown on the carbon fiber substrate resulting in an interconnected network structure, as shown in Figure 7.6a [119]. Vandana et al. reported QDs dispersed on polymers synthesized through a facile hydrothermal method using glucose as the source for quantum dots, as shown in Figure 7.6b [120]. She et al. elaborated on the synthesis of graphene quantum dots (GQDs) and PPy hybrids. The electrochemically reduced GR was decorated on the PPy sphere, as shown in Figure 7.6c [121]. Li et al. reported core-shell@PANI using citric acid as the precursor, which follows pyrolysis and NaOH treatment; further the PANI was composited by *in situ* adsorption polymerization, as shown in Figure 7.6d [122].

4.2.2 Solvothermal/Hydrothermal Process

These synthesis procedures are very frequently employed not only for the synthesis for QDs, but also for a different range of materials having various size ranges and configuration. Mostly, it involves a single-step procedure in which organic precursors are treated in a closed autoclave to prepare the QDs under high pressure and temperature according to the requirement. Here the selection of precursors plays a vital role because the source used may not only contain carbon but may also have doping elements which can impact the synthesized QDs' structure. Citric acid, L-cysteine, melamine, hydrazine, polyethylene glycol-400,(1, 3, 6)-trinitropyrene, and hydrazine hydrate are often used as precursors in the synthesis of CQDs, whereas bromobenzoic acid and citric acid are the different precursors used in synthesis of GQDs. Other than these precursors, CQDs are also prepared via the dehydration of glucose in sulfuric acid and nitric acid [123]. Additionally, the usage of further sulfur and nitrogen doping are observed to achieve better application outcomes of the CQDs or GQDS. For doping purposes, nitrogen (N) and sufhur (S) sources included in the preparation are thiourea, urea, hexamethylenetetramine, ethylenediamine, and diethyl diethanolamine which are used to synthesize N or S-doped QDs [124]. However, rather than focusing only on a one-step approach, researchers have adapted a two-step synthesis approach to mainly obtain doped QDs with improved structure and applicability [125]. For example, GQDs are prepared with citric acid as the precursor under basic conditions of a hydrothermal process, and then the as-synthesized QDs are blended with hydrazine and heated for several hours to prepared N-doped QDs. The hydrothermal technique is not only used to prepare CQDs or GQDs, but also used to prepare CeO2/Ce2O3 quantum dots anchored on reduced graphene oxide (rGO) sheets of different weight fractions, where GO was synthesized initially using natural graphite flakes via employing the modified Hummer's method [126]. Cerium nitrate hexahydrate was the precursor used in this preparation. Similarly, the single-step solvothermal technique was also employed to synthesize paper-like layered CeO2 quantum dots doped Ni-Co hydroxide nanosheets which was used as an electrode in asymmetric SCs [127]. Utilizing citric acid, thiourea, and ceria-like precursors along with hydrothermal method, N and S co-doped GQDs were grown on CeO_2 nanoparticles and this combination of ceria and GQDs was found to provide an increased active surface area, and hence enhanced electrochemical activity [128].

The hydrothermal method is also used to prepare CuO QDs using copper (II) acetate monohydrate as the precursor; and additionally single layer GR was added on the CuO QD surface for better structural stability (unique core-shell structure) and improved performance [129]. Likewise, CuS, SnO_2, $NiCo_2O_4$ QDs were also produced using copper (II) dithiooxamide, stannous chloride dihydrate, nickel acetate tetra-hydrate, cobalt acetate tetra-hydrate precursors, respectively [130–132]. In order to synthesize WS2 and MoS2 QDs, the solvothermal technique has been broadly used [133]. Precursors like sodium molybdate and cysteine were used to synthesize MoS2 QDs, whereas sodium tungstate and L-glutathione were used to prepare WS2 QDs [134,135]. L-glutathione and cysteine are mainly used as the source for sulfide. However, studies have also mentioned the use of thiocarbamide, dibenzyldisulfide, thiourea as the sulfide source in synthesizing WS2 and MoS2 QDs [136]. The key benefits of this method are being less hazardous, easier to conduct, and less expensive. Moreover, controlling and altering the properties as well as composition of the prepared QDs is possible using this method, which are favorable for application purposes [137].

4.2.3 Microwave Synthesis

Compared to the hydrothermal or solvothermal process, microwave-irradiated preparation techniques consume less time and require lesser reaction temperature for synthesizing QDs. By adding water with glucose and ammonia, GQDs were prepared under microwave irradiation conditions at a reaction time of 1 minute [138,110]. Another group has prepared similar GQDs at a reaction time of 5 min under 200°C using acetylacetone along with water in a quartz bowl kept under 800 W power microwave irradiation. Not only because of quicker reaction time and low temperature requirement, but also due to a simple reaction route and overall economical factors, this technique appears to be suitable for large scale production of GQDs as suggested by researchers [139,140]. The microwave synthesis of GQDs and its modification by tuning different parameters like power, time, temperature, etc. have been illustrated in the Figure 7.7 [141–145]. In addition, characteristics of CQDs or GQDs can be controlled and altered via adjusting irradiation power and reaction timing or by varying precursors.

4.3 Quantum Dots and Polymers

Composites synthesized by reducing materials or compounds from their bulk form to their QD structure for SC applications exhibit various advantages such as stable and efficient quantum confinement effects, improved surface area, higher surface area to volume ratio, diffusion length for both electrons and ions, higher concentration of edge atoms or more available edge atoms providing extra active sites, much lesser volume expansion/contraction, etc. [143]. These structural benefits augment the

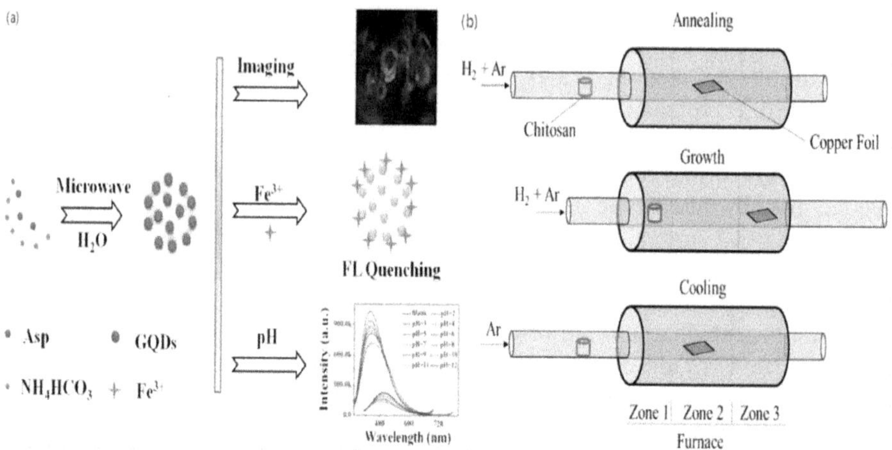

FIGURE 7.7 (a) Schematic illustration of microwave bottom-up route for g-GQDs and b-GQDs: green circles mean carboxyl and carbonyl groups and blue circles indicate hydroxyl groups. (b) Schematic diagram for the synthesis of N-GQDs.

electrochemical activity when used as electrode materials in energy storage devices, especially batteries and SCs.

Apart from the generalized issues, specific issues for specific QDs have been observed in various studies. For instance, ceria QDs are found to be inadequately conductive; so rather than using them individually for SCs, they are necessarily composited with conductive materials to achieve appropriate electrochemical activities [146]. N. Chakrabarty et al. [147] prepared a hybrid composite via mixing reduced graphene oxide (rGO) and ceria which formed conducting pathways and ensures easy movement of charges in the fabricated devices. Interestingly, addition of rGO not only resolved the conductivity issues but also prevented QDs from agglomeration. Similar to ceria, Nb_2O_5 QDs have also exhibited conductivity and aggregation issues. In this case, coating these QDs with biomass-derived nitrogen-rich carbon has alleviated the drawbacks [148]. Additionally, to counter volume expansion drawbacks [149], annealing at higher temperatures during synthesis was also recommended in the study for even distribution of QDs in nitrogen-rich carbon. Few other problems are also observed such as restacking, achieving uniform QDs with narrow size distributions, etc. Reports have further mentioned that such problems can be reduced to some extent by forming composite materials or capping via employing organic groups [150]. Dinari et al. reported PANI/GQDs as an electrode material for SCs; the composite was prepared using citric acid as the carbon source and thiourea as the N, S source for heteroatom doping by an *in situ* electrochemical polymerization procedure. The resulting electrode material exhibited superior electrochemical performance than that of PANI [151]. Mehare et al. reported sucrose-derived quantum dots composited with PANI through electrodeposition. The as-prepared material showed a high specific capacitance of 1512.4 F g^{-1} at 1 A g^{-1} in a three-electrode setup with 1 M H2SO4 as the electrolyte. The assembled asymmetric SC shows a high specific capacitance of 295 F g^{-1} at 1 A g^{-1}. The device exhibited a remarkable energy and power density of 40.86 Wh kg^{-1} and 2000 W kg^{-1}. Kumar et al. synthesized PANI/CQDs composite material using polyethylene glycol as the source for QDs; the nanocomposite was synthesized by chemical oxidation. The assembled device showed a high specific capacitance of 161.3 mF cm^{-2} with a good cycling stability even after 5,000 cycles [152]. Jian et al. reported the electrochemical synthesis of GQDs composited with polymers; the as-prepared material CQDs/PPy composite all-solid-state supercapacitors (ASSSs) showed outstanding electrochemical performance with an areal capacitance of 315 mF cm^{-2} (corresponding to the specific capacitance of 308 F g^{-1}) at a current density of 0.2 mA cm^{-2} and a long cycle life with 85.7% capacitance retention; it exhibited an excellent performance till 2,000 cycles [153]. Shao et al. reported PANI/GQDs/GR co-coated on the commercial compressed non-woven towel through a repeated dyeing and drying method; the material exhibited a high specific capacitance of 195 mF cm^{-2} at 0.1 mA cm^{-2} with a high stability of 96.5% retention after 6,000 cycles [154]. Similarly, Figure 7.8 represents the CV and charge/discharge curves of the electrodes based on PANI composites with GQDs. Figure 7.8d shows the cycle life of the electrodes for 6,000 charge/discharge cycles,

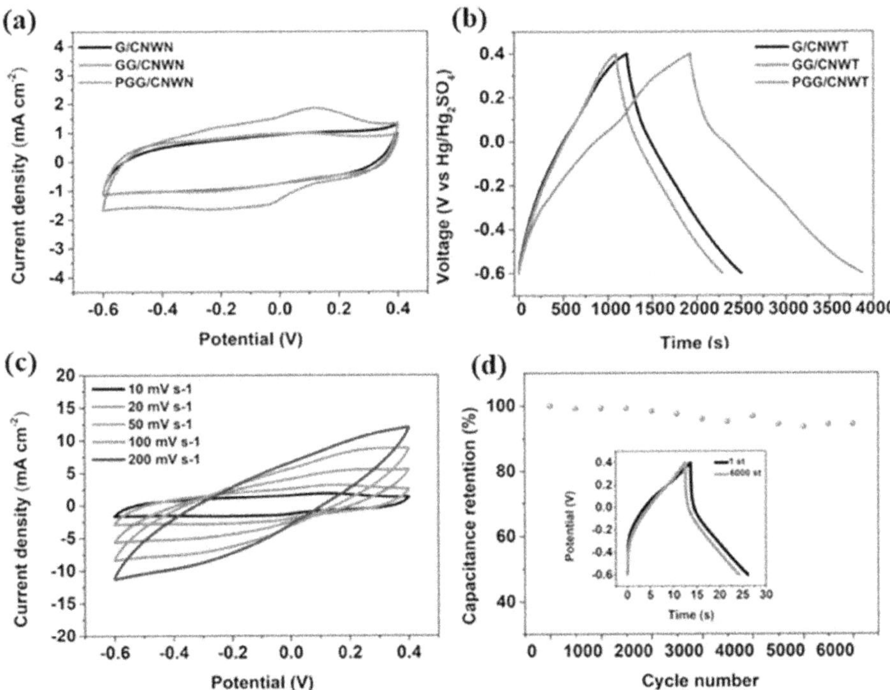

FIGURE 7.8 The electrochemical properties' comparison of G/compressed non-woven towel (G/CNWT), OH-GQDs and rGO co-coated CNWT (GG/CNWT) and polyaniline (PANI) nanoparticles were electrodeposited on the GG/CNWT (PGG/CNWT): (a) CV curves at 10 mV, (b) GCD curves at 0.1 mA cm^{-2}, (c) CV curves of PGG/CNWTs at different scan rates, (d) electrochemical stabilities of PGG/CNWTs at a current density of 5 mA cm^{-2} for 6,000 cycles.

where it is found to be quite stable. Thus, this suggests that polymer/QD composites can be potential electrode materials to acquire high electrochemical performances.

4.4 Conclusions

SCs are found to be very reliable for use in different real-time applications. They are proven to be appropriate candidates for next-generation energy-oriented technologies like solar systems, faster trains, electric vehicles, and wind power equipment, owing to their advantageous properties like excellent power density, longer cycle life, involvement of lesser cost, etc. However, studies are still trying to analyse and improve their properties to make their usage more efficient. In this way, SCs hand in hand with batteries will be able to resolve many energy issues and be able to provide a substantial alternative for energy fuels. In this review, we have concentrated our discussion on different types of quantum dots (QDs) and polymer or nanocomposites employed for SCs along with illustrating their individual performance. Impactful utilization of QDs has actually triggered an interest toward achieving increased efficiency in the performance of SCs. Although not only in the case of the SCs, QDs have also contributed excellently in battery electrode materials, drug delivery, biosensors, bio-imaging, electrochemical, UV, sensor applications, etc. In this review, we have specifically focused our discussion on electrode materials based on quantum dots, polymers, and their composites reported till date. The existing studies helped in unveiling various important insights into their unique properties and minute observations from their utilization in numerous SC configurations were also helpful to understand their functioning. By observing these studies, it has been clearly identified that the polymer nanocomposite with carbon-based QDs are the ones mostly preferred in studies as they have depicted adequate electrochemical performances in terms of power density and cycle life. However, other types of QDs such as transition metal oxide QDs, transition metal dichalcogenide QDs, and polymer QDs are also found to be effectively workable owing to their stable cyclic performances, unique layered structural benefits, and enhanced pathways for mass/ion transfer.

REFERENCES

[1] N. Kamaya, K. Homma, Y. Yamakawa, M. Hirayama, R. Kanno, M. Yonemura, T. Kamiyama, Y. Kato, S. Hama, K. Kawamoto, A. Mitsui. A lithium super Ionic conductor. *Nature Mater.*, 2011, 7: 1–5.

[2] C. A. Vincent. Lithium batteries: A50-year perspective, 1959–2009. *Solid StIonies*, 2000, 134: 159–167.

[3] I-C. Kim, D. Byun, S. Lee, J. K. Lee. Electrochemical characteristics of copper silicide-coated graphite as an anode material of lithium secondary batteries. *Electrochim. Acta*, 2006, 52: 1532–1537.

[4] M. S. Whittingham. Electrical energy storage and intercalation chemistry. *Science*, 1976, 11: 1126–1127.

[5] B. M. L. Rao, R. W. Franeis, H. A. ChristoPher. Lithium-aluminum electrode. *Eleetrochem. Soc.*, 1977, 124: 1490.

[6] M. M. Thackeray, W. I. F. David, P. G. Bruce, J. B. Goodenough. Lithium insertion into manganese spinels. *Mater Res. Bull.*, 1983, 18: 461.

[7] A. Manthiram, J. B. Goodenough. *J. Power Sources,* 1989, 26: 403.

[8] A. K. Padhi, K. S. Nanjundaswamy, J. B. Goodenough. Phospho-olivines aspositive-electrode materials for rechargeable Lithium batteries. *Electrochem. Soc.,* 1997, 144: 1188–1194.

[9] Athula Wijayasinghe. LiFeOz-LiCoOz-NiO cathodes for molten carbonate fuelcells. *J. Electrochem. Soc.,* 2003, 150(5): A558.

[10] K. MitZUshima, P. C. Johnes, P. J. Wiseman, J. B. Goodenough. Li $_x$ CoO2(0<x<1): Anew cathode material for batteries of high energy density. *Mater Res. Bull.,* 1980, 15(6): 783.

[11] G. X. Wang, S. Zhong, D. H. Bradhurst, S. X. Dou, H. K. Liu. Synthesis and characterization ofLiNiOz compounds as cathodes for rechargeable lithium batteries. *J PowerSources,* 1998, 76: 141–146.

[12] A. Urbaro, S. C. Castro, R. Landers, J. Morais, A. D. Siervo, A. Gorenstein, M. H. Tabacniks, M. C. A. Fantini. Electronic structure of Li;NiOy. *Power Sources,* 2001, 108: 239.

[13] A. Rougie, I. Saadoune, P. Gravereau, P. Willmann, C. Delmasa. Effect of cobalt substitution on cationicdistribution in LiNi. yCoyO electrode materials. *Solid State Ionics,* 1996, 90: 83–90.

[14] D. Caurant, N. Baffier, B. Gareia, J. P. Pereira-Ramos. Synthesis by a soft chemistry route andcharacterization of LiNi, Co;. xOz(0Sx<1) cathode materials. *Solid State Ionics,* 1996, 91: 45–54.

[15] J. Molenda, P. Wilk, J. Marzec. Transport properties of the LiNij. yCo, O2 system. *Solid State Ionies,* 1999, 119: 19–22.

[16] T. Ohzuku, A. Ueda, M. Kouguchi, Synthesis and characterization of LiAly/4Nis, O2(R3m)for lithium-ion(shuttlecock) batteries. *Electrochem. Soc.,* 1995, 142(12): 4033–4039.

[17] A. Ott, P. Endres, V. Kleln, et al. Electrochemical performance and chemicalproperties of oxide cathode materials for 4V rechargeable Li-ion batteries. *J. PowerSources,* 1998, 72: 2–7.

[18] Y. Xia, M. Yoshio. An investigation of Lithium Ion insertion into spinel StructureLi-Mn-O compounds. *Electrochem. Soc.,* 1996, 143(3): 825.

[19] W. I. F. David, M. M. Thackeray, L. A. de Picciotto, J. B. Goodenough. Structurerefinement of the spinel-related phases LiMnzO4 and Lio. zMngO4. *Solid State Chem.,* 1987, 67: 316.

[20] A. Mosbah, A. Verbaere, M. Tournous. Phases LiMnOz lrattachees au typespinelle. *Mater. Res. Bull.,* 1983, 18(11): 1375.

[21] James C. Hunter. Preparation of a new crystal form of manganese dioxide: λ-MnO-[]. *Solid State Chem.,* 1981, 39(2): 142–147.

[22] J. M. Tarascon, D. Guyomard. The Li+xMn2Of/C rocking-chair system: A review. *Electrochim. Acta.,* 1993, 38: 1221–1231.

[23] D. H. Jang, Y. J. Shin, S. M. Oh. Dissolution of spinel oxides and capacity losses in4VLi/ Li, Mn, O Coils. *J. Electrochem. Soc.,* 1996, 143(7): 2204–2211.

[24] Z. L. Liu, A. Yu, J. Y. Lee. Synthesis and characterization of LiNi1-x. yCo;Mn, O2 as thecathode materials of secondary Lithium batteries. *J. Power Sources,* 1998, 81–82: 416–419.

[25] T. Ohuzuku, Y. Makimura. Layered Lithium insertion material ofLiCoysNiyzMngsO: For lithium-ion batteries. *Chem. Lett.,* 2001, 30: 642–644.

[26] P. P. Prosini, D. Zane, M. Pasquali. Improved electrochemical performance of aLiFePOx-based composite cathode. *Electrochim. Acta,* 2001, 46(23): 3517–3523.

[27] M. Koltypin, D. Aurbach, L. Nazar, B. Ellis. More on the performance of LiFePO4electrodes-the effect of synthesis route, solution composition, aging, and temperature. *J. Power Sources,* 2007, 174(2): 1241–1250.

[28] N. J. Yun, H. W. Ha, K. H. Jeong, H.-Y. Park, K. Kim. Synthesis and electrochemical properties ofolivine- type LiFePO4/C composite cathode material prepared from a poly-containing precursor. *J. Power Sources*, 2006, 160(2): 1361–1368.

[29] C. M. Julien, A. Mauger, A. Ait-Salah, M. Massot, F. Gendron, K. Zaghib. Nanoscopic scale studies of LiFePO4 ascathode material in lithium-ion batteries for HEV application. *lonics*, 2007, 13: 395–411.

[30] K. Zaghib, N. Ravet, M. Gauthier, F. Gendron, A. Mauger, J. B. Goodenough, C. M. Julien. Optimized electrochemical performance of LiFePO4 at 60°C with purity controlled by sQUID magnetometry. *J Power Sources*, 2006, 163: 560–566.

[31] S. Franger, F. L. Cras, C. Bourbon, H. Rouault. LiFePO4 synthesis routes for enhancedelectro- chemical performance. *Electrochem. Solid-State Lett.*, 2002, 5(10): A231–A233.

[32] J. Barker, M. Y. Saidi, J. L. Swoyer. Lithium iron(I) phospho-olivines prepared by anovel carbothermal reduction method. *Electrochem. Solid-State Lett.*, 2003, 6(3): A53–A55.

[33] Jiangfeng Ni, Masanori Morishita, Yoshiteru Kawabe, Masaharu Watada, Nobuhiko Takeichi, Tetsuo Sakai. Hydrothermalpreparation of LiFePO4 nanocrystals mediated by organic acid. *J. PowerSources*, 2010, 195: 2877–2882.

[34] G. Arnold, J. Garche, R. Hemmer, S. Ströbele, C. Vogler, M. Wohlfahrt-Mehrens. Fine-particle lithium iron phosphate LiFePO4synthesized by a new low-cost aqueous precipitation technology. *J PowerSources*, 2003, 119–121: 247–251.

[35] K. S. Park, K. T. Kang, S. B. Lee, G. Y. Kim, Y. J. Park, H. G. Kim. Synthesis of LiFePO4 with fine particle bycoprecipitation method. *Mater. Res. Bull.*, 2004, 39(12): 1803–1810.

[36] S. L. Bewlay, K. Konstantinov, G. X. Wang, S. X. Dou, H. K. Liu. Conductivity improvements tospray-produced LiFePO: By addition of a carbon source. *Mater. Lett.*, 2004, 58(11): 1788–1791.

[37] T. H. Teng, M. R. Yang, S. H. Wu, Y. P. Chiang. Electrochemical properties ofLiFeo, 9Mgo. PO4/carbon cathode materials prepared by ultrasonic spray pyrolysis. *Solid State Commun.*, 2007, 142: 389–392.

[38] M. S. Islam, D. J. Driscoll, C. A. J. Fisher, P. R. Slater. Atomic-scale investigation of defects, dopants, and lithium transport in the $LiFePO_4$ olivine-type battery material. *Chem. Mater.,* 2005, 17: 5085.

[39] J. R. Ying, M. Lei, C. Y. Jiang, C. Wan, X. He, J. Li, L. Wang, J. Ren. Preparation and characterization of high-densityspherical Lio. 9zCro. oFePOf/C cathode material for lithium ion batteries. *J. PowerSources*, 2006, 158(1): 543–549.

[40] N. Revet et al. *196th Meeting of the Electrochemical Society.* Hawaii, 1999, Abstract 127.

[41] N. Ravet, J. B. Goodenough, S. Besner, et al. *Abstract the Electrochemical Societyand the Electrochemical Society of Japan Meeting Abstracts.* Honoluluhi, 1999, Vol. 9912.

[42] C. John Wen, B. A. Boukamp, R. A. Huggins. Thermodynamic and mass transportproperties of lithium aluminide. *J. Elcetrochem. Soc.*, 1979, 126(12): 2258–2266.

[43] Peter G. Bruce. *Solid State Electrochemistry.* Cambridge University Press, New York, NY 10011-4211, USA, 1995, pp. 199–228.

[44] I. D. Raistrick, R. A. Huggins application of AC technique to the study of lithiumdiffusion in tungsten trioxide thin films. *J Electrochem Soc.*, 1980, 127(2): 343.

[45] S. Y. Lim, W. Shen, Z. Gao. Carbon quantum dots and their applications. *Chem. Soc. Rev.*, 2015, 44(1): 362–381.

[46] L. Tian, Z. Li, P. Wang, et al. Carbon quantum dots for advanced electrocatalysis. *J. Energy Chem.*, 2021, 55: 279–294.

[47] R. Jelinek. *Carbon Quantum Dots.* Springer International Publishing, Cham, 2017: 29–46.

[48] Y. Wang, A. Hu. Carbon quantum dots: Synthesis, properties and applications. *J. Mater. Chem. C*, 2014, 2(34): 6921–6939.

[49] Y. Dong, J. Lin, Y. Chen, et al. Graphene quantum dots, graphene oxide, carbon quantum dots and graphite nanocrystals in coals. *Nanoscale*, 2014, 6(13): 7410–7415.

[50] A. S. Rasal, S. Yadav, A. Yadav, A. A. Kashale, S. T. Manjunatha, A. Altaee, and J.-Y. Chang. Carbon quantum dots for energy applications: A review. *ACS Appl. Nano Mater.*, 2021, 4(7): 6515–6541.

[51] M. J. Molaei. Carbon quantum dots and their biomedical and therapeutic applications: A review. *RSC Advan.*, 2019, 9(12): 6460–6481.

[52] L. Wang, W. Li, L. Yin, et al. Full-color fluorescent carbon quantum dots. *Sci. Advan.*, 2020, 6(40): eabb6772.

[53] D. L. Zhao, T. S. Chung. Applications of carbon quantum dots (CQDs) in membrane technologies: A review. *Water Res.*, 2018, 147: 43–49.

[54] K. A. S. Fernando, S. Sahu, Y. Liu, W. K. Lewis, E. A. Guliants, A. Jafariyan, P. Wang, C. E. Bunker, Y.-P. Sun. Carbon quantum dots and applications in photocatalytic energy conversion. *ACS Appl. Mater. Interfaces*, 2015, 7(16): 8363–8376.

[55] N. Azam, M. Najabat Ali, T. Javaid Khan. Carbon quantum dots for biomedical applications: Review and analysis. *Front. Mater.*, 2021, 8: 700403.

[56] I. Singh, R. Arora, H. Dhiman, R. Pahwa. Carbon quantum dots: Synthesis, characterization and biomedical applications. *Turk. J. Pharm. Sci.*, 2018, 15(2): 219–230.

[57] Y. Zhang, M. Jia, H. Gao, J. Yu, L. Wang, Y. Zou, F. Qin, Y. Zhao. Porous hollow carbon spheres: facile fabrication and excellent supercapacitive properties. *Electrochim. Acta,* 2015, 184: 32–39.

[58] Z. F. Yang, J. R. Tian, Z. F. Yin, C. J. Cui, W. Z. Qian, F. Wei. Carbon nanotube- and graphene-based nanomaterials and applications in high-voltage supercapacitor: A review. *Carbon N Y,* 2019, 141: 467–480.

[59] J. Zhao, Y. F. Jiang, H. Fan, M. Liu, O. Zhuo, X. Z. Wang, Q. Wu, L. J. Yang, Y. W. Ma, Z. Hu. Porous 3D few-layer graphene-like carbon for ultrahigh-power supercapacitors with well-defined structure-performance relationship. *Adv. Mater.,* 2017, 29(11): 1604569.

[60] A. liwak, B. Grzyb, N. Díez, G.y. Gryglewicz. Nitrogen-doped reduced graphene oxide as electrode material for high rate supercapacitors. *Appl. Surf. Sci.,* 2017, 399: 265–271.

[61] R. Pothu, R. Bolagam, Q.-.H. Wang, W. Ni, J.-.F. Cai, X.-.X. Peng, Y.Z. Feng, J.-.M. Ma. Nickel sulfide-based energy storage materials for highperformance electrochemical capacitors. *Rare Metals,* 2020, 40: 353–373. Doi:10.1007/s12598-020-01470-w.

[62] L. L. Zhang, X. S. Zhao. Carbon-based materials as supercapacitor electrodes. *Chem. Soc. Rev.*, 2009, 38: 2520–2531.

[63] X. L. Li, L. J. Zhi. Graphene hybridization for energy storage applications. *Chem. Soc. Rev.*, 2018, 47(9): 3189–3216.

[64] P. B. Geng, S. S. Zheng, H. Tang, R. M. Zhu, L. Zhang, S. Cao, H. G. Xue, H. Pang. Transition metal sulfides based on graphene for electrochemical energy storage. *Adv. Energy. Mater.*, 2018, 8(15): 1703259.

[65] P. H. Yang, W. J. Mai. Flexible solid-state electrochemical supercapacitors. *Nano Energy*, 2014, 8: 274–290.

[66] H. J. Liu, W. L. Zhu, D. F. Long, J. L. Zhu, G. Pezzotti. Porous V2O5 nanorods/reduced graphene oxide composites for high performance symmetric supercapacitors. *Appl. Surf. Sci.,* 2019, 478: 383–392.

[67] R. Ramachandran, Q. Hu, F. Wang, Z. X. Xu. Synthesis of N-CuMe2Pc nanorods/ graphene oxide nanocomposite for symmetric supercapacitor electrode with excellent cyclic stability. *Electrochim. Acta,* 2019, 298: 770–777.

[68] S. H. Ji, N. R. Chodankar, D. H. Kim. Aqueous asymmetric supercapacitor based on RuO2-WO3 electrodes. *Electrochim. Acta,* 2019, 325: 134879 325.

[69] D. W. Wang, L. Feng, L. Min, G. Q. Lu, H. M. Cheng. 3DAperiodic hierarchical porous graphitic carbon material for high-rate electrochemical capacitive energy storage. *Angew. Chem. Int. Ed.*, 2008, 47(2): 373–376.

[70] X. Chen, W.-.D. Oh, P.-.H. Zhang, R. D. Webster, T.-.T. Lim. Surface construction of nitrogen-doped chitosan-derived carbon nanosheets with hierarchically porous structure for enhanced sulfacetamide degradation via peroxymonosulfate activation: Maneuverable porosity and active sites. *Chem. Eng. J.*, 2020, 382: 122908.

[71] B. Duan, X. Gao, X. Yao, Y. Fang, L. Huang, J. Zhou, L. Zhang. Unique elastic N-doped carbon nanofibrous microspheres with hierarchical porosity derived from renewable chitin for high rate supercapacitors. *Nano Energy*, 2016, 27: 482–491.

[72] T. Méndez-Morales, N. Ganfoud, Z. Li, M. Haefele, B. Rotenberg, M. Salanne. Performance of microporous carbon electrodes for supercapacitors: Comparing graphene with disordered materials. *Energy Storage Mater.*, 2019, 17: 88–92.

[73] H. Wang, Z. Xu, A. Kohandehghan, Z. Li, K. Cui, X. Tan, T. J. Stephenson, C. K. King'Ondu, C. M. Holt, B. C. Olsen, Interconnected carbon nanosheets derived from hemp for ultrafast supercapacitors with high energy. *ACS Nano*, 2013, 7(6): 5131–5141.

[74] A. Bagri, C. Mattevi, M. Acik, Y. J. Chabal, M. Chhowalla, V. B. Shenoy. Structural evolution during the reduction of chemically derived graphene oxide. *Nat. Chem.*, 2010, 2(7): 581–587.

[75] S. Kerisit, B. Schwenzer, M. Vijayakumar. Effects of oxygen-containing functional groups on supercapacitor performance. *J. Phys. Chem. Lett.*, 2014, 5(13): 2330–2334.

[76] J. Zhang, H. B. Song, D. W. Zeng, H. Wang, Z.Y. Qin, K. Xu, A. M. Pang, C. S. Xie. Facile synthesis of diverse graphene nanomeshes based on simultaneous regulation of pore size and surface structure. *Sci. Rep-Uk*, 2016, 6: 1–9.

[77] L.-.G. Gong, X.-.X. Qi, K. Yu, J.-.Q. Gao, B.-.B. Zhou, G.-.Y. Yang. Covalent conductive polymer chain and organic ligand ethylenediamine modified MXene-like-{AlW12 O40} compounds for fully symmetric supercapacitors, electrochemical sensors and photocatalysis mechanisms. *J. Mater. Chem. A*, 2020, 8(11): 5709–5720.

[78] X. Li, S. H. Qi, W. C. Zhang, Y.Z. Feng, J. M. Ma. Recent progress on FeS2 as anodes for metal-ion batteries. *Rare Metals,* 2020, 39(11): 1239–1255.

[79] J. Wu, Q.e. Zhang, J. Wang, X. Huang, H. Bai. A self-assembly route to porous polyaniline/reduced graphene oxide composite materials with molecular-level uniformity for high-performance supercapacitors. *Energ. Environ. Sci.*, 2018, 11: 1280–1286.

[80] J. Yan, Q. Wang, T. Wei, L.L. Jiang, M. L. Zhang, X. Y. Jing, Z. J. Fan. Template-assisted low temperature synthesis of functionalized graphene for ultrahigh volumetric performance supercapacitors. *ACS Nano,* 2014, 8(5): 4720–4729.

[81] W. L. Zhang, C. Xu, C. Q. Ma, G. X. Li, Y. Z. Wang, K. Y. Zhang, F. Li, C. Liu, H. M. Cheng, Y. W. Du, N. J. Tang, W. C. Ren. Nitrogen-superdoped 3D graphene networks for high-performance supercapacitors. *Adv. Mater.*, 2017, 29(36): 17 0 16 7 7.

[82] A. Bakandritsos, D. D. Chronopoulos, P. Jakubec, M. Pykal, K. Čépe, T. Steriotis, S. Kalytchuk, M. Petr, R. Zbořil, M. Otyepka. High-performance supercapacitors based on a zwitterionic network of covalently functionalized graphene with iron tetraaminophthalocyanine. *Adv. Funct. Mater.*, 2018, 28(29): 18 01111.

[83] A. Criado, M. Melchionna, S. Marchesan, M. Prato. The covalent functionalization of graphene on substrates. *Angew. Chem. Int. Ed. Engl.*, 2015, 54(37): 10734–10750.

[84] Z. Yu, L. Lei, G. Casillas, Z. Sun, Y. Zheng, G. Ruan, Z. Peng, A. R. O. Raji, C. Kittrell, R. H. Hauge. A seemless three-dimensional carbon nanotube graphene hybrid material. *Nat. Commun.*, 2012, 3: 1–7.

[85] R. V. Salvatierra, D. Zakhidov, J. W. Sha, N. D. Kim, S. K. Lee, A. R. O. Raji, N. Q. Zhao, J. M. Tour, Graphene carbon nanotube carpets grown using binary catalysts for high-performance lithium-ion capacitors. *ACS Nano,* 2017, 11(3): 2724–2733.

[86] J. L. Liu, Y. R. Zhu, X. H. Chen, W. J. Yi. Nitrogen, sulfur and phosphorus tri-doped holey graphene oxide as a novel electrode material for application in supercapacitor. *J. Alloy. Compd.*, 2020, 815: 152328.

[87] J. H. Lee, N. Park, B. G. Kim, D. S. Jung, K. Im, J. Hur, J. W. Choi. Restacking-inhibited 3D reduced graphene oxide for high performance supercapacitor electrodes. *ACS Nano*, 2013, 7(10): 9366–9374.

[88] D. Hulicova-Jurcakova, M. Kodama, S. Shiraishi, H. Hatori, Z. H. Zhu, G. Q. Lu. Nitrogen-enriched nonporous carbon electrodes with extraordinary supercapacitance. *Adv. Funct. Mater.*, 2009, 19(11): 1800–1809.

[89] Z. Liu, L. L. Jiang, L. Z. Sheng, Q. H. Zhou, T. Wei, B. S. Zhang, Z. J. Fan. Oxygen clusters distributed in graphene with "paddy land" structure: Ultrahigh capacitance and rate performance for supercapacitors. *Adv. Funct. Mater.*, 2018, 28(5): 1800–1809.

[90] C. Liu, Z. Yu, D. Neff, A. Zhamu, B. Z. Jang. Graphene-based supercapacitor with an ultrahigh energy density. *Nano Lett.*, 2010, 10(12): 4863–4868.

[91] H. Shang, Z. C. Zuo, H. Y. Zheng, K. Li, Z. Y. Tu, Y. P. Yi, H. B. Liu, Y. J. Li, Y. L. Li. Ndoped graphdiyne for high-performance electrochemical electrodes. *Nano Energy*, 2018, 44: 144–154.

[92] E. N. Gorenskaia, B. C. Kholkhoev, V. G. Makotchenko, M. N. Ivanova, V. E. Fedorov, V. F. Burdukovskii. Hydrothermal synthesis of N-doped graphene for supercapacitor electrodes. *J. Nanosci. Nanotechnol.*, 2020, 20(5): 3258–3264.

[93] Y. J. Wang, L. C. Zhao, H. Peng, X. W. Dai, X. N. Liu, G. F. Ma, Z. Q. Lei. Three-dimensional honeycomb-like porous carbon derived from tamarisk roots via a green fabrication process for high-performance supercapacitors. *Ionics* (Kiel), 2019, 25(9): 4315–4323.

[94] T. E. Balaji, H. T. Das, T. Maiyalagan. Recent trends in bimetallic oxides and their composites as electrode materials for supercapacitor applications. *ChemElectroChem*, 2021, 8: 1723–1746.

[95] S. Uddin, H. T. Das, T. Maiyalagan, P. Elumalai, Influence of designed electrode surfaces on double layer capacitance in aqueous electrolyte: Insights from standard models. *Appl. Surf. Sci.*, 2018, 449: 445–453.

[96] Y. Zhang, L. Li, H. Su, W. Huang, X. Dong. Binary metal oxide: Advanced energy storage materials in supercapacitors. *J. Mater. Chem. A*, 2014, 8: 43–59.

[97] H. T. Das, K. Mahendraprabhu, T. Maiyalagan, P. Elumalai. Performance of solid-state hybrid energy-storage device using reduced graphene-oxide anchored sol-gel derived Ni/NiO nanocomposite. *Sci. Rep.*, 2017, 7: 15342.

[98] H. R. Barai, A. N. Banerjee, S. W. Joo. Improved electrochemical properties of highly porous amorphous manganese oxide nanoparticles with crystalline edges for superior supercapacitors. *J. Ind. Eng. Chem.*, 2017, 56: 212–224.

[99] S. K. Kiran, M. Padmini, H. T. Das, P. Elumalai. Performance of asymmetric supercapacitor using CoCr-layered double hydroxide and reduced graphene-oxide. *J. Solid State Electrochem.*, 2017, 21: 927–938.

[100] E. Duraisamy, H. T. Das, A. S. Sharma, P. Elumalai. Supercapacitor and photocatalytic performances of hydrothermally-derived Co3O4/CoO@carbon nanocomposite. *New J. Chem.*, 2018, 42: 6114–6124.

[101] A. M. Al-Enizi, M. Ubaidullah, J. Ahmed, T. Ahamad, T. Ahmad, S. F. Shaikh, M. Naushad. Synthesis of NiOx@NPC composite for high-performance supercapacitor via waste PET plastic-derived Ni-MOF. *Compos. Part B Eng.*, 2020, 183: 107655.

[102] S. Saini, P. Chand, A. Joshi. Biomass derived carbon for supercapacitor applications: Review. *J. Energy Storage*, 2021, 39: 102646.

[103] M. Shaker, R. Riahifar, Y. Li. A review on the superb contribution of carbon and graphene quantum dots to electrochemical capacitors' performance: Synthesis and application. *Flat. Chem.*, 2020, 22: 100171.

[104] H. T. Das, S. Saravanya, P. Elumalai. Disposed dry cells as sustainable source for generation of few layers of graphene and manganese oxide for solid-state symmetric and asymmetric supercapacitor applications. *ChemistrySelect.*, 2018, 3: 13275–13283.

[105] S. J. Panchu, K. Raju, H. C. Swart, B. Chokkalingam, M. Maaza, M. Henini, M. K. Moodley. Luminescent MoS2 quantum dots with tunable operating potential for energy-enhanced aqueous supercapacitors. *ACS Omega,* 2021, 6: 4542–4550.

[106] H. T. Das, T. Elango Balaji, K. Mahendraprabhu, S. Vinoth. Cost-Effective Nanomaterials Fabricated by Recycling Spent Batteries. In: Makhlouf, A.S.H., Ali, G.A.M. (eds) *Waste Recycling Technologies for Nanomaterials Manufacturing. Topics in Mining, Metallurgy and Materials Engineering.* Springer, Cham, 2021, pp. 147–174.

[107] S. Suriyakumar, P. Bhardwaj, A. N. Grace, A. M. Stephan. Role of polymers in enhancing the performance of electrochemical supercapacitors: A review. *Batter. Supercaps,* 2021, 4: 571–584.

[108] S. S. Shah, H. T. Das, H. R. Barai, A. Aziz. Boosting the electrochemical performance of polyaniline by one-step electrochemical deposition on nickel foam for high-performance asymmetric supercapacitor. *Polymers,* 2022, 14: 270.

[109] V. Subramanian, K. H. Prasad, H. T. Das, K. Ganapathy, S. Nallani, T. Maiyalagan. Novel dispersion of 1D nanofiber fillers for fast ion-conducting nanocomposite polymer blend quasi-solid electrolytes for dye-sensitized solar cells. *ACS Omega,* 2022, 7: 1658–1670.

[110] W. Yin, X. Liu, X. Zhang, X. Gao, V. L. Colvin, Y. Zhang, W.W. Yu. Synthesis of tungsten disulfide and molybdenum disulfide quantum dots and their applications. *Chem. Mater.,* 2020, 32: 4409–4424.

[111] H. R. Barai, A. N. Banerjee, N. Hamnabard, S. W. Joo. Synthesis of amorphous manganese oxide nanoparticles-to-crystalline nanorods through a simple wet-chemical technique using K+ ions as a 'growth director' and their morphology-controlled high performance supercapacitor applications. *RSC Adv.,* 2016, 6: 78887–78908.

[112] S. Kumar, G. Saeed, L. Zhu, K. N. Hui, N. H. Kim, J. H. Lee. 0D to 3D carbon-based networks combined with pseudocapacitive electrode material for high energy density supercapacitor: A review. *Chem. Eng. J.,* 2021, 403: 126352.

[113] A. Aphale, K. Maisuria, M. K. Mahapatra, A. Santiago, P. Singh, P. Patra. Hybrid electrodes by in-situ integration of graphene and carbon-nanotubes in polypyrrole for supercapacitors. *Sci. Rep.,* 2015, 5: 14445.

[114] P. Forouzandeh, V. Kumaravel, S. C. Pillai. Electrode materials for supercapacitors: A review of recent advances. *Catalysts,* 2020, 10: 969.

[115] H. R. Barai, M. Rahman, S. W. Joo. Annealing-free synthesis of K-doped mixed-phase TiO2 nanofibers on Ti Foil for electrochemical supercapacitor. *Electrochim. Acta,* 2017, 253: 563–571.

[116] B. E. Conway. Transition from "supercapacitor" to "battery" behavior in electrochemical energy storage. *In Proceedings of the International Power Sources Symposium,* Cherry Hill, NJ, 25–28 June 1990; IEEE: Piscataway, NJ, 1991; Volume 138, pp. 319–327.

[117] H. R. Barai, A. Banerjee, F. Bai, S. W. Joo. Surface modification of Titania nanotube arrays with crystalline manganese-oxide nanostructures and fabrication of hybrid electrochemical electrode for high-performance supercapacitors. *J. Ind. Eng. Chem.,* 2018, 62: 409–417.

[118] Z. Zhao, Y. Xie. Enhanced electrochemical performance of carbon quantum dots-polyaniline hybrid. *J. Power Sources,* 2017, 337: 54–64.

[119] H. R. Barai, N. S. Lopa, F. Ahmed, N. A. Khan, S. A. Ansari, S. W. Joo, M. Rahman. Synthesis of Cu-Doped Mn3O4@Mn-doped cuo nanostructured electrode materials by a solution process for high-performance electrochemical pseudocapacitors. *ACS Omega,* 2020, 5: 22356–22366.

[120] H. R. Barai, P. Barai, M. Roy, S. W. Joo. Solid-state synthesis of Titanium-Doped Binary Strontium–Copper Oxide as a highperformance electrochemical pseudocapacitive electrode nanomaterial. *Energy Fuels,* 2021, 35: 16870–16881.

[121] H. R. Barai, M.M. Rahman, M. Adeel, S. W. Joo. MnSn(OH)6 derived Mn2SnO4@ Mn2O3 composites as electrode materials for high-performance Supercapacitors. *Mater. Res. Bull.,* 2022, 148: 111678.

[122] S. Najib, E. Erdem. Current progress achieved in novel materials for supercapacitor electrodes: Mini review. *Nanoscale Adv.,* 2019, 1: 2817–2827.

[123] H. R. Barai, M. Rahman, S. W. Joo. Template-free synthesis of two-dimensional titania/titanate nanosheets as electrodes for high-performance supercapacitor applications. *J. Power Sources,* 2017, 372: 227–234.

[124] S. K. Ujjain. Applications of quantum dots in supercapacitors. *Mater. Res. Found.,* 2021, 96: 169–190.

[125] M. Li, T. Chen, J. J. Gooding, J. Liu. Review of carbon and graphene quantum dots for sensing. *ACS Sens.,* 2019, 4: 1732–1748.

[126] P. Tian, L. Tang, K. Teng, S. Lau. Graphene quantum dots from chemistry to applications. *Mater. Today Chem.,* 2018, 10: 221–258.

[127] M. Zvaigzne, I. Domanina, D. Il'gach, A. Yakimansky, I. Nabiev, P. Samokhvalov. Quantum dot–polyfluorene composites for white-light-emitting quantum dot-based LEDs. *Nanomaterials,* 2020, 10: 2487.

[128] B. W. Watson, L. Meng, C. Fetrow, Y. Qin. Core/shell conjugated polymer/quantum dot composite nanofibers through orthogonal non-covalent interactions. *Polymers* 2016, 8: 408.

[129] H. Gordillo, I. Suárez, R. Abargues, P. Rodríguez-Cantó, S. Albert, J. P. Martínez-Pastor. Polymer/QDs nanocomposites for waveguiding applications. *J. Nanomater.,* 2012, 2012: 1–9.

[130] C. Hu, M. Li, J. Qiu, Y. P. Sun. Design and fabrication of carbon dots for energy conversion and storage. *Chem. Soc. Rev.,* 2019, 48: 2315–2337.

[131] S. S. Mittal, G. Ramadas, N. Vasanthmurali, V. S. Madaneshwar, M. S. Kumar, N. K. Kothurkar. Carbon quantum dot-polypyrrole nanocomposite for supercapacitor electrodes. *IOP Conf. Ser. Mater. Sci. Eng.,* 2019, 577: 012194.

[132] D. Arthisree, W. Madhuri. Optically active polymer nanocomposite composed of polyaniline, polyacrylonitrile and greensynthesized graphene quantum dot for supercapacitor application. *Int. J. Hydrogen Energy,* 2020, 45: 9317–9327.

[133] H. Li, X. He, Z. Kang, H. Huang, Y. Liu, J. Liu, S. Lian, A. C. H. Tsang, X. Yang, S.-T. Lee. Water-soluble fluorescent carbon quantum dots and photocatalyst design. *Angew. Chem. Int. Ed.,* 2010, 49: 4430–4434.

[134] Y. Li, Y. Hu, Y. Zhao, G. Shi, L. Deng, Y. Hou, L. Qu. An electrochemical avenue to green-luminescent graphene quantum dots as potential electron-acceptors for photovoltaics. *Adv. Mater.,* 2011, 23: 776–780.

[135] X. Li, M. Rui, J. Song, Z. Shen, H. Zeng. Carbon and graphene quantum dots for optoelectronic and energy devices: A review. *Adv. Funct. Mater.,* 2015, 25: 4929–4947.

[136] Y. Li, Y. Zhao, H. Cheng, Y. Hu, G. Shi, L. Dai, L. Qu. Nitrogen-doped graphene quantum dots with oxygen-rich functional groups. *J. Am. Chem. Soc.,* 2012, 134: 15–18.

[137] S. Anantharaj, M. O. Valappil, K. Karthick, V. K. Pillai, S. Alwarappan, S. Kundu. Electrochemically chopped WS2 quantum dots as an efficient and stable electrocatalyst for water reduction. *Catal. Sci. Technol.,* 2019, 9: 223–231.

[138] M. Jing, C. Wang, H. Hou, Z. Wu, Y. Zhu, Y. Yang, X. Jia, Y. Zhang, X. Ji. Ultrafine nickel oxide quantum dots enbedded with few-layer exfoliative graphene for an asymmetric supercapacitor: Enhanced capacitances by alternating voltage. *J. Power Sources,* 2015, 298: 241–248.

[139] M. O. Valappil, A. Anil, M. Shaijumon, V. K. Pillai, S. Alwarappan. A single-step electrochemical synthesis of luminescent WS2 quantum dots. *Chem. Eur. J.*, 2017, 23: 9144–9148.

[140] M. Vandana, H. Vijeth, S. Ashokkumar, H. Devendrappa. Hydrothermal synthesis of quantum dots dispersed on conjugated polymer as an efficient electrodes for highly stable hybrid supercapacitors. *Inorg. Chem. Commun.*, 2020, 117: 107941.

[141] X. Shi, Z.-Z. Yin, J. Xu, S. Li, C. Wang, B. Wang, Y. Qin, Y. Kong. Preparation, characterization and the supercapacitive behaviors of electrochemically reduced graphene quantum dots/polypyrrole hybrids. *Electrochim. Acta,* 2021, 385: 138435.

[142] L. Li, M. Li, J. Liang, X. Yang, M. Luo, L. Ji, Y. Guo, H. Zhang, N. Tang, X. Wang. Preparation of Core–Shell CQD@PANI nanoparticles and their electrochemical properties. *ACS Appl. Mater. Interfaces,* 2019, 11: 22621–22627.

[143] A. P. P. Alves, R. Koizumi, A. Samanta, L. D. Machado, A. K. Singh, D. S. Galvao, G. G. Silva, C. S. Tiwary, P. M. Ajayan. One-step electrodeposited 3D-ternary composite of zirconia nanoparticles, rGO and polypyrrole with enhanced supercapacitor performance. *Nano Energy,* 2017, 31: 225–232.

[144] D. Vasudevan, R. R. Gaddam, A. Trinchi, I. Cole. Core–shell quantum dots: Properties and applications. *J. Alloys Compd.*, 2015, 636: 395–404.

[145] J. Ju, W. Chen. Synthesis of highly fluorescent nitrogen-doped graphene quantum dots for sensitive, label-free detection of Fe (III) in aqueous media. *Biosens. Bioelectron.*, 2014, 58: 219–225.

[146] H. Duan, T. Wang, X. Wu, Z. Su, J. Zhuang, S. Liu, R. Zhu, C. Chen, H. Pang. CeO2 quantum dots doped Ni-Co hydroxide nanosheets for ultrahigh energy density asymmetric supercapacitors. *Chin. Chem. Lett.*, 2020, 31: 2330–2332.

[147] N. Chakrabarty, A. Dey, S. Krishnamurthy, A. K. Chakraborty. CeO2/Ce2O3 quantum dot decorated reduced graphene oxide nanohybrid as electrode for supercapacitor. *Appl. Surf. Sci.*, 2021, 536: 147960.

[148] G. Oskueyan, M. M. Lakouraj, M. Mahyari. Nitrogen and sulfur Co-Doped graphene quantum dots decorated CeO2 nanoparticles/ polyaniline: As high efficient hybrid supercapacitor electrode materials. *Electrochim. Acta,* 2019, 299: 125–131.

[149] D. Gopalakrishnan, D. Damien, B. Li, H. Gullappalli, V. K. Pillai, P. M. Ajayan, M. M. Shaijumon. Electrochemical synthesis of luminescent MoS2 quantum dots. *Chem. Commun.*, 2015, 51: 6293–6296.

[150] Y. Ko, J. Shim, C.-H. Lee, K. S. Lee, H. Cho, K.-T. Lee, D. I. Son. Synthesis and characterization of CuO/graphene (Core/shell) quantum dots for electrochemical applications. *Mater. Lett.*, 2018, 217: 113–116.

[151] K. Ghosh, S. K. Srivastava. Enhanced supercapacitor performance and electromagnetic interference shielding effectiveness of CuS Quantum dots grown on reduced graphene oxide sheets. *ACS Omega*, 2021, 6: 4582–4596.

[152] P. Siwatch, K. Sharma, S. Tripathi. Facile synthesis of NiCo2O4 quantum dots for asymmetric supercapacitor. *Electrochim. Acta,* 2020, 329: 135084.

[153] J. Geng, C. Ma, D. Zhang, X. Ning. Facile and fast synthesis of SnO2 quantum dots for high performance solid-state asymmetric supercapacitor. *J. Alloys Compd.*, 2020, 825: 153850.

[154] X. Guo, Y. Wang, F. Wu, Y. Ni, S. Kokot. The use of tungsten disulfide dots as highly selective, fluorescent probes for analysis of nitrofurazone. *Talanta*, 2015, 144: 1036–1043.

8 Development of g-C$_3$N$_4$-Based Nanocomposite for Hydrogen Production and Battery Applications

Is Fatimah, Ganjar Fadiilah, Ika Yanti,
Suresh Sagadevan and Won-Chun Oh

1 INTRODUCTION

Renewable energy resources have become a vital issue, play an important role, and determine the sustainability of other sectors. Due to their importance, the development of materials for enhancing the production of renewable energy sources attracted great interest. Within this scheme, the technology for supporting the production of new energy sources such as batteries and new energy sources that are low-cost, reproducible, and high-performance materials such as hydrogen from water splitting are required (Hosseini and Wahid, 2016). g-C$_3$N$_4$, also known as graphitic carbon nitride, is a non-metallic semiconductor material with a unique layered structure is one of the potential materials. During the past few decades, it garnered significant interest due to its potential applications in sensors, photocatalysis, energy storage, membrane, and optoelectronic devices. g-C$_3$N$_4$ has aroused a tremendous interest among researchers. Considering that H$_2$ production and battery are the most important technologies for providing sustainable, clean, and renewable energy in the future, the role of g-C$_3$N$_4$ development for energy inclusive of its use for both purposes is shown in Figure 8.1.

Referring to the chart, it can be observed that an increasing number of publications related to g-C$_3$N$_4$-based material for both usages were recorded during the last 5 years. These phenomena suggest the importance of its research progression in the future.

2 SYNTHESIS AND PHYSICOCHEMICAL CHARACTERISTICS OF G-C$_3$N$_4$

Historically, g-C$_3$N$_4$ was found as Berzelius synthesized the polymeric derivative of C$_3$N$_4$ named as melon by Liebig. The structure of melon is considered as a linear

DOI: 10.1201/9781003561262-8

"

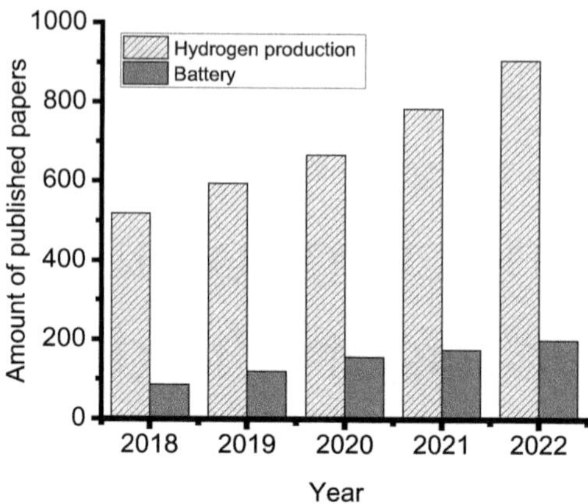

FIGURE 8.1   Popularity of g-C_3N_4 development for H_2 production and battery applications.
Source: SCOPUS database.

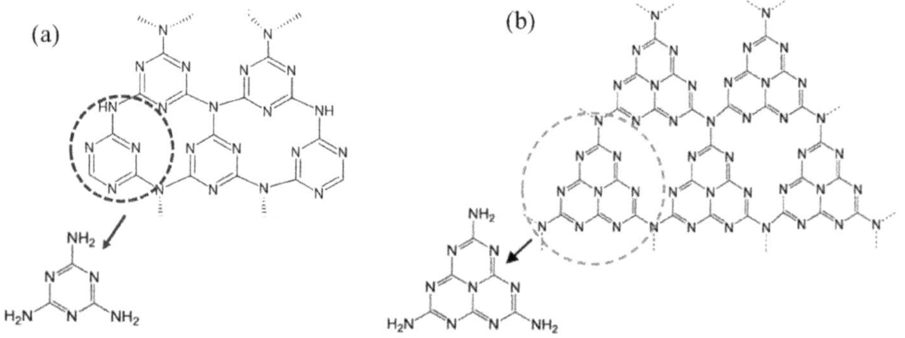

FIGURE 8.2   (a) Triaine and (b) tri-s-triazine (heptazine) structure of g-C_3N_4.

polymer consisting of connected tri-s-triazines via secondary nitrogen, which is similar to that of graphene, where N heteroatoms substitute C atoms in the graphene framework forming p-conjugated systems containing sp^2-hybridized N and C atoms (Figure 8.2). However, with the layer's distance of 0.326 nm, the layer that is slightly higher leads to the stronger interplanar interaction between layers compared to graphite (Guo et al., 2020; H. Zhang et al., 2017).

The beneficial properties of g-C_3N_4 are its physicochemical stability and tunability, and its ability to act as an environmentally friendly semiconductor as a metal-free material. g-C_3N_4 shows thermal stability of mild temperatures, and it decomposes to C and N upon heating above 600 and 900°C at the pressure of 10 and 25 gPa (Fang et al., 2011). Particularly, the bandgap of g-C_3N_4 is about 2.7 eV, with the conduction band (CB) at −1.13 eV and the valence band (VB) at +1.74 eV, and its absorption edge is about 460 nm. With these features, g-C_3N_4 could be

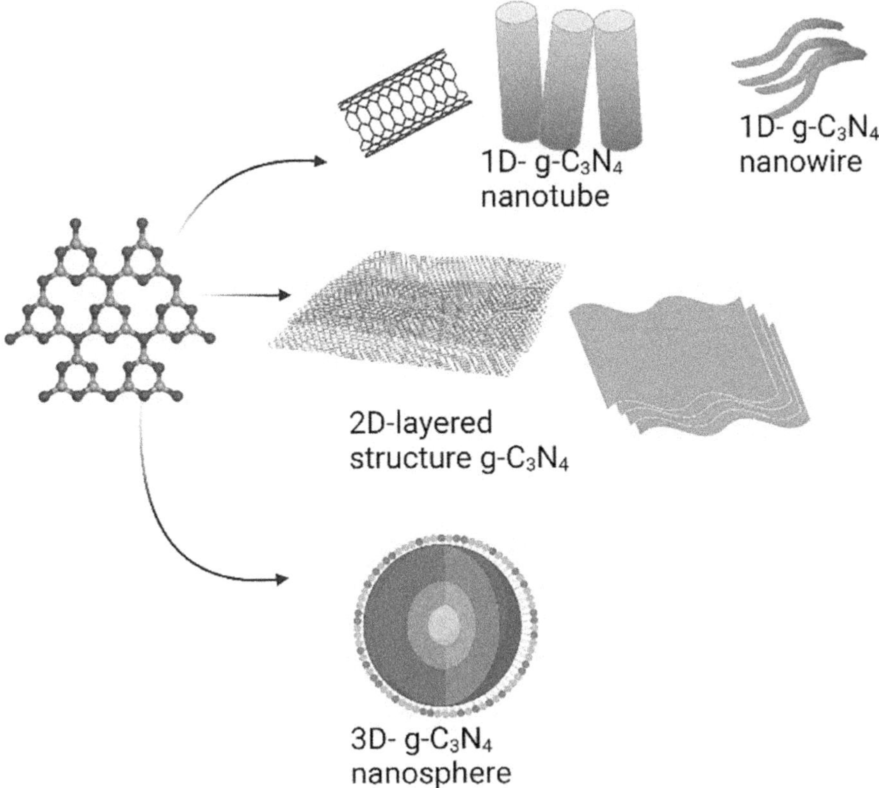

FIGURE 8.3 Various forms of g-C$_3$N$_4$.

utilized as a supporting material for metal or metal oxide materials or nanomaterials. However, besides its potential properties, some drawbacks such as poor specific surface area, low porosity, and high potency in electron-hole recombination for utilization as semiconductors are exhibited by g-C$_3$N$_4$. Therefore, methods to control size morphologies of g-C$_3$N$_4$ were developed. g-C$_3$N$_4$ nanostructures could be present in zero-dimension (0D) to three-dimension (3D), including 0D g-C$_3$N$_4$ quantum dots, 1D g-C$_3$N$_4$ nanotubes, 2D layered structure-g-C$_3$N$_4$, and 3D g-C$_3$N$_4$ nanosphere (Figure 8.3).

Each structure could be obtained by a specific method, but in general, g-C$_3$N$_4$ can be obtained by the pyrolysis of nitrogen-rich precursors at different temperatures according to their specific thermal energy and conversion. Various precursors such as urea, melamine, thiourea, and cyanamide have been reported for the synthesis and their schematic representation is depicted in Figure 8.4.

By the smaller size, g-C$_3$N$_4$ quantum dots (g-C$_3$N$_4$QDs) have a wide band gap that allows for efficient charge carrier separation, which is an important feature to be improved for their enhanced performance in photocatalysis applications. In this way, g-C$_3$N$_4$QDs and its composites with some metal and metal oxides are

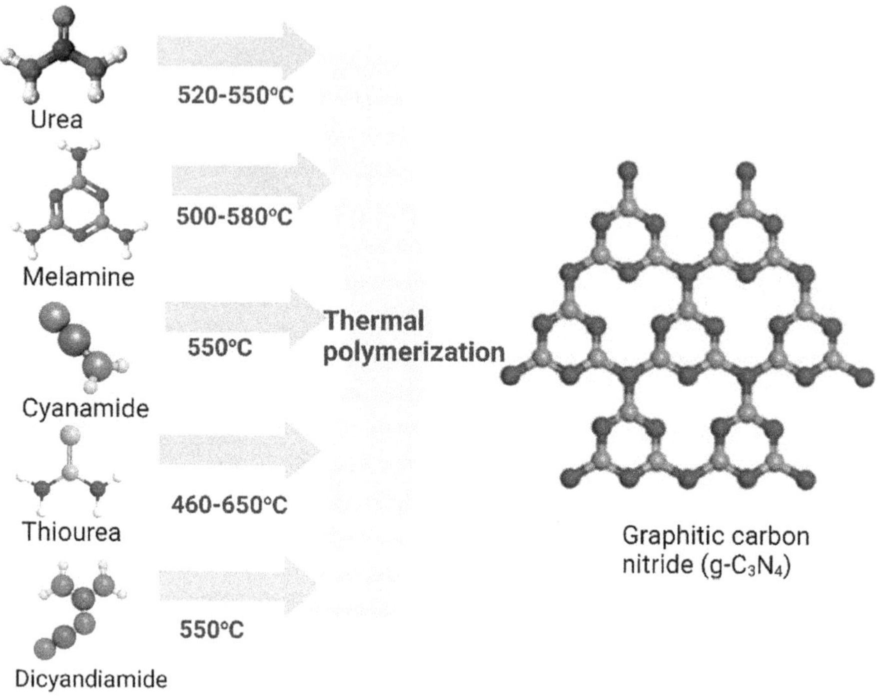

FIGURE 8.4 Schematic representation of g-C$_3$N$_4$ synthesis (Chen et al., 2024).

applicable for hydrogen production via photocatalytic water splitting and also for removing organic-contaminated water. As many other QDs like carbon-based QDs, g-C$_3$N$_4$QDs could be synthesized via top-down and bottom-up routes. The top-down route includes breaking bonding from bulk g-C$_3$N$_4$ structures upon varied intensification methods such as ultrasound irradiation, microwave irradiation, and hydrothermal treatment. Meanwhile, the bottom-up methods utilize carbon and nitrogen precursors from organic compounds under certain conditions. As a carbon precursor, some compounds including sucrose, glucose, lactic acid, citric acid, ascorbic acid, glycerol, ethylene glycol have been reported as ideal precursors for bottom-up synthesis. Furthermore, in the perspective of green chemistry, various plant extracts were used as substitutes for these compounds with low-cost and sustainable properties. Meanwhile as a nitrogen resource, various kinds of nitrogen-rich organic molecules such as N,N-dimethylformamide (DMF), melamine, formamide, guanidine, hydrochloride, urea, dicyandiamide, and organic amines have been employed (Fang et al., 2011). Figure 8.5 represents the summary of synthesis methods of g-C$_3$N$_4$ QDs. As an example, g-C$_3$N$_4$ QDs could be prepared from bulk g-C$_3$N$_4$ directly by a thermal-chemical etching process via three steps: heat etching, acidic cutting, and hydrothermal cutting (Wang et al., 2014).

g-C$_3$N$_4$ in 1D structures including nanotubes, nanochain, nanowires, or other elongated forms of g-C$_3$N$_4$ typically have rapid charge transport due to their high aspect ratios (Figure 8.3).

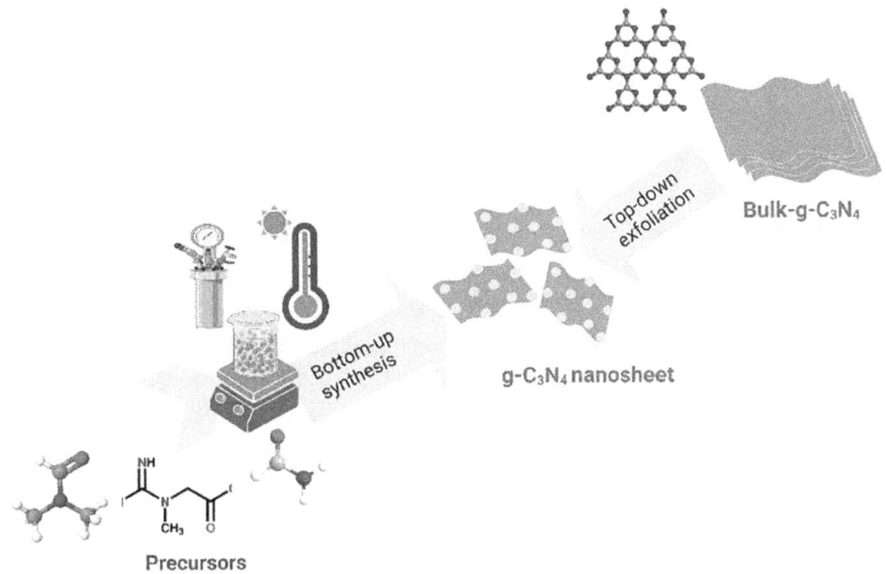

FIGURE 8.5   Synthesis methods of g-C$_3$N$_4$ QDs (Chen et al., 2024).

The unique morphology also provides efficient pathways for charge migration and separation, which is important for electrochemical and photochemical applications. Some strategies were developed to achieve high porosity of g-C$_3$N$_4$ nanotubes; one of these was utilizing 1,5-naphthalene disulfonic acid (NA) through the melamine (MA)-cyanuric acid (CA) precursor self-assembly technique (Zhang et al., 2022). Another development was template-free 1D g-C$_3$N$_4$ in nanochain form onto melamine and ethylene glycol precursors. Physicochemical characterization using photoluminescence (PL) spectroscopy revealed that g-C$_3$N$_4$ with a chain structure provides better separation efficiency of e$^-$-h pairs as represented by an intensely quenched PL spectrum, while bulk g-C$_3$N$_4$ exhibits severe charge recombination (Figure 8.6).

Among all forms of g-C$_3$N$_4$, the 2D g-C$_3$N$_4$ structure or g-C$_3$N$_4$ nanosheets is considered due to its large surface area, high crystallinity, and excellent charge carrier separation performance. Thermal oxidative etching of bulk g-C$_3$N$_4$ was representative of low-cost and simple methods for g-C$_3$N$_4$ nanosheet synthesis. The principle used in this method is the removal of the hydrogen bond from the polymeric unit; and the thickness of the structure was reduced, which leads to a blue-shift of intrinsic absorption edge in the UV-Visible spectrum compared to its bulk form. It leads to an increased band gap energy from the bulk form (2.77 eV) to 2.97 eV in nanosheets. This widening of the band gap reflects quantum confinement which raises and lowers the CB and VB edges, respectively. In addition, electron transport within the sheets was detected and attributed to the I-V curve that, on the contrary, is not represented by the bulk form.

The characteristics of each g-C$_3$N$_4$ form and morphologies could be designed by using a specific synthesis approach. Furthermore, the efficacy and efficiency

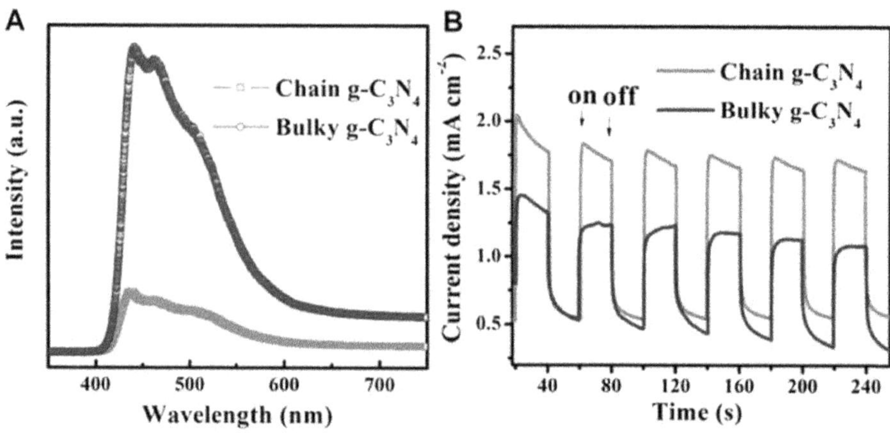

FIGURE 8.6　(a) PL spectra and (b) photocurrent spectra of bulky and chain-structured g-C$_3$N$_4$ (Zhang et al., 2021).

of g-C$_3$N$_4$ and g-C$_3$N$_4$-based nanocomposites for photocatalytic H$_2$ production by water splitting and battery application will be discussed in the next part of this chapter.

3 PHOTOCATALYTIC PROPERTIES G-C$_3$N$_4$ AND G-C$_3$N$_4$-BASED NANOCOMPOSITES

The photocatalytic activity of g-C$_3$N$_4$ is referred to its π-conjugated framework created by sp^2-hybridization of C and N in the structure. As a photocatalyst, the g-C$_3$N$_4$ photocatalyst has a CB of −1.1 eV and a VB of 1.6 eV, which is fit for splitting water to produce H$_2$ and O$_2$. Upon light exposure, absorbing the photons with the energy (hv) higher than the bandgap energy will lead to the excitation of the electrons of g-C$_3$N$_4$ from the VB to CB and leave holes in the VB. Some of the photoexcited charge carriers will be combined rapidly, and some other photogenerated electrons and holes can be transferred to the surface of g-C$_3$N$_4$ to involve the reaction. In the water splitting photoelectron-catalytic system, the photoexcited electrons have the capability to reduce water molecules and at the same time, the photoexcited holes oxidize water for generating O$_2$. For H$_2$ evolution, the CB potential of the photocatalyst should be more negative than the H$_2$ reduction potential, while the VB potential should be more positive than the water oxidation potential for the O$_2$ evolution from water. However, the efficiency of water splitting by g-C$_3$N$_4$ is still low, which is mainly due to limited electron transport along the g-C$_3$N$_4$ sheets, potency of electron-hole pairs recombination, and minimum photo-adsorption in the visible light region. The drawback of g-C$_3$N$_4$ is related to the aromatic structure of g-C$_3$N$_4$ that has delocalized π-electrons and influences the semiconductor properties. As a highly conjugated system, excited electrons are easy to recombine before participating in the reduction-oxidation system. Thus, to mitigate this condition, some modification and anchoring of the existing molecular structure of g-C$_3$N$_4$ could be performed. It consists of doping of g-C_3N_4 with non-metal atoms, surface modification, and combining with other metal oxide semiconductors.

The doping using non-metal atoms is intended to distorting the π-conjugated orbital leading to improved light absorption and efficient charge separation. By adjusting the energy levels of the CB and VB, non-metal doping can promote the generation, migration, and utilization of photogenerated charge carriers, thereby enhancing the overall photocatalytic activity of g-C_3N_4. Alteration is achieved by doping and this is related to changes to the surface and electronic properties due to the substitution of the C/N surface structure, followed by the generation of some defects in g-C_3N_4. It will lead to a better π-electron delocalization and the shifting of VB and CB edge potential (Babu et al., 2023).

For example, due to its strong electronegative nature, F doping into g-C_3N_4 influenced the electronic band structure; it particularly increased the VB and reduced the thermodynamic energy required during H_2 reduction. Besides this, the doping caused the C-F bonding to hold the maximum charge polarization to host materials, followed by the enhancement of the activity of energy-related reactions and improvement of the stability of compounds. The photocatalytic activity of F-g-CN for H_2 evolution under visible irradiation is 11.6 times higher than that of pristine g-C_3N_4 because of accelerated photogenerated charge separation. F-doping also created a mesoporous structure which leads to an increase in the specific surface area (Sun et al., 2022).

B-doping of g-C_3N_4 alters the electronic properties of carbon nitride to be more adaptive in the visible light region. The doping of g-C_3N_4 by P and Br has the potential to improve the carrier separation efficiency. The effectiveness of B-doping was represented by the improvement on the photocatalytic activity which is in line with the prolonged lifetime from 12.3 μS in g-C_3N_4 into 14.4 μS. This feature influences the generated H_2 with a production rate of 1439 and 400 mmol g^{-1} h^{-1} under simulated sunlight and visible light irradiation, respectively. These values are about 2.4 times higher than that of pure g-C_3N_4.

In general, synthesis of non-metal-doped g-C_3N_4 could be classified as (i) *in situ* doping to the precursor of g-C_3N_4 and (ii) doping into g-C_3N_4 powder. Figure 8.7 depicts the different approaches involved in the synthesis of B-g-C_3N_4.

Table 8.1 provides characteristic improvements obtained by non-metal dopants and their method of doping.

Boron (B), Fluorine (F), phosporus (P), and sulfur (S) are some non-metal dopants that have been reported to improve the photocatalytic performance of g-C_3N_4.

Different types of metal oxides, such as TiO_2, MnO, ZnO, WO_3, iron oxide, tin oxide, etc., have been reported to improve the photocatalytic efficiency of g-C_3N_4 by promoting the charge carriers' separation and reducing the electrons-holes recombination. The combination consequently affects charge carrier separation that influences the photocatalysis mechanism.

There are five types of charge carrier separation for g-C3N4–metal oxide photocatalysts:

1. Type I heterojunction,
2. Type II heterojunction,

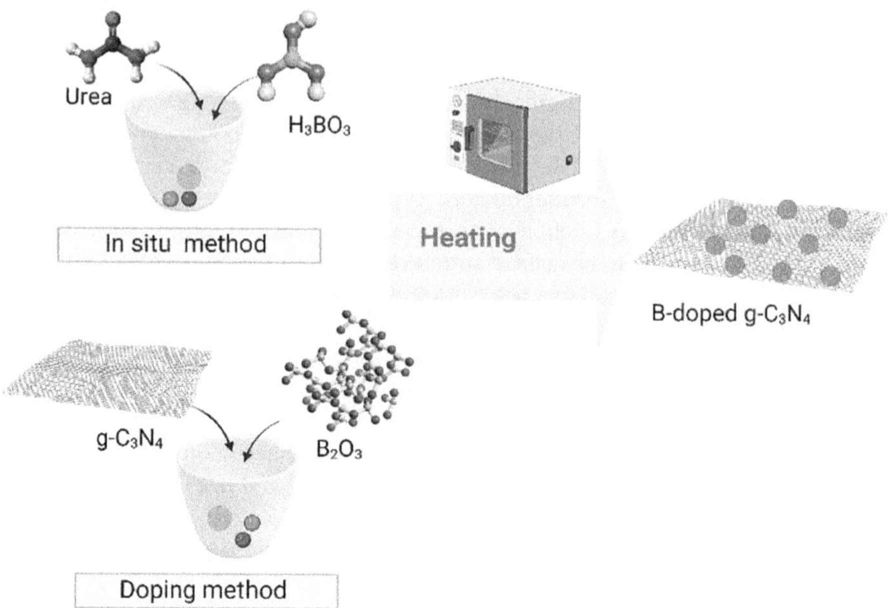

FIGURE 8.7   *In situ* synthesis and doping method for the production of B-g-C₃N₄ (Guo et al., 2019; S. Zhang et al., 2017).

3. Z-scheme heterojunction
4. p-n heterojunction,
5. Schottky junction.

Among those modes, most of g-C₃N₄/metal oxide-based photocatalysts express type II and Z-scheme mechanisms, with schematic representation depicted in Figure 8.8.

In type II heterojunctions, two semiconductors are bound to form a stable heterojunction, and the position of the VB of semiconductor A is higher than that of semiconductor B. The difference in voltages produces the hole migration from the VB of semiconductor B to that of semiconductor A (Figure 8.8). At the same time, electron transfer occurs from the conductance band (CB) of semiconductor A to that of semiconductor B. The condition facilitates electrons to stay in an excited condition for longer times or reduces the rate of the recombination. Contrarily, in the Z-scheme photocatalytic system, the generated electrons in the CB of semiconductor B will be transferred into the VB of semiconductor A and combines with the photogenerated holes. This type of heterojunction correlated with both reducing the recombination by an increase of the electrons and holes separation and improving the reduction-oxidation ability. As an example, the Z-scheme photoelectrocatalysis mechanism for water splitting by TiO₂/ g-C₃N₄ is provided in Figure 8.9.

In detail, Figure 8.9 depicts the scheme of energy in the photocatalytic system. As the absolute electronegativity of TiO₂ and g-C₃N₄ is 5.81 and 4.64 eV, and CB of g-C₃N₄ and TiO₂ is −1.235 and −0.24 eV, respectively, photogenerated electrons

TABLE 8.1

Non-Metal Dopant to g-C$_3$N$_4$ for H$_2$ Production

Dopant	Improvements	Improvements in H$_2$ Production	Synthesis Method	Reference
B	Enhancement on the photoabsorption performance of g-C$_3$N$_4$, prolonging the lifetime of photogenerated electrons, annihilation of charge carrier, increased electron migration	H$_2$ production was 2.4 times higher than that of pure g-C$_3$N$_4$.	Microwave heating of H$_3$BO$_3$, melamine, and urea	(Chen et al., 2018)
P	Enlarged surface area, induces enhanced visible light harvesting, enhanced charge carrier separation, and enhanced proton recombination	The H$_2$ evolution rate of PCN-S-3 is 2.9 times as higher as the CN.	Thermal polycondensation by calcination of g-C$_3$N$_4$ with NaH$_2$PO$_2$	(Lin et al., 2020)
F	Increasing surface area by mesopore formation, vacancy generation, and electronic structure modification.	F-doped g-C$_3$N$_4$ showed that H$_2$ evolution under visible irradiation is 11.6 times higher than that of pristine g-C$_3$N$_4$	Gas phase fluorination of g-C$_3$N$_4$	(Sun et al., 2022)
S	The enlarged specific surface area, enhanced light absorption, and improved separation efficiency of photo-induced charge carriers.	S-doped g-C$_3$N$_4$ yielded 8.4 times higher H2 than that of bulk g-C$_3$N$_4$ in visible light		(Guo et al., 2020)
Br	The enhanced optical absorption and accelerated charge mobility	The optimal activity (48 mol h^{-1}) obtained for Br-g-C$_3$N$_4$ was 2.4 times higher than that of the pure CNU sample (20 mol h^{-1}).	NH$_4$Br impregnation onto g-C$_3$N$_4$ followed by sintering at 550 °C for 2 h with a heating rate of 5.0°C min^{-1} in air	(Lan et al., 2016)

will be transferred from the CB of g-C$_3$N$_4$ to the CB of TiO$_2$ via a heterostructure in between the g-C$_3$N$_4$ and TiO$_2$ interface. Meanwhile, the VB of g-C$_3$N$_4$ and TiO$_2$ are +1.515 and +2.86 eV, respectively. It causes the transfer of photogenerated holes from the VB of TiO$_2$ (+ 2.86 eV) to the VB of g-C$_3$N$_4$ (+ 1.515).

Similar to the non-metal-doping to g-C$_3$N$_4$, metal oxide modifications were prepared by either *in situ* synthesis or metal oxide loading to g-C$_3$N$_4$ (H. Zhang et al., 2017). Figure 8.10 represents the *in situ* synthesis of TiO$_2$/g-C$_3$N$_4$ by using the sol-gel method.

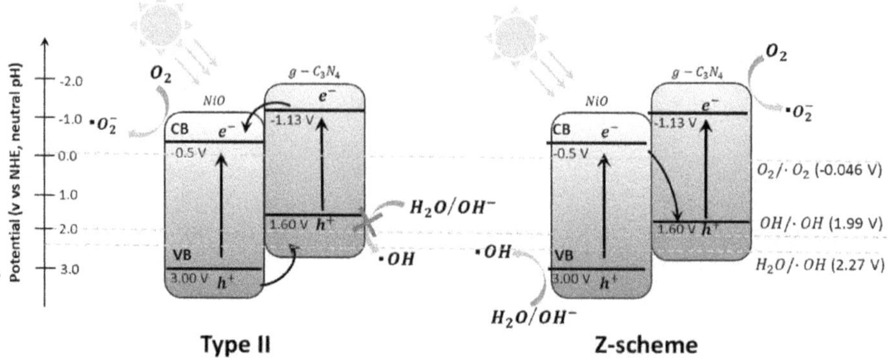

FIGURE 8.8   Type II and Z-scheme of the photocatalysis mechanism.

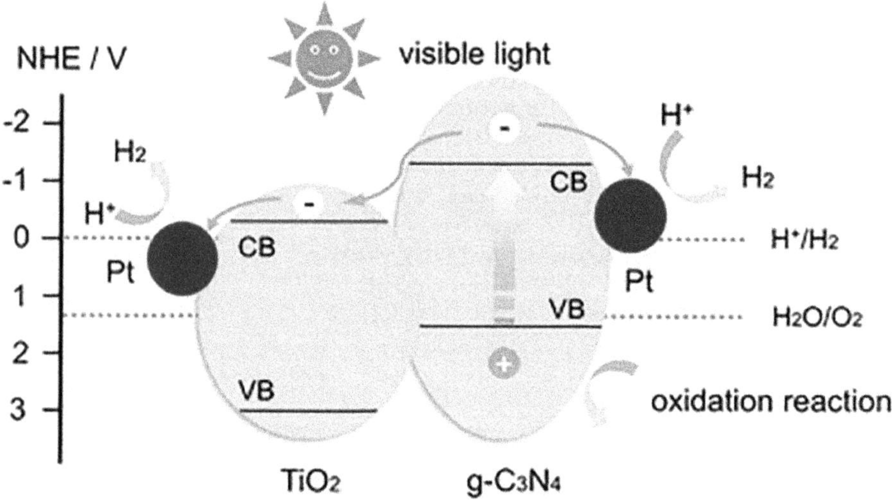

FIGURE 8.9   Schematic representation of the photocatalysis mechanism of TiO_2/g-C_3N_4 (Mary Rajaitha et al., 2020).

From physicochemical characterization using transmission electron microscopy (TEM) analysis, the homogeneous dispersion of TiO_2 onto the g-C_3N_4 nanosheet was observed. The increased absorption of visible light could be attributed to more defects in the g-C_3N_4 phase that are introduced by the TiO_2 nanoparticles; and at the same time, the dispersion influences the increasing specific surface area and porosity. By varying the amount of dispersed TiO_2, it is obtained that the increased H_2 production rate is not linearly correlated with the TiO_2 content. The average H_2 evolution rate observed to be the highest, 40 μmol h^{-1} was obtained by dispersing 200 μL–200 mg as-prepared g-C_3N_4 nanosheets sample, which is about 2.7 times higher than that of the rate expressed by g-C_3N_4. At higher TiO_2 content, an obvious drop of the H_2 evolution rate occurred, which is attributed to the decreasing thermodynamic

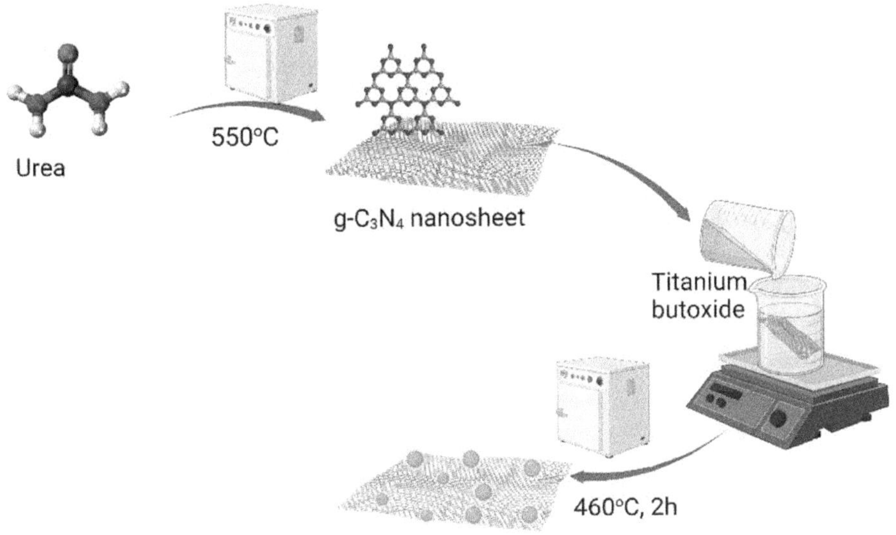

FIGURE 8.10 Schematic diagram of TiO$_2$/g-C$_3$N$_4$ synthesis by using the sol-gel method (H. Zhang et al., 2017).

driving force for H$_2$ evolution caused by the lowered CB of the composite catalyst with more TiO$_2$ (H. Zhang et al., 2017).

The preparation by TiO$_2$ loading onto g-C$_3$N$_4$ was reported by mixing TiO$_2$ or TiOSO$_4$ as the TiO$_2$ precursor with g-C$_3$N$_4$ powder followed by annealing (Alcudia-Ramos et al., 2020; Girish et al., 2022). The increasing band gap energy was relevant with the increasing H$_2$ production of 1042 μmol/g.h, which is 6.42 times higher compared with the H$_2$ production with pure g-C$_3$N$_4$ (Alcudia-Ramos et al., 2020).

By some investigations it can be observed that it is not only charge separation and ion transport properties that influence the effectiveness of the photocatalyst, but other physicochemical characters such as specific surface area affects the performance of the g-C$_3$N$_4$-based composite significantly. As an example, the photocatalytic activity of dispersed Pt onto g-C$_3$N$_4$ for H$_2$ production was determined by the varied synthesis methods; chemical reduction and photo reduction. The dispersion of Pt in the composite provided by chemical reduction increases porosity and specific surface area, and it leads to a remarkable increase in the H$_2$ production rate. It was noted that the H$_2$ production rate by Pt/g-C$_3$N$_4$-nanosheet prepared by chemical reduction was 7862.5 μmol g^{-1} h^{-1}, which is 6.92 times higher than that of the Pt/g-C$_3$N$_4$-nanosheet prepared by photoreduction (1136.8 μmol g^{-1} h^{-1}) (Zhou et al., 2019).

In order to enhance H$_2$ production rate, advanced modifications to g-C$_3$N$_4$ were innovatively conducted by the combination of two metal oxides, metal and metal oxide, and also metal/metal oxide with carbon or graphene. Table 8.2 lists modified g-C$_3$N$_4$ with metal or metal oxides and their H$_2$ production for applications in photocatalytic and photoelectrocatalytic systems.

TABLE 8.2

Some Studies on g-C$_3$N$_4$ Modification with Metal or Metal Oxides and Their H$_2$ Production for Applications in Photocatalytic and Photoelectrocatalytic Systems

Material	H$_2$ Production	Remark	Reference
NiO/ g-C$_3$N$_4$	The H$_2$ production rate was 766 µmol/h/g.	A 127 and 109-fold higher H$_2$ production by a photocatalytic system was shown by NiO/ g-C$_3$N$_4$ compared to NiO and g-C$_3$N$_4$, respectively.	(Rugma et al., 2021)
MoS$_2$/ Ni@ NiO/ g-C$_3$N$_4$	The hydrogen production rate by the photocatalysis system was 7.98 mmol g^{-1} h^{-1}.	The apparent quantum yield of MoS$_2$/Ni@NiO/g-C$_3$N$_4$ was calculated, which was 67.25%.	(Zhang et al., 2024)
SnO$_2$/ g-C$_3$N$_4$	The optimized composite of SnO$_2$/g-C$_3$N$_4$ produced H$_2$ at the rate of 132 µmol/h.		(Zada et al., 2020)
0D/2D SnO$_2$/ g-C3N4	SnO$_2$/g–C$_3$N$_4$ sample is 1389.2 µmol h^{-1} g^{-1}, which is 6.06 times higher than that of g-C3N4 (230.8 µmol h^{-1} g^{-1}).	The reaction system enabled photocatalytic H2 production under visible light.	(Wu et al., 2020)
g-C$_3$N$_4$/ ZnO/Au	The g-C$_3$N$_4$/ZnO/Au composite expressed an optimal photocatalytic H$_2$ production rate of 46.46 µmol·g^{-1}·h^{-1}, representing 25.2-fold and 30.4-fold enhancements compared with a g-C$_3$N$_4$/ZnO composite and bare g-C$_3$N$_4$, respectively.	The reaction system enabled photocatalytic H2 production under visible light.	(Ge et al., 2023)
SnO$_2$– ZnO QDs/ g-C3N4	SnO$_2$–ZnO QDs/g-C$_3$N$_4$ hybrid demonstrated a high rate of hydrogen production (13 673.61 µmol g^{-1})	The reaction system enabled photocatalytic H$_2$ production under visible light. H$_2$ production rates were 1.06 and 2.27 times higher than that of the binary ZnO/ g-C3N4 hybrid (12 785.54 µmol g−1) and pristine g-C3N4 photocatalyst (6017.72 µmol g−1).	(Vattikuti et al., 2018)
Pt/g-C$_3$N$_4$	The H$_2$ production rate by Pt/g-C$_3$N$_4$-nanosheet prepared by chemical reduction was 7862.5 µmol g^{-1} h^{-1}.	The H$_2$ production rate was 6.92 times higher than that of the Pt/g-C$_3$N$_4$-nanosheet prepared by photoreduction (1136.8 µmol g^{-1} h^{-1})	(Zhou et al., 2019)
MoS$_2$/ Fe$_2$O$_3$/ g-C$_3$N$_4$	MoS$_2$/Fe$_2$O$_3$/g-C$_3$N$_4$ composite catalyst had a higher hydrogen production rate than the single-component catalyst g-C$_3$N$_4$, which was 7.82 mmol g^{-1} h^{-1}.	The H$_2$ production rate was about five times higher than the catalyst g-C$_3$N$_4$ (1.56 mmol g^{-1} h^{-1}).	(Zhang et al., 2022)

(Continued)

TABLE 8.2
(Continued)

Material	H$_2$ Production	Remark	Reference
MnO/ g-C$_3$N$_4$	The MnO/ g-C$_3$N$_4$ with 5 wt% MnO shows a superior photocatalytic activity with a hydrogen evolution rate of 559 µmol h^{-1} g^{-1} under visible light.	The H$_2$ production rate was about nine times as that of the bulk g-C$_3$N$_4$.	(Mao et al., 2019)
CdSe QDs/ g-C$_3$N$_4$	Maximum H$_2$ production rate reached by the composite loaded with 13.6 wt % CdSe QDs was 615 µmol·g-1 h-1.	The H$_2$ production rate was approximately 76 times greater than that of pristine g-C$_3$N$_4$ and twice that of CdSe QDs.	(Zhong et al., 2018)

4 G-C$_3$N$_4$ FOR BATTERY APPLICATION

Lithium-ion batteries (LIBs) have recently gained considerable research interest as one of the most promising clean energy sources for many electronic devices owing to their low maintenance cost, low self-discharge rate, high Coulombic efficiency, high energy density, and long cycling life. The recent emergence of high-energy demanding applications requires the electrochemical performance and cost-effectiveness of LIBs. This could be gained by developing low-cost electrode materials with high reversible capacity, cycling stability, and rate capability (Ould Ely et al., 2019). The layered structure of 2D-g-C$_3$N$_4$ is a structural analogue of graphene and received great interest in the field of electrochemical energy storage, in particular, for battery applications. Pure g-C$_3$N$_4$ has been proven as a potential LIB electrode. However, due to some drawbacks that still exist in practical applications, efforts and innovative studies have been directed toward optimizing g-C$_3$N$_4$-based materials. As an N-substituted graphite material with a quite high N doping level, g-C$_3$N$_4$ is an ideal anode material for LIBs in theory. In addition, compared with graphite, the more open structure of g-C$_3$N$_4$ may lead to improved rates of Li$^+$ exchange. However, pure g-C$_3$N$_4$ exhibits poor conductivity and low reversible capacity. Pure g-C$_3$N$_4$ electrode in LIBs showed a relatively low initial discharge capacity at 134.9 mA h g^1 and an irreversible capacity loss up to 90% after only seven cycles (Luo et al., 2019). These parameters were improved by creating porosity and specific morphology of g-C$_3$N$_4$. The porous g-C$_3$N$_4$ nanotubes provide uniform size of pores in the wall and doubly bonded nitrogen at the edges of the pore, which are crucial to Li storage (Pan, 2014).

Metal-doped or metal-supported g-C$_3$N$_4$ is also the strategy for improving cycling performance. Figure 8.11 represents the scheme of lithiation to Sn/g-C$_3$N$_4$.

The porosity of g-C$_3$N$_4$ provides ample opportunities for lithium atoms to penetrate both its interior and exterior, leading to a notable intercalation capacity. Compared to other nanotubes, the energy barrier for lithium atoms to ingress the interior of g-C$_3$N$_4$ nanotubes is significantly lower due to this porous configuration. Initially, as the density of lithium increases, the intercalation energy rises; however,

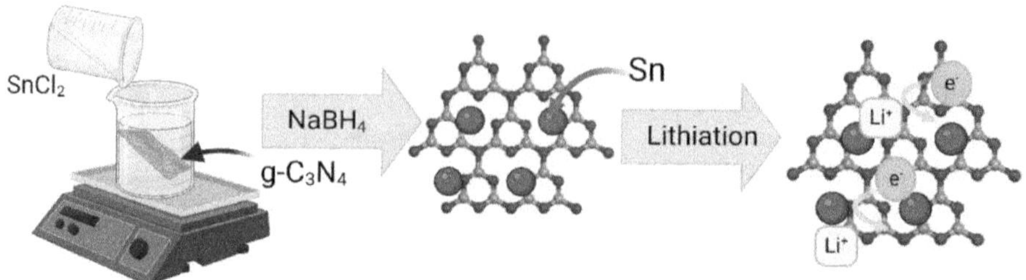

FIGURE 8.11   Schematic representation of the lithiation method to synthesize Sn/g-C$_3$N$_4$ (Pan, 2014).

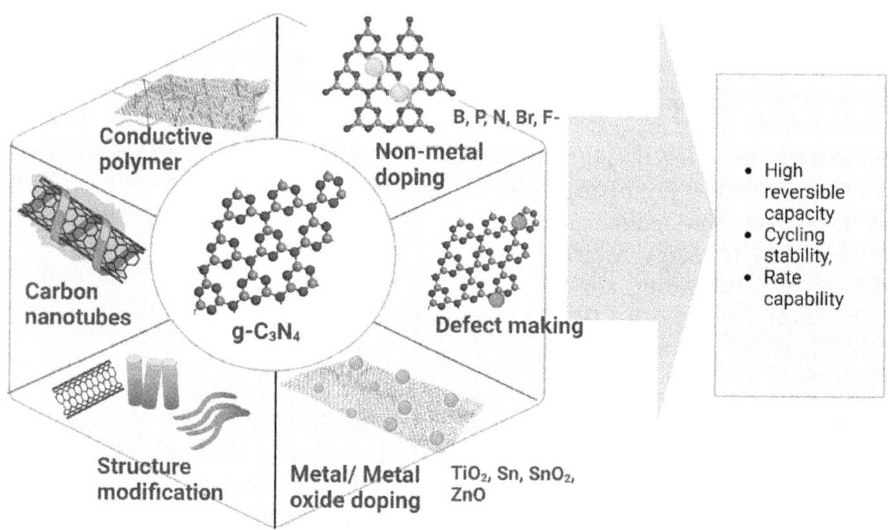

FIGURE 8.12   Scheme of some possible modifications to g-C$_3$N$_4$ for battery applications.

it eventually stabilizes at -0.23 eV at a lithium density of 0.57, with the intercalation process being exothermic. Furthermore, the specific energy rises as lithium density increases and is yet to reach its peak even at a density of 0.57. These computational findings suggest that g-C$_3$N$_4$ nanotubes hold considerable promise as electrode materials for lithium batteries, offering enhanced performance, heightened stability, and rapid response times.

Another procedure to enhance Li$^+$ transport in the electrode system is the use of a solid polymer electrolyte. The interaction of the litium salt with the surface of g-C$_3$N$_4$ promotes the dissociation of the lithium salt that leads to an effective ion transport network in the composite electrolyte (Sun et al., 2019). Figure 8.12 shows some possible modifications to the g-C$_3$N$_4$ structure for battery applications.

5 FUTURE PERSPECTIVES

Based on the studies on the synthesis and modifications of g-C$_3$N$_4$ for H$_2$ production and battery applications, the development of g-C$_3$N$_4$-based material is still demanding. The detailed future challenges to overcome are:

1. Some strategies for the enhancement mechanisms of hydrogen production should be devised by considering some aspects that are low-cost and effective.
2. Not only lifetime, stability, and performance of materials should be taken into consideration, but also the safety procedures and chemicals need to be evaluated.

REFERENCES

Alcudia-Ramos, M.A., Fuentez-Torres, M.O., Ortiz-Chi, F., Espinosa-González, C.G., Hernández-Como, N., García-Zaleta, D.S., Kesarla, M.K., Torres-Torres, J.G., Collins-Martínez, V., Godavarthi, S., 2020. Fabrication of g-C3N4/TiO2 heterojunction composite for enhanced photocatalytic hydrogen production. *Ceram. Int.* 46, 38–45. https://doi.org/10.1016/j.ceramint.2019.08.228

Babu, P., Park, H., Park, J.Y., 2023. Surface chemistry of graphitic carbon nitride: doping and plasmonic effect, and photocatalytic applications. *Surf. Sci. Technol.* 1, 1–50. https://doi.org/10.1007/s44251-023-00026-1

Chen, L., Maigbay, M.A., Li, M., Qiu, X., 2024. Synthesis and modification strategies of g-C3N4 nanosheets for photocatalytic applications. *Adv. Powder Mater.* 3, 100150. https://doi.org/10.1016/j.apmate.2023.100150

Chen, P., Xing, P., Chen, Z., Lin, H., He, Y., 2018. Rapid and energy-efficient preparation of boron doped g-C3N4 with excellent performance in photocatalytic H2-evolution. *Int. J. Hydrogen Energy* 43, 19984–19989. https://doi.org/10.1016/j.ijhydene.2018.09.078

Fang, L., Ohfuji, H., Shinmei, T., Irifune, T., 2011. Experimental study on the stability of graphitic C3N4 under high pressure and high temperature. *Diam. Relat. Mater.* 20, 819–825. https://doi.org/10.1016/j.diamond.2011.03.034

Ge, W., Liu, K., Deng, S., Yang, P., A, L.S., 2023. Z-scheme g-C3N4/ZnO heterojunction decorated by Au nanoparticles for enhanced photocatalytic hydrogen production. *Appl. Surf. Sci.* 607, 155036.

Girish, Y.R., Udayabhanu, Alnaggar, G., Hezam, A., Nayan, M.B., Nagaraju, G., Byrappa, K., 2022. Facile and rapid synthesis of solar-driven TiO2/g-C3N4 heterostructure photocatalysts for enhanced photocatalytic activity. *J. Sci. Adv. Mater. Devices* 7, 100419. https://doi.org/10.1016/j.jsamd.2022.100419

Guo, H., Shu, Z., Chen, D., Tan, Y., Zhou, J., Meng, F., Li, T., 2020. One-step synthesis of S-doped g-C3N4 nanosheets for improved visible-light photocatalytic hydrogen evolution. *Chem. Phys.* 533, 110714. https://doi.org/10.1016/j.chemphys.2020.110714

Guo, X., Rao, L., Wang, P., Zhang, L., Wang, Y., 2019. Synthesis of porous boron-doped carbon nitride: Adsorption capacity and photo-regeneration properties. *Int. J. Environ. Res. Public Health* 16, 581. https://doi.org/10.3390/ijerph16040581

Hosseini, S.E., Wahid, M.A., 2016. Hydrogen production from renewable and sustainable energy resources: Promising green energy carrier for clean development. *Renew. Sustain. Energy Rev.* 57, 850–866. https://doi.org/10.1016/j.rser.2015.12.112

Lan, Z.A., Zhang, G., Wang, X., 2016. A facile synthesis of Br-modified g-C3N4 semiconductors for photoredox water splitting. *Appl. Catal. B Environ.* 192, 116–125. https://doi.org/10.1016/j.apcatb.2016.03.062

Lin, Q., Li, Z., Lin, T., Li, B., Liao, X., Yu, H., Yu, C., 2020. Controlled preparation of P-doped g-C3N4 nanosheets for efficient photocatalytic hydrogen production. *Chinese J. Chem. Eng.* 28, 2677–2688. https://doi.org/10.1016/j.cjche.2020.06.037

Luo, Y., Yan, Y., Zheng, S., Xue, H., Pang, H., 2019. Graphitic carbon nitride based materials for electrochemical energy storage. *J. Mater. Chem. A* 7, 901–924. https://doi.org/10.1039/c8ta08464e

Mao, N., Gao, X., Zhang, C., Shu, C., Ma, W., Wang, F., Jiang, J.X., 2019. Enhanced photocatalytic activity of g-C3N4/MnO composites for hydrogen evolution under visible light. *Dalt. Trans.* 48, 14864–14872. https://doi.org/10.1039/c9dt02748c

Mary Rajaitha, P., Shamsa, K., Murugan, C., Bhojanaa, K.B., Ravichandran, S., Jothivenkatachalam, K., Pandikumar, A., 2020. Graphitic carbon nitride nanoplatelets incorporated titania based type-II heterostructure and its enhanced performance in photoelectrocatalytic water splitting. *SN Appl. Sci.* 2, 1–14. https://doi.org/10.1007/s42452-020-2190-9

Ould Ely, T., Kamzabek, D., Chakraborty, D., 2019. Batteries safety: Recent Progress and current challenges. *Front. Energy Res.* 7, 71. https://doi.org/10.3389/fenrg.2019.00071

Pan, H., 2014. Graphitic carbon nitride nanotubes As Li-ion battery materials: A first-principles study. *J. Phys. Chem. C* 118, 9318–9323. https://doi.org/10.1021/jp4122722

Rugma, T.P., Watts, A., Vijayarajan, V.S., Lakhera, S.K., Neppolian, B., 2021. Synergistic hydrogen evolution activity of NiO/g-C3N4 photocatalysts under direct solar light irradiation. *Mater. Lett.* 302, 130292. https://doi.org/10.1016/j.matlet.2021.130292

Sun, L., Li, Y., Feng, W., 2022. Gas-phase fluorination of g-C3N4 for enhanced photocatalytic hydrogen evolution. *Nanomaterials* 12, 1–13. https://doi.org/10.3390/nano12010037

Sun, Z., Li, Y., Zhang, S., Shi, L., Wu, H., Bu, H., Ding, S., 2019. G-C3N4 nanosheets enhanced solid polymer electrolytes with excellent electrochemical performance, mechanical properties, and thermal stability. *J. Mater. Chem. A* 7, 11069–11076. https://doi.org/10.1039/c9ta00634f

Vattikuti, S.V.P., Reddy, P.A.K., Shim, J., Byon, C., 2018. Visible-light-driven photocatalytic activity of SnO2-ZnO quantum dots anchored on g-C3N4 nanosheets for photocatalytic pollutant degradation and H2 production. *ACS Omega* 3, 7587–7602. https://doi.org/10.1021/acsomega.8b00471

Wang, W., Yu, J.C., Shen, Z., Chan, D.K.L., Gu, T., 2014. G-C3N4 quantum dots: Direct synthesis, upconversion properties and photocatalytic application. *Chem. Commun.* 50, 10148–10150. https://doi.org/10.1039/c4cc02543a

Wu, H., Yu, S., Wang, Y., Han, J., Wang, L., Song, N., Dong, H., Li, C., 2020. A facile one-step strategy to construct 0D/2D SnO2/g-C3N4 heterojunction photocatalyst for high-efficiency hydrogen production performance from water splitting. *Int. J. Hydrogen Energy* 45, 30142–30152. https://doi.org/10.1016/j.ijhydene.2020.08.112

Zada, A., Khan, M., Qureshi, M.N., Liu, S.Y., Wang, R., 2020. Accelerating photocatalytic hydrogen production and pollutant degradation by functionalizing g-C3N4 With SnO2. *Front. Chem.* 7, 1–8. https://doi.org/10.3389/fchem.2019.00941

Zhang, C., Zheng, C., Cao, X., 2024. Preparation of MoS2/Ni@NiO/g-C3N4 composite catalyst and its photocatalytic performance for hydrogen production. *Int. J. Hydrogen Energy* 52, 1078–1086.

Zhang, H., Liu, F., Wu, H., Cao, X., Sun, J., Lei, W., 2017. In situ synthesis of g-C3N4/TiO2 heterostructures with enhanced photocatalytic hydrogen evolution under visible light. *RSC Adv.* 7, 40327–40333. https://doi.org/10.1039/c7ra06786k

Zhang, M., Sun, Y., Chang, X., Zhang, P., 2021. Template-free synthesis of one-dimensional g-c3n4 chain nanostructures for efficient photocatalytic hydrogen evolution. *Front. Chem.* 9, 1–7. https://doi.org/10.3389/fchem.2021.652762

Zhang, S., Gao, L., Fan, D., Lv, X., Li, Y., Yan, Z., 2017. Synthesis of boron-doped g-C3N4 with enhanced electro-catalytic activity and stability. *Chem. Phys. Lett.* 672, 26–30. https://doi.org/10.1016/j.cplett.2017.01.046

Zhang, Y., Wan, J., Zhang, C., Cao, X., 2022. MoS2 and Fe2O3 co-modify g-C3N4 to improve the performance of photocatalytic hydrogen production. *Sci. Rep.* 12, 1–12. https://doi.org/10.1038/s41598-022-07126-2

Zhong, Y., Chen, W., Yu, S., Xie, Z., Wei, S., Zhou, Y., 2018. CdSe quantum dots/g-C3N4 heterostructure for efficient H2 production under visible light irradiation. *ACS Omega* 3, 17762–17769. https://doi.org/10.1021/acsomega.8b02585

Zhou, X., Li, Y., Xing, Y., Li, J., Jiang, X., 2019. Effects of the preparation method of Pt/g-C3N4 photocatalysts on their efficiency for visible-light hydrogen production. *Dalt. Trans.* 48, 15068–15073. https://doi.org/10.1039/c9dt02938a

9 MXenes-Based Energy Storage Applications
Low-Dimensional Structural Design and Functional Expansion

Xiangfeng Shu

1 INTRODUCTION

In recent years, the materials research community has undertaken extensive investigations into two-dimensional (2D) materials characterized by high aspect ratio structures, enlarged lateral dimensions, and atomic/molecular thinness. Notable examples of functional structural materials in this category include graphene [1, 2], transition metal dichalcogenides (TMD) [3], metal organic frameworks (MOFs) [4], silicene [5], phosphorene [6], among others. Atomically thin 2D materials have garnered significant attention for diverse applications spanning energy storage, environmental solutions, and optoelectronics. A prominent group within this realm is the emerging family of 2D layered structures known as transition metal carbides, nitrides, and carbonitrides, collectively referred to as MXenes. Typically, MXenes are fabricated through the selective removal of A layers (comprising group IIIA or IVA elements) from their layered precursors (MAX phases). The general formula for MXene is denoted as $M_{n+1}X_nT_z$ (where $n=1$, 2, or 3), with M representing a transition metal element, X representing carbon and/or nitrogen constituents, and T_z symbolizing surface termination groups such as $-O$, $-OH$, and/or $-F$. Owing to their versatile surface chemistry and customizable composition, MXenes have garnered increasing interest in energy storage domains owing to their exceptional mechanical robustness, high electrical conductivity, superior ion adsorption capabilities, and distinctive topological attributes [7–9].

Nevertheless, the phenomenon of self-accumulation arises as an unavoidable consequence of the van der Waals forces and hydrogen bonding interactions that transpire between adjoining 2D MXene nanosheets, impeding the ingress of ions into the redox sites and consequently constraining the electrochemical efficacy. To mitigate these challenges, a viable approach involves augmenting the interlayer distance through the incorporation of spacers (such as cations, conductive polymers, and so on) to enhance carrier mobility and catalyze electrochemical transformations.

DOI: 10.1201/9781003561262-9

Arising from the intrinsic hydrophilicity of surface-terminated $Ti_3C_2T_x$-MXene, aqueous cations undergo spontaneous insertion into the interlayer interstices. As elucidated by Kurra et al. [10], the proton diffusion coefficient across the 2D slit was determined to be $2.2\times10^{-8}\pm0.4\times10^{-8}$ cm^2 s^{-1}. Notably, their investigation revealed that protons traverse the $Ti_3C_2T_x$ slit approximately three-fold faster than hydrated metal cations (e.g., Li, Na, and Ca++2) and display pronounced potential-dependent characteristics. Consequently, the rate capability of the $Ti_3C_2T_x$-MXene electrode is primarily dictated by ion transport within the 2D slit rather than electron conveyance. Simultaneously, the surface moieties –O, –OH, and/or –F inherently furnish a plethora of active sites, thereby offering substantial potential for surface functionalization and efficient loading of active constituents. Conversely, the ultrathinness of 2D MXenes, with a singular atom or few atomic layers thickness, engenders a plethora of distinctive physicochemical attributes concomitant with metallic conductivity (ranging between 6000 and 8000 S cm^{-1}) facilitated by transition metal carbides/nitrogen. This inherent property endows 2D MXenes with the capacity to conceive functional architectures and devices via diverse methodologies, encompassing the fabrication of thin films endowed with tailored microstructures and adjustable nanochannels.

This review presents a timely and insightful appraisal of the interplay between the synthesis, structural elaboration, and applications of MXenes in the realm of energy storage (refer to Figure 9.1). In its initial segment, a succinct overview is

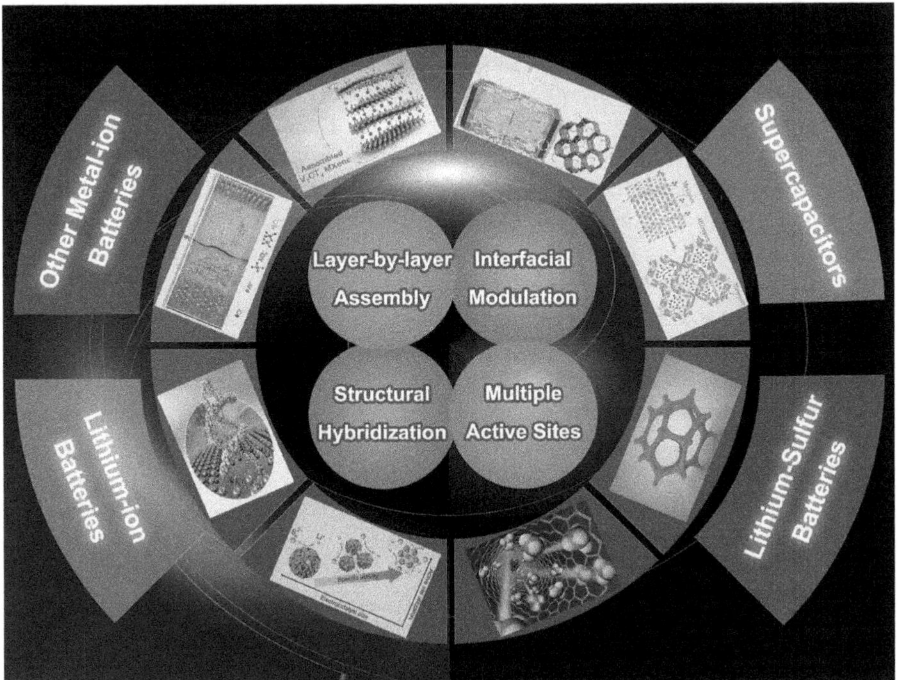

FIGURE 9.1 Structure and applications of MXenes in the realm of energy storage.

extended on the synthesis, layering, and innate characteristics of MXene. Following this, diverse fabrication methodologies for tailoring distinct MXene variants are delineated, encompassing interlayer self-assembly techniques, interfacial modifications, structural amalgamation, and the incorporation of multiple active sites. Subsequent sections delve into the utilization of MXenes in rechargeable batteries and supercapacitors (SCs), highlighting the recent advancements in the development of MXene-centric configurations through the lens of structural design. The critical role of composition and architecture in modulating the pertinent energy storage performance is underscored. Concluding remarks furnish insights into challenges, emerging trends, and prospective avenues within the domain of MXene-based frameworks for energy storage applications. Anticipated outcomes include providing valuable direction for the innovation of advanced MXene structures tailored for energy storage and associated fields.

2 MICROSTRUCTURE AND NANOCHANNEL DESIGN

2.1 INTERLAYER SELF-ASSEMBLY

In the domain of energy storage electrodes, the intricate nanochannels inherent to MXenes serve as pivotal conduits for the dynamic diffusion of ions and the acquisition of the electrolyte. Typically, MXene layers adopt a stacked configuration, wherein the densely packed self-assembled structure constrains the available space within nanochannels, thereby diminishing channel density and impeding ion dynamics diffusion, consequently limiting electrolyte ion acquisition. To circumvent this challenge, one approach involves the manipulation of interlayer van der Waals forces via atomic doping or group modifications. For instance, Jiang et al. harnessed the cooperative effects of covalent and hydrogen bonds between polyvinyl alcohol (PVA) and large-sized MXene (LMX) to fabricate stretchable PVA/LMX films characterized by notable mechanical strength (~280 MPa), electrical conductivity (~6240 S cm^{-1}), and energy storage capacity (~364 F g^{-1}) [11]. Alternatively, an effective strategy entails the introduction of interlayer spacers amidst MXene layers to expand the physical separation, thereby attenuating the influence of van der Waals forces. Illustrative instances encompass the utilization of ions/molecules, quantum dots (QDs), nanoparticles [12], as well as 1D nanotubes or nanowires, and 2D nanosheets. Yang et al. assembled a novel approach utilizing nanoconfined water-induced basal-plane alignment, along with covalent and π-π interplatelet bridging, was employed to fabricate Ti$_3$C$_2$T$_x$-MXene-bridged graphene sheets at room temperature. The resulting hybrid sheets exhibited remarkable isotropic in-plane tensile strength of 1.87 Gpa and moduli of 98.7 Gpa Moreover, the in-plane room temperature electrical conductivity was measured at 1423 S cm^{-1}, while the volumetric specific capacity reached 828 C cm^{-3}. These findings suggest that nanoconfined water-induced alignment can serve as a promising strategy for the fabrication of aligned macroscopic assemblies of 2D nanoplatelets with enhanced properties [13]. Moreover, the inclusion of ethanol in the etching phase enhances the interlayer separation of MXene, consequently streamlining the lift-off procedure [14].

2.2 3D Microstructure Design

Flat configurations comprising horizontally oriented MXene flakes and/or interlayer spacers are commonly characterized by elevated electrode densities; however, they often exhibit restricted high-speed charge transfer capabilities and ion diffusion to redox-active sites. Addressing these challenges necessitates a shift from planar MXene film architectures toward interconnected frameworks featuring open and porous structures. The customized porosity integrated within the cross-linked scaffold functions as an ion reservoir, thereby promoting the facilitated migration of electrolyte ions and mitigating electrode material deformations, thus enabling proficient ion mobility within the conduit.

2.3 In-Plane Microchannel Design

In contrast to interlayer nanochannels and engineered 3D nanochannels, the in-plane nanochannels present within MXene sheets offer a greater abundance of accessible surface sites and demonstrate heightened efficiency in ion/guest transportation. Sikdar et al. present a straightforward strategy to engineer robust 3D freestanding MXene ($Ti_3C_2T_x$) hydrogels with hierarchically porous structures. The tetraamminezinc (II) complex cation ($[Zn(NH_3)_4]^{2+}$) is selected to electrostatically assemble colloidal MXene nanosheets into a 3D interconnected hydrogel framework, followed by a mild oxidative acid-etching process to create nanoholes on the MXene surface. These hierarchically porous, conductive hole-MXene frameworks facilitate 3D transport of both electrons and electrolyte ions [15]. In the pristine layered configuration of MXene, the diffusion of electrolytic ions confronts constraints resulting from the restricted interlayer spacing, contrasting with the enhanced suitability for in-plane rapid ion migration and access to more reactive sites facilitated by the porosity amid curved layers. Further enhancement of ion transmission is achievable through in-plane pore manipulation. Overall, the dimensions and arrangement of nanochannels exhibit a pivotal influence in delineating the propagation distance, governing the accessibility of active sites within MXenes, and supporting electrochemical energy storage and ion conveyance. Moreover, the integration of guest materials introduces modifications to the energy storage mechanism by altering the geometric and chemical properties of MXenes. Leveraging the regulatory effects on transport enables the devising of advanced MXene architectures to elevate electrochemical efficacy. Despite noteworthy strides in the design of microstructures and nanochannels of MXenes, pivotal considerations should be taken into account in the pursuit of high-performance MXenes: (i) judicious selection of appropriate active materials for hybridization with MXene to formulate well-integrated electrodes capable of concurrently achieving remarkable capacity and preserving exceptional flexibility. (ii) Establishment of viable batch fabrication techniques for MXenes to facilitate scalable synthesis and practical implementation. (iii) Thoughtful application design for MXenes to fully harness their inherent capabilities.

The amalgamation of two-dimensional MXene with dimensionally pertinent physicochemical attributes, coupled with its inherent high ionic/electronic conductivity, hydrophilic nature, and facile processability, has sparked a surge in exploration toward the development of MXene-based materials for energy storage purposes. Notably, tailored MXene structures have showcased remarkable prowess in ion/charge storage capabilities and efficacious mitigation of lithium polysulfide (LiPS) shuttle effects. Subsequent sections delineate recent endeavors aimed at crafting bespoke MXene architectures endowed with desirable traits conducive to energy storage applications, spanning metal-ion batteries (MIBs), lithium-sulfur batteries, and SCs.

2.4 METAL-ION BATTERIES

The proliferation of lithium-ion batteries (LIB) within commercial applications stands out as a prominent strategy in addressing the escalating demand for portable electronic devices and electric vehicles. Regrettably, the utilization of graphite as a conventional anode material in LIB systems presents constraints due to its constrained specific capacity (372 mAhg^{-1}) and suboptimal rate capabilities. Historically, an array of alternative anode materials—including MXenes, Si, Sn, TMO, and TMDs—have been explored to enhance LIB performance. Nonetheless, the extensive deployment of these materials encounters multifaceted challenges, prominently characterized by substantial irreversible capacity, inadequate electronic conductivity, and pronounced volume variations during the Li intercalation/deintercalation process. The design of forthcoming materials and architectures necessitates a keen emphasis on augmenting electrical conductivity, fortifying mechanical integrity, and fostering expedited lithium diffusion rates.

Given the abundant natural reserves and cost-effectiveness of sodium and potassium reservoirs, sodium-ion batteries (SIBs) and potassium-ion batteries (PIBs) are being contemplated as promising alternatives to lithium for large-scale energy storage needs. Furthermore, the exploration of rechargeable multivalent ion batteries encompassing magnesium, calcium, aluminum, and zinc variants has garnered significant interest within the research community [16]. Nevertheless, owing to their substantial ionic dimensions, the quest for suitable host materials capable of facilitating the reversible and enduring storage of these alkali metal ions poses a formidable challenge. This segment provides a concise overview of MXenes and their derivative compounds utilized in metal-ion batteries, spanning from anode active materials, matrix components, to current collectors (refer to Table 9.1), elucidating the diverse roles of MXenes within electrodes.

TABLE 9.1

Summary of MXenes for MIBs

Materials	Type	Role	Electrochemical Performance	Ref.
Ti$_3$C$_2$T$_x$	LIB	Active material	163.5 mAh g^{-1}	[17]
Nb$_2$CT$_x$-MXene	LIB	Active material	382 mAh g^{-1}	[18]
3D Hollow Ti$_3$C$_2$T$_x$ Tubes	LIB	Active material	633.2 mAh g^{-1}	[19]
CNF/MXene@Zn	ZIB	Active material	357 mAh g^{-1}	[16]
CoSe$_2$/MXene	AIB	Active material	197 mAh g^{-1}	[20]
Nb$_2$CSe$_2$-MXene/CNTs	LIB	Active material	270 mAh g^{-1} (10 C)	[21]
Cu@Na-MX@Sn	ZMB	Current collector	5000 mAh cm^{-2}	[22]

2.4.1 Active Materials

The functional groups on the surface of MXenes exert a significant influence on their electrochemical characteristics concerning ion storage metrics such as conductivity, ion mobility, and capacity. Nevertheless, pristine MXene nanosheets lacking surface functionalities tend to exhibit enhanced metal capacity. Driven by insights garnered from theoretical predictions and empirical investigations, MXenes have garnered considerable interest as promising candidates for rechargeable MIB anode components. Moreover, multivalent metal ions possessing additional charges boast superior theoretical capacities compared to monovalent counterparts. Notably, recent advancements have seen the pre-embedding of high-valent cations like Mg^{2+} and Al^{3+} within the interlayers of $Ti_3C_2T_x$, facilitating the creation of self-supporting thin film electrodes characterized by expanded ion transport pathways [17]. The incorporation of these cations enhances Coulomb interactions, thereby fortifying lithium transport conduits throughout the lithiation/charging processes.

The surface terminations of MXenes are pivotal in governing their electrochemical characteristics pertinent to ion storage, encompassing attributes like conductivity, ion mobility, and storage capacity. Notably, pristine MXene nanosheets devoid of surface functional groups manifest elevated metal capacity. Ongoing theoretical investigations and empirical investigations are continually unveiling fresh insights, propelling MXenes into the spotlight as promising candidates for rechargeable MIB anode materials. Additionally, multivalent metal ions, which bear supplementary charges, demonstrate superior theoretical capacities relative to their monovalent counterparts. An illustration of this can be seen in recent advancements where high-valent cations like Mg^{2+} and Al^{3+} were introduced within the interlayers of $Ti_3C_2T_x$, thereby facilitating the formation of autonomous thin film electrodes endowed with expanded ion diffusion pathways [17]. The integration of these cations serves to bolster Coulombic interactions, ensuring the integrity of lithium transport conduits throughout the lithiation/charging cycles. Noteworthy work by Yuan et al. [20] delves into elucidating the intricate multi-ion transport dynamics underpinning the battery's electrochemical performance. Specifically, their investigations highlighted the significant impact of $CoSe_2$ lattice distortions induced by $AlCl^{4-}$ ion intercalation on the initial operational stability of the battery. Leveraging MXene as a supporting scaffold presents a compelling solution by curbing $CoSe_2$ growth on its surface, thereby effectively mitigating lattice distortions arising from interactions with aluminum anion complexes. This approach resolves issues related to suboptimal battery reversibility, erratic cycling stability, and diminished Coulombic efficiency [20] (Figure 9.2).

2D materials, including graphene, transition-metal dichalcogenides (TMDs), have been harnessed to optimize the electronic coupling between MXene interlayers. In a comprehensive analysis, Du et al. [23] conducted a systematic investigation into the electrochemical characteristics of the Ti_2CX_2-MXene/graphene heterostructure via first-principles computations. Their findings underscore the pivotal role of graphene in mitigating the restacking propensity of MXene layers, bolstering electrical conductivity, enhancing lithium adsorption affinity (while concurrently preserving high lithium mobility), and augmenting mechanical rigidity. Similarly, Pang et al. [21]. achieved some progress in synthesizing a novel flexible Se-terminated niobium carbide MXene variant (Nb_2CSe_2) characterized by a distinctive "TMD+MXene"

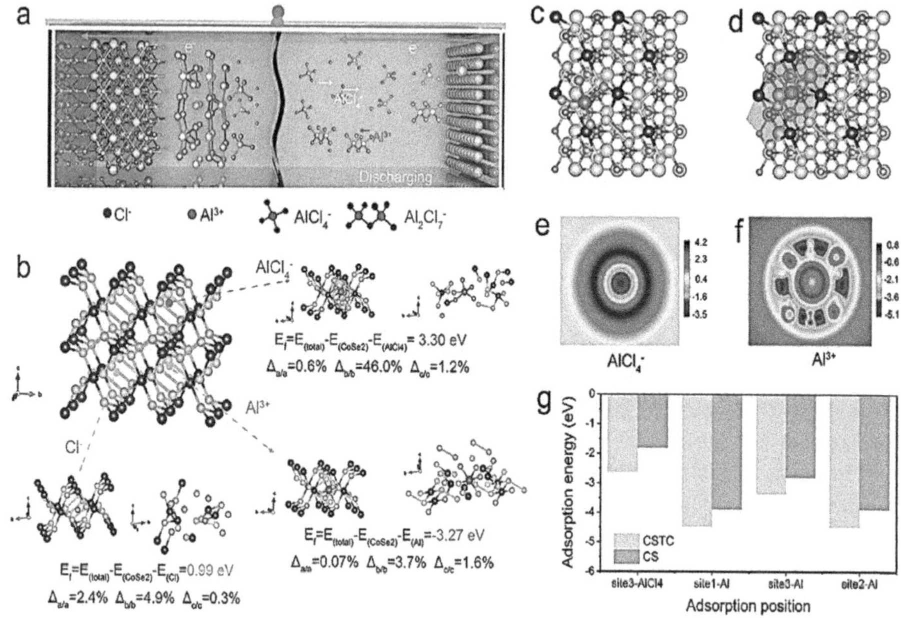

FIGURE 9.2   Theoretical prediction of multiple ion storage in $CoSe_2$. (a) Illustration of the discharging process. (b) Theoretical simulation analysis with $AlCl^{4-}$, Cl^-, and Al^{3+} entering $CoSe_2$. (c,d) The adsorption model of $AlCl^{4-}$ (c) and Al^{3+} (d) on surface Site 3 of $CoSe_2$@$Ti_3C_2(OH)x$. (e,f) The adsorption energy contour map of $(AlCl^{4-})_xCoSe_2$@$Ti_3C_2(OH)_x$ (e) and $(Al^{3+})CoSe_2$@$Ti_3C_2(OH)_x$ (f). (g) Adsorption energy at different sites. [20] Copyright 2023, Wiley.

architecture through a singular "vapor phase activated" methodology devoid of acid-etching. This innovative approach yielded outstanding performance metrics, showcasing a notable capacity of 270 mAh g^{-1} at a high discharge rate of 10 C, indicative of superior rate capability. A prevalent approach in the realm of materials engineering involves the transformation of 2D MXene nanosheets into 3D structures, aiming to optimize surface area exposure and bolster ion diffusion dynamics. A noteworthy paradigm of this methodology is the pioneering work by Gogotsi et al., who successfully fabricated self-supporting and pliable 3D macroporous MXene films employing polymethylmethacrylate (PMMA) as a sacrificial template. Leveraging the intimate interface between the porous framework and the inherent metallic conductivity of MXene yields a resultant composite that exhibits enhanced electrochemical performance metrics, characterized by superior capacity, accelerated rate capabilities, and sustained cycling robustness when deployed as an anode material for sodium-ion storage applications.

2.4.2   Matrix Materials

With their exceptional mechanical robustness, flexibility, superior electrical conductivity, and diverse functional attributes, MXenes have been established as a prime matrix material for integration with other electrochemically active counterparts, particularly for the fabrication of standalone or flexible electrode configurations. Within hybrid constructs, MXene nanosheets serve a dual purpose: firstly, they serve to deter the agglomeration of active nanostructures throughout the intercalation and delayering sequences, and secondly, they facilitate bidirectional electron and ion conveyance at the interfacial boundaries. Moreover, the incorporation of MXenes effectively mitigates the volumetric expansion exhibited by the active materials, thereby enhancing the overall structural integrity of the composite system.

2.4.3 Current Collectors

Conventional lithium-ion battery fabrication processes conventionally rely on copper and aluminum as current collectors for the anode and cathode, respectively. Notably, the inclusion of metal current collectors devoid of any intrinsic capacity contribution imposes superfluous volume and weight burdens, consequently diminishing the overall energy density of lithium-ion batteries. While efforts have been made to introduce carbon-based current collectors as substitutes for traditional metal foils, enhancing their conductivity remains a persistent challenge [24]. Remarkably, employing MXenes films yields reductions in device weight and thickness when juxtaposed against copper and aluminum current collectors, all the while upholding capacity and rate performance standards. By virtue of the favorable wettability of MXenes with liquid metal and the ensuing synergistic interactions, MXenes—when utilized as anode materials in lithium-ion batteries—showcase superior electrochemical performances relative to liquid metal-coated copper foils. Noteworthy contributions from Xu et al. [22] introduce the novel 2D Cu@Na-MX@Sn configuration as a current collector, catering to the augmentation of galvanizing/stripping kinetics and the uniform distribution of electric field and ion flux. The *in situ* formation of a $ZnF+2$ layer during the galvanizing process facilitates the unhindered diffusion of Zn ions while concurrently safeguarding against side reactions, culminating in the achievement of stable zinc plating/stripping operations characterized by a remarkable cumulative area capacity of 5000 mAh cm at high current densities and plating capacities ranging between 10 mA cm^{-2} and 20 mAh cm^{-2}. Despite the relatively modest spotlight on MXenes-based current collectors, their intrinsic engineering versatility portends bespoke applications within the domain of energy storage endeavors.

2.5 LITHIUM-SULFUR BATTERIES

Lithium-sulfur batteries have emerged as a frontrunner among prospective next-generation energy storage platforms, owing to their remarkable attributes including a substantial theoretical energy density of 2600 Wh kg^{-1}, generous specific capacities of 1675 mAh g^{-1} for sulfur and 3860 mAh g^{-1} for lithium, and eco-friendliness [25]. MXenes stand out as a promising contender in lithium-sulfur battery technology, primarily due to their metallic conductivity, abundant surface functional groups, and tunable vacancies or defects that serve to mitigate the deleterious "shuttle effect" and catalyze sulfur redox reactions [26]. Significantly, recent work by Chen et al. underscored that hydroxyl groups on MXenes can be replaced by sulfur and/or sulfide species, leading to robust Ti-S interactions that effectively curb the undesired polysulfide "shuttle effect." [27] The alignment of the d-band center of Ti atoms in

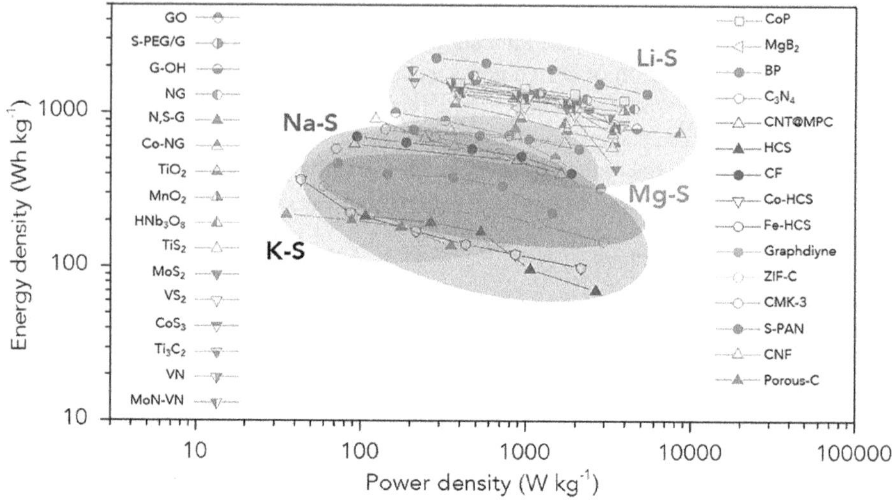

FIGURE 9.3 Summary comparison of power and energy density of metal-sulfur batteries with regard to mass of sulfur cathode. [25] Copyright 2020, Elsevier Inc.

proximity to the Fermi level enhancing the adsorption of LiPS is noteworthy [28]. Despite these advantages, the self-aggregation tendency of MXenes through van der Waals forces can hinder operational efficiency by diminishing active catalytic sites, impeding ion transport and electrolyte infiltration [29, 30]. Consequently, the strategic development of structurally robust MXene-based networks capable of facilitating rapid lithium-ion diffusion and exchange represents a pivotal avenue to drive the sulfur redox conversion process toward achieving peak performance. In Figure 9.3, Qiao and colleagues [25] have succinctly encapsulated a comparative analysis pertaining to the power and energy density of metal-sulfur batteries in relation to the mass of the sulfur cathode.

2.5.1 Sulfur Host

The implementation of MXene encounters persistent challenges associated with suboptimal long-term cycling stability attributable to the pronounced shuttle effect of LiPSs and sluggish reaction kinetics. Addressing these critical concerns, Huang et al. [31] meticulously detailed the fabrication of La$_2$O$_3$-MXene@CNF composites as a dedicated sulfur (S) encapsulating medium, aiming to rectify the aforementioned hurdles. The La$_2$O$_3$ architectural domain within the composite exhibits robust adsorption capabilities, effectively sequestering a surplus of LiPS species for subsequent catalytic processing. Moreover, the presence of an insoluble thiosulfate intermediate, derived from the hydroxyl terminal groups on the MXene surface, significantly expedites the swift conversion of LiPS into lithium sulfide (Li$_2$S) via the "Wackenroder reaction." Remarkably, the as-synthesized composite showcases notable attributes such as a minimal capacity deterioration rate of 0.031% post a single operational cycle, coupled with a strikingly high specific capacity of 857.9 mAh g^{-1} even under exceedingly elevated sulfur loading conditions.

Li et al. [32] focused on WSe_2 as a pivotal exemplar, targeting the tailored design of its electronic structure and catalytic activity model governing the sulfur redox kinetics. Their findings elucidated that the introduction of metal cation dopants with a low electron affinity/ionic radius (E_A/r) ratio notably facilitates the creation of additional Se vacancies and lattice defects, consequently augmenting the pool of active sites and electron accumulation within the material. Notably, the heightened affinity of surface Se sites toward LiPSs enhances LiPS binding, albeit at the expense of reduced competitiveness in the LiPSs/Li_2S interplay on the host surface. This nuanced interplay strengthens the Li-S bond, thereby facilitating enhanced LiPS adsorption while concurrently mitigating the energy threshold for Li_2S nucleation and breakdown. The resultant V-doped WSe_2/MXene catalyst, showcasing the minimal EA/r value, emerges as an exceptionally effective sulfur host, exhibiting a remarkable reversible capacity of 1402.5 mAh g^{-1}, exceptional long-term cycling stability spanning 800 cycles (with $\sim$70% retention rate), and a substantial areal capacity of 6.4 mAh cm^{-2}. In a complementary study by Shen et al. [33], PMMA spheres were ingeniously employed as templates for assembling 2D $Ti_3C_2T_x$ nanosheets into 3D thin-walled spheres to create MXene@$CoSe_2$/NC hollow sphere composites. Among these, the MXene hollow spheres serve as the substrates for sulfur encapsulation. The ensuing MXene@$CoSe_2$/NC electrodes demonstrated outstanding performance, delivering an impressive areal capacity of 3.343 mAh cm^{-2} under a notable sulfur loading of 3.5 mg cm^{-2}. Capitalizing on the unique attributes of the 3D MXenes heterostructure composite material, the electrocatalyst garnered several advantages: firstly, the 3D MXene hollow architecture promotes an enhanced interfacial interaction between the electrode and the electrolyte, facilitating efficient electron transfer processes; secondly, MXene plays a pivotal role in optimizing the chemical bonds within the atomic framework, thereby regulating the excessive dissolution of LiPSs; finally, the heterostructure interface significantly expands the electroactive area, thereby catalyzing chemical reactions within the single-phase electrolyte and driving solid-liquid conversion kinetics in the multi-phase electrocatalytic milieu.

2.5.2 Functional Interlayers

The strategic utilization of 2D MXenes in the fabrication of functional separators/interlayers stands as a pivotal tactic in enhancing the performance of Li-S batteries by impeding the migration of LiPS and finely modulating lithium plating behavior [34]. Functioning as a critical interlayer, MXenes assume a tripartite role: (i) bolstering the cathode's conductivity, (ii) curtailing the diffusion of LiPS, and (iii) expediting the conversion of LiPS. Peng et al. [35] introduced a composite framework with overlapping transmission orbits, labeled as MX@WSSe/PP, as a tailored separator for lithium-sulfur batteries. This innovative separator effectively curbs the shuttle effect of LiPS, ensures optimal utilization of cathode materials, and transforms the deposition mode of Li_2S in a 2D structure, thereby improving the overall reversibility and reaction kinetics of the cell. This design achieved an areal capacity of 9.39 mAh cm^{-2} under high sulfur loading conditions (10.2 mg cm^{-2}) and a reduced electrolyte/sulfur ratio (7.5 μL mg^{-1}). In a complementary advancement, He et al. [36] pioneered the development and synthesis of a composite porous carbon nanofiber imbued with oxygen-enriched Ti_3C_2-MXene nanosheets, serving as a versatile interlayer for Li-S battery systems.

This innovative design leverages an oxygen functionalization strategy that integrates stepwise controllable surface modifications and moderate oxygen adjustments to engender an oxygen-rich MXene surface. Consequently, this tailored approach facilitates enhanced lithium flux diffusion, culminating in a dendrite-free lithium anode and bolstering the overall performance and stability of the Li-S cell.

In the investigation conducted by Tang et al. [37], the Fe_3Se_4/FeSe heterostructure demonstrates a superior capacity to furnish abundant active sites for catalyzing polysulfide species, while the inherent internal electric field significantly bolsters the redox kinetics within the system. Concurrently, the integration of MXene nanosheets facilitates the expeditious transport of electrons and ions, efficiently sequesters polysulfides, thereby fortifying the overall performance of the lithium-sulfur battery. Leveraging these collective advantages manifests the implementation of the Fe_3Se_4/FeSe@MXene-PP separator with a noteworthy initial specific capacity of 1104.2 mAh g^{-1} at 0.2 C, coupled with an exceptional rate capacity of 758.8 mAh g^{-1} at 4 C. Impressively, even under a substantial sulfur loading of 5.8 mg cm^{-2}, the Li-S battery sustains a commendable specific capacity of 862.6 mAh g^{-1} at 0.2 C, showcasing robust cycling stability with a remarkable capacity retention rate of 92.3% following 120 operational cycles. In a holistic perspective, the versatile multifunctional interlayer emerges not only as an adept LiPS capture agent but also as a successful mitigator of dendritic growth in the lithium anode.

2.5.3 Lithium Metal Hosts

Serving as a prospective anode material in lithium-sulfur (Li-S) batteries, lithium metal boasts an impressive theoretical capacity of 3860 mAh g^{-1} and a low redox potential (–3.04V vs. the standard hydrogen electrode). Nevertheless, challenges arise from the unstable interfacial interactions between the Li anode, LiPS, and additives within the organic electrolyte, culminating in substantial volume fluctuations during the repetitive plating and stripping procedures. This phenomenon collaborates to foster the formation of uncontrollable dendrites and a precarious solid electrolyte interface (SEI), consequently inducing issues like compromised Coulombic efficiency, severe lithium powdering, and the potential risk of short-circuit occurrences in Li-S batteries. Addressing the aforementioned hurdles involves a strategic pursuit of stable and conductive substrates to homogenize the electric field and facilitate uniform lithium deposition, a tactic that has demonstrated considerable efficacy. In this context, the utilization of 2D MXenes emerges prominently due to their elevated electronic conductivity, rapid lithium-ion transport capabilities, and an abundance of lithium nucleation sites (represented by surface functional groups). By orchestrating the formation of 3D frameworks endowed with adjustable pores, MXenes effectively modulate lithium deposition processes and dendrite growth, thereby showcasing notable performance in dendrite suppression. Drawing on the profound polarity and intricate surface chemistry intrinsic to 2D MXenes, employing MXenes as sulfur hosts and functional interlayers proves pivotal in proficiently sequestering LiPS throughout Li-S battery operations. Leveraging techniques such as functionalization, hybridization, and the construction of 3D architectures further augments the ability of MXenes to anchor LiPS species, thereby catalyzing enhanced sulfur utilization rates while mitigating the shuttle effect. This culminates in the attainment

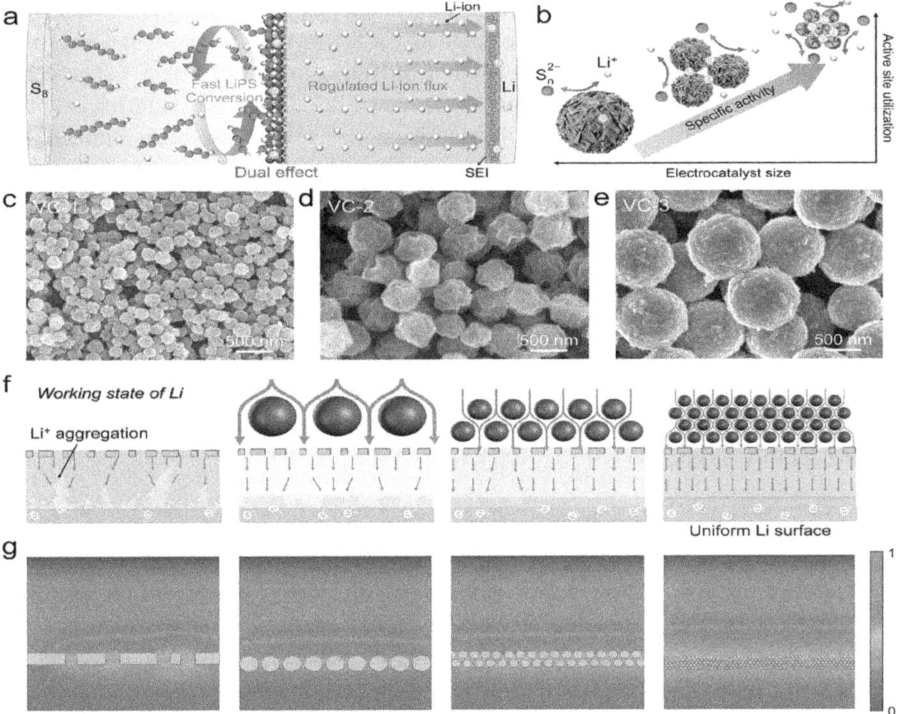

FIGURE 9.4 Schematically illustrating the scale effect of the VC-based separator on sulfur and lithium evolution efficiency. (a) The dual effect of the VC-based separator on sulfur and lithium evolution. (b) The relationship between VC spheres and electrochemical activity. (c–e) SEM images of various VC samples. (f) The impact of VC size on the Li-ion kinetics behavior. (g) COMSOL results of the relationship between the density of the sieve pores and the Li-ion flux. [38] Copyright 2023, Wiley.

of heightened reversible capacity and prolonged cycle stability within Li-S systems. Notably, MXenes also exhibit a capacity to impede the growth of lithium dendrites, thus bestowing improvements upon the lithium anodes featured within Li-S battery configurations.

In the study depicted in Figure 9.4, Song et al. [38] detail the tailored fabrication of V_2C-MXene (VC) spheres by employing melamine formaldehyde (MF) as a template, as illustrated in Figure 9.4a. Through deliberate adjustments of the VC sphere dimensions to enhance the population of active sites, the team achieves tunable reactivity conducive for orchestrating the Li-S redox reaction modulation, as depicted in Figure 9.4b. Notably, the calibration of the VC sphere size serves to expedite the kinetics of the Li_2S conversion reaction and facilitate the screening of structural ions within Li-S batteries, capturing the scale dynamics in Figure 9.4 (c–e). The intricate high-density channel framework sculpted by the VC spheres facilitates a uniform flow of lithium ions, as showcased in Figure 9.4f, g. The lithium-sulfur battery under examination demonstrates an initial discharge capacity of 1206.4 mAh g^{-1} at 0.1 C, coupled with a commendable cycle stability even after 600 charging/discharging cycles, exhibiting a minimal decay rate of 0.04% per cycle at 1 C. Noteworthy is the ability to sustain an exceptional areal capacity of 8.1 mA h cm^{-2} even under a sulfur loading of 8.1 mg cm^{-2} and an electrolyte to sulfur ratio of 4 µL mg S^{-1}. The versatility of MXenes in enhancing the performance of lithium-sulfur batteries is underscored in this investigation. Although significant milestones have been attained, key challenges persist necessitating further exploration for performance enhancement: (i) tailoring the reactivity of MXenes to mitigate the shuttle effect, instigate the conversion of LiPS, and facilitate uniform lithium deposition is imperative for achieving viable Li-S battery technologies. It is crucial (ii) to engineer high-performance multifunctional battery frameworks (MBFs) capable of synergistically enhancing electron/ion transport and electrochemical conversion processes, perhaps through the integration of a multifaceted interlayer that not only curtails LiPS migration but also facilitates unimpeded lithium-ion mobility; (iii) advancing the development of specialized MXenes nanosheet templates to prescribe design approaches that counteract nanosheet restacking and reduce production costs.

2.6 SCs

Beyond rechargeable batteries, SCs stand as an alternative energy storage solution that operates on the principles of electrolyte ion adsorption or pseudocapacitive Faradaic reactions occurring near the surface of electrode materials, fostering the creation of a dual charge layer at the electrode/electrolyte junction to facilitate charge storage [39]. Typically categorized into electric double layer capacitors (EDLCs), pseudocapacitors, and asymmetric SCs predicated on their charge storage mechanisms, SCs center their storage phenomena primarily on the electrode/electrolyte interface within EDLCs and pseudocapacitors. Thus, the specific surface area (SSA) of electrode materials emerges as a critical determinant of SC performance, with a larger SSA offering increased ion adsorption capacity in EDLCs, while nanostructured electrode materials characterized by substantial SSA and active sites prove advantageous in minimizing electrolyte ion transport distances, thereby enhancing the power density of pseudocapacitors. Notably, within this realm, $Ti_3C_2T_x$ stands out as the foremost reported and extensively investigated MXene SC material to date, distinguished by its elevated packing density of up to 4 g cm^{-3} and a surface conducive to redox reactions through protonation [40].

Owing to their stratified architecture, MXenes and their composite derivatives exhibit a capacity to accommodate polar organic compounds and diverse cations (including Li^+, Na^+, K^+, Mg^{2+}, and NH^{4+}), thereby fostering enhanced electron conduction rates and offering significant potential for advancement in this domain. Through the inclusion of guest active species within these composites and the provision of functionalized surfaces that serve as active sites for ion adsorption or redox processes, an augmented volume capacitance is achieved, surpassing that of conventional carbon-based materials. Strategies such as interlayer modulation, surface functionalization, pore design, and amalgamation with alternate materials have been devised to elevate the performance benchmarks of SCs. Subsequent to these improvements, recent investigations delve into the realm of MXenes nanoengineering aimed at fortifying the functionality of conventional SCs and micro-SCs (MSCs).

2.6.1   Active Materials

In the early stages of 2013, the groundbreaking work by the Gogotsi research group [41] unveiled the phenomenon of spontaneous/electrochemical cation intercalation within two-dimensional Ti_3C_2 layers, heralding a new era for the potential utilization of MXene materials in SCs. Leveraging the charge storage pseudocapacitance mechanism with rapid proton transport capabilities, the implementation of a large-pore three-dimensional (3D) MXene electrode architecture distinctly shortens ion diffusion pathways. Noteworthy advancements in the realm of 3D porous structures with optimized loading quality have been demonstrated by Ostrikov et al. [42] The meticulously engineered 3D porous MXene films, distinguished by minimized interlayer repulsion stemming from reduced electronegative functional groups and charge-shielding effects, have yielded a remarkable rate performance of 207.9 F g^{-1} at 10 V s^{-1}, showcasing scalability toward a practical mass loading level of 16.18 mg cm^{-2}. Additionally, Spurling et al. [43] have successfully engineered high-performance thick film electrodes with thicknesses reaching $\sim$ 50 μm, featuring well-defined layered porous architectures achieved by assembling $Ti_3C_2T_x$ 2D nanosheets into interconnected 3D network films. The electrode design, striking a balance between porosity and thickness, embodies high areal loading capabilities of up to approximately 7.2 mg cm^{-2}, a notable material density of around 1440 mg cm^{-3}, and impressive electrochemical performance metrics of 240 F g^{-1} and 140 F cm^{-3}. Consequently, even under demanding conditions such as a high rate of 200 mV s^{-1}, the electrode embodies a notable areal capacitance of $\sim$ 1.4 F cm^{-2}.

The utilization of 0D inorganic particles within composite architectures integrated into layered MXenes has emerged as a promising avenue for the development of sophisticated thin film electrodes. Huang et al. [12] pioneered the synthesis of a multilayer $Ti_3C_2T_x$ superlattice embedded within monolayer Mesoporous Carbon Frameworks (MMCF) via the synergistic co-assembly of $Ti_3C_2T_x$ nanosheets and Fe_3O_4 nanoparticles. The intercalated MMCF framework not only engenders augmented interlayer spacing and creates permeable pathways to facilitate rapid mass transport, but also serves as conductive scaffolds, facilitating efficient electron transfer in the z-direction. The volumetric capacitance of $Ti_3C_2T_x$/MMCF has been reported at an impressive 317 F cm^{-3}. Additionally, Beidaghi et al. [44] proposed a cation-driven assembly approach for the fabrication of highly durable and flexible V_2CT_x/Ti_2CT_x-MXene multilayer films, utilizing chemically labile layered monolayer flakes. Electrodes constructed using assembled V_2CT_x flakes exhibited superior electrochemical performance compared to $Ti_3C_2T_x$ counterparts across various aqueous electrolytes, thereby exhibiting a specific capacitance of up to 1315 F cm^{-3} and retaining approximately 77% of the initial capacitance following a million charge-discharge cycles. 2D materials play a pivotal role in enabling the rapid diffusion of ions through slit-like channels. The quest for superior energy storage devices necessitates the utilization of electrode materials that exhibit heightened electronic conductivity, abundant embedding sites, and long-term stability to optimize the energy density, power density, and operational longevity of the device. To fully exploit the active surface and avert restacking of MXene flakes, novel 2D materials like graphene and borophene have been introduced as interlayer spacers. Concurrently, the elimination of terminal groups (–F/–OH) enhances the availability of pseudocapacitive active sites.

Lee et al. [45] have introduced a pioneering approach, showcasing the fabrication of flexible and high-speed SCs utilizing conductive 2D MXene/Borophene (MxB) electrodes via an electrophoretic deposition methodology. This innovative design minimizes self-weight stacking tendencies and augments layer spacing, culminating in a hybrid electrode that exhibits an impressive specific capacitance of up to 626.7 F g^{-1} at 1 A g^{-1} with a remarkable capacitance retention of 85.14% at 20 A g^{-1}. Furthermore, the device maintains negligible capacitance loss even after 10,000 cycling processes, thereby attaining exceptional ultra-high energy and power densities. The zinc ion migration channels (ZIMCs) devised in this study concurrently facilitate rapid solid-phase zinc diffusion and efficient zinc ion transport, yielding an area-specific capacitance of up to 410 mF cm^{-2} and an energy density reaching 103 μWh cm^{-2} at a power density of 2100 μW cm^{-2}. Ostrikov et al. [46] have introduced a one-step approach for the selenization of MXene, yielding carbon scaffold-anchored $TiSe_2$ nanosheets. This technique engenders a non-stacked structure for $TiSe_2$@ CSA (Carbon-Scaffold-Anchored), characterized by ample interlayer spacing and a profusion of Se vacancies, thus unlocking interlayer and vacancy-mediated ion transport mechanisms essential for rapid lithium-ion storage. The resultant $TiSe_2$@ CSA framework emerges as a promising anode candidate for lithium-ion capacitors (LICs), facilitating swift ion diffusion within the system. $TiSe_2$@CSA exhibits an energy density of 96.7 Wh kg^{-1} at 410 W kg^{-1}, while maintaining a remarkable 31.0 Wh kg^- even under a high power density of 15700 W kg^{-1}. Additionally, Wang et al. [47] have successfully fabricated MXene-based electrodes featuring a hierarchical interconnected architecture spanning micro-meso-macropores, as depicted in Figure 9.5. This intricate material design offers a trifecta of benefits: macroscopic macropores minimize ion transmission distances, mesopores provide a low-impedance conduit for ion transport, and micropores bolster electrochemical reactivity. Subsequently, leveraging the laser engraving process, polyurethane-coated PM (Porous MXene) / CNF foam-based ZIMCs were meticulously assembled.

2.6.2 Matrix Materials

Owing to its inherent mechanical flexibility and complex surface chemistry, MXene emerges as a highly promising material matrix for the development of thin film electrodes in SCs. For instance, Zhang et al. [48] elucidated that flexible films featuring a 2D-3D analogous heterostructure comprising $Ti_3C_2T_x$ layers and 3D cubic Ni-Fe oxide exhibited noteworthy electrochemical performance. This performance superiority can be attributed to the synergistic advantages offered by distinct constituents: the chemically reactive MXene layer fulfills the role of a conductive framework, while the cubic nickel-iron oxide component acts as an active spacer that augments interlayer spacing within the MXene structure, consequently enhancing the capacitance of the composite film.

2.6.3 Micro-Supercapacitors (MSCs)

MSCs are increasingly recognized as a promising energy storage solution tailored for miniaturized and portable electronic devices. Accumulated evidence from prior investigations underscores that both the selection of electrode materials and the methodologies employed for fabrication represent pivotal determinants

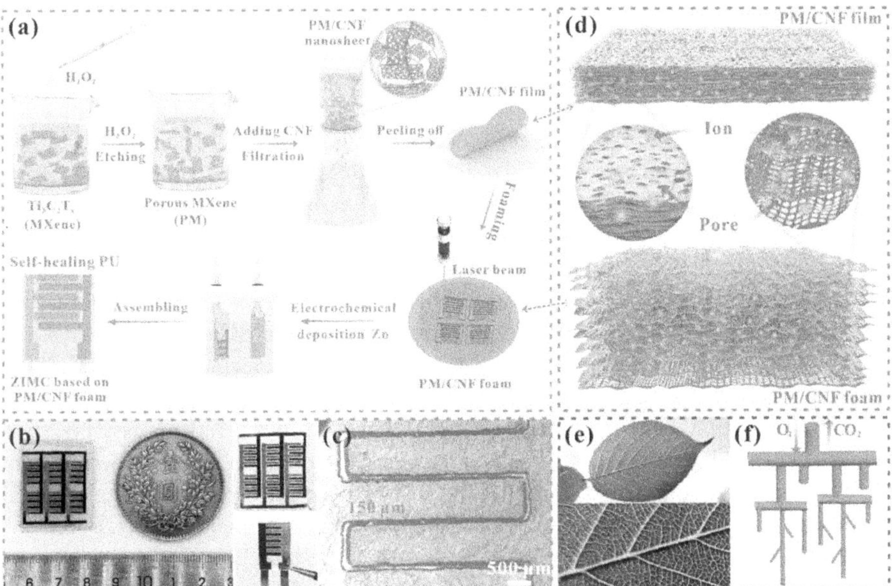

FIGURE 9.5 (a) Schematic diagram of the preparation of PM/CNF foam and the assembly of ZIMC based on PM/CNF foam. (b) Photographs of the PM/CNF foam-based ZIMC array and single forked finger electrode. (c) SEM image of the PM/CNF foam-based ZIMC. (d) Schematic diagram of the electrode structures of PM/CNF film and PM/CNF foam and the corresponding local structure enlargement. (e) The veins of the leaf are interlaced to form a graded network of interconnected porosity for maximum nutrient delivery. (f) Interconnected hierarchical porous structures are the pathways evolved by plants to achieve optimal photosynthesis. [47] Copyright 2022, Wiley.

influencing the efficacy and practical utility of MSCs [49, 50]. Concomitantly, the electrode layout emerges as a critical parameter significantly influencing MSC performance. Amid a myriad of electrode architectures, planar configurations are favored for their integration compatibility within microelectronic systems, primarily due to the enhanced ion diffusion capabilities inherent to planar arrangements vis-à-vis sandwich configurations. Delving into this discourse, we expound upon the recent strides made in harnessing planar configurations of MXenes within the realm of MSCs, illuminating novel pathways for enhanced energy storage advancements.

Wang et al. [51] have documented the preparation of MXene ink featuring an enlarged interlayer spacing achieved through the intercalation of lithium ions, thereby enhancing ion diffusion kinetics. The lithium-ion-incorporated MXene ink was subsequently utilized in the fabrication of MXene-based MSCs through screen printing, enabling a scalable manufacturing process. The unique electrode architecture design contributed to notable performance enhancements in the resulting devices, manifesting as a remarkable area capacitance of 252 mF cm^{-2}, enhanced rate capability with capacitance retention exceeding 80%, and exceptional cycling durability evidenced by maintenance of stability over 10,000 charge-discharge cycles with 98.4% capacitance retention. Noteworthy flexibility in tandem with sustained performance highlights the promising prospects of these MSCs in the realm of wearable smart electronics. Pumera et al. [52] harnessed picosecond pulsed laser technology to engineer high-energy-density MSCs integrated with force-sensing apparatus designed for monitoring radial artery pulses in human subjects. The active electrode materials for the MSCs were formulated using photochemically synthesized spherical nanoparticles derived from laser-induced MXene ($Ti_3C_2T_x$) oxide, uniformly adhered to laser-induced graphene (LIG). Investigation through molecular dynamics simulations and detailed spectroscopic analyses unveiled the underlying cooperative interfacial mechanism characterized by Ti-O-C covalent bonding interactions between MXene and LIG. It was demonstrated that micro-SCs fabricated from the nano-MXene-LIG composite exhibited not only superior mechanical flexibility and durability, but also boasted an ultrahigh energy density reaching 21.16×10^{-3} mWh cm^{-2} and exceptional capacitance (~100 mF cm^{-2} @ 10 mV s^{-1}). Impressively, these devices exhibited prolonged cycle life with a capacity retention rate of 91% even after 10,000 cycles, indicative of their robust and sustainable performance. Zhang et al. [53] introduce a groundbreaking approach for the large-scale production of customizable planar micro-supercapacitors (MSCs) utilizing advanced 3D printing technology with enhanced fidelity. In Figure 9.6, by employing additive-free electrochemically exfoliated graphene inks, the research team overcomes challenges related to structural control in the development of monolithic integrated MSCs (MIMSCs). The resultant MSCs exhibit exceptional engineering specifications, boasting an impressively small engineering footprint of 0.025 cm^2; high areal capacitance of 4900 mF

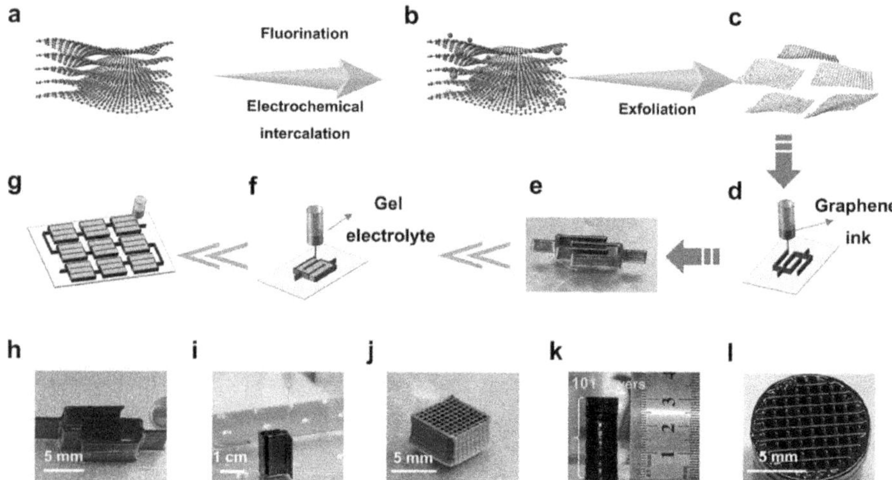

FIGURE 9.6 Schematic diagram of EG-MIMSCs manufacturing. (a–c) Demonstration of the preparation process of EG-NaBF$_4$+NaI (red anion: BF$_4$–; yellow anion: I–; blue molecule: H$_2$O). (d–g) Illustration depicting the manufacturing procedure of 3D-printed EG-MSCs and EG-MIMSCs. (h,i) Photographs of EG-MSCs with different configurations of interdigital shapes and (j–l) lattice structure. [53] Copyright 2024, Wiley.

cm^{-2}; volumetric capacitance of 195.6 F cm^{-3}; and impressive energy density metrics, including an areal energy density of 2.1 mWh cm^{-2} and volumetric energy density of 23 mWh cm^{-3} per single cell, surpassing previous benchmarks for 3D-printed MSCs.

3 SUMMARY AND OUTLOOK

A comprehensive overview of the recent advancements in the multifaceted engineering of MXenes for energy storage applications encompassing MIBs, LIBs, Li-S batteries, and SCs is provided herein. Initially, the foundational principles underpinning MXenes are introduced, elucidating various synthesis methodologies, layering procedures, and their intrinsic properties. Subsequent discussions delve into pertinent strategies pertaining to the design and manufacture of MXenes, alongside addressing key considerations associated with their processing. By capitalizing on their inherent merits, such as metallic conductivity, outstanding hydrophilicity, diverse surface chemistry, and tunable interlayer spacing, MXenes are poised as promising candidates for multifunctional building blocks capable of serving as active materials, conductive elements, additives, hosts, functional interlayers, or matrices, especially when integrated into membrane-like architectures. The burgeoning literature increasingly emphasizes the fabrication of MXenes as standalone electrodes, modifying layers, or functional components within energy storage devices. Noteworthy trends driving this trajectory include: (i) the widespread integration of MXenes and their derivatives into composite films tailored for energy storage purposes; (ii) notable strides in enhancing the synthesis methodologies and solution processing of MXene derivatives to bolster their stability and mechanical attributes in recent years; and (iii) the emergence of diverse technologies aimed at enabling the cost-effective and large-scale production of MXene composite films to enhance device performance outcomes. Collectively, these facets are poised to propel the development of MXene-based composite materials and unlock novel avenues for potential applications spanning ion/molecule separation, environmental remediation, water purification, sensor technologies, and beyond.

REFERENCES

[1] Y. Qi, L. Sun, Z. Liu, Super Graphene-Skinned Materials: An Innovative Strategy toward Graphene Applications, *ACS Nano*, 18 (2024) 4617–4623.

[2] X. Li, Q. Zheng, C. Li, G. Liu, Q. Yang, Y. Wang, P. Sun, H. Tian, C. Wang, X. Chen, J. Shao, Bubble Up Induced Graphene Microspheres for Engineering Capacitive Energy Storage, *Advanced Energy Materials*, 13 (2023) 2203761.

[3] L. Zhang, J. Liu, Y. Zhai, S. Zhang, W. Wang, G. Li, L. Sun, H. Li, S. Qi, S. Chen, R. Wang, Q. Ma, J. Just, C. Zhang, Rational Design of Multinary Metal Chalcogenide Bi0.4Sb1.6Te3 Nanocrystals for Efficient Potassium Storage, *Advanced Materials*, 36 (2024) 2313835.

[4] H. Wu, J. Hao, Y. Jiang, Y. Jiao, J. Liu, X. Xu, K. Davey, C. Wang, S.-Z. Qiao, Alkaline-Based Aqueous Sodium-Ion Batteries for Large-Scale Energy Storage, *Nature Communications*, 15 (2024) 575.

[5] Q. Man, Y. An, H. Shen, C. Wei, S. Xiong, J. Feng, Two-Dimensional Silicene/Silicon and Its Derivatives: Properties, Synthesis and Frontier Applications, *Materials Today*, 67 (2023) 566–591.

[6] R. Herbst-Irmer, D. Stalke, Correspondence to "Structure of Violet Phosphorus and Its Phosphorene Exfoliation", *Angewandte Chemie International Edition*, 63 (2024) e202319571.

[7] Y. Zhou, L. Yin, S. Xiang, S. Yu, H.M. Johnson, S. Wang, J. Yin, J. Zhao, Y. Luo, P.K. Chu, Unleashing the Potential of MXene-Based Flexible Materials for High-Performance Energy Storage Devices, *Advanced Science*, 11 (2023) 2304874.

[8] H. Liu, Z. Xin, B. Cao, B. Zhang, H.J. Fan, S. Guo, Versatile MXenes for Aqueous Zinc Batteries, *Advanced Science*, 11 (2023) 2305806.

[9] Y. An, Y. Tian, H. Shen, Q. Man, S. Xiong, J. Feng, Two-Dimensional MXenes for Flexible Energy Storage Devices, *Energy & Environmental Science*, 16 (2023) 4191–4250.

[10] S. Yadav, N. Kurra, Diffusion Kinetics of Ionic Charge Carriers Across Ti3C2T MXene-Aqueous Electrochemical Interfaces, *Energy Storage Materials*, 65 (2024) 103094.

[11] M. Jiang, D. Jiang, J. Wang, Y. Sun, J. Liu, Stretchable MXene Based Films Towards Achieving Balanced Electrical, Mechanical and Energy Storage Properties, *Chemical Engineering Journal*, 459 (2023) 141527.

[12] X. Huang, X. Lyu, G. Wu, J. Yang, R. Zhu, Y. Tang, T. Li, Y. Wang, D. Yang, A. Dong, Multilayer Superlattices of Monolayer Mesoporous Carbon Framework-Intercalated MXene for Efficient Capacitive Energy Storage, *Advanced Energy Materials*, 14 (2023) 2303417.

[13] J. Yang, M. Li, S. Fang, Y. Wang, H. He, C. Wang, Z. Zhang, B. Yuan, L. Jiang, R.H. Baughman, Q. Cheng, Water-Induced Strong Isotropic MXene-Bridged Graphene Sheets for Electrochemical Energy Storage, *Science*, 383 (2024) 771–777.

[14] P. Liu, R. Pan, B. Li, Z. Su, B. Lin, M. Tong, Mild and Efficient Method for the In Situ Preparation of High-Quality MXene Materials Enabled by Hexafluoro Complex Anion Contained Salts Etching, *Advanced Functional Materials*, 34 (2023) 2308532.

[15] A. Sikdar, F. Héraly, H. Zhang, S. Hall, K. Pang, M. Zhang, J. Yuan, Hierarchically Porous 3D Freestanding Holey-MXene Framework via Mild Oxidation of Self-Assembled MXene Hydrogel for Ultrafast Pseudocapacitive Energy Storage, *ACS Nano*, 18 (2024) 3707–3719.

[16] W. Xu, X. Liao, W. Xu, K. Zhao, G. Yao, Q. Wu, Ion Selective and Water Resistant Cellulose Nanofiber/MXene Membrane Enabled Cycling Zn Anode at High Currents, *Advanced Energy Materials*, 13 (2023) 2300283.

[17] M. Lu, W. Han, H. Li, W. Shi, J. Wang, B. Zhang, Y. Zhou, H. Li, W. Zhang, W. Zheng, Tent-Pitching-Inspired High-Valence Period 3-Cation Pre-Intercalation Excels for Anode of 2D Titanium Carbide (MXene) with High Li Storage Capacity, *Energy Storage Materials*, 16 (2019) 163–168.

[18] Z. Yan, K. Zhu, H. Xu, S. Wei, W. Wen, C. Wang, L. Song, Oxygen-Poor Amorphous Heterostructure Derived from Nb2CTx Toward Superior Lithium-Ion Storage, *Advanced Functional Materials*, 34 (2024) 2402543.

[19] K. Guan, L. Dong, Y. Xing, X. Li, J. Luo, Q. Jia, H. Zhang, S. Zhang, W. Lei, Structure and Surface Modification of MXene for Efficient Li/K-Ion Storage, *Journal of Energy Chemistry*, 75 (2022) 330–339.

[20] Z. Yuan, Q. Lin, Y. Li, W. Han, L. Wang, Effects of Multiple Ion Reactions Based on a CoSe(2) /MXene Cathode in Aluminum-Ion Batteries, *Advanced Materials*, 35 (2023) 2211527.

[21] X. Pang, Z. Lv, S. Xu, J. Rong, M. Cai, C. Zhao, F. Huang, Ultra-Conductive Se-Terminated MXene Nb2CSe2 via One-Step Synthesis for Flexible Fast-Charging Batteries, *Energy Storage Materials*, 61 (2023) 102860.

[22] Y. An, B. Xu, Y. Tian, H. Shen, Q. Man, X. Liu, Y. Yang, M. Li, Reversible Zn Electrodeposition Enabled by Interfacial Chemistry Manipulation for High-Energy Anode-Free Zn Batteries, *Materials Today*, 70 (2023) 93–103.

[23] Y.-T. Du, X. Kan, F. Yang, L.-Y. Gan, U. Schwingenschlögl, MXene/Graphene Heterostructures as High-Performance Electrodes for Li-Ion Batteries, *ACS Applied Materials & Interfaces*, 10 (2018) 32867–32873.

[24] S. Jin, Y. Jiang, H. Ji, Y. Yu, Advanced 3D Current Collectors for Lithium-Based Batteries, *Advanced Materials*, 30 (2018) 1802014.

[25] C. Ye, D. Chao, J. Shan, H. Li, K. Davey, S.-Z. Qiao, Unveiling the Advances of 2D Materials for Li/Na-S Batteries Experimentally and Theoretically, *Matter*, 2 (2020) 323–344.

[26] X. Li, Q. Guan, Z. Zhuang, Y. Zhang, Y. Lin, J. Wang, C. Shen, H. Lin, Y. Wang, L. Zhan, L. Ling, Ordered Mesoporous Carbon Grafted MXene Catalytic Heterostructure as Li-Ion Kinetic Pump toward High-Efficient Sulfur/Sulfide Conversions for Li-S Battery, *ACS Nano*, 17 (2023) 1653–1662.

[27] Z. Ye, Y. Jiang, L. Li, F. Wu, R. Chen, Self-Assembly of 0D-2D Heterostructure Electrocatalyst from MOF and MXene for Boosted Lithium Polysulfide Conversion Reaction, *Advanced Materials*, 33 (2021) 2101204.

[28] X. Wang, D. Luo, J. Wang, Z. Sun, G. Cui, Y. Chen, T. Wang, L. Zheng, Y. Zhao, L. Shui, G. Zhou, K. Kempa, Y. Zhang, Z. Chen, Strain Engineering of a MXene/CNT Hierarchical Porous Hollow Microsphere Electrocatalyst for a High-Efficiency Lithium Polysulfide Conversion Process, *Angewandte Chemie International Edition in English*, 60 (2021) 2371–2378.

[29] Y. Zhang, Z. Cao, S. Liu, Z. Du, Y. Cui, J. Gu, Y. Shi, B. Li, S. Yang, Charge-Enriched Strategy Based on MXene-Based Polypyrrole Layers Toward Dendrite-Free Zinc Metal Anodes, *Advanced Energy Materials*, 12 (2022) 2103979.

[30] W. Peng, J. Han, Y.R. Lu, M. Luo, T.S. Chan, M. Peng, Y. Tan, A General Strategy for Engineering Single-Metal Sites on 3D Porous N, P Co-Doped Ti(3)C(2)T(X) MXene, *ACS Nano*, 16 (2022) 4116–4125.

[31] Z. Huang, Y. Zhu, Y. Kong, Z. Wang, K. He, J. Qin, Q. Zhang, C. Su, Y.L. Zhong, H. Chen, Efficient Synergism of Chemisorption and Wackenroder Reaction via Heterostructured La2O3-Ti3C2Tx -Embedded Carbon Nanofiber for High-Energy Lithium-Sulfur Pouch Cells, *Advanced Functional Materials*, 33 (2023) 2303422.

[32] W. Wang, X. Wang, J. Shan, L. Yue, Z. Shao, L. Chen, D. Lu, Y. Li, Atomic-Level Design Rules of Metal-Cation-Doped Catalysts: Manipulating Electron Affinity/Ionic Radius of Doped Cations for Accelerating Sulfur Redox Kinetics in Li–S Batteries, *Energy & Environmental Science*, 16 (2023) 2669–2683.

[33] T. Li, L. Liang, Z. Chen, J. Zhu, P. Shen, Hollow Ti3C2T MXene@CoSe2/N-Doped Carbon Heterostructured Composites for Multiphase Electrocatalysis Process in Lithium-Sulfur Batteries, *Chemical Engineering Journal*, 474 (2023) 145970.

[34] D. Guo, F. Ming, H. Su, Y. Wu, W. Wahyudi, M. Li, M.N. Hedhili, G. Sheng, L.-J. Li, H.N. Alshareef, Y. Li, Z. Lai, MXene Based Self-Assembled Cathode and Antifouling Separator for High-Rate and Dendrite-Inhibited Li–S Battery, *Nano Energy*, 61 (2019) 478–485.

[35] S. Tian, G. Liu, S. Xu, C. Han, K. Tao, J. Huang, S. Peng, Deposition Mode Design of Li2S: Transmitted Orbital Overlap Strategy in Highly Stable Lithium-Sulfur Battery, *Advanced Functional Materials*, 34 (2023) 2309437.

[36] Y. He, Y. Zhao, Y. Zhang, Z. He, G. Liu, J. Li, C. Liang, Q. Li, Building Flexibly Porous Conductive Skeleton Inlaid with Surface Oxygen-Dominated MXene as an Amphiphilic Nanoreactor for Stable Li-S Pouch Batteries, *Energy Storage Materials*, 47 (2022) 434–444.

[37] X. Zhou, Y. Cui, X. Huang, Q. Zhang, B. Wang, S. Tang, Interface Engineering of Fe3Se4/FeSe Heterostructures Encapsulated in MXene for Boosting LiPS Conversion and Inhibiting Shuttle Effect, *Chemical Engineering Journal*, 457 (2023) 141139.

[38] L. Chen, Y. Sun, X. Wei, L. Song, G. Tao, X. Cao, D. Wang, G. Zhou, Y. Song, Dual-Functional V(2) C MXene Assembly in Facilitating Sulfur Evolution Kinetics and Li-Ion Sieving toward Practical Lithium-Sulfur Batteries, *Advanced Materials*, 35 (2023) 2300771.

[39] Y. Cai, X. Chen, Y. Xu, Y. Zhang, H. Liu, H. Zhang, J. Tang, Ti3C2Tx MXene/Carbon Composites for Advanced Supercapacitors: Synthesis, Progress, and Perspectives, *Carbon Energy*, 6 (2024) 501.

[40] Z. He, L. Yao, W. Guo, N. Sun, F. Wang, Y. Wang, R. Wang, F. Wang, Pseudocapacitance of Bimetallic Solid-Solution MXene for Supercapacitors with Enhanced Electrochemical Energy Storage, *Advanced Functional Materials*, 33 (2023) 2305251.

[41] M.R. Lukatskaya, O. Mashtalir, C.E. Ren, Y. Dall'Agnese, P. Rozier, P.L. Taberna, M. Naguib, P. Simon, M.W. Barsoum, Y. Gogotsi, Cation Intercalation and High Volumetric Capacitance of Two-Dimensional Titanium Carbide, *Science*, 341 (2013) 1502–1505.

[42] J. Kong, H. Yang, X. Guo, S. Yang, Z. Huang, X. Lu, Z. Bo, J. Yan, K. Cen, K.K. Ostrikov, High-Mass-Loading Porous Ti3C2Tx Films for Ultrahigh-Rate Pseudocapacitors, *ACS Energy Letters*, 5 (2020) 2266–2274.

[43] D. Spurling, H. Krüger, N. Kohlmann, F. Rasch, M.P. Kremer, L. Kienle, R. Adelung, V. Nicolosi, F. Schütt, 3D Networked MXene thin Films for High Performance Supercapacitors, *Energy Storage Materials*, 65 (2024) 103148.

[44] A. VahidMohammadi, M. Mojtabavi, N.M. Caffrey, M. Wanunu, M. Beidaghi, Assembling 2D MXenes into Highly Stable Pseudocapacitive Electrodes with High Power and Energy Densities, *Advanced Materials*, 31 (2019) 1806931.

[45] S. T.E, D.T. Tran, S. Jena, Y. Bai, S. Prabhakaran, D.H. Kim, N.H. Kim, J.H. Lee, Flexible 2D Borophene-Stacked MXene Heterostructure for High-*Performance Supercapacitors, Chemical Engineering Journal*, 481 (2024) 148266.

[46] Z. Bo, Z. Zheng, Y. Huang, Z. Huang, P. Chen, J. Yan, K. Cen, H. Yang, K.K. Ostrikov, One-step MXene Selenization-Conversion Into Carbon-Scaffold-Anchored TiSe2 Nanosheets Enabling Interlayer- and Vacancy-Mediated Ion Transport Mechanisms for Fast Lithium-Ion Storage, *Chemical Engineering Journal*, 473 (2023) 145183.

[47] M. Wang, Y. Cheng, H. Zhang, F. Cheng, Y. Wang, T. Huang, Z. Wei, Y. Zhang, B. Ge, Y. Ma, Y. Yue, Y. Gao, Nature-Inspired Interconnected Macro/Meso/Micro-Porous MXene Electrode, *Advanced Functional Materials*, 33 (2023) 2211199.

[48] Y. Fan, J. Li, S. Wang, X. Meng, W. Zhang, Y. Jin, N. Yang, X. Tan, J. Li, S. Liu, Voltage-Enhanced Ion Sieving and Rejection of Pb2+ through a Thermally Cross-Linked Two-Dimensional MXene Membrane, *Chemical Engineering Journal*, 401 (2020) 126073.

[49] J. Orangi, F. Hamade, V.A. Davis, M. Beidaghi, 3D Printing of Additive-Free 2D Ti(3)C(2)T(x) (MXene) Ink for Fabrication of Micro-Supercapacitors with Ultra-High Energy Densities, *ACS Nano*, 14 (2020) 640–650.

[50] C. Lethien, J. Le Bideau, T. Brousse, Challenges and Prospects of 3D Micro-Supercapacitors for Powering the Internet of Things, *Energy & Environmental Science*, 12 (2019) 96–115.

[51] Y. Wang, Y. Yuan, H. Geng, W. Yang, X. Chen, Boosting Ion Diffusion Kinetics of MXene Inks with Water-in-Salt Electrolyte for Screen-Printed Micro-Supercapacitors, *Advanced Functional Materials*, 34 (2024) 2400887.

[52] S. Deshmukh, K. Ghosh, M. Pykal, M. Otyepka, M. Pumera, Laser-Induced MXene-Functionalized Graphene Nanoarchitectonics-Based Microsupercapacitor for Health Monitoring Application, *ACS Nano*, 17 (2023) 20537–20550.

[53] L. Zhang, J. Qin, P. Das, S. Wang, T. Bai, F. Zhou, M. Wu, Z.S. Wu, Electrochemically Exfoliated Graphene Additive-Free Inks for 3D Printing Customizable Monolithic Integrated Micro-Supercapacitors on a Large Scale, *Advanced Materials*, 36 (2024) 2313930.

10 Application of *Nano* Materials in Electrochemistry

Chang-Min Yoon and Zambaga Otgonbayar

1 INTRODUCTION

A *nano* material is defined as a material whose size or dimension falls within the range of 1–100 *nano* meters. In 1914, the Richard Adolf Zsigmondy gave the title "*nano* meter" for the first time and later, Noble Prize winner, Richard Feynman used this term during his annual conference speech in 1959 [1]. Before the 1980s, *nano* technology was mainly a topic of discussion, but researchers have been unanimous in their view that it is a potential field for future development [2–9]. Owing to their unique properties and benefits, *nano* materials are widely used in electrochemical applications such as energy storage, supercapacitors (SCs), batteries (BTs), and solar cells [10–15], owing to their unique properties, such as good electrochemical properties, high surface area, tunable electrical conductivity, and compatibility with other techniques. Figure 10.1 shows a summary of the history and development of *nano* materials and their expansion over the past decade [16–23]. Among the different fields, *nano* materials are widely used in SCs and BTs because of their exceptional properties that significantly enhance energy storage and conversion [24–27]. *Nano* material-based SCs have several advantages, including high energy capacity, short-term energy storage, and extended lifespan [28–30]. Graphite, a common electrode found in Li-ion batteries (LIBs), offers a capacity of 372 mAh g^{-1}, which is insufficient for supercapacitor applications. On the other hand, *nano* materials can provide higher capacitance, making them a more suitable choice for such applications [31]. Another advantage of *nano* materials for SCs is that they enable better loading, thereby maintaining their electrochemical properties with minimal structural changes over the cycle life. In addition to their impressive physicochemical properties, *nano* materials offer a number of advantages that make them particularly well-suited for SCs and BTs [32]. *Nano* materials used in electrochemistry primarily consist of transition metals, sulfides and oxides, carbon, polymers, and noble *nano* materials. The general elemental composition of the *nano* materials is depicted in Figure 10.2.

The novel electrochemical sensors and the electrodes are classified into four types based on the *nano* materials chosen: metal, metal-compound, carbon, and *nano* composite-based. The properties of *nano* materials used in electrodes and sensors are contingent on their precise composition, shape, and size, which also determine

DOI: 10.1201/9781003561262-10

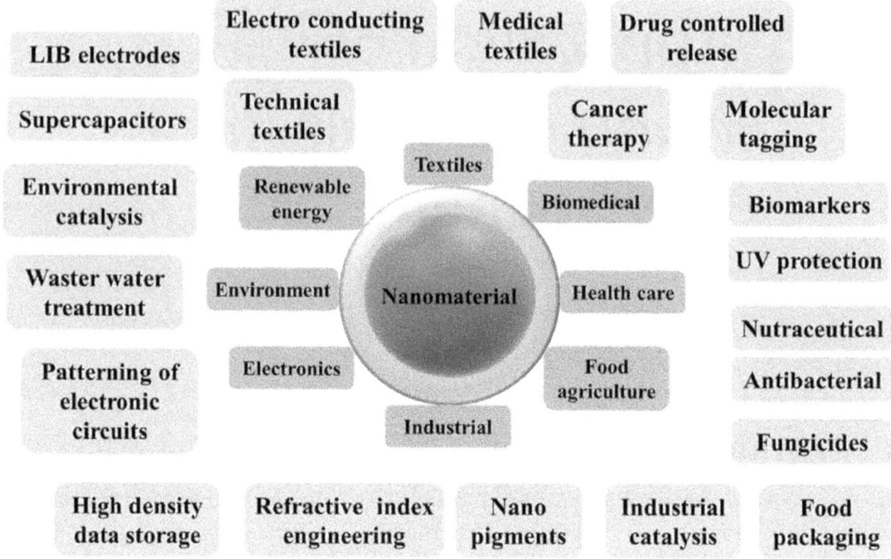

FIGURE 10.1 Key applications of *nano* materials.

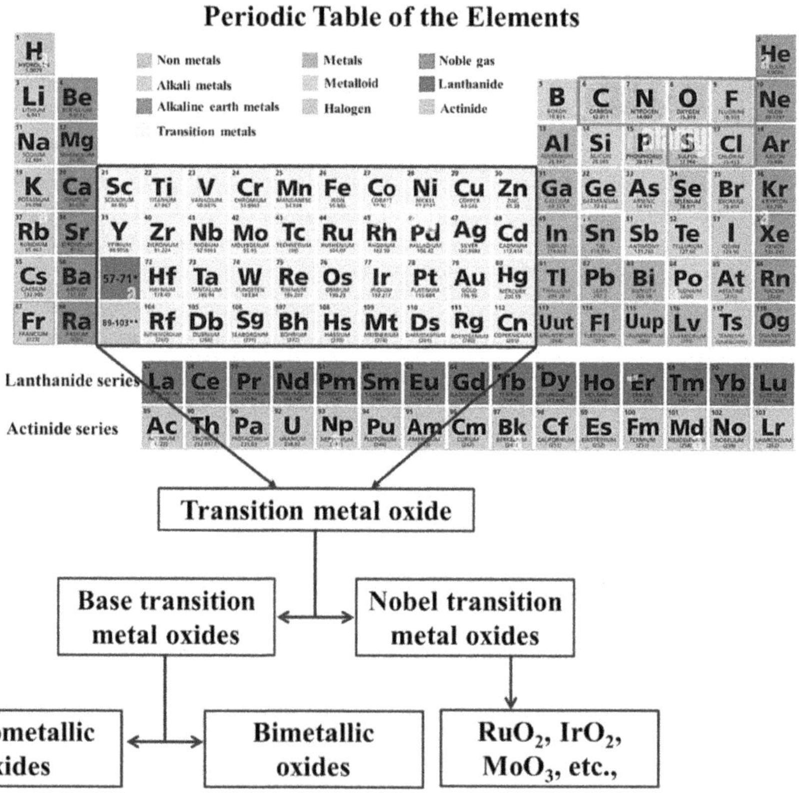

Periodic Table of the Elements

FIGURE 10.2 General elemental composition of *nano* materials.

TABLE 10.1

Brief Description of Keywords and Symbols Related to the Nano Materials

No	Term	Description
1	Aspect ratio	The proportion of the longer axis to the shorter axis in length
2	*Nano* scale	Range of sizes from 1 to 100 nm
3	*Nano* technology	Technology to develop objects by controlling factors at the *nano* scale to achieve unique properties.
4	*Nano* manufacturing	Bottom-up and top-down methods are used.
5	*Nano* material	Materials with a size between 1 and 100 nm.
6	Nano composite	Matter with a structure of having repetitive distances between the different phases.
7	*Nano* object	Materials with a confinement of one, two, and three dimensions at the *nano* scale.
8	*Nano* particle	Ultrafine units with dimensions measured in nm
9	*Nano* sphere	Spherical particles with a diameter measured in nm
10	*Nano* rod	Rod-shaped particle with one dimension measured in nm
11	*Nano* fiber	Fiber with the diameter in the range of nm
12	*Nano* wire	Wire with length in the range of nm
13	*Nano* tube	Microscopic tube with a dimension measured in nm

TABLE 10.2

Quantitative Data Collected from Web of Science

No	Application	Number of Research Articles Published by *Year*						Total Articles
		2018	2019	2020	2021	2022	2023	
1	*Nano* materials in "Sensor"	2,415	2,878	3,482	3,919	4,574	3,621	20,889
2	*Nano* materials in "Energy Storage and Conversion"	1,616	1,832	2,162	2,537	3,102	2,492	13,743
3	*Nano* materials in "Electrocatalysis"	296	300	414	491	606	450	2,557
4	*Nano* materials in "Supercapacitor"	697	809	904	999	1,168	994	5,571
5	*Nano* materials in "Solar cells"	1,022	1,054	1,242	1,375	1,545	1,209	7,447

their impact on health and the environment. The definitions and explanations of the terms related to *nano* materials are listed in Table 10.1.

In the last 6 years, there have been a total of 50,207 publications on *Nano* materials for sensor, energy storage, electrocatalysis, SC, and solar cells applications reported in Web of Science, searched with keywords "*Nano* materials." Table 10.2 presents a quantitative analysis of published papers on *nano* materials and their applications in the field of electrochemistry based on data from primary sources between January 2018 and September 2023.

2   SYNTHESIS METHODS

Top-down and bottom-up techniques are the primary methods for *nano* material synthesis. Top-down methods involve reducing the size of the raw and bulk materials using laser ablation, ball milling, etching, electroplating, and sputtering. In contrast, bottom-up methods involve crystallization of small organic and inorganic molecules and atoms to form *nano* materials of varying sizes. The most commonly used bottom-up methods include chemical vapor deposition (CVD), hydrothermal, and sol-gel methods. CVD is widely recognized as an effective method for producing high-quality *nano* materials on a variety of substrates. Plasma-enhanced pulsed laser deposition (PEPLD) is another technique that combines the traditional PLD configuration with low-temperature electrically generated oxygen plasma. PEPLD offers the advantage of improved stoichiometric control because adjusting parameters such as plasma density and electron temperature during deposition can significantly affect the properties of the deposited materials. The first "top-down" method is ball milling, which is a cost-effective method that allows the preparation of *nano* materials in different phases and the reduction of bulk materials to the *nano* scale [33]. The second top-down method is electrospinning, is primarily used to prepare *nano* fibrous and membranes. As a part of the top-down approach, sputtering is used to prepare *nano* materials after blasting a solid object with high energy particles using plasma or gas. In this method, intense gas ions are blast the target surface, forming *nano* materials through sputtering The latest method, laser ablation, uses a laser beam to create *nano* materials to vaporizing the precursor material without using any chemicals or stabilizers. Figure 10.3 shows a schematic of the top-down synthesis approach.

The next synthesis approach is bottom-up, which builds up the *nano* material from the atomic level to cluster into *nano* particles. The first method is CVD, which involves creating a vapor-phase material to produce a thin coating on a substrate. The following

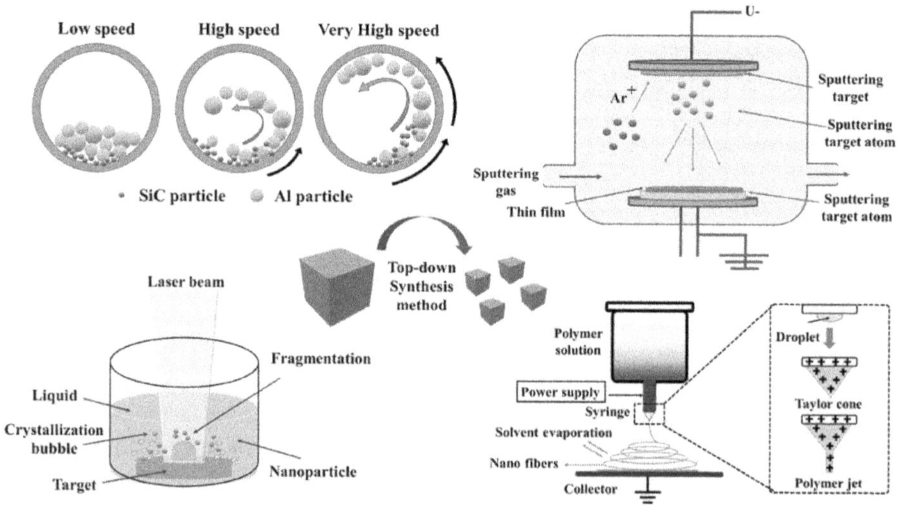

FIGURE 10.3   Schematic illustration of the top-down synthesis.

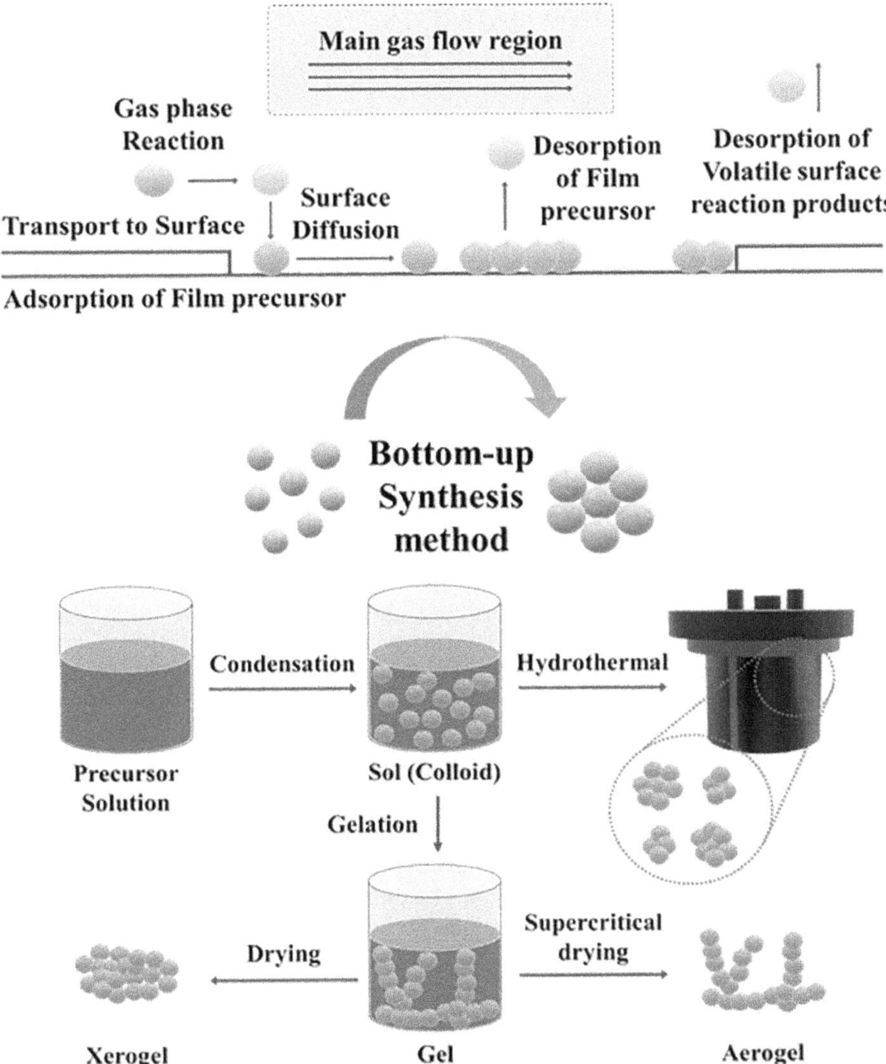

FIGURE 10.4 The schematic illustration of the bottom-up synthesis method.

criteria must be met for this method: high chemical purity, nonhazardous composition, good stability during evaporation, extended shelf life, and affordability with no residual impurities from decomposition [34]. A well-known bottom-up method is the hydrothermal method, which synthesizes *nano* materials through a heterogeneous reaction in a liquid medium under high pressure and temperature. The advantage of the hydrothermal synthesis is its ability to synthesize *nano* materials with various morphologies and sizes (Figure 10.4). Another method is the sol-gel method that is based on a wet-chemical process for forming *nano* material-based metal oxides (MOs), offering an eco-friendly approach for creating homogeneous *nano* materials [35].

3 PROPERTIES OF *NANO* MATERIALS

As mentioned earlier, *nano* materials are an extraordinary class of materials characterized by at least one dimension falling within the range of 1–100 nm. These materials exhibit unique properties owing to their *nano* scale size and are described in detail in the following subsections.

- *Mechanical properties*: *nano* materials have fascinating unique properties compared to their bulk counterparts owing to their small size, surface effects, and quantum effects. The key mechanical properties of *nano* materials include enhanced strength, hardness, fracture strength, elastic modulus, stress, and strain.
- *Thermal properties*: *nano* materials exhibit distinct thermal properties than their bulk counterparts due to their size, surface defect, and the unique surface area. The key thermal properties of *nano* material include thermal conductivity, melting point, heat capacity, and thermal expansion.
- *Electrical properties*: *nano* materials can have outstanding electrical properties that can differ significantly. The key optical properties of the *nano* materials include conductivity, semiconducting behavior, and resistivity.
- *Optical properties*: *nano* materials exhibit unique and interesting optical properties with controllable factors. The key optical properties of *nano* materials are absorption, scattering, emission, and photoluminescence, which can be modified by precisely controlling the surface, shape, surface functionalization, and size of *nano* materials.
- *Magnetic properties*: *nano* materials have unique and exceptionally intriguing magnetic properties, because of their magnetic domain. The key magnetic properties of the *nano* materials include super-para-magnetism and magnetostriction.
- *Catalytic properties*: *nano* materials exhibit excellent catalytic properties, enabling sustainable and efficient chemical processes. The key catalytic properties of *nano* materials include high surface area, reactive sites, surface energy, and improved selectivity, which are strongly affected by the increase in the reaction rate and production efficiency.
- *Large surface area*: *nano* materials have a large surface area to volume ratio. The key advantages of the large surface area of *nano* materials include enhanced reactivity, high efficiency catalytic properties, unique quantum effect, adsorption site, and tailored functionality.

The above key properties of *nano* material can be tuned by the precise control of size, shape, crystallinity, composition and alloying, surface energy and state. Therefore, *nano* materials provide immense potential across diverse application fields, fostering innovation and sustainable development, including electrochemistry.

4 KEY APPLICATIONS OF *NANO* MATERIALS

4.1 Electrochemical Sensors

The unique compatibility and tunable properties of *nano* materials enable their use in *bio*-electrochemical and medical applications. The compatibility of *nano* material

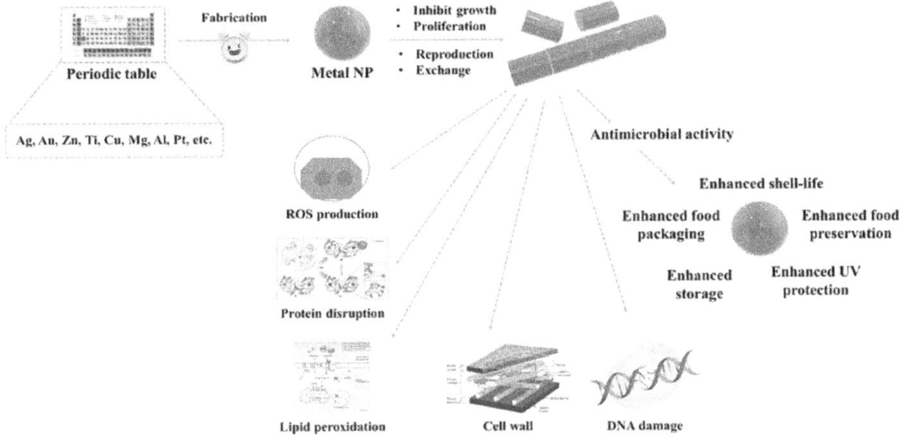

FIGURE 10.5   Metal-based *nano sensors* and their future used in biofield.

opens the way to revolutionize *bio*-electrochemical sensing, through "*nano sensors*". The main applications of *nano* materials in *bio*-electrochemical applications are sensing, *bio*-receptor immobilization, and signal amplification [36]. To be specific, the *nano* materials, like gold, carbon, and polymer-based *nano* materials are promising candidates for constructing biosensor electrodes (Figure 10.5). These materials present the high bio-receptor signal, selectivity, and sensitivity. The modest emergence of the *nano* materials propounds the design of the new electrode, building efficient interaction and electron transfer between the analyte and the electrode. *Nano* materials with controlled assemblies have emerged as electrode materials, improving bio-electrocatalytic performance. Four different types of *nano* materials with distinct properties have been used to construct sensing electrodes, each providing a unique sensing response [37–40]. First "*nano sensor*" is metal-based; it shows high electrochemical properties derived from the free movement of valence electrons in the metal lattice under an electric field. Metals, including silver, gold, and platinum, are promising sources of high detection signals owing to their excellent conduction mechanisms which are influenced by the lattice structure, surface deviations, and impurities like defects and vibrations. In addition, it is possible to improve and activate electronic conduction and mechanisms by adding other atoms to the lattice to control its defects and vibrations [41]. Therefore, it can be concluded that the modification of the lattice and its configuration is a direct approach for improving the sensing capability of metal-based sensors. *Nano sensors* are composed of MOs, and metal sulfides exhibit high sensing capabilities because of their high electron mobility, ionic conductivity, active sites, and bandgap [42].

The second "*nano sensor*" is carbon-based, mainly composed of graphite, graphene, polymers, and carbon *nano* tubes, with unique physicochemical properties associated with changes in sp, sp^2, and sp^3 hybridization, resulting in better electrical conductivity. The novelty of the carbon-based *nano* sensor is its high sensing capability and and providing accurate outcomes in a short term. Graphene, composed of π bond and sp^2 hybridized carbon atoms, is the potential carbon *nano* material for the electrode. The next material is carbon *nano* tubes, which are a candidate

TABLE 10.3

An Overview of the Electrochemical and Biosensor Systems and Sensing Results

No	*Nano* particles	Samples	Analytes	Linear Range	Detection	Ref
1	Ag	Nature water	Arsenic	13.3–375.2	0.24	[52]
2	Ag-GO	Bio samples	–	1.3–200 nm	32.5	[53]
3	Ag-Fe$_3$O$_4$	Waste water	–	0.01–1 ppb	0.00097	[54]
4	Pt-PPy	Ground water	Mercury	5–500 nM	0.27	[55]
5	Fe$_2$O$_3$-rGO	Tap-water	Nitrite	50 nM–0.78 mM	15	[56]
6	CoO-rGO	Vegetables	Carbo furan	0.2–70 mM	4.2	[57]
7	Ag-CQD	Mercury	Water	–	0.02	[58]

material that corrects the heterogeneous electron transfer between the electrode and the detection medium, thus improving the selectivity of the biosensor and electrode. Abovementioned carbon *nano* sensors have quick and accurate sensing properties that originated from the rapid electron transfer with physical aspects. In detail, the π–π bond of carbon *nano* materials enables the free movement of electric charge carriers in the entire lattice, which enhances the speed of electron transport. In addition, carbon-based *nano sensors* can be easily modified with the surface functional groups [-COOH and -OH] which are the actives sites, accelerating the electrochemical reaction [43–47]. The third "*nano sensor*" is composite-based, mainly consisting of carbon and another *nano* material which enable addressing the instability of the carbon-based electrode. Even though carbon-based *nano* materials had advantages, it has the drawback of agglomeration resulting from the van der Waals interaction between sheets. To address this complexity, the junction between two materials with distinct properties can be the driving force to attain the free charge carrier region, resulting in better sensing capability. Therefore, it can be concluded that *nano* materials with complex structures are emerging as better alternatives to overcome the instability of carbon electrodes, thus becoming a hot topic in the field [48–51]. An overview of the electrochemical and biosensor systems and their eco-friendly monitoring results are listed in Table 10.3.

4.2 ENERGY STORAGE AND CONVERSION

Energy storage and conversion is the pivotal field in this modern society of economic growth. Thousands of research projects have been focused on using non-renewable energy sources, but there are projects rarely focused on renewable energy sources that can meet rising demands. Recently, scientists and technologists have been focusing on renewable energy sources that can supply large scale markets for high energy applications, including electronic devices and autonomous driving applications used in daily life. Thus, it is becoming possible to meet the conditions for reducing the production of toxic gases. The key factors sustaining high performance in energy storage device is materials used in electrodes and the sorption and reaction occurred on the electrode surface. Due to their unique properties and controllable morphology, the *nano* materials become promising materials for use in electrodes, thus improving its practical potential. However, there are some challenges that need to be addressed, including the intrinsic and extrinsic properties of the *nano* materials to achieve the momentous improvement with the control of synthesis procedure from the viewpoint of chemistry, *nano* technology, and engineering. Although *nano*-sized materials hold great promise for applications in energy storage, research is still in its initial stage and has not fully utilized its potential. The types of studied *nano* materials are listed as follows: carbon-based *nano* materials, metals, MOs, polymer-based *nano* materials, hybrid *nano* materials, and nitride *nano* materials, are widely used in the energy storage and conversion fields [59–64].

The first *nano* material is "*transition metal and metal oxides*" that exhibits variable number of valence electrons and chemical stability. Metal oxides (MOs), including MnO_2, RuO_2, Fe_2O_3, and NiO, have been used as electrode materials; not only single MOs but also bimetallic MOs have been studied owing to their high electronic conductivity and stability for cyclability [65]. The second is "*polymer-based*, " which includes good redox and charge/discharge properties. Polymers such as polyaniline are the most noble materials used in energy storage devices owing to their flexibility and conductivity; however, they can be stored in air, thus bringing up the challenge of improving their properties by combining them with polyindole, resulting in hybrid polymer *nano* materials with improved redox properties and better thermal stability [66,67]. The third material is "*carbon-based*" that has high specific surface area (SSA) and more reactive sites than the MOs [68]. The well-known material is "*graphene (G)*" which has a single sheet-like structure with a high surface area and charge/discharge rate. The charge storage rates of the G and GO are found to be *ca.* 185 F g^{-1} and these are enhanced by combining with other MOs and other materials, which are called "*hybrid nano* materials." These materials benefits from fast electron transfer between the electrode and electrolyte, taking advantage of G and MO. The final conventional material is nitride, which possesses both long-term chemical stability and electrochemical properties, making it a potential candidate for electrode material. The summary of the capacitance values of the studied materials are listed in Table 10.4.

TABLE 10.4
The List of the Studied Nano Materials and the Specific Capacitance Values

No	Electrode Material	Specific Capacitance Value (F g^{-1})	Ref
1	MnO_2-V_2O_5//AC (activated carbon)	61	[69]
2	$Ni(OH)_2$ *nano* sphere//AC	120	[70]
3	$NiCo_2O_4$//AC	135	[71]
4	$Ni_{0.67}Co_{0.33}$ Se//RGO	176	[72]
5	HPCNTs//HPCNTs	139	[73]
6	CNT/graphene//Mn_3O_4/graphene	72.6	[74]
7	Co_2AlO_4@MnO_2 *nano* sheet//Fe_3O_4 *nano* flakes	99.1	[75]
8	$ZnCo_2O_4$@MnO_2//Fe_2O_3	161	[76]
9	Ni-Co-BH-G//CCN	340	[77]
10	Ni-Mn LDH/rGO//AC	82.26	[78]
11	RuO_2-NPG//$Co(OH)_2$-NPG	350	[79]

4.3 ELECTROCATALYSIS

Nano materials hold a significant place in the field of electrocatalysis because of their exceptional properties and high efficiency. The focus of this field is to determine the optimal design and characterization of functional *nano* materials to enhance their electrocatalytic performance as electrode materials in alternative energy devices. This includes studying model electrode *nano* materials and establishing a theoretical basis for their electrocatalytic activities in specific reactions. Advanced *nano* materials (ADNs), high-entropy *nano* materials, and electrospinning *nano* materials are the main materials for electrocatalysis that have received the most attention. In the following paragraph, the properties of each material are briefly explained. ADNs have properties which significantly enhanced the electrocatalysis field with efficiency. The quickest and most effective method to improve the properties is the control of their size, shape, and intrinsic structure. In addition, ADNs possess large surface areas for increased reaction rates, along with favorable transport properties for enhanced electron and mass transfer, resulting in superior electrocatalytic performance.

High-entropy *nano* materials (HEMs) have multiple compositions and unique entropy states, offering adjustable activities and stability [80–82]. The high-entropy state refers to a high degree of disorder at the atomic level, which results from the equal or near-equal molar ratios of multiple elements in the material (Figure 10.6). The unique properties of HEMs make them promising materials for electrocatalytic applications. In particular, their multi-element composition can lead to a rich variety of active sites, and the high-entropy state can result in a unique electronic structure, leading to improved resistance to deactivation, thereby enhancing catalyst stability. HEMs can be synthesized using various methods including mechanical alloying, sputtering, and electrodeposition. These methods aid in precise control over the composition and structure of the materials, enabling the optimization of their electrocatalytic properties.

Electrospun *nano* materials (ESMs), have large surface areas, unique chemical structures, easily tunable compositions, and excellent electron and mass transport properties, thus enabling its use in efficient energy storage and conversion applications (Figure 10.7). The high surface area of ESMs is the key factor that provides

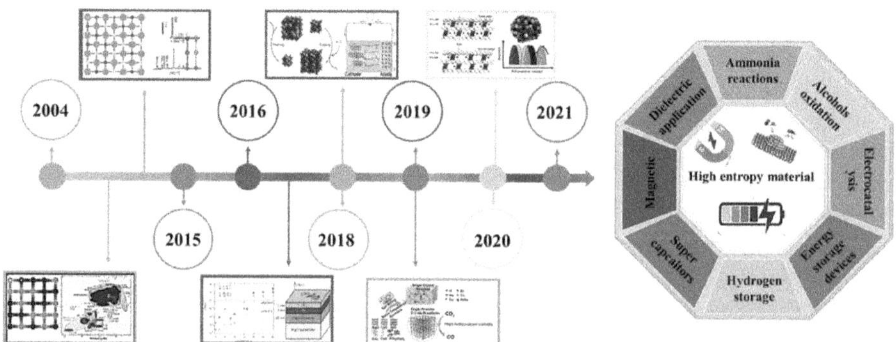

FIGURE 10.6 Roadmap of HEMs for recent energy-related applications.

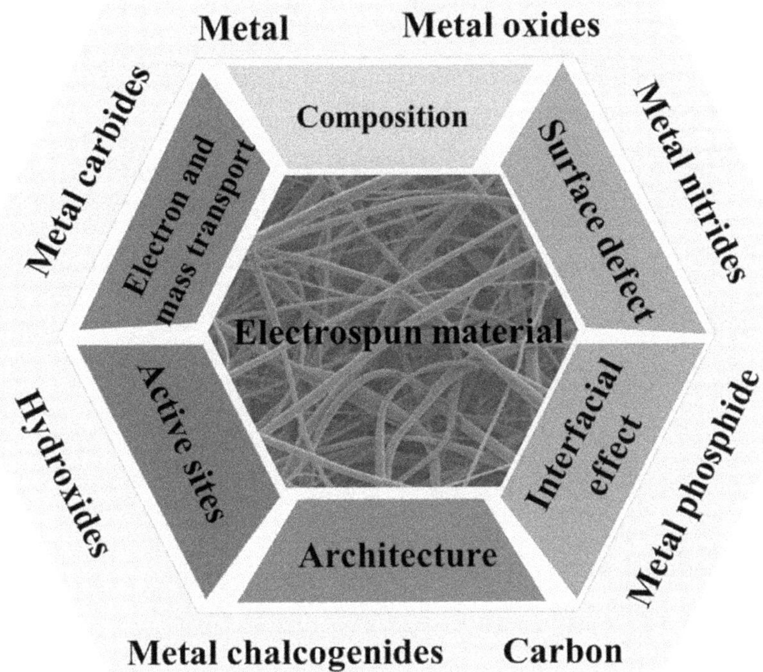

FIGURE 10.7 Schematic illustration of the electrospun *nano* materials for their applications in electrocatalysis.

more active sites for electrocatalytic reactions, leading to improved reaction rates and efficiency. In addition, the ESMs have excellent electron and mass transport properties, which are particularly important in electrocatalysis where the rate of the reaction is often limited by the rate of electron transfer or mass transport [83,84].

4.4 SCs

As an energy storage system, SCs have received enormous attention owing to the large specific capacity, higher energy density, long cycle lifetime, and low voltage limits [85]. SCs can store 100 times more energy per unit volume than the electrocatalytic capacitors, with high power density and charge/discharge rates. The usage field of SCs may include various fields that has high necessity of using clean and renewable energy sources [86]. According to the published articles, the first patent of SCs was in 1957, and the first application field of SCs was hybrid electric vehicles. Since then, the SCs have become the supplementing device for the field of energy storage. In 2014, the US Department of Energy certified the SCs as a comparable energy storage device with BTs for future society and the rapid growth of industrial technologies. Since then, researchers, organizations, and government agencies have paid considerable attention to the development of new SCs for practical applications. The important parameters and the features of SCs are illustrated in Figure 10.8. Despite

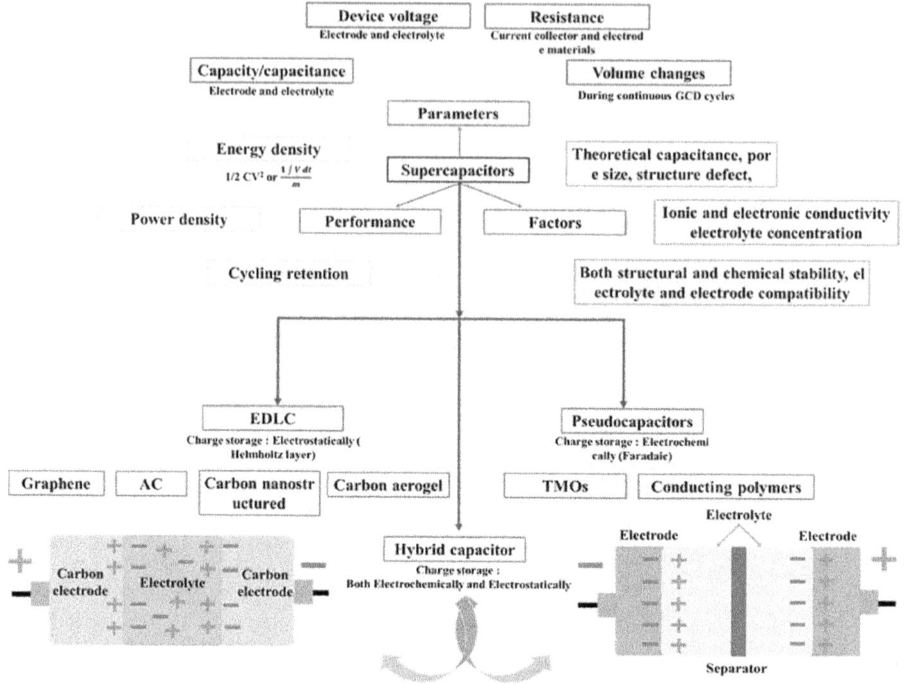

FIGURE 10.8　SC performance determinants and parameters and classification of SCs.

these advantages, manufacturing SCs with low production costs while maintaining their advantages remains a challenge. Researchers have been studying to find a way to reduce the production costs, and the most effective way to reach the goal was fabricating the *nano* material electrode. The highest cost was originally derived from the electrode fabrication process [87]. Substantial progress has recently been made in enhancing the theoretical and practical capabilities of SCs. Understanding SC design is important to enhance the functionality and potential, as the electrode design of SCs significantly impact on the energy storage mechanism, and this is thoroughly discussed [88].

For designing electrodes for SCs, the natural properties of the *nano* material with the choice of electrolytes and their compatibility are important. The classification of the electrolyte for SCs is discussed. An electrolyte is the parameter for balancing charge in SCs and it forms an electrical connection between the cathode and anode. An electrolyte is the most important part responsible for charge transfer and balance between SC electrodes [89–95]. Three different electrolytes have been selected as the SC electrolyte, including aqueous, ionic, organic, and solid-state. Each electrolyte has its advantages but has limitations to be used in SCs for a long term [96–98]. Solid-state electrolytes (SSEs) are a commonly utilized type of electrolyte within the electrolyte family; however, they are known to have relatively low ionic conductivities. The development of electrolytes that are compatible with electrode materials is crucial for achieving a high capacitance, high rate performance, complete

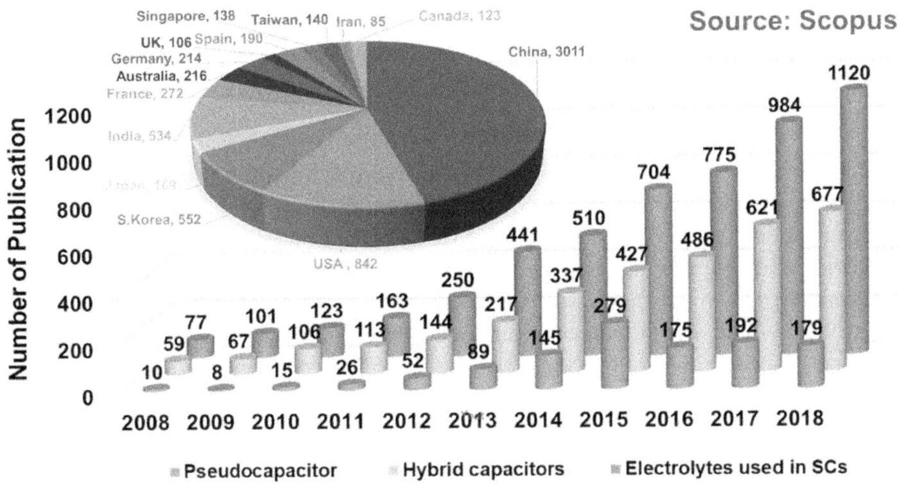

FIGURE 10.9 The total number of publications for SCs, electrolytes used in SCs, and the country distribution of publications.

Source : Scopus

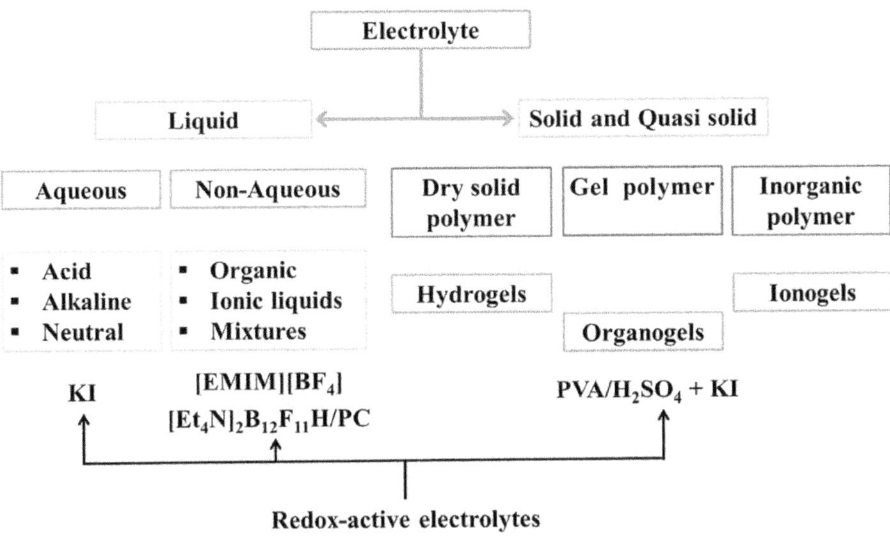

FIGURE 10.10 Types of electrolytes and the example of redox-active electrolytes.

cyclability, and the ability to operate at high voltages within broad operational windows. Figure 10.9 illustrates the number of articles on electrodes and electrolytes accompanied by a map representing the distribution of electrolytes in relation to countries. In contrast, Figure 10.10 presents the classifications of electrolytes used in SCs along with their subcategories [99,100]. In addition, SCs fall into three main categories, like electrostatic double layer capacitors (EDLCs), pseudo capacitors (PCs), and hybrid capacitors (HCs). The fundamental history of each capacitor is summarized in the following subcategories.

4.4.1 EDLCs and the *Nano* Materials

EDLCs, the member of the SC family, is unlike traditional capacitors that rely on the electrostatic field between conductive plates; EDLCs store energy through an electrochemical process. The concept of EDLC was first reported in the nineteenth century by von Helmholtz [101] and later modified by Gouy et al. The fundamental working mechanism of the EDLC is based on the charge/discharge process in porous electrodes, which are immersed in an electrolyte. With the application of voltage, the ions of the electrolyte move toward the electrodes of opposite charges, forming a "double layer" of a charge at the electrode/electrolyte interface. The accumulation of the charges at the interface results in energy storage. The physical nature of charge storage contributes to the long cycle life with minimal material degradation over time. The advantages of EDLCs include rapid charge/discharge, durability, high energy density, and zero redox reactions. A concise and detailed description of the advantages is listed: For rapid charge/discharge, EDLCs can handle the rates without any degradation, thus making them suitable for practical applications that require quick energy release [102–104]. For durability, physical mechanisms ensure a long lifespan. For high energy density, compared with BTs, they offer a higher energy density than traditional capacitors. In addition, zero chemical reactions occur in the EDLC during the energy storage process. It can be concluded that EDLCs can provide a balanced and rapid energy release from capacitors and the energy density of BTs, making them valuable candidates for SCs. Conventional capacitors store energy by accumulating charges on two parallel metal electrodes separated by a dielectric medium, where a potential difference exists between them. The capacitance C of the parallel-plate capacitor is given by:

$$C = \varepsilon \varepsilon \tau A / d \tag{1}$$

where A is the area of the metal plate, ε is the permittivity of the medium, and d is the distance between metal plates. As mentioned before, the carbon family is the promising candidate for electrode materials of EDLCs. For EDLCs, the carbon and carbon derivatives are well-suited electrodes due to their novel advantages, including the no-redox reaction involved, superior capacitive performance, and charge/discharge profile [105]. Researchers have explored and reported on various carbon-based compounds, such as graphene and carbon *nano* tubes, in scientific literature. Figure 10.11 shows different types of allotropes of carbon and carbon derivatives that can be used for such purposes.

Among the different carbon-based materials, activated carbon (AC) has a noteworthy specific surface area (SSA). However, it also has a large number of micropores, which impede ion transmission and prevent the formation of an effective double electric layer. Although activated carbon fibers (ACFs) exhibit better performance, their capacitance density is relatively low. The cost of carbon aerogels (CAs) is extremely high, and large-scale production is challenging. Many researchers

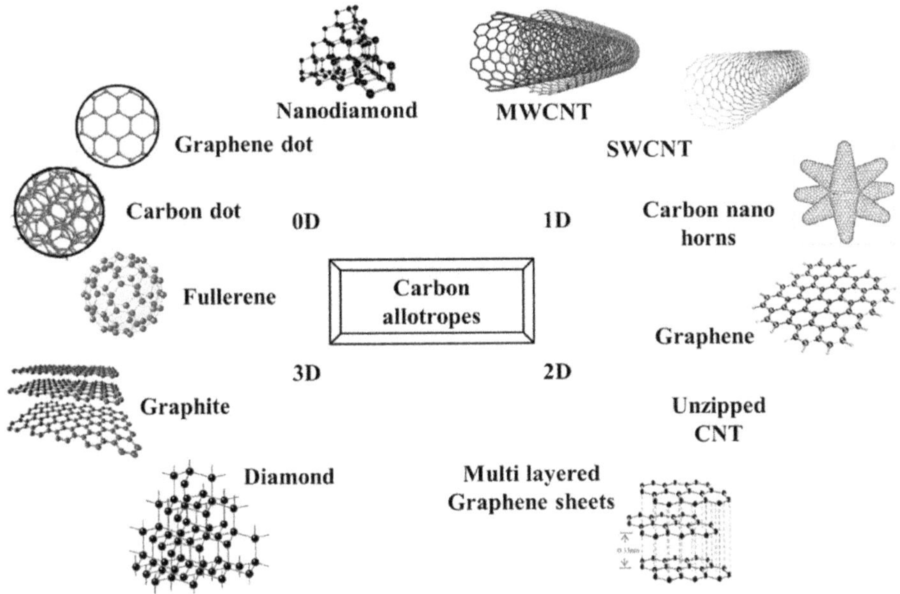

FIGURE 10.11 The allotropes of carbon and their structure.

have proposed new techniques to produce CAs as substitutes for supercritical drying. However, the performance of the resulting CA deviates from that of the CA produced using a supercritical gel.

The first carbon material is activated carbon (AC) that offers controllable properties by its chemical and physical properties. In addition, the incorporation with other metals and MOs is easy and enable the improvement of their properties. The specialty of AC, compared with other carbon, is low cost, cycle-stability, high corrosion resistance, and bio-waste. The second carbon material is graphene, which comprises a single layer of carbon atoms. The surface area, high conductivity, and distribution of the porous structure are crucial factors for graphene, which makes it a good candidate for electrode materials. However, electrodes made of graphene do not operate in their full potential. To utilize the maximum potential of graphene, it is necessary to prevent restacking of the graphene sheets [103]. The third most widely studied carbon material is carbon *nano* tubes (CNTs), which consist of carbon atoms arranged in a hexagonal lattice to form a cylindrical structure. Specifically, CNTs have two distinct structures: single-wall and multi-wall. Owing to their structure, CNTs have a unique porous structure with interconnected porosity, which allows for continuous charge distribution. Owing to their morphology, incorporation with other electrode materials can be easily conducted, except for the reduction of the surface area. The carbon-based electrode material for EDLCs and their responsive results are listed in Table 10.5.

4.4.2 Pseudocapacitors and the *Nano* Materials

PCs, or pseudocapacitors, is the energy storage device based on the Faradaic charge transfer processes occurring on the surface of electrodes. Owing to the

TABLE 10.5

Summary of Various Carbon-Based Electrode Materials for EDLCs

No	Electrode Material	Electrolyte	Cell Voltage (V)	Power Density (W kg^{-1})	Energy Density (Wh kg^{-1})	Specific Capacitance (F g^{-1})	Ref
1	AC	TEABF$_4$/AN	2.5	681.0	20.0	80.0	[106]
2	TiO$_2$/AC	1M Na$_2$SO$_4$	1.2	4200	28.0	282.0	[107]
3	AC on rGO	1M Na$_2$SO$_4$	1.6	469.2	11.9	541.0	[108]
4	Miler straw AC	2M KOH	1	101.0	18.7	144.0	[109]
5	AC from organic	6M KOH	0.8	400.0	20.9	59.0	[110]
6	rGO	2M KOH	1	–	–	114.0	[111]
7	PANI/rGO	1M H$_2$SO$_4$	0.8	368.0	24.5	299.0	[112]
8	N-doped Graphene	EMIMBF$_4$	4.1	1000	82.0	239.0	[113]
9	N-doped carbon aerogel	6M KOH	1	–	–	208.0	[114]
10	Hierarchical porous carbon	6M KOH	0.9	–	10.5	384.0	[115]
11	Graphene flakes	γ-Butyrolactone/[EMIM][TFSI]	3.0	7500	23.5	101.0	[116]

electrochemical energy storage mechanism, it was positioned between the traditional capacitors and the conventional BTs. Total capacitance of the PCs is sum of the electronic charge transfer and double layer component at the interface of the electrode/electrolyte. Although the final outcome of the PCs is similar to that of EDLCs, the power density of the PCs is lower than EDLCs, despite their high capacitance. In spite of charge transfer consisting of a redox reaction, the electrodes of PCs tend to swell and lose their cycle lifetime and the stability. Metals, metal oxides, and polymers are the main electrode materials of PCs, owing to their conductivity. A summary of the *nano* materials studied for PCs is listed in Table 10.6.

According to previously reported articles, some PCs exhibited higher capacitance than the EDLC, because of absence of charge storage in the double layer type. Currently, the researchers are focused on forming the sustainable electrode material for PCs, using the MOs and conducting polymers (CPs).

CPs are beneficial for creating electrodes because they exhibit tunable redox behavior and conductivity that can be chemically modified. These traits allow the CP electrodes to store more ions by exchanging them with the polymer backbone during oxidation and releasing them into the electrolyte during reduction. The bulk of the CPs undergoes these reactions, resulting in an extremely reversible discharge process that does not involve any structural changes, such as

TABLE 10.6

The Summary of the Electrode Material for PCs

No	Electrode Material	Electrolyte	Cell Voltage (V)	Power Density (W kg⁻¹)	Energy Density (Wh kg⁻¹)	Specific Capacitance (F g⁻¹)	Ref
1	Cr_2O_3 NPs	6M KOH	1	312.5	12.5	340	[117]
2	Ni_3BHT	1M $LiPF_6$	1.7	–	–	227	[118]
3	$Co_3V_2O_8$	–	–	–	–	790	[119]
4	MoS_2/rGO	1M H_2SO_4	1.5	370	23	361	[120]
5	NiCoHC/CDs	6M KOH	1.5	–	–	1964.6	[121]
6	$VO(OH)_2$/CNT	1M $LiClO_4$/PC	2.2	63.65	32	265	[122]
7	$CuFeS_2$	1M LiOH	1.4	1566	4.74	95	[123]
8	PHATN	6M KOH	1	–	–	686	[124]
9	THPP-PA-Mn	1M Na_2SO_4	1	1000	12.6	90.9	[125]
10	$MnCo_2O_4$ HSs/NF	–	–	250.1	37.1	648.4	[126]
11	Fe-Al co-doped Ni-based	6M KOH	0.5	370	30.27	495.6	[127]

phase transitions. However, the preparation of CPs requires specialized procedures, and expensive chemicals and reagents. Additionally, polymers are relatively expensive electrode materials, which restricts their practical application in SCs. Consequently, CPs have not been extensively explored and accepted as materials. The background history of CPs is summarized in the following sections. The first discovery of CPs was in 1963, and since 1976, MacDiarmid has begun extensive research on its possible applications [128]. The working mechanism of CPs involves oxidation-reduction reactions during charging and discharging, leading to the rapid production of n-type or p-type doping on the polymer film [129]. However, the poor mechanical stability of CP-based PCs often leads to degradation after 1,000 cycles because of repeated intercalation and depletion of ions during charging and discharging. The commonly used CPs include polyaniline, polythiophene, polypyrrole, polyvinyl alcohol, poly (3,4-ethylene dioxythiophene), polyacetylene, poly (4-styrene sulfonate), and polyphenylene-vinylene.

The first CP was polyaniline (PANI), discovered in the mid-nineteenth century by Henry Letheby, who studied the electrochemical and chemical oxidation products of aniline in acidic media. The theoretical specific capacitance of PANI is 233–1220 F g⁻¹ [130]. The properties of PANI electrode materials are determined by their microscopic morphologies and sizes. Various PANI electrode materials with different morphologies have been synthesized, including PANI *nano* fibers, *nano* wires, *nano* tubes, hollow spheres, *nano* belts, and octahedra. The summary of the PANI-based electrode and their properties are listed in Table 10.7. The second most promising CP is polythiophene (PTh), a modification of the thiophene used as an electrode material

TABLE 10.7

The List of PANI-Electrode Material for PCs

No	Electrode Material	Scan Rate	Specific Capacitance (F g^{-1})	Cycle Time	Ref
1	Urchin-like PANI	10 mV s^{-1}	435.0	1,000 (93%)	[131]
2	Cabbage-like PANI	10 mV s^{-1}	584.0	3,000 (91%)	[132]
3	Feather fan-like PANI	50 mV s^{-1}	403.5	2,000 (93%)	[133]
4	PANI-Graphene	2 mV s^{-1}	233.5	1,500 (99%)	[134]
5	Sandwich like PANI/ graphene	50 mV s^{-1}	900.0	5,000 (-)	[135]
6	PANI/GO	5 mV s^{-1}	475.0	2,000 (90%)	[136]
7	PVA/Graphene/PANI	30 mV s^{-1}	90.0	88,000 (75.6%)	[137]
8	rGO/Fe$_3$O$_4$/PANI	100 mV s^{-1}	283.4	5,000 (78%)	[138]
9	Cotton-like PANI/CNT		144.0	2,000 (92%)	[139]

in SCs and related electrode materials. Although the n-doped form of PTh typically has a lower mass than the p-doped form, its conductivity is relatively low. PT derivatives are generally stable and can be doped in air and moisture; however, this limits their use as additive anode materials owing to their high self-discharge in devices and short cycle times [132]. Polypyrrole (PPy) is a highly flexible CP that has been widely studied as a cathode material for electrochemical processing. Despite its high density, which gives it a high capacitance of 400–500 Fcm^{-3}, it cannot be n-doped like thiophene derivatives, which limits its use. To address this limitation, Guo et al. designed flexible solid-state symmetric SCs by combining 3D interconnected PPy *nano* structures with a porous H$_2$SO$_4$-PVA electrolyte. The resulting SCs achieved a high Cs value of 505 mF cm^{-2} (gravimetric Cs = 168 F g^{-1}) at 2.0 mA cm^{-2}, as reported in a study by Guo et al. [140].

The second sustainable electrode material for PCs is MOs owing to their low cost, environmental friendliness, and accessibility. MO-based electrodes possess the characteristics of both carbon and Faradaic materials, allowing them to store energy electrostatically and undergo faster redox reactions because of their diverse oxidation states. When selecting MOs for electrochemical applications, it is essential to consider the following aspects: high electro-conductive MOs, having changeable oxidation states which enables the fast reversible reactions and the layered structure which allows the reciprocal transformation of O$_2$ to OH- of the electrolyte. Researchers have primarily focused on studying MOs such as RuO$_2$, NiO, MnO$_2$, V$_2$O$_5$, and CoO as electrode materials owing to their remarkable attributes, including large theoretical capacitances (1360 F g^{-1}). However, their high cost restricts their practical applications. Consequently, scientists are investigating more cost-effective and efficient alternatives. The specific details of some MOs are summarized. RuO$_2$ is the first MO, which is currently highly favored due to its impressive properties, such as the potential to deliver a high theoretical capacitance of approximately 2000 F g^{-1}, good conductivity, and stability in solution [141]. Moreover, RuO$_2$ exhibits a relatively constant and considerable capacitance (600–1000 F g^{-1}) over a 1.4 V range, depending on the preparation, measurement, and the use of support. However, the actual capacitance often falls short of the theoretical value due to the high crystallinity and high power demands. Nonetheless, RuO$_2$/H$_2$SO$_4$ aqueous solution system has been successful, as demonstrated by Fugare et al., who used ultrasonic spray pyrolysis to create RuO$_2$ samples in a H$_2$SO$_4$ solution. The optimized electrode exhibited a maximum capacitance of 2192 F g^{-1} at 2 mV s^{-1}, exceeding the theoretical value. Despite the high cost and toxicity of RuO$_2$, its low porosity, and other limitations, it is predominantly used in device and military applications. The most commonly used second electrode material for PCs is MnO$_2$, which offers several advantages such as a wide range of resources, low cost, environmental friendliness, multiple oxidation valence states, and high theoretical capacity (1370 mAh g^{-1}) in neutral electrolytes [142]. However, MnO$_2$ powders prepared using different methods exhibit significantly varying capacity values, ranging from 100 to 250 mAh g^{-1}. Several methods, including the sol-gel approach, dip coating, drop coating, hydrothermal methods, electrochemical deposition, and combustion methods have been used to prepare MnO$_2$ films. However, the cycle stability of materials synthesized by these methods is not as good as that of pure carbon-based materials. To addres this, scientists are focusing on developing *nano* structured carbon/MnO$_2$ composites with uniform morphologies and large exposed surface areas to achieve high capacitance performance and excellent electrochemical stability. Nickel oxide is a highly recommended electrode material for SCs owing to its widespread availability, affordability, environmental friendliness, and low toxicity [143]. With a high theoretical capacitance of 2584 F g^{-1} in the potential window of 0.5 V, NiO is the strong candidate. However, the actual capacitance of NiO is lower than its theoretical value, which is attributed to its p-type semiconductor characteristics and poor contact between the substrate and the active material. To overcome these limitations, researchers have prepared *nano* structured NiO using SiO$_2$/C core-shell microspheres as templates. This approach produced hierarchical NiO/carbon hollow spheres with an inner cavity that acted as a reservoir for electrolyte ions, ensuring a stable supply of the electrolyte during rapid charging and discharging. Additionally, the multi-shell hollow spheres increased the number of active sites for redox reactions. NiO microspheres fabricated *via* precipitation on a carbon template have been reported to exhibit a high capacitance of 1063 F g^{-1} at 2 A g^{-1}. Although NiO foam is used in SC applications owing to its three-dimensional layered structure and high electrical conductivity, its performance per unit area remains limited. However, foamed Ni-based electrodes have demonstrated impressive cycle stabilities and high capacity retention.

Except for some specific MOs, ferric oxides and their derivatives, which are TMOs, have been extensively studied in the field of electrochemistry [144]. Ferric oxide is one of the most abundant elements on Earth and can exist in various valence states including Fe^0/Fe^{2+}, Fe^0/Fe^{3+}, and Fe^{2+}/Fe^{3+}, making it an ideal material for use in PCs. The main reason for Fe-based materials being suitable materials for PCs is the presence of multiple valence states in the ferrium (Fe^{2+} and Fe^{3+}). The charge

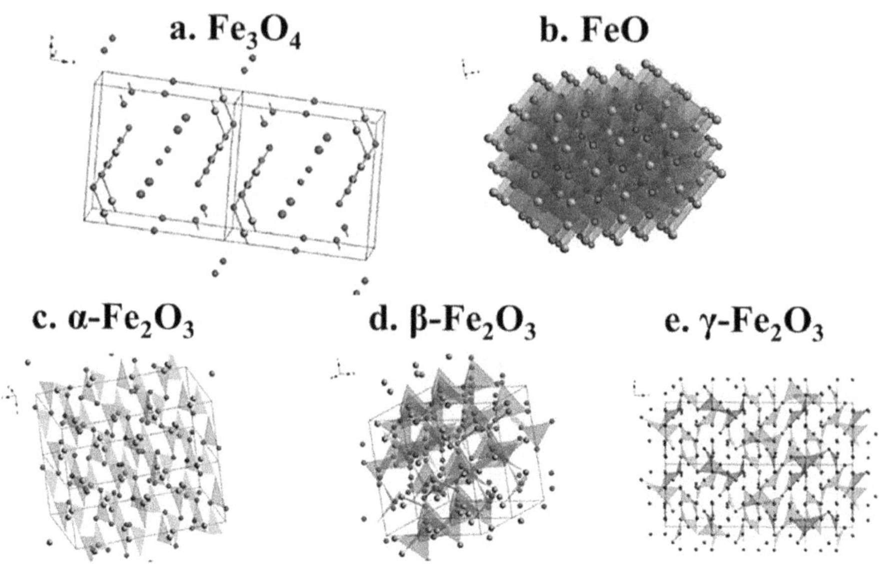

FIGURE 10.12　The multiple phase structures of Fe_2O_3.

storage process in Fe-based oxides during redox reactions in aqueous electrolytes can be divided into two distinct stages:

$$FeO_x + OH^- \leftrightarrow FeOOH + e^- \tag{2}$$
$$FeOOH + OH^- \leftrightarrow FeO_x + H_2O + e^- \tag{3}$$

The Fe-based electrode undergoes an oxidation reaction during charging, where Fe ions transitions from a lower valence state (Fe^{2+}) to a higher valence state (Fe^{3+}) and vice versa during discharge. This reaction involves both charge transfer and ion adsorption, allowing the SCs to store electrical energy and release it. This chemical reaction defines the energy storage mechanism of the Fe-based electrode. The standard form of Fe is Fe_2O_3, which has numerous phase structures that require specific synthesis conditions to attain various electrochemical properties. Among these phases, α-Fe_2O_3 is more structurally stable than β-Fe_2O_3 and γ-Fe_2O_3 because of its corundum crystal structure (Figure 10.12). This crystal structure enables α-Fe_2O_3 to undergo phase transformations effortlessly, such as transitioning from the α- to the β-phase, when heated to 500°C in air, making it an ideal electrode material with a long lifetime.

Nano materials exhibit various characteristics including crystal structure, form, alignment, and imperfections, which significantly influence their electrochemical efficiency. Methods, such as molecular defect design and polyatomic co-doping, can improve the capacitive performance of Fe-based materials. Defects in the crystal lattice, such as missing and additional atoms, can be used to generate new electrically active sites. In ionic crystals, defect movement facilitates electrical conduction, and point defects can create local states that are distinct from those of Bloch waves. These states may supply electrons by heating the

TABLE 10.8

Summary of the Fe-Based Electrodes and Electrochemical Results in Various Electrolytes

No	Electrode Material	Electrolyte	Capacitance Performance	Cycle Time	Energy Density (Wh kg^{-1})	Power Density (W kg^{-1})	Ref
1	α-Fe$_2$O$_3$.*Nano* flake	KOH	164 F g^{-1} 4A g^{-1}	10,000 (92%)	21.5	158.2	[145]
2	α-Fe$_2$O$_3$.*Nano* rods	3M KOH	1253 F g^{-1} at 3mV s^{-1}	3,000 (93%)	85.7	1.4	[146]
3	α-Fe$_2$O$_3$@CeO$_2$	2M Na$_2$SO$_4$	168 F g^{-1} at 5 mV s^{-1}	2,000 (50%)	15.62	781	[147]
4	Fe$_2$O$_3$@ACC	3M LiNO$_3$	2775 mF cm^{-2} at 1mA cm^{-2}	10,000 (92%)	9.2	12	[148]
5	α-Fe$_2$O$_3$ NDs/ RGO	2M KOH	347.4 F g^{-1} at 1A g^{-1}	2,500 (93.6%)	21.6	750	[149]
6	Fe$_3$O$_4$/C	–	278 F g^{-1} at 1A g^{-1}	5,000 (96.5%)	17.3	700	[150]
7	Fe$_3$O$_4$/N-CNTs	–	314 C g^{-1} at 1A g^{-1}	5,000 (92%)	25.3	699	[151]
8	Fe$_3$O$_4$-C-C	6M KOH	1153 F g^{-1} at 2 A g^{-1}	8,000 (96.7%)	17	400	[152]
9	Graphene/ Fe$_3$O$_4$	1M Na$_2$SO$_4$	268 F g^{-1} at 2mV s^{-1}	10,000 (98.9%)	87.6	394	[153]
10	Fe$_3$O$_4$/PANI/DE	–	178.7 F g^{-1} at 1A g^{-1}	1,000 (73%)	–	–	[154]
11	Fe$_3$O$_4$/@CF/C	1M KOH	355 F g^{-1} at 3.73 A g^{-1}	8,000 (93.39%)	1.8	32.4	[155]
12	FeS-CNF	6M KOH	503 F g^{-1} at 2A g^{-1}	1,000 (90%)	–	–	[156]
13	FeS$_2$/rGO	1M Na$_2$SO$_4$	12.41 mF cm^{-2} at 5 mV s^{-1}	10,000 (90%)	–	–	[157]
14	Fe$_3$C/Fe@NC	6M KOH	1695 F g^{-1} at 2 A g^{-1}	4,000 (91.4%)	72 and 35.8	0.83	[158]
15	ZnFe$_2$O$_4$/gCN	6M KOH	103 F g^{-1} at 10A g^{-1}	5,000 (94%)	–	–	[159]
16	NiFe$_2$O$_4$/MoS$_2$	Alkaline solution	506 F g^{-1} at 1A g^{-1}	3,000 (90.7%)	–	–	[160]
17	FeCo$_2$O$_4$/PPy	2M MOH	2269 F g^{-1}	5,000 (90.2%)	68.8 and 52	0.82 and 15.5	[161]

conduction band, leading to electrical conductivity in ionic crystals. The results and compositions of the Fe-based electrodes are presented in Table 10.8. From the analysis results of Fe$_2$O$_3$-based electrode, it can be inferred that Fe$_2$O$_3$-based composites have improved electrochemical performance compared to that of pristine Fe$_2$O$_3$ [162].

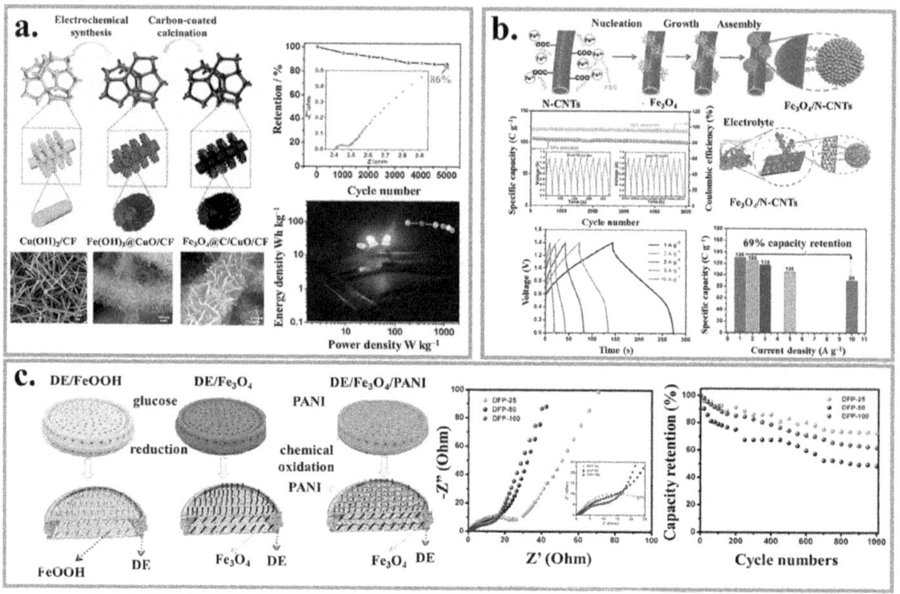

FIGURE 10.13 (a) Analysis result of the Fe_3O_4@C/CuO/CF [163], (b) the formation mechanism of the Fe_3O_4/N-CNTs nano composite and cyclic stability along with coulombic efficiency of the asymmetric supercapacitor device, rate capability [164], (c) schematic illustration and the electrochemical result of the formation process of DE/Fe_3O_4/PANI [165].

Fe_3O_4, an Fe-based electrode, has received considerable attention in the field of electrochemistry. Despite its promising electrical conductivity, its magnetic properties lead to aggregation on a large scale, decreasing the surface area of the active material and the available reaction sites. To overcome this limitation, various strategies have been explored to enhance the conductivity and prevent its aggregation.

The first strategy is encapsulation, which prevents aggregation and enhances the electrode conductivity. Conductive polymers (CPs) are widely used for encapsulation because they offer an additional active surface area and contact area with the electrolyte, resulting in improved energy storage properties. Zhang et al. employed hollow CuO *nano* tubes as scaffolds to deposit Fe_3O_4 *nano* particles onto copper-foam substrates, resulting in the formation of Fe_3O_4@C/CuO/CF. This composite boasted increased active sites for the Fe_3O_4 active material and a high specific capacitance of 1250 F g^{-1} at 5 mA cm^{-2} [as depicted in Figure 10.13]. The next study was conducted on using PANI CPs to create the porous structure with prevention of aggregation. The incorporation of PA improved the electron migration efficiency owing to its high electrical conductivity.

The incorporation of Fe_3O_4 into carbon materials is a common strategy to create electrodes with excellent conductivity and stability (Figure 10.14). This method effectively prevents the loss of Fe_3O_4 NPs during cycling, thereby enhancing the cycling stability and long-term durability of the electrode. This ensures sustained performance in energy storage applications. By significantly improving the conductivity

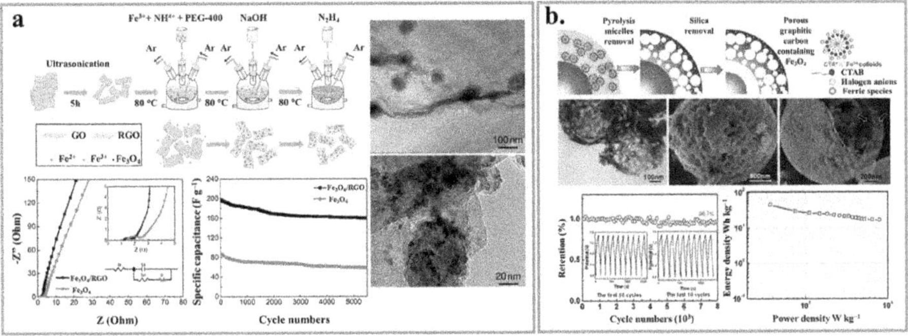

FIGURE 10.14 (a) Schematic illustration of the synthesis and the electrochemical result of Fe_3O_4/RGO [166], (b) synthesis procedure and the cyclic test of the double-shelled C-C-Fe_3O_4 hollow spheres [167].

of Fe_3O_4 NPs, suppressing aggregation, and preserving high surface area and active sites of Fe_3O_4, this process enables efficient electron and ion exchange between the electrode and electrolyte. As a result, the efficiency and performance of electrochemical reactions are improved.

Electrochemical research has focused on iron compounds including halides, sulfides, nitrides, and phosphides. The introduction of Fe to nitrides, sulfides, and phosphides alters their bonding structure, resulting in unique physicochemical properties. Transition metal nitrides (TMNs) possess an electronic structure with high conductivity, chemical stability, and mechanical stability, making them promising candidates for energy storage. Iron nitrides, owing to their unique properties, are of interest in energy storage research. TMNs can form eutectic complexes, crystalline ions, and transition metals (Figure 10.15). However, their formation is challenging due to unfavorable thermodynamic conditions. Researchers have addressed this challenge by reducing the grain size of iron nitride and using carbon materials, such as reduced graphene oxide and carbon *nano* tubes, to support TMN *nano* particles. These materials suppress TMN dispersion or aggregation, thus improving electrochemical performance. To achieve optimal performance, fine-tuning the electronic configuration of TMNs and facilitate efficient charge transfer and interfacial interactions during electrochemical reactions are crucial. *Nano* engineering TMNs aims to expand their SSA, increase the number of active centers, promote ion mass transport, and shorten the electrolyte diffusion distance. Two synthetic methods have been introduced to prepare TMNs: physical and chemical. Physical methods offer advantages, such as high crystallinity, simplified operation, and improved efficiency. These methods enable the production of iron nitride with enhanced structural properties, making it a promising material for various application. Unlike physical methods, chemical methods involve the reaction of iron precursors, such as iron chloride, with nitrogen sources, such as urea and ammonia, under a high-temperature atmosphere to produce iron nitride. Chemical methods provide several advantages over physical methods, as they enable greater control over the composition and structure of Fe_2N, allowing for the incorporation of impurities and facilitating the direct synthesis of

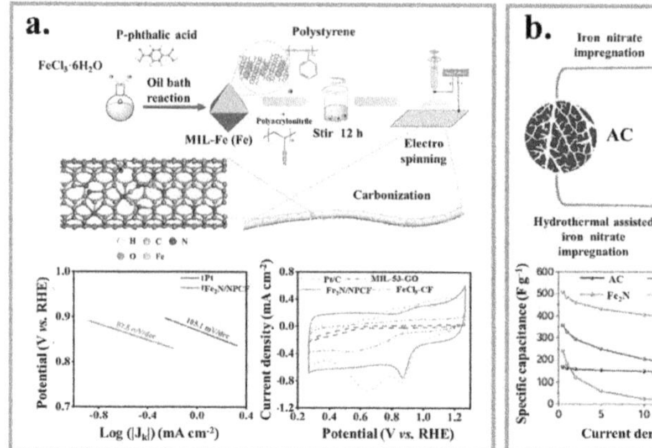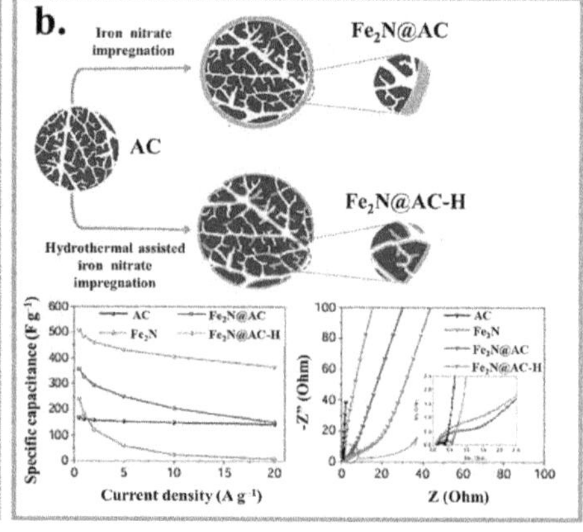

FIGURE 10.15 (a) The process of fabricating Fe$_2$N/N-doped porous carbon fibers (Fe$_2$N/NPCF) using electrospinning technique roadmap [168] and (b) the synthetic route of Fe$_2$N@AC and Fe$_2$N@AC-H with electrochemical results [169].

iron-based nitride materials. By controlling the reaction conditions and incorporating suitable precursors or additives, it is possible to achieve precise control over the composition, morphology, and properties of a material.

The materials commonly used in electrochemistry include single MOs. Binary and ternary MOs have been extensively studied in this field, particularly for their potential energy storage applications. Binary transition metal oxides (BTMOs) have gained recognition owing to their high reversible storage capacity, structural stability, and electronic conductivity, making them novel electrode materials for SCs. On the other hand, ternary MOs have also shown great promise in electrochemical energy storage. For example, mixed ternary metal (Co/Zn/Cu) metal organic frameworks (MOFs) have been developed for SC applications. These binder-free Cu–Zn$_{1.5}$Co$_{1.5}$O$_4$ petal-like composite electrodes exhibited high specific capacitance. Compared with single MOs, binary and ternary MOs often exhibit enhanced properties. For instance, binary MOs like Ir (1-x)Mx where M = Pd, Rh, Ru, and ternary MOs like Ir (1-x-z)MxMz' where M,M' = Pd, Rh, and Ru have been studied [170]. These binary and ternary MOs frequently exhibit superior performances compared to their single MO counterparts. This was likely due to the synergistic effects of the different metals in the oxide structure, which enhanced the electrochemical properties. However, ternary composite materials are more sensitive to high current rates (above 372 mA g^{-1}) than binary composites, owing to the presence of graphitic carbon. This indicates that although single composite materials have their own advantages, binary and ternary composite materials often provide improved performance owing to the synergistic effects of their components.

4.4.3 Hybrid Capacitor

A hybrid SC was created with the aim of increasing the energy density to 20–30 Wh kg^{-1}. Its mechanism involves the combination of an EDLC and a PC, which can be configured in a symmetric or asymmetric manner. Asymmetric hybrid SCs, which consist of two electrodes made of different materials, exhibit better electrochemical properties than individual electrodes and are, therefore, more commercially available. For instance, an asymmetric SC can be fabricated by combining AC as the anode and $Li_4Ti_5O_{12}$ as the Faradaic cathode with an organic electrolyte. These hybrid devices, which can be either internal or external, are formed by a combination of materials, electrodes, or the entire SCs and BT at either the material or device level. The performance of hybrid capacitors is influenced by the electrodes and electrolyte, and there are three categories of hybrid electrodes: composite electrodes composed of carbon incorporated with CPs or MOs; hybrid SCs with two different materials with redox properties; and SCs with a BT-type material electrode and an SC electrode that exhibits better performance. The following sections discuss in detail the mechanisms of metal-ion capacitors, including lithium-ion capacitors (LICs), sodium-ion capacitors (SICs), and potassium-ion capacitors (KICs), as well as their recent electrochemical properties.

The first LIC capacitor was fabricated with a *nano* structured anode and AC cathode [171,172]. It exhibits high energy and high power properties due to the physical adsorption and desorption of ions using a hybrid electrolyte. During charging and discharging, anions are adsorbed and desorbed from the high surface area anode, while Li ions are intercalated and deintercalated at the cathode. The electrolyte, which can be either aqueous or non-aqueous, comprises a Li salt, a solvent to dissolve the components, and additives that enhance compositional features. Aqueous electrolytes have a potential window of up to 1.2 V and include compounds such as alkaline sulfates (Li_2SO_4), nitrates ($LiNO_3$), and hydrates (LiOH).

A sodium-ion capacitor (SIC) is a hybrid device with two electrodes, one exhibiting capacitive properties and the other BT-like properties, an electrolyte, and a separator. Unlike traditional BTs and SCs, SICs store energy through a combination of capacitive and Faradaic reactions [173–175]. The cathode stores charge through the electrochemical double layer capacitance (EDLC) mechanism, whereas the anode undergoes a Faradaic reaction in the bulk electrode material. Advanced electrode materials for SICs are designed to balance the capacitive and Faradaic behavior, enhance cathode capacity, and improve the power and energy density of SICs. Therefore, determining the capacitive and Faradaic contributions is crucial for SICs. Potassium-ion capacitors (KICs) have also been discussed as alternatives to LICs and SICs because of the abundance of sodium on Earth (176). Although K^+ ions are larger than Li^+ ions, they can be stored in BT-type materials through intercalation reactions. In 2012, Chen et al. introduced the first non-aqueous KIC using 1M $NaClO_4$ in a propylene carbonate electrolyte, which utilized an interpenetrating network composite of layered V_2O_5 *nano* wires and carbon *nano* tubes synthesized through a simple hydrothermal process.

4.5 BTs

A BT is a power source that converts chemical energy into electricity and typically comprises one or more cells to facilitate this process. The term "battery" often

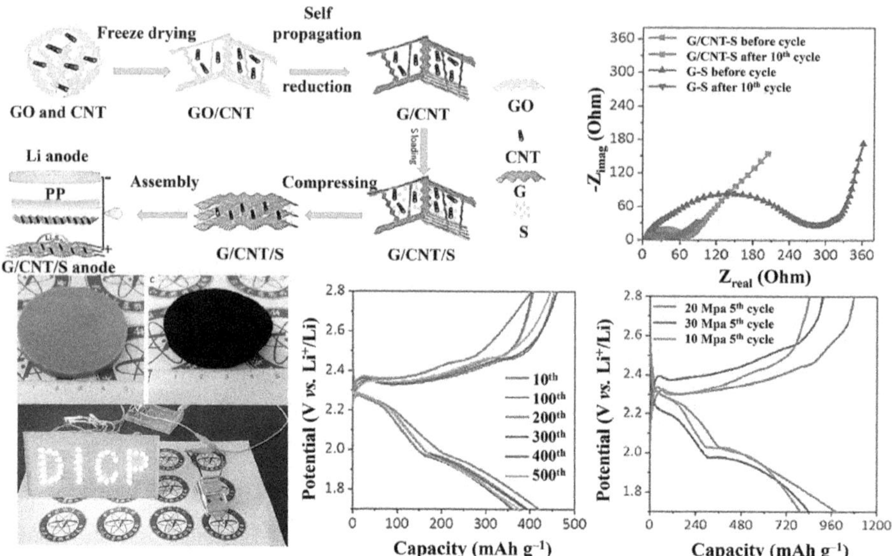

FIGURE 10.16 Schematic of the synthesis of G/CNT-S//G/CNT cathode integrated G/CNT-S host with G/CNT interlayer for Li-S batteries and the photograph of the prepared aerogel with the photograph of the "DICP" letters composed of 42 LED arrays, powered by two serially connected Li-S batteries [177].

refers to a group of two or more galvanic cells that can convert energy. Various types of BTs are available, including lithium-ion (Li-ion), sodium-ion (Na-ion), and lithium-sulfur (Li-S). Li-ion batteries (LIBs) are widely used because of their high energy density, small memory effect, and low self-discharge rate. Na-ion batteries, which are more abundant and potentially more affordable than LIBs, are promising alternatives. However, challenges, such as the lack of a well-established raw material supply chain, still need to be addressed. Li-sulfur batteries, while being studied, have limited cycle life and safety issues. The best battery type depends on the specific requirements of the application, as each type has its own advantages and disadvantages. Lithium-sulfur batteries are a promising technology owing to the high theoretical capacity of sulfur, which can undergo multi-electron redox reactions. However, poor sulfur conductivity and polysulfide shuttle effects lead to a low performance (Figure 10.16). To address this, graphene aerogels with high conductivity and SSA have been developed. Their porous structure allowed uniform sulfur loading and improved electron transport. Additionally, CAs with high SSAs and electrical conductivities have been used to increase sulfur loading and achieve high capacity. These advancements have led to improved rate performances and cycle stabilities.

The next type of BT is the Na-ion BT, which utilizes sodium ions as charge carriers. Although Na-ion BTs operate on the same principle as lithium-ion batteries (LIBs), they differ in that they employ sodium instead of lithium as the cathode material (Figure 10.17). The use of sodium in these BTs is appealing because of its

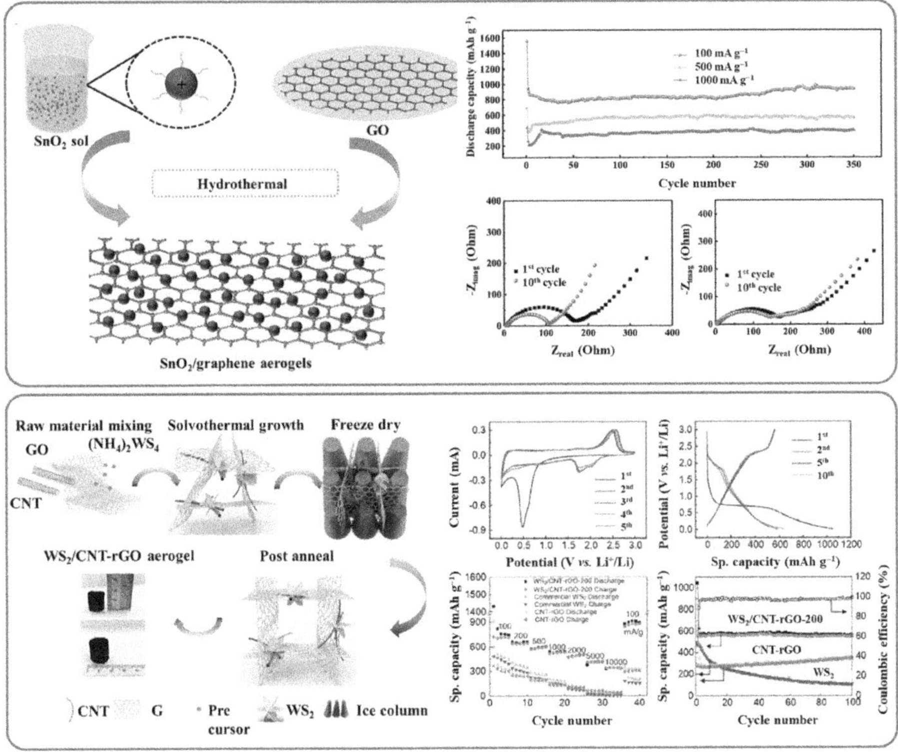

FIGURE 10.17 Schematic illustration of the synthesis of SnO$_2$/graphene aerogels *nano* composites and the electrochemical performance for LIBs [178], schematic diagram of WS$_2$/CNT-rGO aerogel 3D ordered microchannel *nano* architecture synthesis process and the electrochemical performance of WS$_2$/CNT-rGO-200 aerogel nano composites as the anode materials of LIBs [179].

abundance, particularly in saltwater, and because it eliminates the need for materials such as Co, Cu, or Ni. Iron-based materials work well in Na-ion BTs as opposed to transition metals because of the larger ionic radius of sodium, which prevents ion exchange and results in slower ion movement inside the crystal lattice. MOs/ bimetal oxides, such as SnO$_2$, CoO, and MnO$_2$, can enhance the capacity of Na-ion BTs. Additionally, graphene aerogels contribute to increased capacity owing to their high porosity, which provides space for MO alloying reactions and reduces the volume expansion during cycling. Metals and bimetallic oxides offer high theoretical capacities and fast redox reactions, whereas Mn$_3$O$_4$-GO hybrids promote rapid ion diffusion. Na-ion BTs are gaining attention as potential alternatives to LIBs owing to their potential to address environmental concerns, high cost, and uneven distribution of materials.

LIBs have gained widespread popularity owing to their high energy density and long cycle life. To improve their performance, various *nano* materials have been incorporated into BT designs. Graphene-coated *nano* particles and

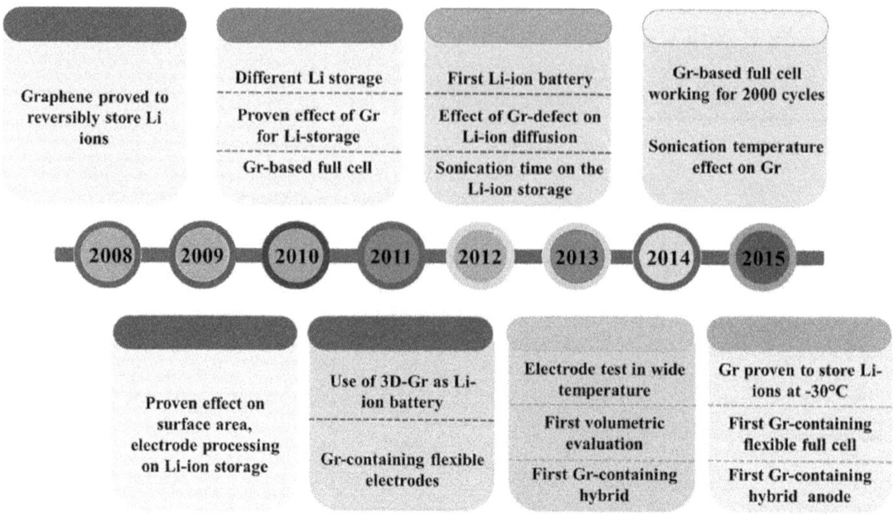

FIGURE 10.18 A comprehensive overview of key discoveries regarding graphene and graphene-based materials as anode materials for LIBs between 2008 and 2015. /Gr- Graphene /

nano composites are among the most commonly used *nano* materials in LIBs (Figures 10.18 and 10.19). *Nano* structured lithium titanate and SnO_2-based *nano* materials have also been employed to enhance the charge/discharge capability and increase the upper temperature limit of BTs. However, despite the significant progress in using graphene in BTs, its practical application remains limited due to the lack of clear focus and appropriate performance metrics. In particular, volumetric capacity, which is more critical than gravimetric capacity for practical applications, is often overlooked. Although the high porosity of graphene can improve ion conduction, its superior performance is typically attributed to its enhanced electronic conductivity. However, this parameter has not been widely reported and has never been adopted as a standard metric for fair comparison with state-of-the-art materials.

5   CONCLUSION

The integration of *nano* materials into BTs and SCs represents a significant advancement in the field of energy storage. The unique properties of *nano* materials, such as their large surface areas relative to their volumes, have been exploited to enhance the performance of these devices. In BTs, this has led to improvements in the energy density, power density, and recharge rates. The use of *nano* materials in SCs increases their available surface area for charge storage, thereby enhancing their energy density. Furthermore, the exploration of *nano* materials in the development of new types of BTs and SCs promises even greater advancements in energy storage technology. As research in this field continues, we anticipate the emergence of increasingly efficient and powerful energy storage devices that will drive technological innovation.

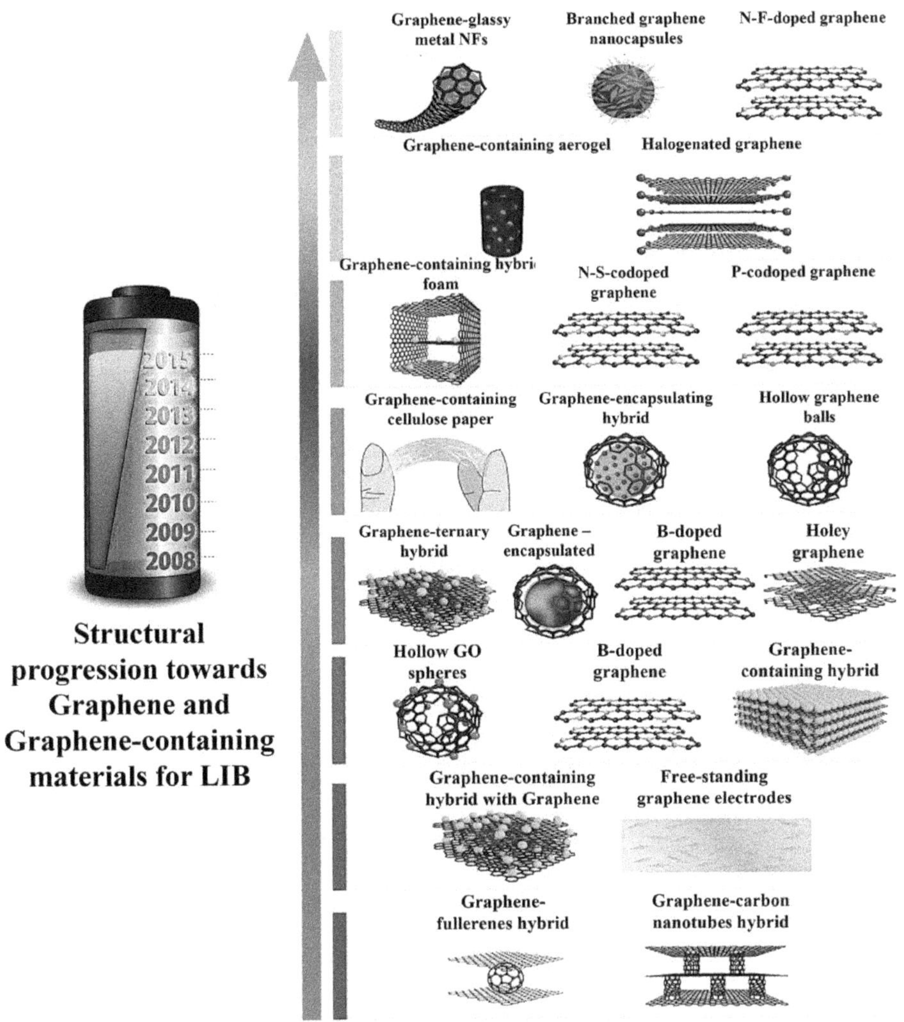

FIGURE 10.19 The development of graphene and graphene-containing materials as LIB anodes presented between 2008 and 2015.

REFERENCES

1. F. J. Heiligtag, M. Niederberger, The fascinating world of nanoparticle research, *Mater. Today*, 2013, 16, 262–271.
2. P. Walter, E. Welcomme, P. Hallégot, N. J. Zaluzec, C. Deeb, J. Castaing, P. Veyssière, R. Bréniaux, J.-L. Lévêque, G. Tsoucaris, Early use of PbS nanotechnology for an ancient hair dyeing formula, *Nano Lett.*, 2006, 6, 2215–2219.
3. S. Zhuang, E. S. Lee, L. Lei, B. B. Nunna, L. Kuang, W. Zhang, Synthesis of nitrogen-doped graphene catalyst by high-energy wet ball milling for electrochemical systems, *Int. J. Energy Res.*, 2016, 40, 2136–2149.
4. R. Kaur, A. Marwaha, V. A. Chhabra, K.-H. Kim, S. K. Tripathi, Recent developments on functional nanomaterial-based electrodes for microbial fuel cells, *Renew. Sustain. Energy Rev.*, 2020, 119, 109551.

5. N. Baig, I. Kammakakam, W. Falath, Nanomaterials: A review of synthesis methods, properties, recent progress, and challenges, *Mater. Adv.*, 2021, 2 (6), 1821–1871.

6. V. Prakash Sharma, U. Sharma, M. Chattopadhyay, V. N. Shukla, Advance applications of nanomaterials: A review, *Mater. Today: Proc.*, 2018, 5 (2, Part 1), 6376–6380.

7. Q. Cao, Q. Fan, Q. Chen, C. Liu, X. Han, L. Li, Recent advances in manipulation of micro- and nano-objects with magnetic fields at small scales, *Mater. Horiz.*, 2020, 7 (3), 638–666.

8. V. K. Pawar, Y. Singh, J. G. Meher, S. Gupta, M. K. Chourasia, Engineered nanocrystal technology: In-vivo fate, targeting and applications in drug delivery, *J. Control. Release*, 2014, 183, 51–66.

9. Y. Wang, M. Yao, R. Ma, Q. Yuan, D. Yang, B. Cui, C. Ma, M. Liu, D. Hu, Design strategy of barium titanate/polyvinylidene fluoride-based nanocomposite films for high energy storage, *J. Mater. Chem. A*, 2020, 8 (3), 884–917.

10. J. Liu, J. Wang, C. Xu, H. Jiang, C. Li, L. Zhang, J. Lin, Z.X. Shen, Advanced energy storage devices: Basic principles, analytical methods, and rational materials design, *Adv. Sci.*, 2017, 5 (1), 1700322.

11. J. R. Miller, P. Simon, Materials science: Electrochemical capacitors for energy management, *Science*, 2008, 321, 651–652.

12. D. Tie, S. Huang, J. Wang, J. Ma, J. Zhang, Y. Zhao, Hybrid energy storage devices: Advanced electrode materials and matching principles, *Energy Stor. Mater.*, 2019, 21, 22–40.

13. S. Chowdhury, R. Balasubramanian, Three-dimensional graphene-based macrostructures for sustainable energy applications and climate change mitigation, *Prog. Mater. Sci.*, 2017, 90, 224–275.

14. B. Wang, T. Ruan, Y. Chen, F. Jin, L. Peng, Y. Zhou, D. Wang, S. Dou, Graphene-based composites for electrochemical energy storage, *Energy Stor. Mater.*, 2020, 24, 22–51.

15. Q. A. Sial, M. S. Javed, Y.-J. Lee, H. Seo, Flexible and transparent graphene-based supercapacitors decorated with nanohybrid of tungsten oxide nanoflakes and nitrogen-doped-graphene quantum dots, *Ceram. Int.*, 2020, 46, 23145–23154.

16. J. Zhang, J. Jiang, H. Li, X. Zhao, A high-performance asymmetric supercapacitor fabricated with graphene-based electrodes, *Energy Environ. Sci.*, 2011, 4, 4009–4015.

17. J. Cao, J. Li, L. Li, Y. Zhang, D. Cai, D. Chen, W. Han, Mn-doped Ni/Co LDH nanosheets grown on the natural N-dispersed PANI-derived porous carbon template for a flexible asymmetric supercapacitor, *ACS Sustain. Chem. Eng.*, 2019, 7, 10699–10707.

18. S. Tajik, Y. Orooji, Z. Ghazanfari, F. Karimi, H. Beitollahi, R. S. Varma, H. W. Jang, M. Shokouhimehr, Nanomaterials modified electrodes for electrochemical detection of Sudan I in food, *J. Food Meas. Char.*, 2021, 15 (4), 3837–3852.

19. M. Tayyab, Y. Liu, S. Min, R. M. Irfan, Q. Zhu, L. Zhou, J. Lei, J. Zhang, Simultaneous hydrogen production with the selective oxidation of benzyl alcohol to benzaldehyde by a noble-metal-free photocatalyst VC/CdS nanowires, *Chin. J. Catal.*, 2022, 43 (4), 1165–1175.

20. M. Z. Iqbal, M. M. Faisal, S. R. A. M. Alzaid, A facile approach to investigate the charge storage mechanism of MOF/PANI based supercapattery devices, *Solid State Ionics*, 2020, 354, 115411.

21. J. Cherusseri, D. Pandey, J. Thomas, Symmetric, asymmetric, and battery-type supercapacitors using two-dimensional nanomaterials and composites, *Batteries Supercaps*, 2020, 3 (9), 860–875.

22. S. C. Pang, M. A. Anderson, T. W. Chapman, Novel electrode materials for thin-film ultracapacitors: comparison of electrochemical properties of sol-gel-derived and electrodeposited manganese dioxide, *J. Electrochem. Soc.*, 2000, 147 (2), 444.

23. M. Z. Iqbal, M. M. Faisal, S. R. Ali, Integration of supercapacitors and batteries towards high-performance hybrid energy storage devices, *Int. J. Energy Res.*, 2021, 45 (2), 1449–1479.

24. B. L. Vijayan, I. I. Misnon, G. M. Anilkumar, C. C. Yang, R. Jose, Void-size-matched hierarchical 3D titania flowers in porous carbon as an electrode for high-density super capacitive charge storage, *J. Alloys Compd.*, 2021, 858, 157649.
25. J. Wang, Q. Li, C. Peng, N. Shu, L. Niu, Y. Zhu, To increase electrochemical performance of electrode material by attaching activated carbon particles on reduced graphene oxide sheets for supercapacitor, *J. Power Sources*, 2020, 450, 227611.
26. H. Gao, S. Xin, J. B. J. C. Goodenough, The origin of superior performance of $Co(OH)_2$ in hybrid supercapacitors, *Chem*, 2017, 3(1), 26–28.
27. B. Conway, W. J. Pell, Double-layer and Pseudo capacitance types of electrochemical capacitors and their applications to the development of hybrid devices, *J. Solid State Electrochem.*, 2003, 7, 637–644.
28. D. Lin, Y. Liu, Y. Cui, Reviving the lithium metal anode for high-energy batteries, *Nature Nanotech.*, 2017, 12 (3), 194–206.
29. M. Thiruvengadam, G. Rajakumar, I.-M. Chung, Nanotechnology: current uses and future applications in the food industry, *3 Biotech*, 2018, 8 (1), 1–13.
30. K. Velusamy, S. Periyasamy, P. S. Kumar, G. Rangasamy, J. M. N. Pauline, P. Ramaraju, S. Mohanasundaram, D. V. Nguyen Vo, Biosensor for heavy metals detection in wastewater: A review, *Food Chem. Toxicol.*, 2022, 168, 113307.
31. M. I. A. Abdel Maksoud, R. A. Fahim, A. E. Shalan, M. Abd Elkodous, S. O. Olojede, A. I. Osman, C. Farrell, A. H. Al-Muhtaseb, A. S. Awed, A. H. Ashour, D. W. Rooney, Advanced materials and technologies for supercapacitors used in energy conversion and storage: A review, *Environ. Chem. Lett.*, 2021, 19, 375–439.
32. M. Ramya, P. S. Kumar, G. Rangasamy, V. Uma shankar, G. Rajesh, K. Nirmala, A. Saravanan, A. Krishnapandi, A recent advancement on the applications of nanomaterials in electrochemical sensors and biosensors, *Chemosphere*, 2022, 308 (2), 136416.
33. C. A. Tao, J. F. Wang. Synthesis of Metal organic frameworks by Ball milling, *Crystals*, 2021, 11 (1), 15.
34. K. A. Shah, B. A. Tali, Synthesis of carbon nanotubes by catalytic chemical vapour deposition: A review on carbon sources, catalysts and substrates, *Mater. Sci. Semicond. Process.*, 2016, 41, 67–82.
35. S. Cao, C. Zhao, T. Han, L. Peng, Hydrothermal synthesis, characterization and gas sensing properties of the WO_3 nanofibers, *Mater. Lett.*, 2016, 169, 17–20.
36. U. Yogeswaran, S. Thiagarajan, S. M. Chen, Nanocomposite of functionalized multiwall carbon nanotubes with nafion, nano platinum, and nano gold biosensing film for simultaneous determination of ascorbic acid, epinephrine, and uric acid, *Anal. Biochem.*, 2007, 365 (1), 122–131.
37. C. Yang, C. B. Jacobs, M. D. Nguyen, M. Ganesana, A. G. Zestos, I. N. Ivanov, A. A. Puretzky, C. M. Rouleau, D. B. Geohegan, B. J. Venton Carbon nanotubes grown on metal microelectrodes for the detection of dopamine, *Anal. Chem.*, 2016, 88, 645–652.
38. J. Yang, B. Dou, R. Yuan, Y. Xiang, Proximity binding and metal ion-dependent DNAzyme cyclic amplification-integrated aptasensor for label-free and sensitive electrochemical detection of thrombin, *Anal. Chem.*, 2016, 88, 8218–8223.
39. M. Zahran, A. H. Marei, Innovative natural polymer metal nanocomposites and their antimicrobial activity, *Int. J. Biol. Macromol.*, 2019, 136, 586–596.
40. W. S. Zheng, Z. L. Guo, Z. Z. Huang, J. Q. Zhuang, W. S. Yang, In-situ preparation of size-tunable gold nanoparticles in porous resorcinol–formaldehyde resin, *Colloids Surf., A*, 2015, 484, 271–277.
41. G. Zhu, P. Gai, L. Wu, J. Zhang, X. Zhang, J. Chen, beta-Cyclodextrin-platinum nanoparticles/graphene nanohybrids: Enhanced sensitivity for electrochemical detection of naphthol isomers, *Chem. Asian J.*, 2012, 7, 732–737.

42. W. Zheng, Q. Li, L. Su, Y. Yan, J. Zhang, L. Mao, Direct electrochemistry of multi-copper oxidases at carbon nanotubes noncovalently functionalized with cellulose derivatives, *Electroanalysis*, 2006, 6, 587–594.

43. P. Veerakumar, S. Chen, R. Madhu, S. Liu, Palladium nanoparticle incorporated porous activated carbon: electrochemical detection of toxic metal ions, *ACS Appl. Mater. Interfaces*, 2016, 8, 1319–1326.

44. M. Amal Raj, S. Abraham John, Fabrication of electrochemically reduced graphene oxide films on glassy carbon electrode by self-assembly method and their electrocatalytic application, *J. Phys. Chem. C*, 2013, 117 (8), 4326–4335.

45. P. Ramnani, A. Mulchandani, Carbon nanomaterial-based electrochemical biosensors for label-free sensing of environmental pollutants, *Chemosphere*, 2016, 143, 85–98.

46. P. Niu, M. Gich, C. Navarro-Hernandez, P. Fanjul-Bolado, A. Roig, Screen-printed electrodes made of a bismuth nanoparticle porous carbon nanocomposite applied to the determination of heavy metal ions, *Microchim. Acta*, 2016, 183, 617–623.

47. H. Moradpour, H. Beitollahi, F. G. Nejad, A. Di Bartolomeo, Glassy carbon electrode modified with N-doped reduced graphene oxide sheets as an effective electrochemical sensor for amaranth detection, *Materials*, 2022, 15 (9), 3011.

48. A. Gholami, F. Farjami, Y. Ghasemi, The development of an amperometric enzyme biosensor based on a polyaniline-multiwalled carbon nanocomposite for the detection of a chemotherapeutic agent in serum samples from patients, *J. Sens.*, 2021.

49. M. Govindhan, A. Chen, Simultaneous synthesis of gold nanoparticle/graphene nanocomposite for enhanced oxygen reduction reaction, *J. Power Sources*, 2015, 274, 928–936.

50. M. Govindhan, M. Amiri, A. Chen, Au nanoparticle/graphene nanocomposite as a platform for the sensitive detection of NADH in human urine, *Biosens. Bioelectron.*, 2015, 66, 474–480.

51. H. Dai, N. Wang, D. Wang, H. Ma, M. Lin, An electrochemical sensor based on phytic acid functionalized polypyrrole/graphene oxide nanocomposites for simultaneous determination of Cd(II) and Pb(II), *Chem. Eng. J.*, 2016, 299, 150–155.

52. R. A. Dar, N. G. Khare, D. P. Cole, S. P. Karna, A. K. Srivastava, Green synthesis of a silver nanoparticle-graphene oxide composite and its application for As(iii) detection, *RSC Adv.*, 2014, 4, 14432.

53. H. H. Chen, J. F. Huang, EDTA assisted highly selective detection of As($^{3+}$) on Au nanoparticle modified glassy carbon electrodes: Facile in situ electrochemical characterization of Au nanoparticles, *Anal. Chem.*, 2014, 86, 12406–12413.

54. H. Cui, W. Yang, X. Li, H. Zhao, Z. Yuan, An electrochemical sensor based on a magnetic Fe_3O_4 nanoparticles and gold nanoparticles modified electrode for sensitive determination of trace amounts of arsenic(iii), *Anal. Methods*, 2012, 4, 4176–34181.

55. M. R. Mahmoudian, W. J. Basirun, Y. Alias, A sensitive electrochemical Hg^{2+} ions sensor based on polypyrrole coated nanospherical platinum, *RSC Adv.*, 2016, 6, 36459–36466.

56. S. Radhakrishnan, K. Krishnamoorthy, C. Sekar, J. Wilson, S. J. Kim, A highly sensitive electrochemical sensor for nitrite detection based on Fe_2O_3 nanoparticles decorated reduced graphene oxide nanosheets, *Appl. Catal., B*, 2014, 1489, 22–28.

57. M. Wang, J. Huang, M. Wang, D. Zhang, J. Chen, Electrochemical nonenzymatic sensor based on CoO decorated reduced graphene oxide for the simultaneous determination of carbofuran and carbaryl in fruits and vegetables, *Food Chem.*, 2014, 151, 191–197.

58. S. L. Ting, A. Ananthanarayanan, K. C. Leong, P. Chen, Graphene quantum dots functionalized gold nanoparticles for sensitive electrochemical detection of heavy metal ions, *Electrohim. Acta*, 2015, 172, 7–11.

59. A. González, E. Goikolea, J. A. Barrena, R. Mysyk, Review on supercapacitors: Technologies and materials renew, *Sust. Energ. Rev.*, 2016, 58, 1189–1206.

60. K. Gopalakrishnan, A. Govindaraj, C. N. R. Rao, Extraordinary supercapacitor performance of heavily nitrogenated graphene oxide obtained by microwave synthesis, *J. Mater. Chem. A,* 2013, 1 (26), 7563.

61. M. A. A. Mohd Abdah, N. H. N. Azman, S. K. Y. Sulaiman, Review of the use of transition-metal-oxide and conducting polymer-based fibres for high-performance supercapacitors, *Mater. Des.*, 2020, 186, 108199.

62. D. Zhang, Y. Zhang, Y. Luo, P. K. Chu, Highly porous honeycomb manganese oxide@carbon fibres core–shell nanocables for flexible supercapacitors, *Nano Energy*, 2015, 13, 47–57.

63. K. Wu, J. Zhao, X. Zhang, H. Zhou, M. Wu, Hierarchical mesoporous MoO_2 sphere as highly effective supercapacitor electrode, *J. Taiwan Inst. Chem. Eng.*, 2019, 102, 212–217.

64. R. Atchudan, T. N. J. I. Edison, S. Perumal, P. Thirukumaran, R. Vinodh, Y. R. Lee, Green synthesis of nitrogen-doped carbon nanograss for supercapacitors, *J. Taiwan Inst. Chem. Eng.*, 2019, 102, 475–486.

65. W. H. Low, P. S. Khiew, S. S. Lim, C. W. Siong, E. R. Ezeigwe, Recent development of mixed transition metal oxide and graphene/mixed transition metal oxide based hybrid nanostructures for advanced supercapacitors, *J. Alloys Compd.*, 2019, 775, 1324--356.

66. W. Wu, D. Niu, J. Zhu, Y. Gao, D. Wei, C. Zhao, C. Wang, F. Wang, L. Wang, L. Yang, Hierarchical architecture of Ti_3C_2@PDA/$NiCo_2S_4$ composite electrode as high-performance supercapacitors, *Ceram. Int.,* 2019, 45 (13), 16261–16269.

67. Y. Huang, H. Li, Z. Wang, M. Zhu, Z. Pei, Q. Xue, Y. Huang, C. Zhi, Nanostructured polypyrrole as a flexible electrode material of supercapacitor, *Nano Energy*, 2016, 22, 422–438.

68. Y. J. Peng, T. H. Wu, C. T. Hsu, S. M. Li, M. G. Chen, C. C. Hu, Electrochemical characteristics of the reduced graphene oxide/carbon nanotube/polypyrrole composites for aqueous asymmetric supercapacitors, *J. Power Sources*, 2014, 272 (Supplement C), 970–978.

69. H. Shang, Z. Zhang, C. Liu, X. Zhang, S. Li, Z. Wen, S. Ji, J. Sun, MnO_2@V_2O_5 microspheres as cathode materials for high performance aqueous rechargeable Zn-ion battery, *J. Electroanal. Chem.*, 2021, 890, 115253.

70. R. Zhang, Q. Tu, X. Li, X. Sun, S. Liu, L. Chen, Template-free preparation of α-$Ni(OH)_2$ nanosphere as high-performance electrode material for advanced supercapacitor, *Nanomaterials,* 2022, 12, 2216.

71. Q. Li, C. Lu, C. Chen, L. Xie, Y. Liu, Y. Li, Q. Kong, H. Wang, Layered $NiCo_2O_4$/reduced graphene oxide composite as an advanced electrode for supercapacitor, *Energy Storage Mater.*, 2017, 8, 59–67.

72. Q. Pei, M. Lu, Z. Liu, D. Li, X. Rao, X. Liu, S. Zhong, Improving the $Na_{0.67}Ni_{0.33}Mn_{0.67}O_2$ Cathode material for high-voltage cyclability via Ti/Cu Co-doping for sodium-ion batteries, *ACS Appl. Energy Mater.*, 2022, 5 (2), 1953–1962.

73. J. G. Wang, H. Liu, X. Zhang, X. Li, X. Liu, F. Kang, Green synthesis of hierarchically porous carbon nanotubes as advanced materials for high-efficient energy storage, *Small*, 2017, 14 (13), 1703950.

74. D. Gangaraju, V. Sridhar, I. Lee, H. Park, Graphene – carbon nanotube – Mn_3O_4 mesoporous nano-alloys as high capacity anodes for lithium-ion batteries, *J. Alloy. Comp.*, 2017, 699, 106–111.

75. F. Li, H. Chen, X. Y. Liu, S. J. Zhu, J. Q. Jia, C. H. Xu, F. Dong, Z. Q. Wen, Y. X. Zhang, Low-cost high-performance asymmetric supercapacitors based on Co_2AlO_4@MnO_2 nanosheets and Fe_3O_4 nanoflakes, *J. Mater. Chem. A*, 2016, 4, 2096–2104.

76. M. Priya, V. K. Premkumar, P. Vasantharani, G. Sivakuamr, Structural and electrochemical properties of $ZnCo_2O_4$ nanoparticles synthesized by hydrothermal method, *Vacuum*, 2019, 167, 307–312.

77. A. Kytsya, V. Berezovets, Yu. Verbovytskyy, L. Bazylyak, V. Kordon, I. Zavaliy, V. A. Yartys, Bimetallic Ni-Co nanoparticles as an efficient catalyst of hydrogen generation via hydrolysis of $NaBH_4$, *J. Alloy. Compds.*, 2022, 908, 164484.

78. L. Sun, Y. Zhang, Y. Zhang, H. Si, W. Qin, Y. Zhang, Reduced graphene oxide nanosheet modified NiMn-LDH nanoflake arrays for high-performance supercapacitors, *Che. Comm.*, 2018, 54, 10172–10175.

79. X. Chen, J. Song, Y. Xing, Y. Qin, J. Lin., X. Qu, B. Sun, S. Du, D. Shi, C. Chen, D. Sun, Nickel-decorated RuO_2 nanocrystals with rich oxygen vacancies for high-efficiency overall water splitting, *J. Clloid. Interface Sci.*, 2023, 630 (Part A), 940–950.

80. B. He, Y. Zu, Y. Mei, Design of advanced electrocatalysts for the high-entropy alloys: Principle, progress, and perspective, *J. Alloys Compd.*, 2023, 958, 170479.

81. Y. Yao, Q. Dong, A. Brozena, J. Luo, J. Miao, M. Chi, C. Wang, I.G. Kevrekidis, Z.J. Ren, J. Greeley, G. Wang, A. Anapolsky, L. Hu, High entropy nanoparticles; Synthesis-structure-property relationships and data driven discovery, *Science*, 2022, 376, 6589.

82. S. A. Lee, J. Bu, J. Lee, H. W. Jang, High-entropy nanomaterials for advanced electro-catalysis, *Small*, 2023, 3 (5), 2200109.

83. J. Han, L. Xiong, X. Jiang, X. Yuan, Y. Zhao, D. Yang, Bio-functional electro spun nanomaterials: From topology design to biological applications, *Progress in Polymer Sci.*, 2019, 91, 1–28.

84. L. Zhang, H. Zhao, S. Xu, Q. Liu, T. Li, Y. Luo, S. Gao, X. Shi, A. M. Asiri, X. Sun, Recent advances in 1D electrospun nanocatalysts for electrochemical water splitting, *Science*, 2020, 2 (2), 20000048.

85. X. Zhang, Q. Lu, H. Liu, K. Li, M. Wei, Nature-inspired design of NiS/carbon micro-spheres for high-performance hybrid supercapacitors, *Appl. Surf. Sci.*, 2020, 528, 146976.

86. I. Hussain, C. Lamiel, S. Sahoo, M. Ahmad, X. Chen, M. S. Javed, N. Qin, S. Gu, Y. Li, T. Nawaz, Factors affecting the growth formation of nanostructures and their impact on electrode materials: A systematic review, *Mater. Today Phys.*, 2022, 27, 100844.

87. Z. Sun, Y. Wang, Z. Chen, X. Li, Min-max game based energy management strategy for fuel cell/supercapacitor hybrid electric vehicles, *Appl. Energy*, 2020, 267, 115086.

88. B. Sahoo, V. Pandey, A. Dogonchi, P. Mohapatra, D. Thatoi, N. Nayak, M. Nayak, A state-of-art review on 2D material-boosted metal oxide nanoparticle electrodes: Supercapacitor applications, *J. Energy Storage*, 2023, 65, 10733.

89. C. Zhong, Y. Deng, W. Hu, J. Qiao, L. Zhang, J. Zhang, A review of electrolyte materials and compositions for electrochemical supercapacitors, *Chem. Soc. Rev.*, 2015, 44, 7484–7539.

90. B. Pal, S. Yang, S. Ramesh, V. Thangadurai, R. Jose, Electrolyte selection for superca-pacitive devices: a critical review, *Nanoscale Adv.*, 2019, 1, 3807–3835.

91. A. Sumboja, J. Liu, W. G. Zheng, Y. Zong, H. Zhang, Z. Liu, Electrochemical energy storage devices for wearable technology: a rationale for materials selection and cell design, *Chem. Soc. Rev.,* 2018, 47, 5919–5945.

92. C. W. Huang, C. A. Wu, S. S. Hou, P. L. Kuo, C. Te Hsieh, H. Teng, Gel electro-lyte derived from poly(ethylene glycol) blending poly(acrylonitrile) applicable to roll-to-roll assembly of electric double layer capacitors, *Adv. Funct. Mater.,* 2012, 22, 4677–4685.

93. Y. J. Kang, S. J. Chun, S. S. Lee, B. Y. Kim, J. H. Kim, H. Chung, S. Y. Lee, W. Kim, All-Solid-state flexible supercapacitors fabricated with bacterial nanocellulose papers, carbon nanotubes, and triblock-copolymer ion gels, *ACS Nano*, 2012, 6, 6400–6406.

94. I. Osada, H. De Vries, B. Scrosati, S. Passerini, Ionic-Liquid-Based polymer electro-lytes for battery applications, *Angew. Chem. - Int. Ed.*, 2016, 55, 500–513.

95. J. Zhang, B. Sun, X. Huang, S. Chen, G. Wang, Honeycomb-like porous gel polymer electrolyte membrane for lithium ion batteries with enhanced safety, *Sci. Rep.*, 2014, 4, 6007.

96. J. Zhou, Y. Yin, A. N. Mansour, X. Zhou, Experimental studies of mediator-enhanced polymer electrolyte supercapacitors, *Electrochem. Solid-State Lett.*, 2011, 14, A25–A28.
97. G. Ma, J. Li, K. Sun, H. Peng, J. Mu, Z. Lei, High performance solid-state supercapacitor with PVA-KOH-K$_3$[Fe(CN)$_6$] gel polymer as electrolyte and separator, *J. Power Sources*, 2014, 256, 281–287.
98. F. Yu, M. Huang, J. Wu, Z. Qiu, L. Fan, J. Lin, Y. Lin, A redox-mediator-doped gel polymer electrolyte applied in quasi-solid-state supercapacitors, *J. Appl. Polym. Sci.*, 2014, 131, 39784.
99. Z. Wu, L. Li, J. M. Yan, X. B. Zhang, Materials design and system construction for conventional and new-concept supercapacitors, *Adv. Sci.*, 2017, 4, 1600382.
100. L. Zhang, X. S. Zhao, Carbon-based materials as supercapacitor electrodes, *Chem. Soc. Rev.*, 2009, 38, 2520–2531.
101. J. R. Miller, R. A. Outlaw, B. C. Holloway, Graphene double-layer capacitor with ac line-filtering performance, *Science*, 2010, 329, 1637–1639.
102. Y. Sun, H. Wang, W. Wei, Y. Zheng, L. Tao, Y. Wang, et al. Sulfur-rich graphene nanoboxes with ultra-high potassiation capacity at fast charge: Storage mechanisms and device performance, *ACS Nano*, 2020, 15, 1652–1665.
103. B. K. Saikia, S. M. Benoy, M. Bora, J. Tamuly, M. Pandey, D. Bhattacharya, A brief review on supercapacitor energy storage devices and utilization of natural carbon resources as their electrode materials, *Fuel*, 2020, 282, 118796.
104. M. Mirzaeian, Q. Abbas, D. Gibson, M. Mazur, Effect of nitrogen doping on the electrochemical performance of resorcinol-formaldehyde based carbon aerogels as electrode material for supercapacitor applications, *Energy*, 2019, 173, 809–819.
105. M. Mirzaeian, Q. Abbas, D. Gibson, M. Mazur, Effect of nitrogen doping on the electrochemical performance of resorcinol-formaldehyde based carbon aerogels as electrode material for supercapacitor applications, *Energy*, 2019, 173, 809–819.
106. M. Usha Rani, K. Nanaji, T. N. Rao, A. S. Deshpande, Corn husk derived activated carbon with enhanced electrochemical performance for high-voltage supercapacitors, *J. Power Sources*, 2020, 471, 228387.
107. B. L. Vijayan, I. I. Misnon, G. M. Anilkumar, C. C. Yang, R. Jose, Void-size-matched hierarchical 3D titania flowers in porous carbon as an electrode for high-density supercapacitive charge storage, *J. Alloys Compd.*, 2021, 858, 157649.
108. J. Wang, Q. Li, C. Peng, N. Shu, L. Niu, Y. Zhu, To increase electrochemical performance of electrode material by attaching activated carbon particles on reduced graphene oxide sheets for supercapacitor, *J. Power Sources*, 2020, 450, 227611.
109. Y. Ding, T. Wang, D. Dong, Y. Zhang, Using biochar and coal as the electrode material for supercapacitor applications, *Front. Energy Res.*, 2020, 7, 1–11.
110. A. A. Mohammed, C. Chen, Z. Zhu, Low-cost, high-performance supercapacitor based on activated carbon electrode materials derived from baobab fruit shells, *J. Colloid Interface Sci.*, 2019, 538, 308–319.
111. S. Siyahjani, S. Oner, H. Diker, B. Gultekin, C. Varlikli, Enhanced capacitive behaviour of graphene based electrochemical double layer capacitors by etheric substitution on ionic liquids, *J. Power Sources*, 2020, 467, 228353.
112. N. Macherla, K. Singh, M. S. Santosh, K. Kumari, R. G. R. Lekkala, Heat assisted facile synthesis of nanostructured polyaniline/reduced crumbled graphene oxide as a high-performance flexible electrode material for supercapacitors, *Colloids Surf. A Physicochem. Eng. Asp.*, 2021, 612, 125982.
113. M. Mirzaeian, Q. Abbas, D. Gibson, M. Mazur, Effect of nitrogen doping on the electrochemical performance of resorcinol-formaldehyde based carbon aerogels as electrode material for supercapacitor applications, *Energy*, 2019, 173, 809–819.
114. X. Xu, J. Yang, X. Zhou, S. Jiang, W. Chen, Z. Liu, Highly crumpled graphene-like material as compression-resistant electrode material for high energy-power density supercapacitor, *Chem. Eng. J.*, 2020, 397, 25525.

115. Y. Wu, J. Cao, X. Zhao, Q. Zhuang, Z. Zhou, Y. Huang, et al., High-performance electrode material for electric double-layer capacitor based on hydrothermal pre-treatment of lignin by ZnCl$_2$, *Appl. Surf. Sci.*, 2020, 508, 144536.

116. M. Akbari, S. Bellani, V. Pellegrini, R. Oropesa-nuñez, A. Esau, D. Rio, et al., Scalable spray-coated graphene-based electrodes for high-power electrochemical double-layer capacitors operating over a wide range of temperature, *Energy Stor. Mater.*, 2021, 34, 1–11.

117. I. Shafi, E. Liang, B. Li, Ultrafine chromium oxide (Cr_2O_3) nanoparticles as a pseudo-capacitive electrode material for supercapacitors, *J. Alloys Compd.*, 2021, 851, 156046.

118. H. Banda, J.-H. Dou, T. Chen, N. J. Libretto, M. Chaudhary, G. M. Bernard, et al., High-capacitance pseudocapacitors from Li$^+$ ion intercalation in nonporous, electrically conductive 2D coordination polymers, *J. Am. Chem. Soc.*, 2021, 143, 2285–2292.

119. S. E. Arasi, R. Ranjithkumar, P. Devendran, M. Krishnakumar, A. Arivarasan, Studies on electrochemical mechanism of nanostructured cobalt vanadate electrode material for pseudocapacitors, *J. Energy Storage*, 2021, 41, 102986.

120. Q. Liu, H. Zhu, Q. Ma, M. Liu, B. Wang, C. Tang, et al., Ultrathin MoS_2 nanosheets hybridizing with reduced graphene oxide for high-performance pseudocapacitors, *FlatChem*, 2020, 26, 100212.

121. W. Luo, W. Zeng, H. Quan, M. Pan, Y. Wang, D. Chen, Carbon dots decorated NiCo hydroxycarbonate hierarchical nanoarrays on carbon cloth with high areal capacitance as pseudocapacitor electrode, *J. Alloys Compd.*, 2021, 868, 159048.

122. M. Chen, Y. Zhang, Y. Liu, J. Zheng, C. Meng, Applied surface science a novel intercalation pseudocapacitive electrode material : $VO(OH)_2$ / CNT composite with cross-linked structure for high performance flexible symmetric supercapacitors, *Appl. Surf. Sci.*, 2019, 492, 746–755.

123. S. Sahoo, P. Pazhamalai, V. K. Mariappan, G. K. Veerasubramani, N. J. Kim, S. J. Kim, Hydrothermally synthesized chalcopyrite platelets as an electrode material for symmetric supercapacitors, *Inorg. Chem. Front.*, 2020, 7, 1492–1502.

124. J. C. Russell, V. A. Posey, J. Gray, R. May, D. A. Reed, H. Zhang, et al., High-performance organic pseudocapacitors via molecular contortion, *Nat. Mater.*, 2021, 20, 1136–1141.

125. Z. Cheng, Y. Qiu, G. Tan, X. Chang, Q. Luo, L. Cui, Synthesis of a novel Mn(II)-porphyrins polycondensation polymer and its application as pseudo-capacitor electrode material, *J. Organomet. Chem.*, 2019, 900, 120940.

126. Z. Zhu, F. Gao, Z. Zhang, Q. Zhuang, Q. Liu, H. Yu, et al., In-situ growth of $MnCo_2O_4$ hollow spheres on nickel foam as pseudocapacitive electrodes for supercapacitors, *J. Colloid Interface Sci.*, 2021, 587, 56–63.

127. Z. Hou, T. Liu, M. U. Tahir, S. Ahmad, X. Shao, C. Yang, Applied surface science facile conversion of nickel-containing electroplating sludge into nickel- based multi-level nano-material for high-performance pseudocapacitors, *Appl. Surf. Sci.*, 2021, 538, 147978.

128. A. Gabe, M. J. M. Lopez, D. S. Torres, E. Morallon, D. C. Amoros, Synthesis of conducting polymer/carbon material composites and their application in electrical energy storage, *Hybrid Polymer Composite Mater.*, 2017, 8, 173–209.

129. J. Li, J. Qiao, K. Lian, Hydroxide ion conducting polymer electrolytes and their applications in solid supercapacitors: A review, *Energy Storage Mater.*, 2020, 24, 6–21.

130. M. Gandara, E. S. Gonzalves, Polyaniline supercapacitor electrode and carbon fiber graphene oxide: electroactive properties at the charging limit, *Electrochim. Acta*, 2002, 345, 136197.

131. Y. Wang, S. Xu, W. Liu, H. Cheng, S. Chen, X. Liu, J. Liu, Q. Tai, C. Hu, Facile fabrication of urchin-like polyaniline microspheres for electrochemical energy storage, *Electrochim. Acta*, 2017, 254, 25–35.

132. C. Hu, S. Chen, Y. Wang, X. Peng, W. Zhang, J. Chen, Excellent electrochemical performances of cabbage-like polyaniline fabricated by template synthesis, *J. Power Sources*, 2016, 321, 94–101.

133. T. Li, X. Wang, P. Liu, B. Yang, S. Diao, Y. Gao, Synthesis of feather fan-like PANI electrodes for supercapacitors, *Synthetic Metals*, 2019, 258, 116194.

134. D. W. Wang, F. Li, J. Zhao, W. Ren, Z. G. Chen, J. Tan, Z. S. Wu, I. Gentle, G. Q. Lu, H. M. Cheng, Fabrication of Graphene/Polyaniline composite paper via In situ Anodic electro polymerization for High performance flexible electrode, *ACS Nano*, 2009, 3 (7), 1745–1752.

135. N. Chen, L.Ni, J. Zhou, G. Zhu, Q. Kang, Y. Zhang, S. Chen, W. Zhou, C. Lu, J. Chen, X. Feng, X. Wang, X. Guo, L. Peng, W. Ding, W. Hou, Sandwich-like holey graphene/PANI/graphene nanohybrid for ultrahigh-rate supercapacitor, *ACS Appl. Energy Mater.*, 2018, 1 (10), 5189–5197.

136. T. W. Chang, L. Y. Lin, P. W. Peng, Y. X. Zhang, Y. Y. Huang, Enhanced electrocapacitive performance for the supercapacitor with tube-like polyaniline and graphene oxide composites, *Electrochim. Acta*, 2018, 259, 348–354.

137. A. Khosrozadeh, G. Singh, Q. Wang, G. Luo, M. Xing, Supercapacitor with extraordinary cycling stability and high rate from nano-architectured polyaniline/graphene on Janus nanofibrous film with shape memory, *J. Mater. Chem. A*, 2018, 6 (42), 21064–21077.

138. S. Mondal, U. Rana, S. Malik, Reduced graphene oxide/Fe$_3$O$_4$/polyaniline nanostructures as electrode materials for an all-solid-state hybrid supercapacitor, *J. Phys. Chem. C*, 2017, 121 (14), 7573–7583.

139. D. Xie, G. Fu, Y. Ding, X. Kang, W. Cao, Y. Zhao, Preparation of cotton-shaped CNT/PANI composite and its electrochemical performances, *Rare Metals*, 30 (Suppl.1) 2011, 94–97.

140. A. Gök, M. Omastova, A. G. Yavuz, Synthesis and characterization of polythiophenes prepared in the presence of surfactants, *Synth. Met.*, 2007, 157, 23–29.

141. A. Yavuz, N. Ozdemir, H. Zengin, Polypyrrole-coated tape electrode for flexible supercapacitor applications, *Int. J. Hydrog. Energy*, 2020, 45, 18876–18887.

142. X. Deng, X. Bai, Z. Cai, M. Huang, X. Chen, B. Huang, Y. Chen, Renewable carbon foam/δ-MnO$_2$ composites with well-defined hierarchical microstructure as supercapacitor electrodes, *J. Mater.Res. Technol.*, 2020, 9, 8544–8555.

143. M. Chen, Q. Ge, M. Qi, X. Liang, F. Wang, Q. Chen, Cobalt oxides nanorods arrays as advanced electrode for high performance supercapacitor, *Surf. Coat. Technol.*, 2019, 360, 73–77.

144. L. Sun, Y. Sun, Q. Fu, C. Pan, Facile preparation of NiO nanoparticles anchored on N/P-co doped 3D carbon nanofibers network for high-performance asymmetric supercapacitors, *J. Alloys Compd.*, 2021, 888, 161488.

145. T. Feng, G. Liu, G. Li, Y. Li, J. Liang, K. Wang, Engineering iron-rich nanomaterials for supercapacitors, *Chem. Eng. J.*, 2023, 473, 145045.

146. A. A. Yadav, Y. M. Hunge, S. Ko, S. W. Kang, Chemically synthesized iron-oxide-based pure negative electrode for solid-state asymmetric supercapacitor devices, *Mater. (Basel)*, 2022, 15 (17), 6133.

147. M. Aalim, M. A. Shah, Role of oxygen vacancies and porosity in enhancing the electrochemical properties of Microwave synthesized hematite (α-Fe$_2$O$_3$) nanostructures for supercapacitor application, *Vacuum*, 2023, 210, 111903.

148. M. Mazloum-Ardakani, F. Sabaghian, M. Yavari, A. Ebady, N. Sahraie, Enhance the performance of iron oxide nanoparticles in supercapacitor applications through internal contact of α-Fe$_2$O$_3$@CeO$_2$ core-shell, *J. Alloys Compd.*, 2020, 819, 152949.

149. J. Li, Y. Wang, W. Xu, Y. Wang, B. Zhang, S. Luo, X. Zhou, C. Zhang, X. Gu, C. Hu, Porous Fe$_2$O$_3$ nanospheres anchored on activated carbon cloth for high-performance symmetric supercapacitors, *Nano Energy*, 2019, 57, 379–387.

150. C. Wu, Z. Zhang, Z. Chen, Z. Jiang, H. Li, H. Cao, Y. Liu, Y. Zhu, Z. Fang, X. Yu, Rational design of novel ultra-small amorphous Fe_2O_3 nanodots/graphene heterostructures for all-solid-state asymmetric supercapacitors, *Nano Res.*, 2020, 14 (4), 953–960.

151. C. Sun, W. Pan, D. Zheng, G. Guo, Y. Zheng, J. Zhu, C. Liu, Advanced asymmetric supercapacitors with a squirrel cage structure Fe_3O_4@carbon nanocomposite as a negative electrode, *RSC Adv.*, 2021, 11 (62), 39399–39411.

152. X. Zhang, W. Yang, A. Liu, Z. Guo, J. Mu, J. Hou, H. Che, Anchoring mesoporous Fe_3O_4 nanospheres onto N-doped carbon nanotubes toward high-performance composite electrodes for supercapacitors, *Ceram. Int.*, 2020, 46 (14), 22373–22382.

153. X. Li, L. Zhang, G. He, Fe_3O_4 doped double-shelled hollow carbon spheres with hierarchical pore network for durable high-performance supercapacitor, *Carbon*, 2016, 99, 514–522.

154. S. Sheng, W. Liu, K. Zhu, K. Cheng, K. Ye, G. Wang, D. Cao, J. Yan, Fe_3O_4 nanospheres in situ decorated graphene as high-performance anode for asymmetric supercapacitor with impressive energy density, *J. Colloid Interface Sci.*, 2019, 536, 235–244.

155. X. Wang, D. Jiang, C. Jing, X. Liu, K. Li, M. Yu, S. Qi, Y. Zhang, Biotemplate synthesis of Fe_3O_4/polyaniline for supercapacitor, *J. Energy Storage*, 2020, 30, 101554.

156. J. Zhu, Q. Zhang, L. Chen, S. Yang, P. Zhao, Q. Yan, Controlled synthesis of Fe_3O_4 microparticles with interconnected 3D network structures for high-performance flexible solid-state asymmetric supercapacitors, *J. Alloys Compd.*, 2022, 918, 165731.

157. Z. Wang, C. Liu, G. Shi, G. Wang, H. Zhang, Q. Zhang, X. Jiang, X. Li, F. Luo, Y. Hu, K. Yi, Preparation and electrochemical properties of electrospun FeS/carbon nanofiber composites, *Ionics*, 2020, 26 (6), 3051–3060.

158. B. Balakrishnan, S. K. Balasingam, K. Sivalingam Nallathambi, A. Ramadoss, M. Kundu, J. S. Bak, I. H. Cho, P. Kandasamy, Y. Jun, H.-J. Kim, Facile synthesis of pristine FeS_2 microflowers and hybrid rGO-FeS_2 microsphere electrode materials for high performance symmetric capacitors, *J. Ind. Eng. Chem.*, 2019, 71, 191–200.

159. L. Hou, W. Yang, X. Xu, B. Deng, Z. Chen, S. Wang, J. Tian, F. Yang, Y. Li, In-situ activation endows the integrated Fe_3C/Fe@nitrogen-doped carbon hybrids with enhanced pseudocapacitance for electrochemical energy storage, *Chem. Eng. J.*, 2019, 375, 122061.

160. B. Palanivel, S. D. Mudisoodum perumal, T. Maiyalagan, V. Jayarman, C. Ayyappan, M. Alagiri, Rational design of $ZnFe_2O_4$/g-C_3N_4 nanocomposite for enhanced photo-Fenton reaction and supercapacitor performance, *Appl. Surface Sci.*, 2019, 498, 143807.

161. Y. Zhao, L. Xu, J. Yan, W. Yan, C. Wu, J. Lian, Y. Huang, J. Bao, J. Qiu, L. Xu, Y. Xu, H. Xu, H. Li, Facile preparation of $NiFe_2O_4$/MoS_2 composite material with synergistic effect for high performance supercapacitor, *J. Alloys Compd.*, 2017, 726, 608–617.

162. X. He, Y. Zhao, R. Chen, H. Zhang, J. Liu, Q. Liu, D. Song, R. Li, J. Wang, Hierarchical $FeCo_2O_4$@polypyrrole core/shell nanowires on carbon cloth for high-performance flexible all-solid-state asymmetric supercapacitors, *ACS Sustain. Chem. Eng.*, 2018, 6 (11), 14945–14954.

163. D. Zhang, Y. Shao, X. Kong, M. Jiang, X. Lei, Hierarchical carbon-decorated Fe_3O_4 on hollow CuO nanotube array: fabrication and used as negative material for ultrahigh-energy density hybrid supercapacitor, *Chem. Eng. J.*, 2018, 349, 491–499.

164. X. Zhang, W. Yang, A. Liu, Z. Guo, J. Mu, J. Hou, H. Che, Anchoring mesoporous Fe_3O_4 nanospheres onto N-doped carbon nanotubes toward high-performance composite electrodes for supercapacitors, *Ceram. Int.*, 2020, 46 (14), 22373–22382.

165. X. Wang, D. Jiang, C. Jing, X. Liu, K. Li, M. Yu, S. Qi, Y. Zhang, Bio template synthesis of Fe_3O_4/polyaniline for supercapacitor, J. *Energy Storage*, 2020, 30, 101554.

166. Y. Jiang, J. Han, X. Wei, H. Zhang, Z. Zhang, L. Ren, Magnetite nanoparticles in-situ grown and clustered on reduced graphene oxide for supercapacitor electrodes, *Mater.* (Basel), 2022, 15 (15), 5371.

167. X. Li, L. Zhang, G. He, Fe_3O_4 doped double-shelled hollow carbon spheres with hierarchical pore network for durable high-performance supercapacitor, *Carbon*, 2016, 99, 514–522.
168. H. Lu, Y. Jiang, G. Xiao, J. Hu, L. Yang, X. He, X. Xiang, M. Li, W. Sun, Z. Lu, Z. Zhu, Y. Qiao, Nitrogen-doped porous carbon fiber with enriched $Fe_{(2)}N$ sites: Synthesis and application as efficient electrocatalyst for oxygen reduction reaction in microbial fuel cells, *J. Colloid Interface Sci.*, 2022, 616, 539–547.
169. A. Śliwak, A. Moyseowicz, G. Gryglewicz, Hydrothermal-assisted synthesis of an iron nitride–carbon composite as a novel electrode material for supercapacitors, *J. Mater. Chem. A*, 2017, 5 (12), 5680–5684.
170. X. Ling, G. Zhang, Z. Long, Z. Lu, Z. He, J. Li, Y. Wang, D. Zhang, Core–shell structure γ-MnO_2-PANI carbon fiber paper-based flexible electrode material for high-performance supercapacitors, *J. Indus. Eng. Chem.*, 2021, 99, 317–325.
171. M. R. Thalji, G. A. M. Ali, P. Liu, Y. L. Zhong, K. F. Chong, $W_{18}O_{49}$ nanowires-graphene nanocomposite for asymmetric supercapacitors employing $AlCl_3$ aqueous electrolyte, *Chem. Eng. J.*, 2021, 409, 128216.
172. Y. Ouyang, Y. Chen, J. Peng, J. Yang, C. Wu, B. Chang, Nickel sulfide/activated carbon nanotubes nanocomposites as advanced electrode of high-performance aqueous asymmetric supercapacitors, *J. Alloys Compd.*, 2021, 885, 160979.
173. Q. Zhao, D. Yang, A. K. Whittaker, X. S. Zhao, A hybrid sodium-ion capacitor with polyimide as anode and polyimide- derived carbon as cathode, *J. Power Sources*, 2018, 396, 12–18.
174. S. Chen, J. Wang, L. Fan, R. Ma, E. Zhang, Q. Liu, An ultrafast rechargeable hybrid sodium-based dual-ion capacitor based on hard carbon cathodes, *Adv. Energy Mater.*, 2018, 1800140, 1–8.
175. G. Subburam, K. Ramachandran, S. A. El-khodary, B. Zou, J. Wang, L. Wang, J. Qiu, X. Liu, D. H. J. Ng, J. Lian, Development of porous carbon nanosheets from polyvinyl alcohol for sodium-ion capacitors, *Chem. Eng. J.*, 2021, 415, 129012.
176. H. D. Pham, N. R. Chodankar, S. D. Jadhav, K. Jayaramulu, A. K. Nanjundan, D. P. Dubal, Large interspaced layered potassium niobate nanosheet arrays as an ultra-stable anode for potassium ion capacitor, *Energy Storage Mater.*, 2021, 34, 475–482.
177. H. Shi, X. Zhao, Z. S. Wu, Y. Dong, P. Lu, J. Chen, W. Ren, H. M. Cheng, X. Bao, Free-standing integrated cathode derived from 3D graphene/carbon nanotube aerogels serving as binder-free sulfur host and interlayer for ultrahigh volumetric-energy-density lithium single bond sulfur batteries, *Nano Energy*, 2019, 60, 743–751.
178. L. Fan, D. Xiong, X. Li, Enhanced lithium/sodium storage of SnO_2/Graphene aerogels nanocomposites, *Mater. Chem. Phy.*, 2018, 238, 121870.
179. Y. Wang, D. Kong, W. Shi, B. Liu, G. J. Sim, Q. Ge, H. Y. Yang, Ice templated free-standing hierarchically WS_2/CNT-rGO Aerogel for high-performance rechargeable Lithium and Sodium ion batteries, *Adv. Energy Mater.*, 2016, 21 (6), 1601057.

11 Electric Double-Layer Capacitor Using Carbon Materials

Toyoda Masahiro

1 INTRODUCTION

1.1 FEATURES OF ELECTRIC DOUBLE LAYER CAPACITORS (EDLCs)

The electric double-layer capacitor is an energy storage device classified as a capacitor and is also called an EDLC, which is an acronym for "Electric Double-Layer Capacitor." Compared to rechargeable batteries, which are known as energy storage devices, EDLCs exhibit a low energy density (the amount of energy that can be stored per unit weight or unit volume), but an improved output density (the instantaneous amount of power that can be extracted per unit weight or unit volume) and have extremely low performance degradation due to repeated charge/discharge cycles at high currents. In addition, the performance degradation due to repeated charging and discharging at high currents is extremely small, and the device has a long service life.

Figure 11.1 shows the relationship between energy density and output density of typical energy storage devices. EDLCs complement the characteristics of the so-called "capacitor," such as aluminum electrolytic capacitor and ceramic capacitor, and "secondary batteries," such as lithium-ion secondary batteries. Rechargeable batteries are suitable for applications that require more energy, while EDLCs are

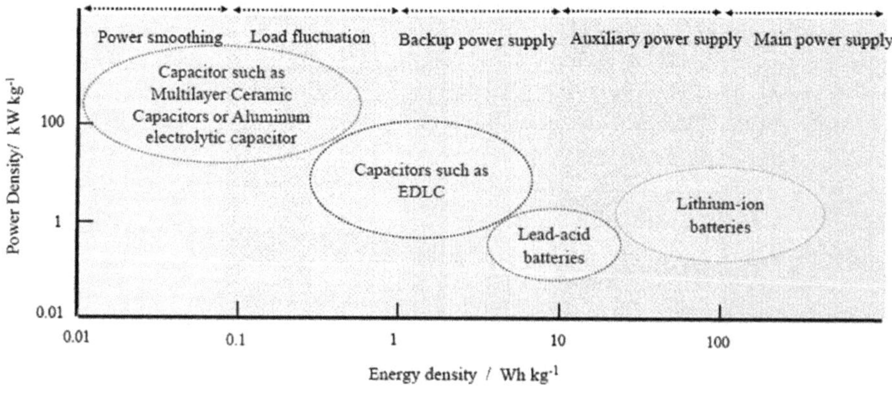

FIGURE 11.1   Energy applications [1].

DOI: 10.1201/9781003561262-11

suitable for applications that require instantaneous charging/discharging, charging/discharging at high currents, and durability for repeated use.

In addition, an EDLC is capable of complete discharge. Compared to rechargeable batteries, which cannot release all of their energy, EDLCs are characterized by a large ratio of the amount of energy that can be extracted to the amount of stored energy. The voltage fluctuation is in proportion to changes in the amount of energy retained, which is the same as in a rechargeable battery. However, since the voltage fluctuation of an EDLC, which can discharge down to 0 V, is also large, voltage stabilization by a power converter (DC/DC converter) is required depending on the load. On the other hand, this characteristic has the advantage that the remaining charge can be accurately determined by measuring the terminal voltage.

The main features can be summarized as follows:

- Capable of millions of charge/discharge cycles (long life)
- Capable of rapid charging and discharging at high power density
- Low loss during charging and discharging (low internal resistance)
- Complete discharge is possible (no limit on the depth of discharge)
- No heavy metals in constituent materials
- High safety in the event of an anomaly; will not fail even in the event of an external short circuit

1.2 STRUCTURE OF EDLCs

An EDLC is commercialized in a wide variety of sizes and shapes, ranging from small products with a capacitance of 1 F or less, such as surface-mount type and coin type, to radial lead type, laminated type, and large screw terminal type, with some large products having a capacitance of over 2000 F in one cell.

The electrode material used in EDLCs is activated carbon, which has a large specific surface area. Electric charge is stored by utilizing the phenomenon of charge orientation over an extremely short distance at the interface between the electrolyte and electrode (electric double layer), as shown in Figure 11.2. Unlike rechargeable batteries that store energy through chemical reactions, these batteries store energy only through physical adsorption of ions on the activated carbon surface, resulting in a long service life with almost no degradation of the constituent materials. Taking advantage of this feature, EDLCs are being used as energy storage devices in applications and locations where the periodic replacement of parts is difficult (or significantly increases costs), in order to reduce the frequency of replacement and to be maintenance-free.

Electrolytes used in EDLCs can be broadly classified into aqueous, organic, and ionic liquids. Typical examples of organic electrolytes used in many products are propylene carbonate-based (PC-based) and acetonitrile-based (AN-based) electrolytes, which have different solvents. AN-based electrolytes have a low volatilization temperature, which limits the ambient temperature of use and may generate highly toxic cyanide gas if ignited.

A large EDLC is often used in a bank structure with multiple cells connected in series or parallel to obtain the necessary voltage and capacitance for the equipment in which they are mounted. The standard lineup of EDLC manufacturers includes

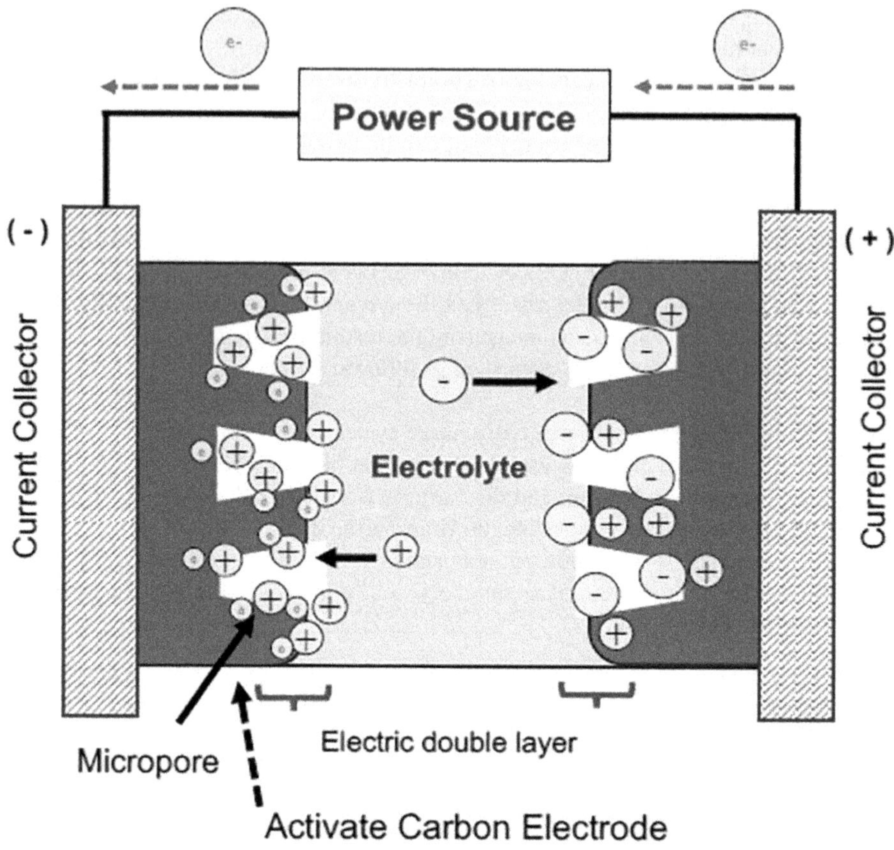

FIGURE 11.2 Illustration of the charge process of EDLCs (electric double-layer capacitors).

capacitor modules that bank multiple cells and incorporate a "balancing circuit" that suppresses the voltage fluctuation of each cell. Some manufacturers also offer modules with special specifications that provide an enhanced resistance to vibration and shock.

1.3 Applications of EDLCs

Applications of small and large capacitors (EDLCs) differ greatly. Small and medium-sized products, which have been established in the market for some time, are widely used in cell phones, smartphones, audio-visual equipment, toys, and game consoles. Major applications include real-time clocks and memory backup power supplies. They are also used as auxiliary power supplies for printers and projectors in office automation equipment to improve performance, such as faster startup and power saving properties. In recent years, they have also been adopted as backup power supplies for smart meters, and in the future, they will be targeted for peak power assist applications in multifunctional smartphones and for camera flash applications, and the market is expected to expand through the development of new products. On the other hand, the main applications for large products are energy

TABLE 11.1

Major Practical Applications of EDLCs [1]

Main Applications	Equipment Used	Main Purpose and Effect
Electricity storage	Road studs (solar power + LED)	Wiring-less and maintenance-free properties
	Traffic information collection terminal (solar power generation + sensors)	Wiring-less and maintenance-free properties
	LED streetlights (solar power generation + LED)	Wiring-less and maintenance-free properties
Power stabilization	Large-scale instantaneous power loss countermeasure device	Measures against momentary power shortages (crisis management)
Power supply assist	Digital multifunction devices (copiers)	Energy-saving properties, reduction of warm-up time
Backup power supply	Distribution line failure of the monitoring equipment	Signal transmission in the event of power failure, maintenance-free properties (crisis management)
	Vending machine for disaster countermeasures and backup power source in the case of power failure	Backup power source in the event of power outage and storage of manual power generated (crisis management)
	Various in-vehicle systems	Increased power supply redundancy and toughness improvement
Energy regeneration	Hybrid transfer crane	Downsized diesel engines and energy-saving properties
	Electric forklift truck	Battery-assisted operation for improved uptime
	Hybrid construction heavy equipment	Downsized diesel engines and energy-saving properties
	Transportation equipment	Fuel consumption improvement and CO_2 reduction

storage/stabilization, power assist, backup power supply, and energy regeneration. In particular, with the growing awareness of energy demands, the development of "energy-saving devices" that take advantage of the characteristics of EDLCs is being promoted in various fields. In many cases, EDLCs are used by themselves as energy storage devices, and in many cases, they are combined with rechargeable batteries to complement each other's characteristics. Table 11.1 summarizes major examples of practical application of large products, but there are many other fields where practical application is being promoted and research is being actively conducted.

2 PROPERTIES OF EDLCS

2.1 Properties Required of Electrode Materials for EDLCs

The energy E stored in a capacitor is half the product of the square of the charging voltage V and the capacitance C ($E = CV^2/2$). Therefore, to improve the energy density of EDLCs, it is necessary to increase the maximum voltage that can be applied to the EDLC (withstand voltage) and the electric double-layer capacitance of the electrode [3–6]. The former is governed by the electrolyte's electrolysis starting voltage or potential window. Therefore, for EDLCs intended for energy storage, electrolytes using organic solvents (organic electrolytes) are exclusively used. For example, compared to aqueous electrolytes such as dilute sulfuric acid that have a withstand voltage of only about 1 V (equivalent to the electrolysis of water), electrolytes using propylene carbonate (PC) as the solvent have a withstand voltage of about 3 V. On the other hand, the pore structure (specific surface area) of the electrode material is important for double-layer capacitance properties since the charge/discharge cycle of the double layer is performed by the physical adsorption and desorption of electrolyte ions on the electrode surface.

In the electrolyte used for EDLCs, the concentration of the electrolyte salt is relatively high (about 1 M) to ensure sufficient ionic conductivity. Therefore, in the case of EDLCs, only the capacitance of the Helmholtz layer, which corresponds to the unimolecular adsorption layer of ions, needs to be considered [6]. In this case, the electric double-layer capacitance C is ideally expressed by the following equation:

$$C = \frac{\varepsilon_0 - \varepsilon_r}{\delta} S \tag{1}$$

In this formula, ε_0 is the dielectric constant of vacuum, ε_r is the relative permittivity of the bilayer, δ is the thickness of the bilayer layer, and S is the effective electrode surface area where ions can be adsorbed and desorbed. The above equation indicates that if the specific surface area where ions can be adsorbed and desorbed is large, the bilayer capacitance also increases proportionally. Therefore, an EDLC requires an electrode material with a large surface area and sufficient electronic conductivity and electrochemical stability to function adequately as an electrode. The material that satisfies these conditions is carbon nanoporous material such as activated carbon. Practical activated carbon is used in commercially available EDLCs, taking manufacturing costs into consideration. Since the pore structure of carbon nanopores (activated carbon) is closely related to the effective surface area of electrodes, many researchers are interested in the correlation between pore structure and capacitance, and active research is being conducted.

2.2　PORE STRUCTURE OF ACTIVATED CARBON AND ELECTRIC DOUBLE-LAYER CAPACITANCE

In Japan and abroad, the search for the optimal pore structure for EDLCs was conducted to increase the effective electrode surface area of activated carbon [3–6]. The search can be roughly divided into two categories: "enlarge the pores of activated carbon to a size where electrolyte ions can be easily adsorbed" and "increase the specific surface area of activated carbon as much as possible."

According to the IUPAC definition, pores of porous materials are classified into micropores that have a pore diameter (width) of 2 nm or less, mesopores that have a pore diameter between 2 nm and 50 nm, and macropores that are larger than 50 nm (recently, IUPAC has defined pores smaller than 100 nm as nanopores [7]). Activated

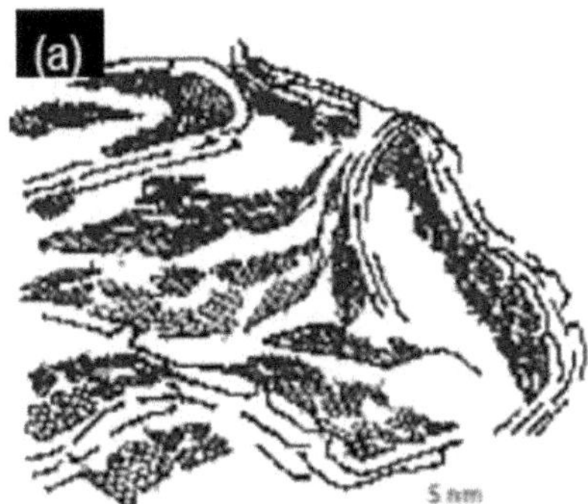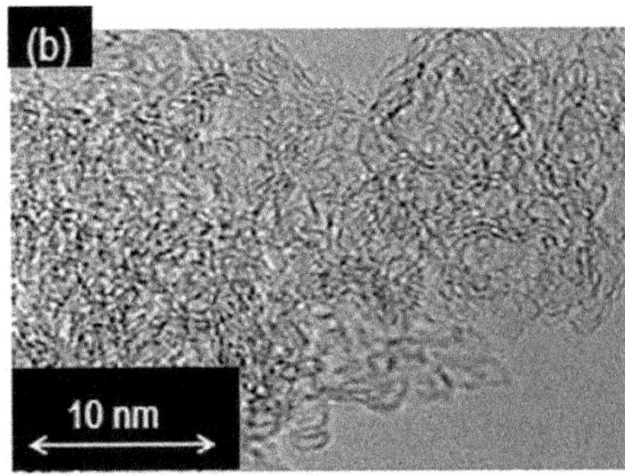

FIGURE 11.3 (a) Structural model of activated carbon (Prof. T. Kyotani, Tohoku University, Japan) and (b) TEM image of activated carbon.

carbon is made by gasifying raw carbon matrices with water vapor, carbon dioxide, or chemicals such as KOH and phosphoric acid at 800°C to develop pores [3–6]. The process of pore formation by gasification is called "activation," and only carbon nanopores produced by activation are technically considered "activated carbon." In general, micropores are selectively generated by activation. Activated carbon is therefore a large porous material with highly developed micropores and a high surface area. Activated carbon is an aggregate of interconnected small pieces of stacked carbon hexagonal mesh surfaces (micrographite) as observed in the TEM images. As shown by Kyotani et al., its micropores correspond to slit-like distorted gaps between the small pieces (Figure 11.3 (a))[6]. Figure 11.3 (b) shows its TEM observation.

On the other hand, electrolyte ions in organic electrolytes for EDLCs are approximately 1 nm in size, depending on the type, and are close to the size of micropores. For example, the ion diameter of tetraethylammonium cation $((C_2H_5)_4N^+)$, a typical electrolyte cation, is 0.7 nm, and that of tetrafluoroborate (BF_4^-), an anion, is about 0.5 nm [8, 9]. From this fact, it was suspected that electrolyte ions could not be smoothly adsorbed in the micropores, and research development focused on mesopores. This is related to the "expansion of activated carbon pores to a size where electrolyte ions can be easily adsorbed."

The "ion sieving" effect is a phenomenon in which the micropores are smaller than the electrolyte ions, making adsorption and desorption difficult and preventing the development of capacitance [3,4,9,10]. A study on mesoporous carbon nanopores, i.e., carbon nanopores containing many mesopores, revealed that the presence of mesopores suppresses the ion sieving effect caused by micropores [9].

However, the development of mesopores leads to an increase in the voids in the electrode, resulting in a decrease in the electrode bulk density. The decrease in electrode bulk density leads to a decrease in volume specific capacitance (specific capacitance normalized by electrode volume), which is important for energy storage applications [3–6].For EDLCs intended for energy storage, the volumetric energy density (energy density normalized by volume) is important. Since the volumetric energy density is proportional to the volume specific capacitance of the electrode, if the withstand voltage is the same, mesoporous carbon, which tends to have a small

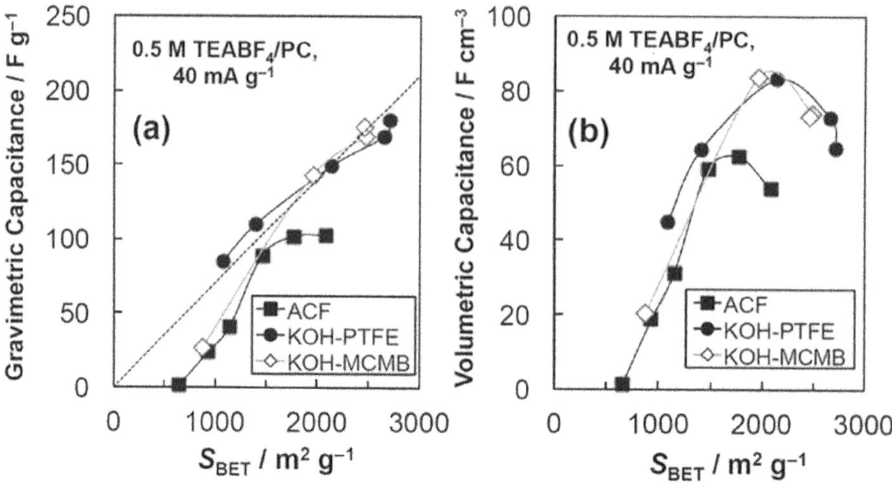

FIGURE 11.4 Correlation between specific surface area (BET) and (a) gravimetric capacitance or (b) volumetric capacitance of various activated carbons in 0.5 M $(C_2H_5)4NBF_4$/propylene carbonate solution [2,6].

Dotted line in (a) corresponds to the specific capacitance per surface area of 7 $\mu F\ cm^{-2}$. ACF: activated carbon fiber, KOH-PTFE: KOH-activated porous carbon derived from polytetrafluoroethylene, KOH-MCMB: KOH-activated mesophase carbon microbeads.

volume specific capacitance, often does not provide an essential solution for improving the energy density of EDLCs.

Another research direction, high surface area, also involves the issue of specific volume capacitance. The correlation between the surface area (Brunauer-Emmett-Teller (BET) specific surface area) and bilayer capacitance of various activated carbons evaluated by the author is shown in Figure 11.4 [2,3–6]. Except in the region of low specific surface area, where the effect of ion sieving is strongly expressed, BET specific surface area and specific capacitance by weight are roughly proportional to each other. This indicates that the electric double-layer capacitance of activated carbon follows equation (1). However, the volume specific capacitance has a convex curve with respect to the BET specific surface area, and a maximum value exists. This is because when pores are developed by activation, the surface area increases, but at the same time the pore width widens, resulting in a trade-off relationship between the increase in specific capacitance by weight and the decrease in bulk density. Therefore, it is difficult to improve the volume specific capacitance by simply increasing the surface area. Therefore, it is necessary to consider an ideal carbon nanopore material in which a single carbon hexagonal network (graphene) is stacked at a minimum distance for ion adsorption. Assuming that the proportionality constant $(= \varepsilon_0\ \varepsilon\ r\ /\ \delta)$ in Eq. (1), i.e., the area specific capacitance is 7–8 $\mu\ F\ cm^{-2}$, referring to the value for activated carbon, the volume specific capacitance of the graphene stack with slit pores of 0.8 nm is calculated to be about 120–140 F cm⁻³ (Figure 11.5) [3–6]. This value is only about twice that of the typical coconut shell activated carbon for EDLCs. Therefore, it is fundamentally difficult to improve the

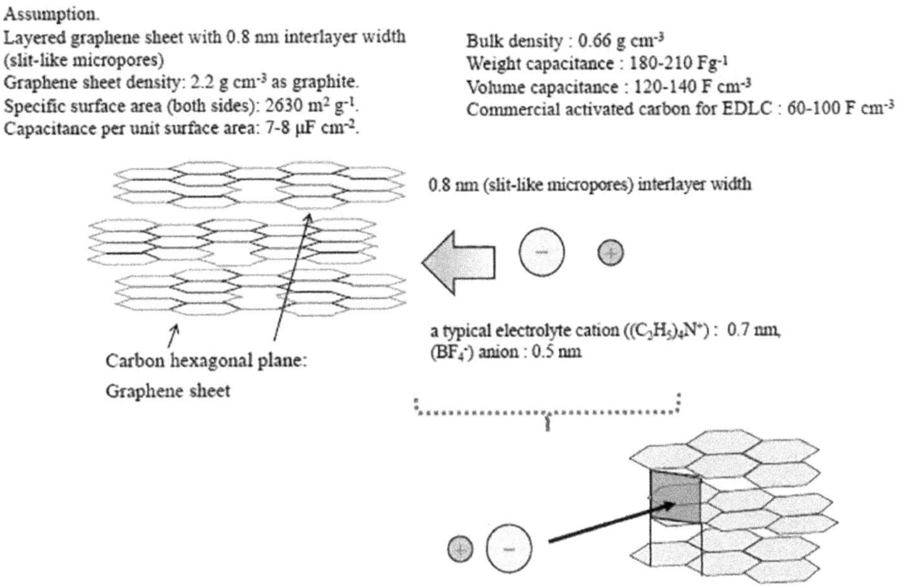

FIGURE 11.5 Gravimetric capacitance and volumetric capacitance of nanoporous carbon [2].

energy density of EDLCs only by controlling the pore structure of carbon nanopores or at least activated carbon.

On the other hand, there is an aqueous EDLC that uses sulfuric acid with a concentration of about 40% as an electrolyte. Compared to the aqueous EDLC, the aforementioned organic EDLC has the advantage of having a higher usable voltage. This is because organic electrolytes have a wider range of electrochemical decomposition voltages than aqueous electrolytes, with a withstand voltage of about 3.0 V compared to about 0.8 V for aqueous electrolytes. On the other hand, aqueous EDLCs are safer, since they do not use a flammable electrolyte, and from the manufacturer's standpoint, they have the process advantage of not having to worry about water contamination. The use of different types of EDLCs is considered to take advantage of their respective characteristics.

It was reported that the electrochemical treatment (anodic oxidation) of carbon fibers in acidic solution can refine carbon fibers from sub-micro to nano size [11]. Figure 11.6 shows the results of SEM observation of pitch-based carbon fibers after anodic oxidation in an acidic solution followed by thermal treatment expansion. After heat treatment and expansion, a single fiber changed significantly to a bundle shape (Figure 11.6(a)). Figure 11.6(b) shows the TEM observation results of sub-micro to nano size miniaturization. Despite miniaturization to nanometer or sub-micrometer size, c-axis stacking was observed and graphite crystallinity was maintained. In addition, microscopic to nano-sized pores were observed in the cross-section of bundle-like fibers (Figure 11(c)). It was reported that a high capacity of 200F/g could be achieved when the obtained microfibers were used for EDLC electrodes, despite

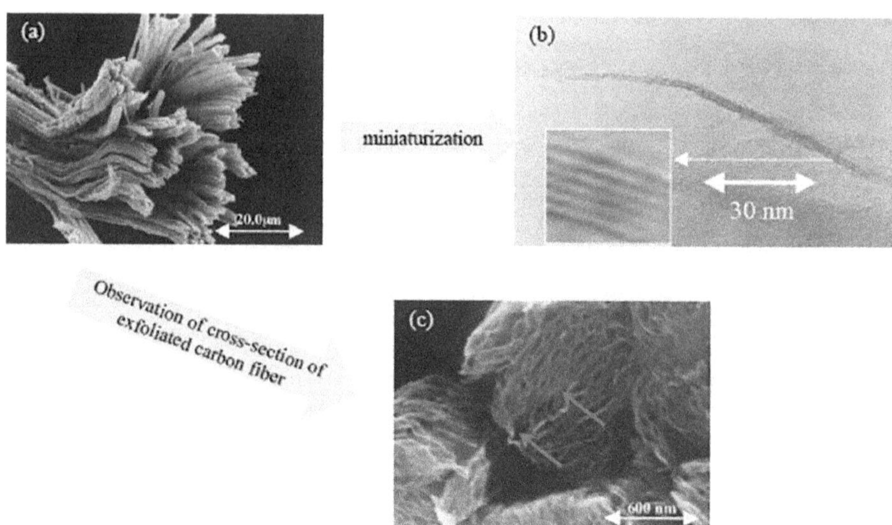

FIGURE 11.6 Microscopic observation in an exfoliated carbon fiber. (a) After exfoliation and (b) TEM observation of miniaturized exfoliated carbon fiber and cross-sectional morphology of the exfoliated carbon fiber [11].

their small specific surface area [12,13]. There were challenges in raising the voltage. However, the voltage could not be increased, and higher capacitance could not be obtained.

In recent years, research has also been conducted on diamond electrochemical capacitors by increasing the power density, especially by using an aqueous electrolyte with excellent ion transfer properties, and by increasing the energy density by using boron-doped diamonds to widen the potential window. In fact, there is a report that a cell voltage of 1.8 V can be applied by using boron-doped diamonds [14].

2.3 Research Directions Required of EDLCs

There is a limit to improving energy density by optimization of the pore structure alone. Therefore, research and development focusing on other factors while keeping pore structure control in mind is extremely important nowadays. For example, carbon nanotubes (CNTs), graphene, and other nanocarbons are considered.

3 NANOCARBONS FOR EDLCS

Nanocarbons are nanomaterials that have attracted a great deal of attention in recent years, but their definitions differ considerably among researchers. For example, Inagaki et al. define nanocarbons as "not only nanosized carbon, but also nanostructured carbon materials whose structure is controlled at the nanoscale" [15]. The characteristics of nanocarbons as capacitor electrodes are described according to this definition.

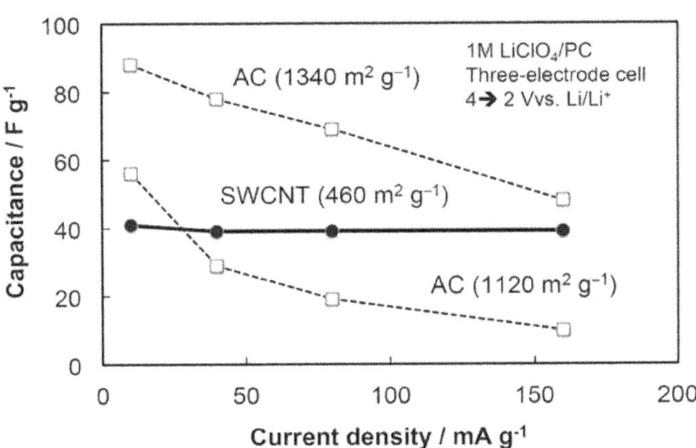

FIGURE 11.7 Rate performance (dependence of capacitance on current density) of single-walled carbon nanotube (SWCNT, HiPcoTM BuckytubeTM) electrode and activated carbon (AC) composite electrode [2].

The nanocarbon that first attracted attention as an electrode material for EDLCs was carbon nanotubes (CNTs), a type of ideal carbon nanopore material with tubular nanopores. In the early days of research, it was reported that the capacitance of CNTs greatly exceeded that of activated carbon; but in reality, CNTs are not necessarily superior to activated carbon in terms of capacitance. This is basically because they do not have the same specific surface area as activated carbon. In the case of single-walled CNTs (SWCNTs), the theoretical surface area is 2630 m² g⁻¹ if the inner cylinder part is also open-pore; but in practice, it is often less than 1000 m² g⁻¹ because of the bundle structure. However, single-walled CNT electrodes are academically very interesting because of their higher area-normalized capacitance (area specific capacitance) compared to activated carbon [16]. In addition, EDLCs using CNT electrodes are characterized by excellent high-speed charge/discharge properties (Figure 11.7) [16–18]. The high-speed response of CNTs is due to the fact that unlike activated carbon, electrolyte ions do not need to diffuse and swim in complex pore spaces. In addition, CNT electrodes are also an electrode material that can contribute to higher voltages in EDLCs, as discussed later. Therefore, even if CNTs do not have much advantage over activated carbon in terms of capacitance, they may enhance the basic properties of EDLCs. Fullerene cannot be used as an electrode material as it is, because it does not exhibit electronic conductivity. However, fullerene soot, a byproduct of fullerene production, is a fragment aggregate that strongly reflects the properties of fullerenes and is at the same time an electronic conductor. Despite its undeveloped micropores and small specific surface area, this soot exhibits a volume specific capacitance that is comparable to that of activated carbon [19]. The cause of this unique property is not clear, but it is considered to be due to the five-membered ring in the soot.

There have been many reports on graphene as an EDLC electrode material.

However, it should be noted that in many cases, the properties of graphene are similar to those of graphene laminate (multilayer graphene) or expanded graphite, which is a classical carbon material, rather than graphene. According to Ruoff, a capacitance of about 100 F g^{-1} has been reported [20]. However, since there are no measurement data for activated carbon electrodes under the same conditions, it is difficult to determine the relative superiority of the two electrodes over activated carbon electrodes.

Miller et al. clarified that the characteristic of Chemical Vapor Deposition single-layer graphene wall electrodes is not capacitance but fast response, and that input-output characteristics (rate characteristics) comparable to those of an electrolytic capacitor can be obtained [21]. Single-layer graphene wall is not expected to have volumetric capacitance due to its low bulk density. Cheng et al. have succeeded in increasing capacitance by incorporating CNTs as a filler between graphene layers to solve the problem of graphene's tendency to stack, which reduces the real surface area [22]. In addition, as with the aforementioned CNTs, there have been studies on graphene exhibiting a higher voltage in EDLCs. Carbon nanoporous materials other than activated carbon can also be called nanocarbons according to the previous definition, if they are conceived and controlled at the nanoscale. The carbon nanopores are filled with carbon by using a porous material such as mesoporous silica or zeolite as a template, and the template then removed after the filling process. Most of the carbon nanopore materials prepared by the template method are mainly mesopore-based, which often have issues with volume specific capacitance as EDLC electrode materials. However, zeolite-templated carbon (ZTC) is micropore-based and has both a volume specific capacitance higher than that of activated carbon and excellent rate characteristics [23].

The excellent rate characteristics despite the microporous nature of ZTC are thought to be due to its uniform pore structure.

It has been reported by Shiraishi et al. that carbon nanopores can be prepared by the halogenation of carbides [1]. In these carbon nanopores, it has been reported that the area specific capacitance increases significantly when the pore width of the micropore approaches the size of the electrolyte ion [24]. Shiraishi et al. carefully evaluated steam-activated carbon to clarify whether this phenomenon is a universal property applicable to carbon nanopores in general. As a result, it was confirmed that narrower pore widths of the micropores resulted in relatively higher area specific capacities [25]. However, considering that activated carbons with narrower pore widths generally have lower rate characteristics and that the crystal structure of the pore sidewalls, which is unique to carbon nanopores, may affect the specific surface area capacitance, further research is needed to increase the potential of narrower pore widths (lower activation). To further explore the application of nanocarbon materials in EDLCs and to obtain higher capacities, it is important to further understand the factors that influence specific surface area capacitance.

4   CONCLUSION

In recent years, a remarkable improvement in performance has been achieved through the elucidation of charge storage mechanisms and the development of advanced nanostructured carbon materials. The discovery that ion desolvation occurs in pores smaller than the solvated ions led to the development of high-capacity electrochemical double-layer capacitors using carbon electrodes with sub-nanometer pores, opening the door to the design of high energy density devices using a variety of electrolytes. By combining the latest generation of nanostructured lithium electrodes with quasi-capacity nanomaterials such as oxides, nitrides, and polymers, the energy density of electrochemical capacitors approaches that of batteries. The use of CNTs, for example, has further advanced micro-electrochemical capacitors and enabled the fabrication of flexible and adaptable devices.

REFERENCES

1. https://www.chemi-con.co.jp/products/edlc/knowledge.html
2. S. Shiraishi and Y. Hatakeyama, Electrode Carbon Material for Electric Double Layer Capacitor, *Vacuume and Surface Sci.*, 62, 703 (2019).
3. S. Shiraishi and Y. Hatakeyama, Electrode Carbon Material for Electric Double Layer Capacitors, *Hyomen*,40, 13 (2002). (in Japanese).
4. S. Shiraishi, Development of Novel Carbon Electrode for Electrochemical Energy Storage. Nano-sized Carbon and Classic Carbon Electrodes for Capacitors, *J. Jpn. Inst. Energy*, 81, 788 (2002). (in Japanese).
5. S. Shiraishi, Heat-Treatment and Nitrogen-Doping of Activated Carbons for High Voltage Operation of Electric Double Layer Capacitor, *Key Eng. Mater.*, 497, 80 (2012).
6. S. Shiraishi, *Encyclopedia of Applied Electrochemistry*, edited by R. F. Savinell, K. Ota and G. Kreysa (Springer, New York, 2014), p. 1.
7. M. Thommes, K. Kaneko, A.V. Neimark, J. P. Olivier, F. Rodriguez-Reinoso, J. Rouquerol and K. S. W. Sing, Physisorption of Gases, with Special Reference to the Evaluation of Surface Area and Pore Size Distribution (IUPAC Technical Report), *Pure Appl. Chem.*, 87, 1051 (2015).
8. M. Ue, Mobility and Ionic Association of Lithium and Quaternary Ammonium Salts in Propylene Carbonate and γ-Butyrolactone, *J. Electrochem. Soc.*, 141, 3336 (1994).
9. S. Shiraishi, H. Kurihara, L. Shi, T. Nakayama and A. Oya, Electric Double-Layer Capacitance of Meso/Macroporous Activated Carbon Fibers Prepared by the Blending Method : I. Nickel-Loaded Activated Carbon Fibers in Propylene Carbonate Solution Containing, *J. Electrochem. Soc.*, 149, A855 (2002).
10. L. Eliad, G. Salitra, A. Soffer and D. Aurbach, Ion Sieving Effects in the Electrical Double Layer of Porous Carbon Electrodes: Estimating Effective Ion Size in Electrolytic Solutions, *J. Phys. Chem.*, B 105, 6880 (2001).
11. M. Toyoda and M. Inagaki, Exfoliation of Carbon Fibers, *J. Phy and Chem.*, 65, 109 (2004).
12. M. Toyoda, Y. Tani and Y. Soneda, Exfoliated Carbon Fibers as an Electrode for Electric Double Layer Capacitor in a 1 Mole dm-3 H2SO4 Electrolyte, *Carbon*, 42, 2833 (2004).
13. Y. Soneda, J. Yamashita, M. Kodama, H. Hatori, M. Toyoda and M. Inagaki, Pseudo-Capacitance on Exfoliated Carbon Fibers in Sulfuric Acid Electrolyte, *Appl. Phys.*, A 82, 575 (2006).,
14. P. Simon and Y. Gogotsi, Materials for Electrochemical Capacitors, *Nat. Mater.*, 7, 845 (2008).
15. M. Inagaki and L.R. Radovic, Nanocarbons, *Carbon*, 40, 2279 (2002).
16. S. Shiraishi, H. Kurihara, K. Okabe, D. Hulicova and A. Oya, Electric Double Layer Capacitance of Highly Pure Single-Walled Carbon Nanotubes in Propylene Carbonate Electrolytes, *Electrochem. Commun.*, 4, 593 (2002).
17. D. N. Futaba, K. Hata, T. Yamada, T. Hiraoka, Y. Hayamizu, Y. Kakudake, O. Tanaike, H. Hatori, M. Yumura and S. Iijima, Shape-Engineerable and Highly Densely Packed Single-Walled Carbon Nanotubes and their Application as Super-Capacitor Electrodes, *Nat. Mater.*, 5, 987 (2006).

18. Y. Honda, T. Haramoto, M. Takeshige, H. Shiozaki, T. Kitamura and M. Ishikawa, Aligned MWCNT Sheet Electrodes Prepared by Transfer Methodology Providing High-Power Capacitor Performance, *Electrochem. Solid-State Lett.*, 10, A106 (2007).

19. M. Egashira, S. Okada, Y. Korai, J. Yamaki and I. Mochida :, Toluene-insoluble Fraction of Fullerene-Soot as the Electrode of a Double-Layer Capacitor, *J. Power Sources*, 148, 116 (2005).

20. M. D. Stoller, S. Park, Y. Zhu, J. An and R. S. Ruoff, Graphene-Based Ultracapacitors, *Nanolett.*, 8, 3498 (2008).

21. J. Miller, R. A. Outlaw and B. C. Holloway, Graphene Double-Layer Capacitor with ac Line-Filtering Performance, *Science*, 329, 1637 (2010).

22. Q. Cheng, J. Tang, J. Ma, H. Zhang, N. Shinya and L.-C. Qin, Graphene and Carbon Nanotube Composite Electrodes for Supercapacitors with ultra-High Energy Density, *Phys. Chem.*, 13, 17615 (2011).

23. H. Itoi, H. Nishihara, T. Kogure and T. Kyotani, Three-Dimensionally Arrayed and Mutually Connected 1.2-nm Nanopores for High-Performance Electric Double Layer Capacitor, *J. Am. Chem. Soc.*, 133, 1165 (2011).

24. J. Chmiola, G. Yushin, Y. Gogotsi, C. Portet, P. Simon and P. L. Taberna, Anomalous Increase in Carbon Capacitance at Pore Sizes Less Than 1 Nanometer, *Science*, 313, 1760 (2006).

25. S. Shiraishi, Dependence of Electric Double Layer Capacitance on Electrolyte Ion for Carbon Electrolyte Interface, *Tanso*, 237, 237-241 (2007). (in Japanese).

<h1>12 Advanced Hybrid Nanomaterials for Energy Storage</h1>

Yujie Ding and Zhuanzhe Zhao

1 INTRODUCTION

Energy is undoubtedly the source of human progress. Traditional sources of fossil energy, such as coal, oil, and gas, have played an irreplaceable role in the modernization of humanity, but they are not renewable and are in danger of being exhausted [1]. Renewable and environmentally friendly sources such as wind, water, and solar energy are being rapidly developed to solve the above problems [2]. However, the use of these energy sources is obviously limited by time periods and geographical locations and does not guarantee continuous use at any time and in any place; so we need energy storage devices to store and convert these clean energy sources.

There are three types of energy storage: heat storage, electricity storage, and hydrogen storage [3]. In the field of electricity storage, there are three types of storage, namely, physical energy storage, electromagnetic storage, and electrochemical energy storage. Electromagnetic energy storage is a new type of energy storage technology, which includes supercapacitor energy storage and superconductive energy storage, which are emerging energy storage technologies. Compared with other energy storage technologies, electromagnetic energy storage has obvious advantages in charging and discharging speed and conversion efficiency and plays a significant role in some specific fields. Electrochemical energy storage, including lithium, lead-acid, liquid current, and sodium-sulfur batteries, is the most widely used and developed energy storage technology [4]. Lithium batteries can be broadly classified into two categories: lithium metal batteries and lithium-ion batteries. Lithium metal batteries are usually non-rechargeable and contain lithium in a metallic state. Lithium-ion batteries do not contain lithium in a metallic state and are rechargeable. The latter is being developed as the primary power battery due to its ability to be recharged repeatedly.

Over the last decade or so, energy storage technology has seen a rapid growth in performance due to the discovery of new electrode materials, particularly nanomaterials. Nanomaterials have become increasingly important in energy storage due to their size finiteness and their unusual electrical, mechanical, and surface properties. However, single nanomaterials do not have all the superior properties required for a particular use, which makes them uncompetitive in applications [5]. Ordinary lithium-ion

DOI: 10.1201/9781003561262-12

batteries and supercapacitor systems suffer from a number of drawbacks due to the single nanostructure of the electrode material. For example, carbon nanomaterials have good rate properties and high cyclic stability, but their specific capacitance is relatively low due to their double-layer capacitive properties. At the same time, transition metal oxides or conductive polymers have a high energy density, but poor cyclic stability and a low rate. The creation of heterostructures is considered an effective way to simultaneously achieve high charging quality and increased utilization of the electrode material; these heterostructures demonstrate potential synergies between each structure, resulting in low internal resistance, high capacitance, excellent stability, and remarkable rate capability. This design meets the main criteria for an ideal electrode material, including high specific surface area, controlled porosity, high electrical conductivity, good mechanical stability, high thermal stability, excellent chemical stability, inexpensive raw material and production costs, and environmental friendliness [6].

In recent years, with the advancement of nanotechnology, scientists and engineers have shown great enthusiasm and interest in learning to hybridize different kinds of nanomaterials to accommodate the physicochemical properties of hybrid nanomaterials [7]. It has been reported that the hybridization of such emerging nanomaterials exhibits better physicochemical properties than any other reported single nanoparticles (NPs) [8]. Hybrid nanomaterials can be defined as a combination of two or more constituents at the nanoscale, integrating both the inherent physical and chemical properties of the constituent materials and benefiting from increased synergistic effects between the materials compared to single nanomaterials, promising to yield better physicochemical properties that can help in energy storage applications. Hybrid nanomaterials offer the flexibility to change their properties by simply modifying their structure and morphology. This gives us customized materials with excellent physicochemical properties such as high thermal stability, mechanical strength, electrical conductivity, optical properties, and controlled wetting properties [9]. Thus, hybrid nanomaterials have great potential for technological innovation and are particularly important for the development of electrochemical conversion and storage devices.

2   PREPARATION OF HYBRID MATERIALS FOR ENERGY STORAGE

There are a number of methods for the preparation of hybrid materials for energy storage, which are classified into physical and chemical methods according to the presence or absence of chemical reactions during synthesis. Hybrid nanomaterials can be synthesized by both physical and chemical routes. In general, the physical synthesis route provides a simpler method to hybridize nanomaterials without going through any rigorous and complex steps. In contrast, the chemical synthesis route requires some chemical processing that must be tightly controlled [10]. For the preparation of hybrid nanofluids, hybrid nanomaterials will be synthesized and then dispersed into the base fluid.

2.1   PHYSICAL METHODS

Physical methods are categorized into mechanical mixing and stirring and self-assembly methods.

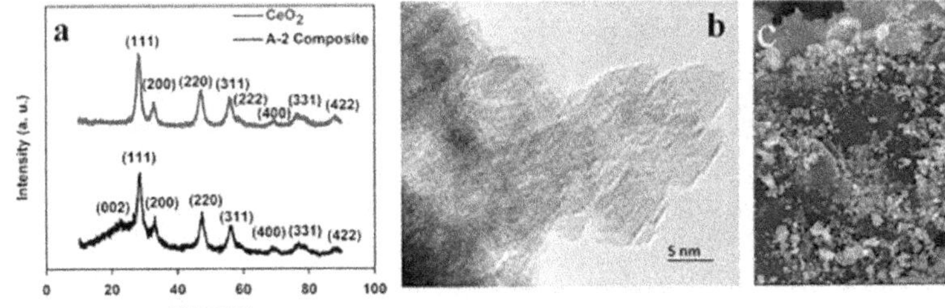

FIGURE 12.1   (a) X-ray diffractogram of CeO_2 NPs and A-2 composite, (b) TEM image of CeO_2 NPs, and (c) FESEM image of 10 wt% CeO_2/AC composite.

2.1.1 Mechanical Mixing and Stirring Method

The mechanical mixing and stirring method involves the use of mechanical devices, such as magnetic stirrers and ultrasound equipment, to facilitate the dispersion of nanomaterials. The use of these mechanical devices prevents the prepared nanomaterials from aggregating or clumping.

Bhat et al. [11] prepared CeO_2/activated carbon-based composites with high scalability using a simple mechanical mixing method. The samples were characterized by X-ray diffraction, transmission electron microscopy, and field emission scanning electron microscopy (Figure 12.1). The results showed that the specific capacitance of the composite mixture with 10 wt% CeO_2 in the two-electrode configuration was 162 F g^{-1}. The test element maintained a specific capacitance of 86% even at a high flux density and exhibited an excellent cycle stability. At high current densities, the power density of the composite electrode reached 3500 W kg^{-1}, making it a good electrode material for SCs.

Heteroatom doping alters the electronic and electrical properties of conducting materials. To understand the effect of nitrogen doping on SC performance, Ghosh et al. [12] synthesized nitrogen-doped graphene/$CuCr_2O_4$ nanocomposites (NCs) for high-performance SCs with a specific capacitance of 530.6 F g^{-1} at 0.5 A g^{-1} current density. The composites were prepared by a simple one-step ultrasound-assisted mechanical mixing method using pre-synthesized $CuCr_2O_4$ and nitrogen-doped graphene with scope for scaling up of the manufacturing process (Figure 12.2).

2.1.2 Self-Assembly Method

In order to produce electrostatic self-assembly interactions, two materials are required: a polyelectrolyte, which is usually soluble in water or has a strong adsorption capacity, and a material with an opposite charge to the polyelectrolyte. When these two nano-sized materials are combined, the polyelectrolyte tends to adsorb it due to the interaction between the two opposite charges, resulting in a stable NC. The method is simple and allows the preparation of stable hybrid nanomaterials that do not contain any chemical bonds [13].

MXenes (transition metal carbides and/or nitrides) are promising future materials for energy storage due to their excellent electrical conductivity and surface chemistry. However, processing MXenes into separate films leads to the restacking and

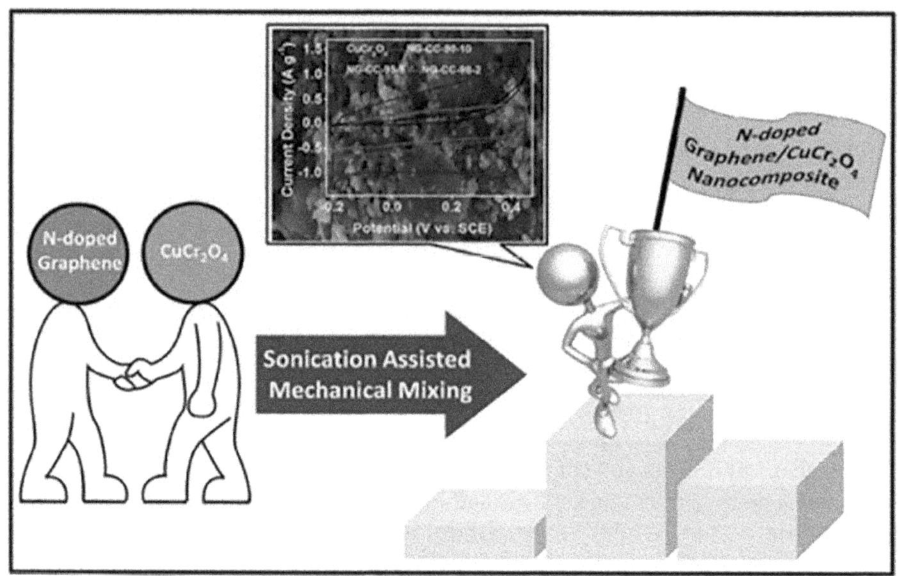

FIGURE 12.2 Roadmap of synthesis by ultrasound-assisted mechanical stirring.

aggregation of 2D nanosheets, which limits the ion dynamics inside the material. Current solutions make it difficult to make thin films that satisfy both rate performance and flexibility. Here, Tang et al. [14] proposed a new chitosan-induced self-assembly strategy to build MXene into flexible Ti_3C_2Tx@Chitosan films with 3D ordered and porous structures (Figure 12.3). The interconnected 3D network provides fast transport channels for electrons and electrolyte ions. As a result, the film exhibits an excellent rate performance and low mass loading dependence. The Ti_3C_2Tx@Chitosan film with a high mass loading of 4 m g^{-2} has a capacitance of 245.2 F g^{-1} at 2 V s^{-1} with a capacitance retention of 57.1% and a 400-fold increase in scan rate. The cross-linked chitosan and MXene nanosheets were interconnected and interwoven, which enhanced the mechanical strength and flexibility of Ti_3C_2Tx@Chitosan. In addition, by applying Ti_3C_2Tx@Chitosan films to assemble flexible symmetric supercapacitors (SCs), an ultrahigh power density of 143.2 μWh cm^{-2} can be obtained. This work develops a simple method for assembling 2D MXene into 3D highly flexible porous films as state-of-the-art electrodes for high-rate pseudocapacitor energy storage.

2.2 Chemical Methods

Chemical methods are categorized as hydrothermal, solvent-thermal, *in situ* growth, and heat treatment methods.

2.2.1 Hydrothermal Method

The hydrothermal method is one of the commonly used methods for the synthesis of hybrid nanomaterials, which mainly depends on the solubility of its aqueous solution in hot water and at high temperatures. To operate, it requires an autoclave (sturdy container) filled with the solution to withstand high temperature and pressure. Typically,

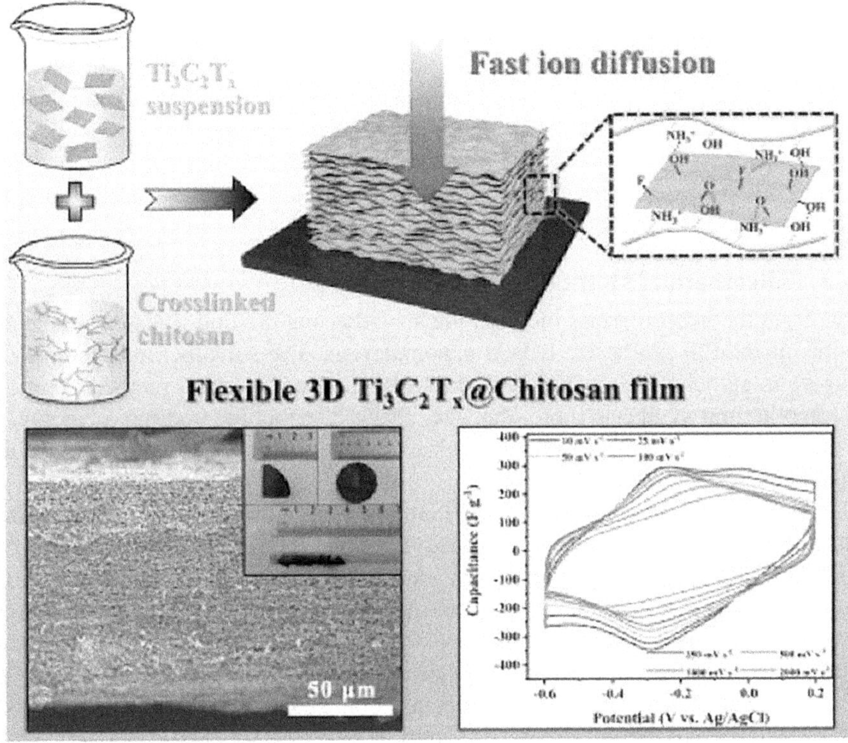

FIGURE 12.3 Schematic of self-assembly of three-dimensionally ordered porous $Ti_3C_2Tx@$ Chitosan membranes.

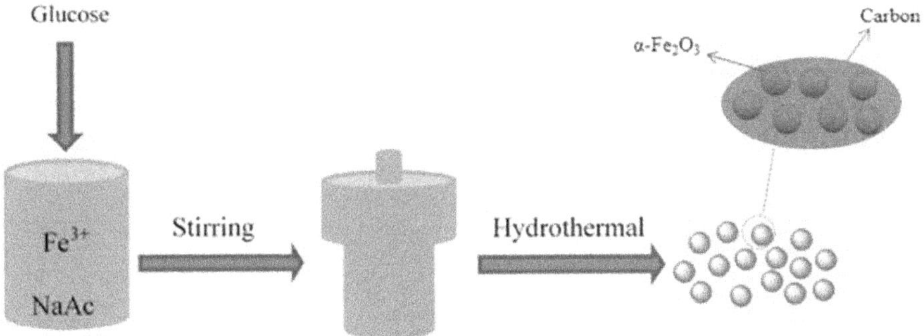

FIGURE 12.4 Schematic illustration of the fabrication of the $\alpha\text{-}Fe_2O_3@C$ composite.

the precursor is placed in an autoclave at a certain temperature for a specific time, followed by any necessary post-processing, such as washing and drying [15].

Zhu and Li et al. [16] prepared $Fe_2O_3@C$ NCs (Figure 12.4) under alkaline conditions using $FeCl_3$ centroid $6H_2O$ as the iron source and glucose as the carbon source. The morphology and structure of the products were characterized using transmission electron microscopy, high resolution transmission electron microscopy, field emission scanning electron microscopy, X-ray diffraction, Raman spectroscopy, Fourier transform infrared spectroscopy, and thermogravimetric analysis. The prepared α-Fe$_2$O$_3$@C NCs were used as supercapacitor electrode materials. The synergistic combination of the capacitance of a carbon-electrical double layer and α-Fe$_2$O$_3$ pseudo-capacitance makes this NC a versatile substrate for high-performance SCs. It is hoped that the synthesis method developed in this work can be used to prepare other metal oxide/carbon composites.

2.2.2 Solvothermal Method

Apart from the hydrothermal method, the solvothermal method is also a simple and feasible method to synthesize hybrid nanomaterials. The solvothermal method also requires an autoclave to operate, which is like the hydrothermal method. Through the solvothermal synthesis route, the size, shape distribution, and nanostructure of the product can be simply tuned by varying the reaction temperature, reaction time, surfactant, and nanomaterial precursor.

Kumar et al. [17] prepared Fe$_3$S$_4$/rGO composites by a one-pot solvent heat method and achieved good energy storage properties (Figure 12.5). The composites were characterized by XRD, SEM, and TEM. The electrochemical specific capacitance of the composite reached 560 F/g at a current density of 1 A/g. This may be attributed to the synergistic effect of Fe$_3$S$_4$ and graphene oxide (GO). The composite has good multiplicative performance and cycling stability with a capacitance loss of 9.0% after 2,000 cycles. The high supercapacitance performance of the composite opens a new platform for the fabrication of high-performance electrode materials for energy storage.

2.2.3 *In Situ* Growth

In addition to the hydrothermal method, *in situ* growth methods offer a promising route for the preparation of hybrid nanomaterials for energy storage. In this case, growth refers to a synthesis process performed on a reaction mixture at the same location without separation or alteration of the original conditions. It should be noted that the *in situ* method facilitates the uniform growth of particles on the substrate surface as it prevents the introduction of impurities and the synthesis process is usually carried out under mild conditions [18].

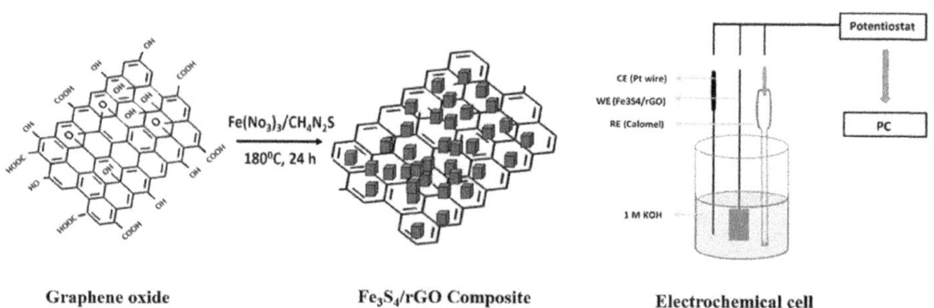

FIGURE 12.5 Scheme for synthesis of the Fe$_3$S$_4$/rGO composite for high-performance SCs.

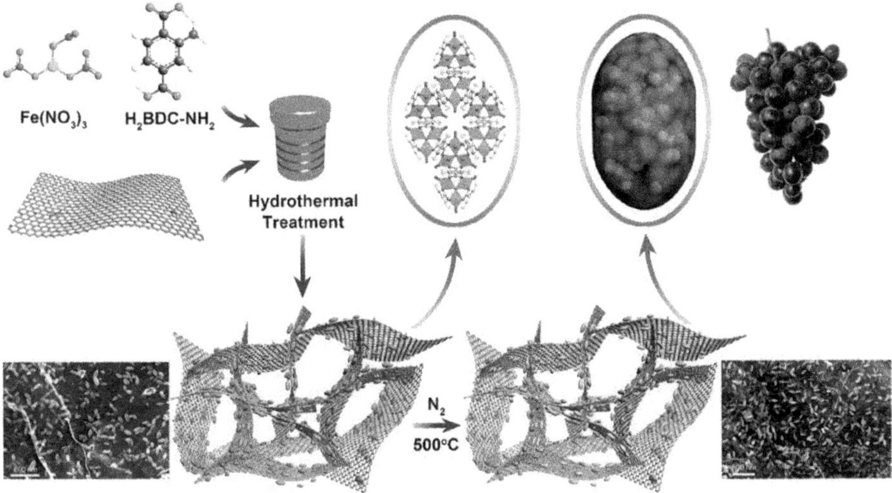

FIGURE 12.6 Schematic illustration of the formation process of Fe_3O_4@C/NG and the corresponding morphology's SEM image

Qi et al. [19] effectively utilized an *in situ* synthesis method to synthesize Fe_3O_4@C/NG hybrid nanomaterials by embedding uniformly dispersed ultrafine Fe_3O_4 on metal-organic framework (MOF)-derived composite nitrogen-doped graphene (Figure 12.6). New synergistic functions can be achieved in this hybrid layered structure. GO layers with functional groups can alleviate the aggregation problem during MOF growth. Moreover, the morphology, size, and carbon content of Fe-based MOFs can be easily controlled by adjusting the concentration of the precursor solution. The synthesized Fe_3O_4@C/NG exhibits significantly improved long-term stability and excellent storage capacity. The excellent lithium storage performance is mainly attributed to the unique 3D hierarchical nanostructure, which has a synergistic effect in preventing NP aggregation, mitigating large volume variations, and facilitating electron and electrolyte transport during cycling. Furthermore, this simple synthesis strategy and the hybrid concept of intelligent combinations of graphene and MOFs for the development of multifunctional materials will lead to even more specific designs that will broaden the range of applications in catalysis, gas storage and separation, and drug delivery.

Wu et al. [20] successfully synthesized $CaCl_2$@zeolitic imidazolate framework-8(Zn) using an *in situ* growth method, as shown in Figure 12.7. This unprecedented $CaCl_2$@ZIF-8(Zn) amido adsorbent not only inherits the advantages of ZIF-8(Zn) and $CaCl_2$, but also has only a slight decrease in specific surface area and pore volume due to the *in situ* growth of halides around the pores of ZIF-8(Zn) and its backbone surfaces, which are not found in previous MOF-derived halide composite adsorbents. Notably, preliminary experimental results showed that the $CaCl_2$@ ZIF-8(Zn) composites exhibited excellent ammonia adsorption performance in saturated ammonia water, higher than that of ZIF-8(Zn) in water, due to the hydrophobicity of the ZIF-8(Zn) skeleton surface. The comprehensive study points out that the

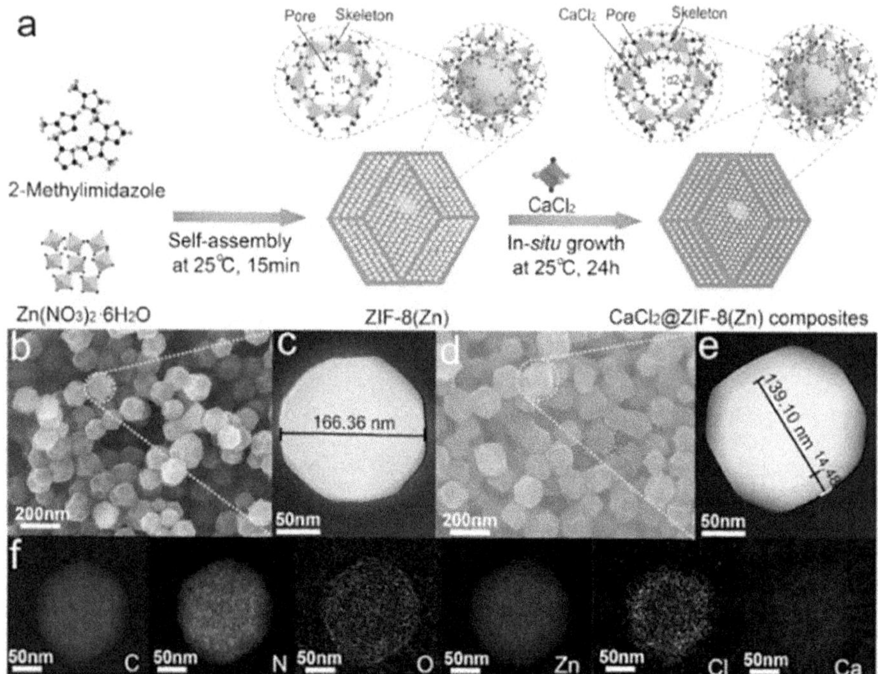

FIGURE 12.7   Synthesis and characterization of MOFs and their composites. (a) Schematic illustration of the new fabrication process of $CaCl_2$@ZIF-8(Zn) composites. (b–e) FE-SEM and FE-TEM images of pure ZIF-8(Zn) and $CaCl_2$@ZIF-8(Zn) composites. (f) Elemental mapping with EDS showing the homogeneous distribution of $CaCl_2$ in the ZIF-8(Zn) matrix.

incorporation of halides in MOFs has significant ammonia adsorption properties, and the solar seasonal thermal battery (STB) enabled by the MOF composite-ammonia work pair has the special function of achieving seasonal heating. They are combined with the halides grown *in situ* on the surface of the MOFs and cascade to utilize the solar heat in the thermal energy storage stage in summer, and upgrade to supply the heat in the energy release stage in winter, which are of high energy efficiency. They are suitable for both seasonal thermal energy storage and space heat management, with high energy efficiency and high adaptability to working conditions.

2.2.4   Thermal Treatment Method

The thermal reduction method is another method used to prepare hybrid nano-materials. This method is simpler, more convenient, and economical compared to other techniques. This method is also one of the favorable methods for obtaining high-purity NPs (e.g., graphene, zinc oxide, etc.) [21,22]. It has also been reported that the thermal reduction method using microwave irradiation is more effective, and the reaction time can be significantly reduced.

Cao et al. [23] used amorphous P-doped NiCo-MOF as a template and is simultaneously carbonized and sulfided by heat treatment to obtain amorphous P-NiCoS@C nanomaterials (Figure 12.8), which were used as cathode materials

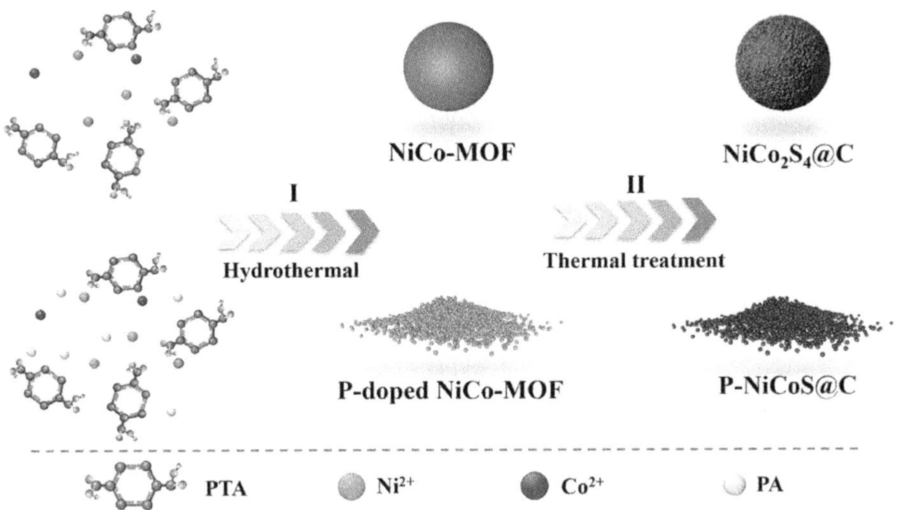

FIGURE 12.8   The preparation diagram of NiCo$_2$S$_4$@C and P-NiCoS@C hybrid materials.

for hybrid supercapacitors (HSCs). The unique structural characteristics of NPs and the advantages of multi-component hybrid materials make them have a large specific surface area, abundant reaction sites, and satisfactory reactivity. These features provide P-NiCoS@C with superior charge storage capability with a specific capacitance up to 1168.2 C g^{-1} at 1 A g^{-1}. In addition, the assembled P-NiCoS@C// AC HSC achieved an excellent energy density of 55.8 W h kg^{-1} at a power density of 800.0 W kg^{-1} and showed a stability of 75.7% after 10,000 cycles. This study opens new perspectives for obtaining electrode materials with high charge storage efficiency.

Hybrid nanomaterials are synthesized using various physical and chemical synthesis methods. It is worth noting that physical synthesis methods are much simpler than chemical synthesis methods because of their stringent requirements and complex steps. In the physical method, the mechanical hybrid method uses a mechanical device to evenly disperse the nanomaterial to prevent its accumulation or agglomeration. In addition, the self-assembly method is a simple method that does not require any chemical bonding, and an electrostatic self-assembly interaction occurs between the two materials, resulting in stable NCs. In addition, the hybridization of nanomaterials also adopts the hydrothermal method and the dissolved heat synthesis route. In these technologies, the reactor is filled with a solution that can withstand high temperatures and pressures. However, in the solvothermal synthesis route, the size, shape distribution, and nanostructure of the product can be adjusted simply by changing the time and temperature of the reaction, the surfactant used, and the nanomaterial precursor. On the other hand, the *in situ* growth method is a promising approach to obtain hybrid nanomaterials for energy storage. This method is performed at the same location of the reaction mixture, without the need to isolate or change the original conditions, facilitating the uniform growth of particles on the substrate surface and preventing the formation of impurities. Finally, the heat treatment method discussed is also a simpler and more economical way to prepare high-purity hybrid nanomaterials. This method uses microwave irradiation, which is more effective and significantly shortens the reaction time.

3 HYBRID NANOMATERIALS IN ENERGY STORAGE

3.1 ELECTROCHEMICAL ENERGY STORAGE

The development of energy storage technologies in the field of renewable energy has become particularly important in the process of dealing with the current energy crisis and combating global warming. To date, the major problems with most electrode materials include poor volume expansion, insufficient layer spacing, inability to be treated in solution due to surface oxidation or defects, and low electrical conductivity, which have hindered their wide application in electrochemical energy storage and conversion [24,25]. The use of suitable active materials, such as hybrid nanomaterials that are able to encounter all of these major problems, is an essential part of high-performance, cost-effective, and environmentally friendly energy storage devices. Hybrid nanomaterials are expected to provide a larger specific surface area, which facilitates the provision of a greater number of active sites for enhanced energy storage capability [26].

3.1.1 SCs

Supercapacitors (SCs) are advanced electrochemical storage devices. SCs are of great interest in academia and industry due to their advantages in terms of high energy density, fast charge and discharge, low cost, and long cycle life [27]. The development of SCs plays an important role in meeting the growing demand for energy, and recently, SCs have gained attention as energy storage devices such as batteries in their design and fabrication. SCs utilize high surface area electrode materials and thin dielectrics to achieve higher capacitance compared to conventional capacitors [28].

There are two main types of SCs: electric double layer supercapacitors (EDLCs) and reversible fast Faraday supercapacitors (FS or pseudocapacitors) [29]. SCs are known as a type of electrical energy storage device that can store electrical energy by creating a double layer of electricity at the interface between electrodes [30]. It is different from the ordinary capacitors because of its very high capacitance. The energy storage mechanism in SCs is due to electrostatic interactions between the polarized surfaces of the porous carbon electrodes and the electrolyte that forms the double electric layer, which results in a fast charging and discharging response time of the device [31,32]. In addition, SCs offer longer charging cycles, allowing them to be facilitated in hybrid vehicles and backup power stations. Despite their higher power density compared to batteries, the commercialization of SCs remains a great challenge due to their low energy density, which is affected by their electrode capacitance and battery operating voltage [33].

Beka et al. [34] reported the use of two-dimensional nanosheet-like nanostructured nickel-cobalt-based MOFs (Ni-Co-MOFs) in combination with rGO hybrid electrode materials for SCs by simple precipitation at room temperature. They observed that the hybrid surface morphology of Ni-Co-MOFs and rGO hybrid electrode material

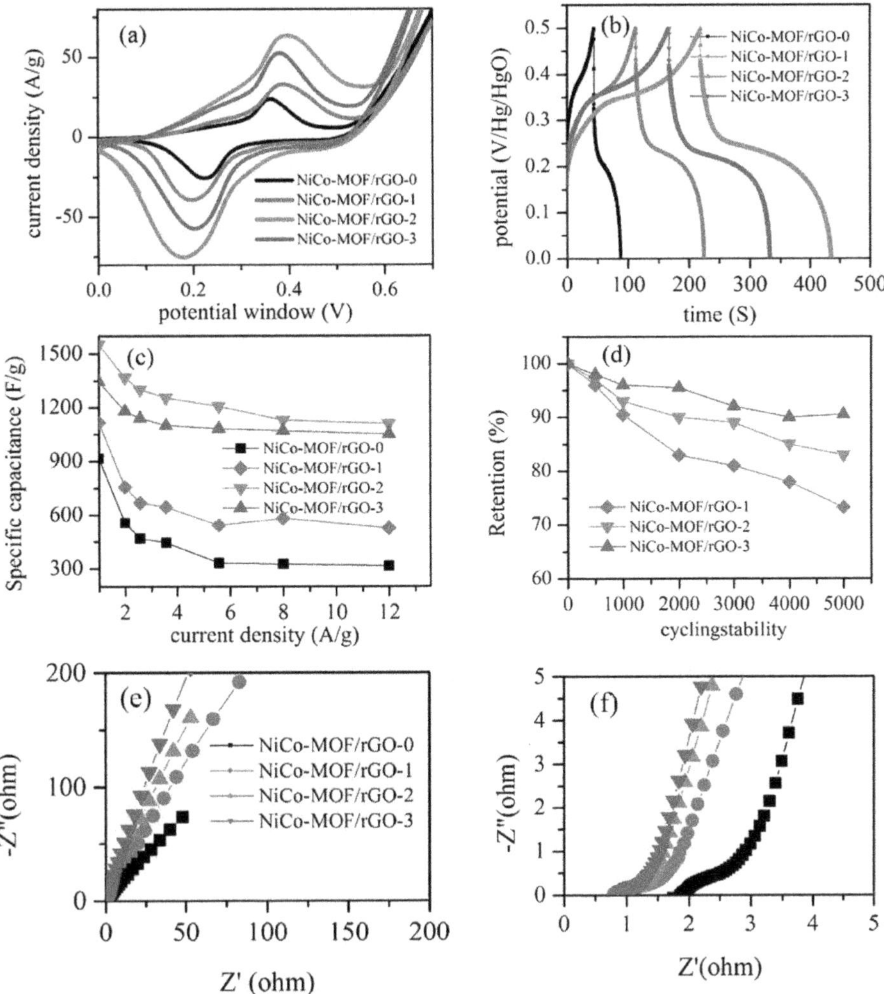

FIGURE 12.9 Electrochemical test characters: (a) relative CV plots at a scan rate of 1 mV s^{-1}, (b) relative galvanic charge plots at 4 mA g^{-1}, (c) relative specific capacitance as a function of current density, (d) cycling stability test, (e) and (f) electrochemical impedance spectroscopy plots at low and high frequency regions, respectively.

composites provides more specific surface area, which leads to faster ion transport in electrochemical testing, as well as lower resistance and improved collector electrodes. They reported the effect of different concentrations of rGO on the electrochemical performance of Ni-Co-MOFs electrodes. Ni-Co-MOFs/rGO-2 in combination with rGO and Ni-Co-MOFs showed excellent electrochemical performance, as shown in Figure 12.9 (a–f). At a lower current density of 1 A/g, the Cs value is as high as 1553 F/g, which may be due to the positive synergistic effect between Ni-Co-MOFs and different concentrations of rGO electrode materials. At a current density of 1 A/g, the Cs value of the Ni-Co-MOFs electrode is 901 F/g. The optimized Ni-Co-MOFs/

rGO composite electrode showed a better stability of 83.6% after 5,000 cycles. After testing the electrochemical properties, they fabricated an asymmetric device with an energy density of up to 44 Wh/kg at a power density of 3168 W/kg.

3.1.2 Lithium-Ion Batteries

For decades, lithium-ion batteries (LIBs) have attracted much attention as the most important electrochemical energy storage system for various applications due to their high Coulomb efficiency, high energy density, low cost, high stability, excellent electrical conductivity, and small size [35]. LIBs are rechargeable batteries whose charging and discharging process typically relies on an electrochemical redox reaction to release lithium ions from the conditioning electrode. Throughout the charging and discharging process, Li^+ will be at both the positive (e.g., $LiCoO_2$, $LiCoO_2$) and negative electrodes, enabling an ion insertion/de-insertion process [30]. With the advantages of high energy density, low self-discharge rate, and long cycling stability, LIBs have become the most commonly used power source for small portable electronic devices, electric vehicles, and hybrids [36]. However, the conventional graphite anode in LIBs is of 372 mAh g^{-1} theoretical capacity. Its limitations and poor multiplicative performance have restricted LIBs' widespread application [37]. Therefore, the development of high-power, high-energy-density energy storage devices is imperative to meet the growing energy demand.

In general, these advances in energy storage will depend on the innovation of the system, i.e., the development of an electrode made of emerging nanomaterials capable of charging and discharging at all current rates [38]. In recent years, great efforts have been made to develop new hybrid electrodes from nanomaterials due to the better electrochemical properties of tailored nanomaterials. The diffusion of lithium ions is mainly influenced by the ion transport pathway and active sites on the electrode surface. Therefore, the use of NC electrodes will not only open the reaction mechanism, but is also expected to improve their electrochemical performance by generating higher energy storage capacity, superior charge/discharge capacity, and good long-term cycling stability due to short ion diffusion pathways and effective contact zone efficiency between the active substance and the electrolyte [38].

To improve electrochemical performance, Ahmed and colleagues [39] used H_2O_2-treated MXene as an anode electrode material for lithium-ion batteries. This hybrid nanomaterial (TiO_2/Ti_2C) derived from MXene (Ti_3C_2Tx) had a specific capacity of 297 mAh/g at a current density of 1000–5000 mA/g. The partially oxidized MXene showed an excellent cycling stability with a specific capacity of 280 mAh/g at 1000 mAh/g (Figure 12.10). In this study, H_2O_2 treatment and titanium dioxide formation led to the opening of or the formation of an increased MXene layer, which resulted in a larger surface area for lithium ions, thereby improving the performance of lithium-ion batteries. In addition, the partially oxidized MXene has high multiplicity and excellent cycling stability and is expected to be a viable candidate for lithium-ion batteries.

3.2 Thermal Energy Storage

Thermal energy storage systems are capable of storing the thermal energy supplied to them, transferring a complete storage cycle of charge, storage, and discharge, so

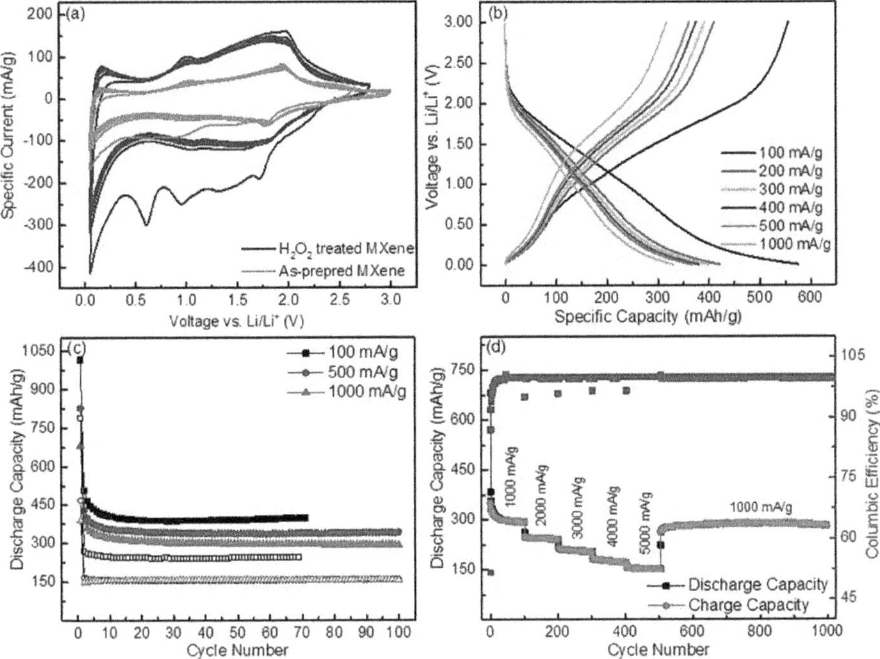

FIGURE 12.10   (a) The cyclic voltammetry (CV) curves and (b) charge/discharge curves for the treated MXene with H_2O_2. (c) Cyclic stability performance at 100 mA/g, 500 mA/g, and 1000 mA/g. (d) Rate-ability assessment of H_2O_2-treated MXene ($\approx$5 min) at various current densities.

that the stored energy can be used at a later date. Thermal energy storage systems are available in sensible and latent forms, with sensible storage systems storing thermal energy only by increasing the temperature of the storage medium, using rock or water as the thermal energy storage medium, while latent thermal storage systems use phase change materials (PCMs) to store thermal energy. Between these two forms of energy storage, latent heat storage using PCMs is widely used for thermal energy storage because of the high latent heat during phase change, the high energy storage density, and the low temperature variation due to the iso-thermal nature of the work, which not only reduces the mismatch between energy demand and supply, but is also an environmentally friendly technology. Thanks to all these features, latent heat storage using PCMs has become one of the most efficient ways of storing thermal energy and has proved to be an ideal means of storing thermal energy for a wide range of applications, such as solar energy storage, heat recovery from industrial waste, climate-controlled intelligent buildings, electrical appliances, and temperature-controlled textiles [40]. In spite of the desirable properties of PCMs for storing thermal energy, it has a major disadvantage in that during solid-liquid phase change processes there is a leakage problem, which in turn increases the thermal resistance, thus limiting its practical applications. Therefore, improving the stability and thermal conductivity of PCMs is the key to develop efficient thermal storage systems.

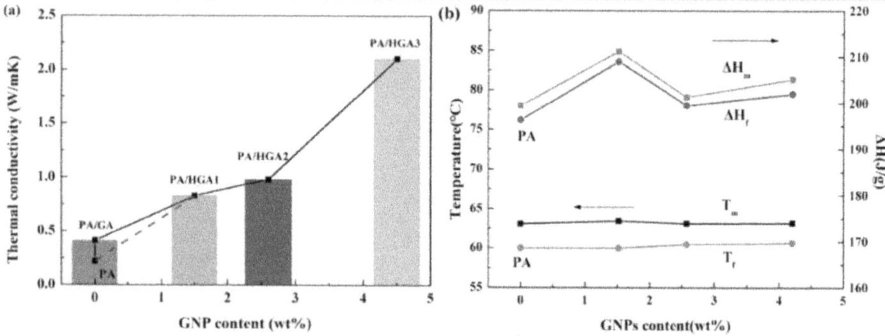

FIGURE 12.11 (a) Thermal conductivity of PA and PA/HGA, (b) the phase-change temperature and latent heat of PA and PA/HGA at different load percentages of GNPs.

In order to improve the thermal properties of PCMs, Shang et al. [41]. successfully synthesized well-defined hybrid graphene aerogels (HGAs) with a three-dimensional porous structure through the hydrothermal reaction of GO and graphene nanoparticles (GNPs) with the help of GO as a dispersant. The uniformly dispersed GNPs along the graphene interconnection network acted as a thermally conductive filler and support material. Palmitic acid (PA) was impregnated into the HGAs by vacuum force. It was found that the thermal conductivity of PA/HGA was enhanced without affecting its thermal storage capacity. The thermal conductivity of PA/HGA containing 2 wt% GNPs was increased by 2.1 W/mK compared to PA, while the latent heat was also increased by 206.2 J/g, as shown in Figure 12.11. PA/HGA with good thermal properties has potential applications in thermal energy storage.

3.3 Hydrogen Storage

With the promotion of the energy transition, new energy storage technology has gradually become the key to promote the development of renewable energy. Hydrogen energy storage technology refers to a kind of energy storage method in which electricity is converted into hydrogen for storage, and then hydrogen is converted into electricity when needed. As an important technology in the field of new energy storage, it has the advantages of high energy storage density, flexible storage method, and high utilization of renewable energy. In hydrogen energy storage, hydrogen is produced by direct methods (e.g., photovoltaic conversion) or by electrolysis, stored for a certain period of time, and then oxidized or subjected to other chemical reactions to restore the energy used. Hydrogen is the result of a chemical reaction, but not a source of energy. Electricity has been the main energy source for many social energy technologies for decades, and hydrogen has complementary properties to electricity [42]. Hydrogen storage materials can be divided into three main categories: (1) adsorbents (e.g., MOFs and carbon-based materials); (2) chemical hydrides (e.g., ammonia boranes and boron hydrides); (3) metal hydrides (e.g., binary hydrides, intermetallic hydrides, and complex metal hydrides) [43,44]. Among these materials, researchers have efficiently stored hydrogen by adding new elements, phase control,

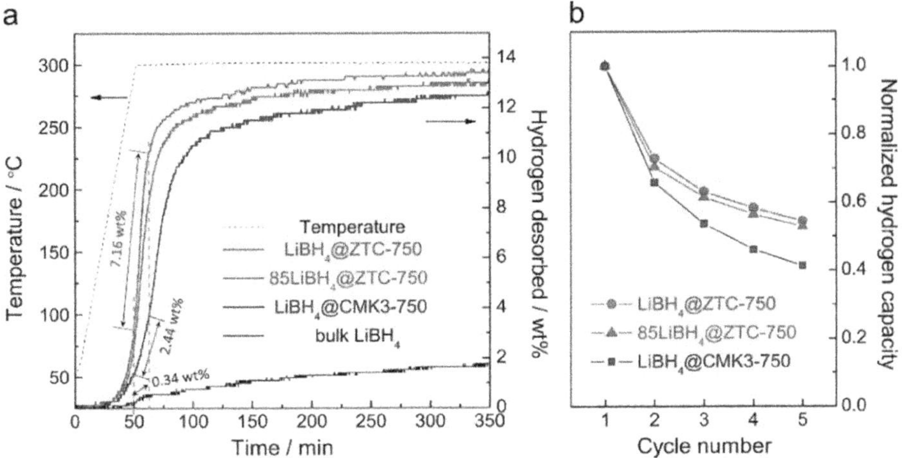

FIGURE 12.12 Hydrogen desorption curves (a) and normalized cyclic hydrogen storage capacities (b) of NCs at 300°C under 0.02 bar H_2; the hydrogenation was carried out at 260°C and 120 bar H_2 for 12 h.

and microstructure modification to obtain high hydrogen storage capacity, operating temperature, hydrogenation/dehydrogenation kinetics, and cycling performance [45,46].

Shao et al. [47] investigated the hydrogen storage properties of nano-confined $LiBH_4$ in zeolite template carbon ($LiBH_4$@ZTC) and ordered mesoporous carbon ($LiBH_4$@CMK3). These samples, prepared at a compaction pressure of 750 megapascals, were designated as $LiBH_4$@ZTC-750 and $LiBH_4$@CMK3–750. From the desorption kinetics shown in Figure 12.12 (a), it can be seen that the average dehydrogenation rate of the densified sample $LiBH_4$@ZTC-750 is 21 times higher than that of bulk $LiBH_4$ and 2.9 times that of the $LiBH_4$@CMK3, and the amount of H_2 obtained is also more than 13.4 wt%. In addition, as shown in Figure 12.12 (b), $LiBH_4$@ZTC-750 has a higher retention rate of hydrogen storage capacity than the $LiBH_4$@CMK3–750. The authors also note that the interface between the nano-enclosed hydride and the scaffold also plays an important role in the hydrogen storage properties.

Recently, Boateng et al. [48] reported a simple one-pot synthesis method of novel three-dimensional rGO and expanded graphite (EG) NCs decorated with Pd NPs for hydrogen storage. Figures 12.13(a–c) show FE-SEM images of rGO, EG, and Pd/rGO-EG NCs, respectively. As can be seen from Figures 12.13(d and e), the electrochemically active surface area and the density of surface oxygen functional groups can be controlled by varying the amount of rGO and EG materials during the synthesis process, which plays an important role in their hydrogen storage performance. In addition, the integration of rGO and EG forms a NC that provides a three-dimensional structure with wide layer spacing and minimal restacking. This also allows the decorated Pd NPs to be evenly distributed on the graphene scaffold, which greatly improves their surface-active sites in electrochemical hydrogen storage.

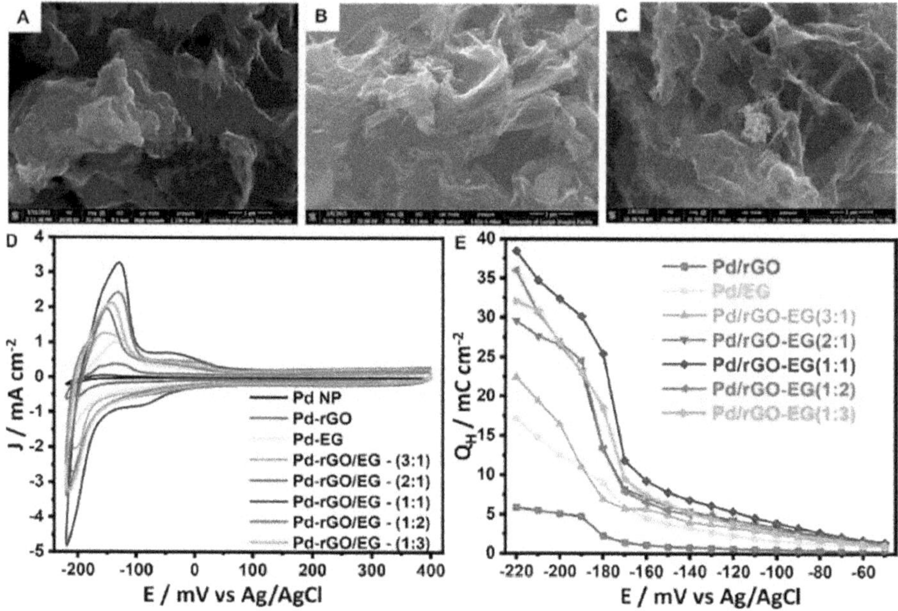

FIGURE 12.13 FE-SEM images of (a) reduced GO, (b) expanded graphite, and (c) three-dimensional (3D) reduced graphene oxide (rGO) and expanded graphite (EG) NC decorated with Pd NPs as hydrogen storage media (Pd/rGO-EG); (d) cyclic voltammogram, (e) overall hydrogen desorption charges, Q_H, at 10 mV s^{-1} in 0.5 M H_2SO_4, respectively.

4 PROBLEMS AND PROSPECTS FOR HYBRID MATERIALS

A full understanding of the new aspects of hybrid nanomaterials is still hampered by the large number of reports in recent years. These reports have only focused on analyzing relevant effects that can improve electrochemical performance and life cycle. The lack of a detailed analysis of the unique physical and mechanical properties of each hybrid electrode compared to other electrodes has led to assessments that do not comprehensively and clearly highlight its properties. Understanding the material properties of the electrode is particularly important as it has a significant impact on the perceived changes in the hybrid structure between the two components, which in turn determines the impact on battery performance. Undoubtedly, this issue poses a significant challenge in the development of advanced hybrid systems. Cost remains a serious concern in the commercialization of hybrid batteries and other approaches. Some of the materials used in hybrid structures, particularly transition metal oxides and carbon-based nanomaterials, are small-scale and expensive materials that are difficult to synthesize in large quantities and in cost-effective ways. In addition, production methods in this field are more complex and costly, requiring expensive equipment and advanced technologies and harsh synthesis conditions, thereby making commercial applications difficult. Despite the large number of experimental studies carried out to improve the structural stability of hybrid nanomaterials, there are few reports on the processes controlling the formation of such structures and, in particular, on how to control the interfacial problems that arise during the assembly of hybrid electrodes. These contribute to maintaining the stability of the structure and making it unrecoverable under a wide range of experimental conditions. Indeed, the main experiments on the organization of hybrid nanostructures seem to rely heavily on self-assembly processes and random placement. Another one that has received less attention recently is the directed assembly technique, which allows controlling the directional growth of its components, such as MOFs [49,50], and core-shell array architectures [51,52]. Interestingly, the control of the assembly process is expected to lead to more promising results for hybrid electrodes, and therefore, it is necessary to further promote this directed assembly process.

More research is needed to develop hybrid electrode materials in the future, and it is important to understand the chemical fundamentals of the electrodes and materials produced in different hybrid electrode systems. Not only the physical properties of hybrid electrode morphologies, but also the details of the electrochemical mechanisms or electrocatalytic activity in LIB and FS systems need to be better understood. A better understanding of these issues will help to achieve an optimal hybrid nanostructure designed for each system, leading to better performance. In addition, research should be conducted to find new fabrication methods to achieve higher levels of organization and efficient and economical electrodes. Replacing transition metal oxides or non-durable materials with other cheaper and more environmentally friendly materials can also significantly reduce the cost of synthesis. The size, shape, orientation, porosity, and thickness of hybrid nanostructures as well as the ratio of active components need to be further optimized to exploit the full potential of hybrid electrodes. In this chapter, in particular, some hybrid nanostructures show promising properties for applications in other systems, but the number of studies is still insufficient or limited. It is clear that an in-depth investigation of these missing hybrid nanomaterials can provide an excellent approach. Future research should focus on the fabrication and design of new morphologies or different nanostructures to maximize the inherently effective properties of nanomaterials, rather than focusing on recent research on improvements or the search for new high-performance materials.

REFERENCES

1. Khezri, M.; Karimi, M. S.; Mamkhezri, J.; Ghazal, R.; Blank, L., Assessing the Impact of Selected Determinants on Renewable Energy Sources in the Electricity Mix: The Case of Asean Countries. *Energies* 2022, *15* (13), 4604.
2. Baskutis, S.; Baskutiene, J.; Navickas, V.; Bilan, Y.; Cieglinski, W., Perspectives and Problems of Using Renewable Energy Sources and Implementation of Local "Green" Initiatives: A Regional Assessment. *Energies* 2021, *14* (18), 5888.
3. Rohit, A. K.; Devi, K. P.; Rangnekar, S., An Overview of Energy Storage and its Importance in Indian Renewable Energy Sector: Part I-Technologies and Comparison. *Journal of Energy Storage* 2017, *13*, 10–23.
4. Aneke, M.; Wang, M., Energy Storage Technologies and Real Life Applications – A State of the Art Review. *Applied Energy* 2016, *179*, 350–377.
5. Devendiran, D. K.; Amirtham, V. A., A Review on Preparation, Characterization, Properties and Applications of Nanofluids. *Renewable and Sustainable Energy Reviews* 2016, *60*, 21–40.

6. Zhi, M. J.; Xiang, C. C.; Li, J. T.; Li, M.; Wu, N. Q., Nanostructured Carbon-Metal Oxide Composite Electrodes for Supercapacitors: A Review. *Nanoscale* 2013, *5* (1), 72–88.

7. Zhou, Z. P.; Gong, J. W.; Guo, Y. C.; Mu, L. Q., Multifunctional Hybrid Nanomaterials for Energy Storage. *Journal of Nanomaterials* 2019, *2019*, 3013594.

8. Devendiran, D. K.; Amirtham, V. A., A Review on Preparation, Characterization, Properties and Applications of Nanofluids. *Renewable & Sustainable Energy Reviews* 2016, *60*, 21–40.

9. Meroni, D.; Ardizzone, S. Preparation and Application of Hybrid Nanomaterials. *Nanomaterials* [Online], 2018.

10. Liu, Y.; Yu, J.; Guo, D.; Li, Z.; Su, Y., Ti3C2Tx MXene/Graphene Nanocomposites: Synthesis and Application in Electrochemical Energy Storage. *Journal of Alloys and Compounds* 2020, *815*, 152403.

11. Aravinda, L. S.; Bhat, K. U.; Bhat, B. R., Nano CeO2/Activated Carbon Based Composite Electrodes for High Performance Supercapacitor. *Materials Letters* 2013, *112*, 158–161.

12. Sarkar, S.; Akshaya, R.; Ghosh, S., Nitrogen Doped Graphene/CuCr2O4 Nanocomposites for Supercapacitors Application: Effect of Nitrogen Doping on Coulombic Efficiency. *Electrochimica Acta* 2020, *332*, 135368.

13. Amadi, E. V.; Venkataraman, A.; Papadopoulos, C., Nanoscale Self-Assembly: Concepts, Applications and Challenges. *Nanotechnology* 2022, *33* (13), 132001.

14. Ding, Y.; Liu, Y. Q.; Sun, X. Y.; Yao, Y. Q.; Yuan, B. L.; Huang, T. T.; Tang, J., Three-Dimensional Ordered and Porous Ti3C2Tx@Chitosan Film Enabled by Self-Assembly Strategy for High-Rate Pseudocapacitive Energy Storage. *Chemical Engineering Journal* 2022, *442*, 136255.

15. Venkateshalu, S.; Grace, A. N., MXenes-A New Class of 2D Layered Materials: Synthesis, Properties, Applications as Supercapacitor Electrode and Beyond. *Applied Materials Today* 2020, *18*, 100509.

16. Zhu, M. Y.; Kan, J. R.; Pan, J. M.; Tong, W. J.; Chen, Q.; Wang, J. C.; Li, S. J., One-Pot Hydrothermal Fabrication of α-F2O3@C Nanocomposites for Electrochemical Energy Storage. *Journal of Energy Chemistry* 2019, *28*, 1–8.

17. Karuppasamy, M.; Muthu, D.; Haldorai, Y.; Kumar, R. T. R., Solvothermal Synthesis of Fe3S4@graphene Composite Electrode Materials for Energy Storage. *Carbon Letters* 2020, *30* (6), 667–673.

18. Liu, X. W.; Wang, F. Y.; Zhen, F.; Huang, J. R., *In situ* Growth of Au Nanoparticles on the Surfaces of Cu2O Nanocubes for Chemical Sensors with Enhanced Performance. *RSC Advances* 2012, *2* (20), 7647–7651.

19. Qi, L.; Xin, Y.; Zuo, Z.; Yang, C.; Wu, K.; Wu, B.; Zhou, H., Grape-Like Fe3O4 Agglomerates Grown on Graphene Nanosheets for Ultrafast and Stable Lithium Storage. *ACS Applied Materials & Interfaces* 2016, *8* (27), 17245–17252.

20. Wu, S.-F.; Wang, L.-W.; An, G.-L.; Zhang, B., Excellent Ammonia Sorption Enabled by Metal-Organic Framework Nanocomposites for Seasonal Thermal Battery. *Energy Storage Materials* 2023, *54*, 822–835.

21. Kong, L. J.; Sun, P. Q.; Liu, J. C.; Lin, Y. X.; Xiao, C.; Bao, C.; Zheng, K.; Xue, M.; Zhang, X.; Liu, X. L.; Tian, X. Y., Superhydrophobic and Mechanical Properties Enhanced the Electrospinning Film with a Multiscale Micro-Nano Structure for High-Efficiency Radiation Cooling. *Journal of Materials Chemistry A* 2024, *12* (13), 7886–7895.

22. Sengupta, I.; Chakraborty, S.; Talukdar, M.; Pal, S. K.; Chakraborty, S., Thermal Reduction of Graphene Oxide: How Temperature Influences Purity. *Journal of Materials Research* 2018, *33* (23), 4113–4122.

23. Cao, W.; Chen, N.; Zhao, W.; Xia, Q.; Du, G.; Xiong, C.; Li, W.; Tang, L., Amorphous P-NiCoS@C Nanoparticles Derived from P-Doped NiCo-MOF as Electrode Materials for High-Performance Hybrid Supercapacitors. *Electrochimica Acta* 2022, *430*, 141049.

24. Chaudhari, N. K.; Jin, H.; Kim, B.; Baek, D. S.; Joo, S. H.; Lee, K., MXene: An Emerging Two-Dimensional Material for Future Energy Conversion and Storage Applications (vol 5, pg 24564, 2017). *Journal of Materials Chemistry A* 2018, *6* (4), 1865–1865.

25. Boota, M.; Pasini, M.; Galeotti, F.; Porzio, W.; Zhao, M.-Q.; Halim, J.; Gogotsi, Y., Interaction of Polar and Nonpolar Polyfluorenes with Layers of Two-Dimensional Titanium Carbide (MXene): Intercalation and Pseudocapacitance. *Chemistry of Materials* 2017, *29* (7), 2731–2738.

26. Sethi, M.; Bantawal, H.; Shenoy, U. S.; Bhat, D. K., Eco-Friendly Synthesis of Porous Graphene and its Utilization as High Performance Supercapacitor Electrode Material. *Journal of Alloys and Compounds* 2019, *799*, 256–266.

27. Yuan, W.; Cheng, L.; Wu, H.; Zhang, Y.; Lv, S.; Guo, X., One-Step Synthesis of 2D-Layered Carbon Wrapped Transition Metal Nitrides from Transition Metal Carbides (MXenes) for Supercapacitors with Ultrahigh Cycling Stability. *Chemical Communications* 2018, *54* (22), 2755–2758.

28. Arico, A. S.; Bruce, P.; Scrosati, B.; Tarascon, J.-M.; van Schalkwijk, W., Nanostructured Materials for Advanced Energy Conversion and Storage Devices. *Nature Materials* 2005, *4* (5), 366–377.

29. Qorbani, M.; Esfandiar, A.; Mehdipour, H.; Chaigneau, M.; Irajizad, A.; Moshfegh, A. Z., Shedding Light on Pseudocapacitive Active Edges of Single-Layer Graphene Nanoribbons as High-Capacitance Supercapacitors. *ACS Applied Energy Materials* 2019, *2* (5), 3665–3675.

30. Pang, J. B.; Mendes, R. G.; Bachmatiuk, A.; Zhao, L.; Ta, H. Q.; Gemming, T.; Liu, H.; Liu, Z. F.; Rummeli, M. H., Applications of 2D MXenes in Energy Conversion and Storage Systems. *Chemical Society Reviews* 2019, *48* (1), 72–133.

31. Yu, A. P.; Chen, Z. W.; Maric, R.; Zhang, L.; Zhang, J. J.; Yan, J. Y., Electrochemical Supercapacitors for Energy Storage and Delivery: Advanced Materials, Technologies and Applications. *Applied Energy* 2015, *153*, 1–2.

32. Piñeiro-Prado, I.; Salinas-Torres, D.; Ruiz-Rosas, R.; Morallón, E.; Cazorla-Amorós, D., Design of Activated Carbon/Activated Carbon Asymmetric Capacitors. *Frontiers in Materials* 2016, *3*, 16.

33. Inagaki, M.; Konno, H.; Tanaike, O., Carbon Materials for Electrochemical Capacitors. *Journal of Power Sources* 2010, *195* (24), 7880–7903.

34. Beka, L. G.; Bu, X.; Li, X.; Wang, X.; Han, C.; Liu, W., A 2D Metal-Organic Framework/Reduced Graphene Oxide Heterostructure for Supercapacitor Application. *RSC Advances* 2019, *9* (62), 36123–36135.

35. Zhao, S.; Nivetha, R.; Qiu, Y.; Guo, X., Two-Dimensional Hybrid Nanomaterials Derived from MXenes (Ti3C2Tx) as Advanced Energy Storage and Conversion Applications. *Chinese Chemical Letters* 2020, *31* (4), 947–952.

36. Goodenough, J. B.; Park, K. S., The Li-Ion Rechargeable Battery: A Perspective. *Journal of the American Chemical Society* 2013, *135* (4), 1167–1176.

37. Yang, H. X.; Xie, Y.; Zhu, M. M.; Liu, Y. M.; Wang, Z. K.; Xu, M. H.; Lin, S. L., Hierarchical Porous MnCo2O4 Yolk-Shell Microspheres from MOFs as Secondary Nanomaterials for High Power Lithium Ion Batteries. *Dalton Transactions* 2019, *48* (25), 9205–9213.

38. Jiang, C.; Hosono, E.; Zhou, H., Nanomaterials for Lithium Ion Batteries. *Nano Today* 2006, *1* (4), 28–33.

39. Ahmed, B.; Anjum, D. H.; Hedhili, M. N.; Gogotsi, Y.; Alshareef, H. N., H2O2 Assisted Room Temperature Oxidation of Ti2C MXene for Li-Ion Battery Anodes. *Nanoscale* 2016, *8* (14), 7580–7587.

40. Song, S. K.; Qiu, F.; Zhu, W. T.; Guo, Y.; Zhang, Y.; Ju, Y. Y.; Feng, R.; Liu, Y.; Chen, Z.; Zhou, J.; Xiong, C. X.; Dong, L. J., Polyethylene glycol/halloysite@Ag Nanocomposite PCM for Thermal Energy Storage: Simultaneously High Latent Heat and Enhanced Thermal Conductivity. *Solar Energy Materials and Solar Cells* 2019, *193*, 237–245.

41. Shang, Y.; Zhang, D.; An, M. R.; Li, Z., Enhanced Thermal Performance of Composite Phase Change Materials Based on Hybrid Graphene Aerogels for Thermal Energy Storage. *Materials* 2022, *15* (15), 5380.

42. Koohi-Fayegh, S.; Rosen, M. A., A Review of Energy Storage Types, Applications and Recent Developments. *Journal of Energy Storage* 2020, *27*, 101047.

43. Schlapbach, L.; Zuttel, A., Hydrogen-Storage Materials for Mobile Applications. *Nature* 2001, *414* (6861), 353–358.

44. Ronnebro, E. C. E.; Majzoub, E. H., Recent Advances in Metal Hydrides for Clean Energy Applications. *MRS Bulletin* 2013, *38* (6), 452–461.

45. Ding, X.; Chen, R. R.; Zhang, J. X.; Cao, W. C.; Su, Y. Q.; Guo, J. J., Recent Progress on Enhancing the Hydrogen Storage Properties of Mg-Based Materials Via Fabricating Nanostructures: A Critical Review. *Journal of Alloys and Compounds* 2022, *897*, 163137.

46. Yu, X. B.; Tang, Z. W.; Sun, D. L.; Ouyang, L. Z.; Zhu, M., Recent Advances and Remaining Challenges of Nanostructured Materials for Hydrogen Storage Applications. *Progress in Materials Science* 2017, *88*, 1–48.

47. Shao, J.; Xiao, X.; Fan, X.; Huang, X.; Zhai, B.; Li, S.; Ge, H.; Wang, Q.; Chen, L., Enhanced Hydrogen Storage Capacity and Reversibility of LiBH4 Nanoconfined in the Densified Zeolite-Templated Carbon with High Mechanical Stability. *Nano Energy* 2015, *15*, 244–255.

48. Boateng, E.; van der Zalm, J.; Chen, A., Design and Electrochemical Study of Three-Dimensional Expanded Graphite and Reduced Graphene Oxide Nanocomposites Decorated with Pd Nanoparticles for Hydrogen Storage. *The Journal of Physical Chemistry C* 2021, *125* (42), 22970–22981.

49. Zhu, G. L.; Wen, H.; Ma, M.; Wang, W. Y.; Yang, L.; Wang, L. C.; Shi, X. F.; Cheng, X. W.; Sun, X. P.; Yao, Y. D., A Self-Supported Hierarchical Co-MOF as a Supercapacitor Electrode with Ultrahigh Areal Capacitance and Excellent Rate Performance. *Chemical Communications* 2018, *54* (74), 10499–10502.

50. Ramachandran, R.; Xuan, W.; Zhao, C.; Leng, X.; Sun, D.; Luo, D.; Wang, F., Enhanced Electrochemical Properties of Cerium Metal-Organic Framework Based Composite Electrodes for High-Performance Supercapacitor Application. *RSC Advances* 2018, *8* (7), 3462–3469.

51. Wang, C.; Wu, L.; Wang, H.; Zuo, W.; Li, Y.; Liu, J., Fabrication and Shell Optimization of Synergistic TiO2-MoO3 Core-Shell Nanowire Array Anode for High Energy and Power Density Lithium-Ion Batteries. *Advanced Functional Materials* 2015, *25* (23), 3524–3533.

52. Tang, X.; Jia, R.; Zhai, T.; Xia, H., Hierarchical Fe3O4@Fe2O3 Core–Shell Nanorod Arrays as High-Performance Anodes for Asymmetric Supercapacitors. *ACS Applied Materials & Interfaces* 2015, *7* (49), 27518–27525.

Index

For Product Safety Concerns and Information please contact our EU
representative GPSR@taylorandfrancis.com
Taylor & Francis Verlag GmbH, Kaufingerstraße 24, 80331 München, Germany